THE CITY & GUILDS TEXTBOOK

LEVEL 2 NVQ DIPLOMA
PLUMBING AND HEATING

About City & Guilds

City & Guilds is the UK's leading provider of vocational qualifications, offering over 500 awards across a wide range of industries, and progressing from entry level to the highest levels of professional achievement. With over 8500 centres in 100 countries, City & Guilds is recognised by employers worldwide for providing qualifications that offer proof of the skills they need to get the job done.

Equal opportunities

City & Guilds fully supports the principle of equal opportunities and we are committed to satisfying this principle in all our activities and published material. A copy of our equal opportunities policy statement is available on the City & Guilds website.

Copyright

The content of this document is, unless otherwise indicated, © The City and Guilds of London Institute 2012 and may not be copied, reproduced or distributed without prior written consent.

First edition 2012

Reprint 2012 (twice), 2013 (twice), 2014, 2016, 2017, 2018 (twice)

ISBN 978 0 85193 209 5

Series design by Purpose

Typesetting by Paul Sloman Design

Printed in the UK by Cambrian Printers Ltd

British Library Cataloguing in Publication Data

A catalogue record is available from the British Library.

Publications

For information about or to order City & Guilds support materials, contact 0844 543 0000 or centresupport@cityandguilds.com. Calls to our 0844 numbers cost 7 pence per minute plus your telephone company's access charge.

Every effort has been made to ensure that the information contained in this publication is true and correct at the time of going to press. However, City & Guilds' products and services are subject to continuous development and improvement and the right is reserved to change products and services from time to time. City & Guilds cannot accept liability for loss or damage arising from the use of information in this publication.

City & Guilds
1 Giltspur Street
London EC1A 9DD

cityandguilds.com
publishingfeedback@cityandguilds.com

I would like to thank the following people for their help, patience and support during the writing of this book. My good friend Mark Sainsbury, who started the ball rolling with his 'go-for-it' attitude. Assistant Principal of Stockport College, Ian Burns who gave me every support and encouragement. My friends in the plumbing department at Stockport College, Sharmy, John, Tony. G and Bradders for their encouragement and good wishes.

To my lovely wife I say a huge thank you for putting up with the late nights and my constant talk about plumbing.

And finally, to my late father Eric Maskrey RP, who, if he were alive, I would say just this. Thanks Dad. This is your book just as much as it is mine. I love you and miss you terribly.

Mike Maskrey

Stockport College 2012

CONTENTS

Acknowledgements		vi
About the author		viii
Foreword		ix
Introduction – how to use this textbook		x
001	Understand and carry out safe working practices In building services engineering	2
002	Understand how to communicate with others Within building services engineering	80
003	Understand how to apply environmental protection methods within building services engineering	104
004	Understand how to apply scientific principles within mechanical engineering services	152
005	Understand and carry out site preparation and pipework fabrication techniques for domestic plumbing and heating systems	208
006	Understand and apply domestic cold water system installation and maintenance techniques	278
007	Understand and apply domestic hot water system installation and maintenance techniques	346
008	Understand and apply domestic central heating system installation and maintenance techniques	392
009	Understand and apply domestic rainwater system installation and maintenance techniques	460
010	Understand and apply domestic above ground drainage installation and maintenance techniques	486
Test your knowledge answers		546
Glossary		555
Index		566

ACKNOWLEDGEMENTS

City & Guilds would like to sincerely thank the following.

For invaluable plumbing knowledge and expertise

Martin Biron, Richard Saunders and John Mys.

For their help with taking pictures

Jules Selmes and Adam Giles (photographer and assistant); Martin Biron and the staff at the College of North West London and the following models: Vivian Chioma, Jennifer Close, Peko Gayle-Reveault, Adam Giles, Michael Maskrey, Nahom Sirane, Zhaojie Yu; Michael Maskrey and the staff at Stockport College and the following models: Michael Maskrey, Jordan Taylor.

Picture credits

Every effort has been made to acknowledge all copyright holders as below and the publishers will, if notified, correct any errors in future editions.

Ace Fixings p256; **AEC** p29; **Albion Water Heaters** p360; **Alamy/AKP Photos** p28(br), **Alamy/ allOver photography** p24 **Alamy/allOver photography** p65, **Alamy/Apex News and Pictures Agency** p33, **Alamy/The Art Gallery Collection** p73, **Alamy,/Simon Belcher** p64, **Alamy/Prisma Bildagentur AG** p23, **Alamy,/Bubbles Photolibrary** p71, **Alamy/DWD-photo** p16, **Alamy, David J. Green - lifestyle themes** p41(tl), **Alamy, Robert Harding Picture Library Ltd** p43, **Alamy/ Russell Kord** p66, **Alamy, mediablitzimages** (uk) p71(tr), **Alamy/Parleycoot Landscape & Travel Images** p19, **Alamy/Andrew Paterson** p28(tl), **Alamy/Andrew Paterson** p47 (ml), **Alamy/Pixellover** p24(tl), **Alamy/Photos 12** p11, **Alamy/ RM 8**, **Alamy/Robert Harding Picture Library** p53 9tl), **Alamy/TAS** p17; **ANT Hire** p542; **Aqualisa** p324 **Architab** p93; **Arctic Products** p170; **Armada Seamless Gutters** p481, p484; **Arctic Products** p218; **Astroflame** p269; **Auctiva** p322; **Automated Cable Solution** p213; **Avenue Supplies** p528; **Bananastock** p43; **B&Q** p65(m), p106, p501; **Bargain Tools** p217; **Bath Superstore** p504; **Bathroom Emporium** p495(br), p495(bl), p502 **Bathworks** p499, p502, p503; **Bella Bathrooms** p325(t), p501(tl), p501(tr), p501(br); **Better Bathrooms** p492, p499 **Bosch** p220, p221(b) **Brannan Thermometers** p173, p174; **Brass Die Casting** p238; **Brett Martin** p475, p476, **Brand X** p56, p57; **Calor Gas Ltd** p59; **Ceramics & Bathrooms** p496; **Charlie Mullins ix Choiceful** p302; **City & Guilds** p4, p8, p12(m), p20, p25, p35, p36, p37(m), p39, p44, p66, p68, p468; **Comstock** p6(ml); **Corbis Images** p40, p71(br), p72, **Corbis/Schenectady Museum; Hall of Electrical History Foundation** p29 (br), **Corbis/Sean Sexton Collection** p29(br); **Cosi** p420; **Crown Copyright** p127; **Cylinder Store** p350, p354, p355(b); **Dan Hall** p94; **Digitalvision** p75(ml); **Direct Heating Supplies** p444; **DM Tools** p212, p213(m), p475(r); **Drain Depot** p475(r), p477, p514; **Drainage Online** p247, p477; **Draper Tools** p210, p211, p213(b), p214(b), p217; **Dreamstime** p354; **Dundee Wharf** p377 (b); **Dynapipe** p327; **Essex Exports** p211; **Facelift** p12(t), p74(m), p75(m); **Fancy** p8 (br); **Fascias**.com p469; **Fast Fix Direct** p243, p244; **Fotalia** p61; **Frog Bathrooms** p500(m); **Garden Hot Tubs** p500(r); **Gas Products** p216, p238; **Gas Register** p238; **Georg Fischer** p243; **Getty/Altrendo** p55(tl), **Getty/Jeffrey Coolidge** p28(tr) **Getty Images** p6(br), **Getty Images** p28 (bl), **Getty Images** p29 (tr), **Getty Images** p47 (tm), **Getty/FPG** p9, **Getty/FilmMagic** p37(tl), **Getty/David Gould** p47(tl), **Man Utd via Getty Images** p53(tr), **Getty/Moviepix** p34, **Getty/Peter Nicholson** p41, **Getty/Photo & Co** p 55(tr), **Getty/Plush Studios** p24, **Getty/Sports Illustrated** p51, **Getty/Superstock** p27, **Getty/Jason Tanner/UNHCR** p39, **Getty/Tetra images** Cover, **Getty/Mark Weiss** p67, **Getty/WireImage** p32; **Getty/WireImage** p59; **Goodshot** p23(mr), p73(tr); **Green Gates** p527; **Greened House** p435; **Grohe** p324, p496; **Groundforce** p76; **Gutter UK** p472; **GW Supplies** p413 **Hatria** p495; **Havelock Controls** p439; **Health & Safety Executive** p8(t), p13(t), p13(b), p15, p21, p22, p27; **Heat & Plumb** p320, p434(l), p493(tl), p503; **Heating & Plumbing** p452; **Heating Powerflush** p455; **Hepworth** p519; **HETAS** p419; **Direct Heating Supplies** p444; **HIE** p238, p253; **Holloways of Ludlow** p492; **Home Supply** p320; **Homebase** p230(bl), p248;

Honeywell p439; Horne p380; HunterJones p531; Hunter Plastics p465, p468, p531; Hydraulic Pumps Motors p215(b); Imagesource p25; IMG Europe p444; Irwell Street Metal Co Ltd p135 Irwin p221; istock p5, p10, p16, p84, p87(t), p87(m), p87(b), p89, pp98(tl), p98(ml), p98(bl), p98(tr), p98(mr), p111, p115, p116(t), p116(m), p116(b), p128; Iwan Barn p4; Jack Sealey Ltd p32, p33; JHM Butt Co p211(m); JMS Plant Hire p331; Jo Edkins Minerals p155; John Guest p253; Johnson Valves p291; JSP Ltd p31, p32(m), p32(b); Jules Selmes Photography p231, p232, p234, p235, p245, p246, p322, p336, p337, p338, p339, p534, p536, p537; Just Loos p505; Knighton Tools p215; KTS/Pierre K. Roberge p161(b); Ladder Store p478; Ladders Direct p63, p65(l); Legrand Electric p206(tl) Lincoln Electric & Harris Products p56; Mac Building Products p254; Machine Mart Ltd p36(m), p36(b) Maintain Your Building p482; Martindale Electric p52; Mary Evans Photolibrary p29(tl); Met Office p462; Metabo p219; Meteor Electrical p47; Michael Maskrey pviii, p18, p119, p227(t); Mighty Oak Trading p542; Milano p436; Mira Showers p375(bl); Monument Tools p60, p212; My Tool Store p220; Myson 435(br), p435(bl), p437 Next Bathrooms p498, p500; Northstar Energy p120; Not Just Taps p503; NTMZJ p216(m); Oil Heating Services p112; Osma p473; P Munn Midland Corrosion p162(l), p162(r), p163(l), p163(r); Partridge Hadleigh p318; Pegasus Whirlpool Baths p502(t); Pegler p233, p315; Pimlico Plumbers p96; Pipetech p290; Plannet Plumbing p23; Plumb Click p438; Plumb Nation p363, p375(tr); p375(br), p411, p384(t); Plumb World p404, p432, p440, p443, p496, p497, p499(tr), p506(m); Plumbing For Less p247; Plumbing Pages p431(l), p431(m), p431(r) Plumbing Supply Services p253; Plumbing World 237; Plumbworld p214(mb), p342, p444; Polar Commercial p218; Porcher US p504; Product Serve p238, p248, p332(t); Professional Building Services p247; Professional Building Supplier p527; PTS p30(m); PVC Guttering p247; Q-Fonic p211(b); QVS Direct p362 (m), p362(b); Radiators 4 You p434(r); Rain Clear p471, p475(r); Rain Guard p475(r); Rapid Tools p210, p210(t), p214(t), p538; Recycle Now p143; Regin Products p268; Rex Features/ W.Disney/Everett p18(tl), Rex/Warner Br/Everett p15(tl), Rex/CBS/Everett p61(tm), Rex/CBS/Everett p61(mr), Rex/Warner Br/Everett p15(tr), Rex/Focus/Everett/ p13 (mr), Rex/20thC.Fox/Everett p63(tr), Rex/Ken McKay p63 (mr), Rex/Garo/Phanie p69, Rex/NBCUPhotobank p61(tl), Rex/NBCUPhotobank p61(tr), Rex/Weinstein/Everett p13(mr); RIBA p250; Richards UK p505(mr); Ridgid Tools p216(b), p217(mt), p235; Roof Line Replacement p472; Rothenburger p217; RP Media p426; RS Online p249; Rutland Plastics p292; Scale Master p325; Screwfix p212(m), p212(b), p213(t), p213(mt), p217(t), p218(m), p218(mt), p218(b), p220, p221, p237, p347, p254, p255, p256, p306, p312, p313, p491, p493(b) Shutterstock p30; Stockbyte p75(mr); Stuart Turner p325(m); Sun Flow p423(t); T Glynes p250; TD Online p350(b); Thannet Tool Supplies p215(m); The Blue Book p505, p506; Thomas Meldrum p211; TIK Products p541; Tiles & Bathrooms Online p499(br); TJ Builders p472; Tool Net p214(m); Tool Station p230(br), p238, p247, p248, p383(r), p414, p470, p517; Tooled Up p213(mt), p215(mb), p216, p217, p336(t); Tools 4 Trade p212; Trade Counter Direct p541; Trade Plumbing p504(r), p504(l); Tradelink Plumbing p494(r), p494(m), p494(l), p495, p498; Trading Depot p238; p253, p526, p527; Transtools p214; Triton p324; True Rooms p321, p504(m); Tycowaterworks p290; UK Bathroom Store p497; UK Bathrooms p492(t), p492(b); UKHPS p319, p407; Unendlich Viel Energie p125; US Government (Public Domain) p155; Victorian Bathrooms p496; Vokera p371(t), p372; Warm Rooms p436(tr); Water Heaters Direct p377(t); WD Bathrooms p503; Weller Soldering p215(mt); Wickes p8(b), p238, p247, p254, p516, p517, p518; Wiha Tools p213(t); William Padden p224(t), p224(b), p225(t), p226(tl), p227, p228, p229, p240, p241, p242, p232, p252, p257, p260, p261, p262, p263, p266, p274, p275, p279, p280, p282, p284, p285, p286, p288, p289, p291, p292, p294, p295, p298, p300, p302, p303, p304, p305, p306, p308, p311, p312, p313, p314, p315, p316, p317, p318, p319, p321, p323, p327, p329(t), p329(b), p347, p348, p350, p351, p352, p355, p356, p357, p358, p359, p362(t), p364(l), p364(r), p365, p366, p367, p369, p370, p317, p373, p376, p377, p380, p383, p383, p384, p385, p386, p391, p397, p398, p399, p400, p401, p403, p404, p405, p407, p408, p409, p410, p411, p141, p415, p416, p418, p420, p421, p423, p425, p427, p428, p429, p430, p431, p432, p433, p435(t), p438, p439, p463, p460, p474, p475, p476, p490, p491, p492, p493, p497, p498, p505, p508, p510, p511, p512, p513, p514, p515, p516, p519, p520, p521, p522, p523, p524, p528, p529, p530, p532, p539, p506, p540; Wood Burners p420; WRAP p133, p136, p139; WRAS p293; Free Signage p21, p37(b); Reece Safety p53(t); Safety Photo p50, p51; Snickers Direct p30(b); Ultimate Handyman p206; Valiant p109; Yorkshire Fittings p236, p237, p417; Youngman Group p65(r).

Illustrations by **William Padden**.

ABOUT THE AUTHOR

My father was quite simply the best plumber I have ever seen. It was his enthusiasm for the trade that he loved that rubbed off on me at a very early age. I was working with him at weekends and school holidays from the age of 10!

In 1977, aged 16, fresh from school and armed with the little knowledge I had gained from my father, I started as an apprentice at a local plumbing firm in my home city of Nottingham where I gained a superb background of plumbing both industrial and domestic. In 1982, I joined my father's small plumbing firm where I stayed, until his retirement in 1999 at the ripe old age of 73. He sadly passed away in 2006.

In 1988, I started teaching part-time at the Basford Hall College (now New College Nottingham), the same college where I did my own training, initially teaching Heating and Ventilation and, soon after, plumbing at both craft and advanced craft and later NVQ Level 2 and Level 3. In July 2000, I moved to Stockport, where my teaching career continued at Stockport College on a full-time basis … and I'm still here!

To the readers of this book, I say simply, you are beginning a journey into a trade, which brings so much satisfaction when you get it right. To be a good plumber requires three D's – Desire, Discipline and Dedication. Be the best plumber that you can possibly be and always strive to achieve excellence.

FOREWORD

Congratulations on choosing a career in plumbing – you are now well on your way to a skilled and varied profession that will provide you with a job for life.

I'm a great believer in first rate training and that is exactly what this qualification offers. It is a respected hands-on training route into plumbing that will require a lot of skill, dedication and attention to detail. This book will guide you through your training with clear text, step-by-step procedures and detailed illustrations.

I qualified as a City & Guilds Plumber and Advanced Plumber more years ago than I care to remember, and it gave me the grounding and the skill set to launch Pimlico Plumbers. The skilled Apprentices that City & Guilds have provided us with over many years are invaluable.

If you put in enough hard work and practise the techniques detailed in this book you will become part of the next generation of skilled workers that are key to the future success of the plumbing industry. Best of luck!

Charlie Mullins

Charlie Mullins started his plumbing business from scratch and built it into a multi-million pound enterprise. He started as a plumbing apprentice at 15 and qualified as a City & Guilds Plumber and Advanced Plumber by the time he was 19. He then went alone and six years later Pimlico Plumbers was launched. It has grown into the most high-profile plumbing company in Europe, employing 200 people.

INTRODUCTION – HOW TO USE THIS TEXTBOOK

Welcome to your City & Guilds Level 2 NVQ Diploma in Plumbing and Heating textbook. It is designed to guide you through your Level 2 qualification and be a useful reference for you throughout your career.

Each chapter covers a unit from the 6189 Level 2 qualification. Each chapter covers everything you will need to understand in order to complete your written or online tests and prepare for your practical assessments.

Throughout this textbook you will see the following features:

KEY POINT
The thread direction refers to the direction in which fittings (such as hoses and regulators) screw on to their bottles. Fuel gases have left-hand threads so that their fittings cannot be mistakenly used with right-hand thread oxygen cylinders, and vice versa.

Key Point – These are particularly useful hints that may assist in you in revision for your tests or to help you remember something important.

Air admittance valve
Allows air into a stub stack to prevent the loss of trap seals.

Definitions – Words in colour in the text are explained in the margin to aid your understanding. They also appear in the glossary at the back of the book.

SUGGESTED ACTIVITY...
Who is the Local Authority in your area? Check out its website and see what services it offers to the construction industry.

Suggested activity – These hints suggest that you try an activity to help you practice and learn.

SmartScreen Unit 001 handout 1

SmartScreen – these icons refer to City & Guilds SmartScreen resources/activities. Ask your tutor for your log-in details.

At the end of every chapter are some 'Test your knowledge' questions. These questions are designed to test your understanding of what you have learnt in that chapter. This can help with identifying further training or revision needed. You will find the answers at the end of the book.

The Level 2 Plumbing qualification no longer includes an optional unit on sheet lead weathering. This will be covered separately in a continuing professional development qualification for plumbers and other trades involved with roofwork. City & Guilds Support Materials for this unit can be found at: www.smartscreen.co.uk/print

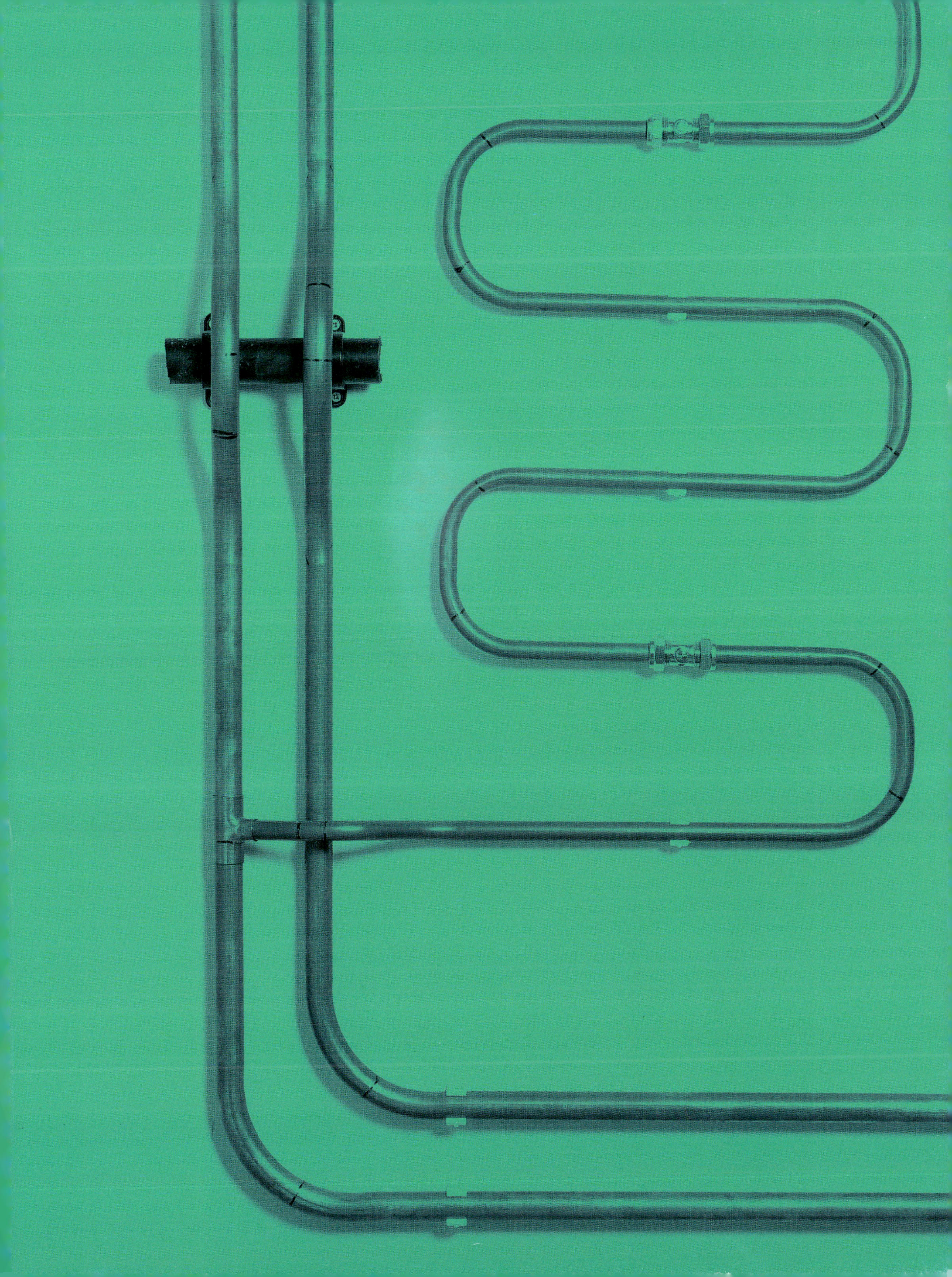

001
UNDERSTAND AND CARRY OUT SAFE WORKING PRACTICES IN BUILDING SERVICES ENGINEERING

Plumbers that work on construction sites are at risk from hazards and accidents every day. Construction is one of the UK's largest industries and arguably the most dangerous. In the past 25 years, nearly 3000 people have been killed on construction sites or as a direct result of construction work. In recent years, there has been a drop in fatality figures, and with proper health and safety management, this figure can be reduced further. The key factor to remember is that health and safety is everyone's responsibility.

IN THIS CHAPTER, YOU WILL COVER:

A Health and safety legislation in the construction industry
- **A1** The Health & Safety at Work Act 1974
- **A2** The Construction (Design and Management) Regulations 2007
- **A3** The Personal Protective Equipment at Work Regulations 1992
- **A4** The Control of Substances Hazardous to Health (COSHH) Regulations 2002
- **A5** Reporting of Injuries, Diseases and Dangerous Occurrences Regulations 1995 (RIDDOR)
- **A6** The Electricity at Work Regulations 1989
- **A7** The Work at Height Regulations 2005
- **A8** Manual Handling Operations Regulations 2005
- **A9** The Safety Signs and Signals Regulations 1996
- **A10** The Control of Lead at Work Regulations 2002
- **A11** The Control of Asbestos Regulations 2006
- **A12** Building services specific legislation
- **A13** The legal status of health and safety publications
- **A14** Who enforces the health and safety regulations?

B General and personal construction site safety
- **B1** Accident prevention and reporting
- **B2** Safety signs
- **B3** Working with hazardous substances
- **B4** Personal protective equipment
- **B5** Manual handling
- **B6** First-aid provision in the workplace

C Working safely with equipment on site
- **C1** Safety with electricity
- **C2** Safety with gas heating equipment
- **C3** Fire safety
- **C4** Working at height
- **C5** Working in trenches and excavations
- **C6** Working in confined spaces

Test your knowledge

A HEALTH AND SAFETY LEGISLATION IN THE CONSTRUCTION INDUSTRY

Hazards encountered by plumbers in particular include asbestos, strained muscles, broken bones, falls, slips, trips and noise. Diseases they risk contracting include dermatitis, asbestosis and emphysema.

In many instances, when the work is subcontracted on a construction project, there is confusion as to who is responsible for safety. However, legislation is very clear that everyone has duties and responsibilities regarding health and safety: the worker, each contractor, the Architect, the client, and the owner of the building being built.

In the first section of this chapter, we will look at some of the many pieces of legislation surrounding health and safety in the construction industry.

A1 THE HEALTH & SAFETY AT WORK ACT 1974

The Health & Safety at Work etc Act 1974 (HASAWA) is the principal piece of legislation covering work-based health and safety in the UK. This act lays down the principles for the management of health and safety at work. There are more specific pieces of legislation and codes of practice for various areas, such as the Control of Substances Hazardous to Health (COSHH) Regulations 2002 and the Personal Protective Equipment (PPE) Regulations 1992.

The HASAWA relates to all people at work (except domestic servants in private employment), whether they are employers, employees or the self-employed. It is specifically aimed at people and their activities at work rather than premises or processes. It includes provisions for both the protection of people at work and members of the general public who may be at risk as a consequence of the workplace activities. The main objectives of the HASAWA are to:

1. secure the health, safety and welfare of all people at work
2. protect others from the risks arising from work activities
3. control the obtaining, keeping and use of explosives and highly flammable substances
4. control emissions into the atmosphere of noxious or offensive substances.

Sections 2, 3, 7 and 8 of the Health & Safety at Work Act 1974 cover more general duties that relate directly to you, your employer and the general public. We will briefly cover these sections here.

Section 2 – The general duties of the HASAWA 1974

Section 2 of the HASAWA deals specifically with the general duties of the employer towards their employee, stating that the employer must, as far

Legislation
A law or group of laws that have come into force. Health and safety legislation for the plumbing industry includes the Health & Safety at Work Act and the Electricity at Work Regulations.

SmartScreen Unit 001 handout 1

as possible, ensure the health, safety and welfare of those in their employment. More specifically, this ensures the following:

- The plant and systems must be safe and without risk to health.
- There is no risk to health in connection with the use, handling, storage and transport of articles and substances.
- Information, instruction and supervision with regard to health and safety at work of employees is available.
- The working environment for employees is safe, without risk to health and adequate with regards to facilities and arrangements for their welfare at work.
- The place of work is maintained in a safe condition and without risk to health, and the means of access to it and egress from it are safe and without risk.

This legislation also states that employers must have a health and safety policy and if the company has more than five employees, that policy must be written down. It must be revised as necessary at regular intervals and all employees must have access to and be informed of any changes made to the policy.

Health and safety notice

KEY POINT
Every employer must consult with health and safety representatives. These people are appointed by employees of an organisation to act on their behalf. Their role is to make and maintain arrangements that will enable the employer and employees to promote and develop health and safety measures and to check their effectiveness.

Section 3 – The general duties of employers and the self-employed to people other than their employees

The HASAWA states that every employer must ensure, as far as is reasonably practicable, that people not in their employment who may be affected by their work are not exposed to risks to their health and safety. These duties also apply to the self-employed.

Every employer and self-employed person must give information to those people who are not in their employ about the way in which aspects of their work might affect the health and safety of others.

Additional employer responsibilities
In addition, the Health & Safety at Work Act 1974 tells us that all employers must:

- carry out risk assessments of all the company's work activities
- identify and implement adequate control measures
- inform all employees of the risk assessments and associated control measures
- review the risk assessments at regular intervals

- make a record of the risk assessments if five or more operatives are employed.

Section 7 – The general duties of employees at work

While at work, it is the duty of every employee to take reasonable care for the health and safety of him or herself and others. It is also essential that every employee cooperates with his or her employer to enable any duty or requirement to be performed or complied with.

Section 8 – Duty not to interfere with or misuse anything provided

As an employee, you must never intentionally or recklessly interfere with or misuse anything provided in the interests of health, safety or welfare. For example, a fire extinguisher must never be tampered with, and never modify a safety helmet by cutting out air holes.

A2 THE CONSTRUCTION (DESIGN AND MANAGEMENT) REGULATIONS 2007

The Construction (Design and Management) Regulations make up the principal piece of health and safety legislation specifically written for the construction industry. This legislation came into force on 6 April 2007, replacing and updating the previous Regulations.

The main aim of these regulations is to integrate health and safety into the management of large construction projects and to encourage everyone involved to work together to:

- improve the planning and management of projects at all stages
- identify hazards early on, eliminate or reduce these at the design planning stage, and manage the remaining risks throughout the project
- target effort where it can have the most beneficial effect on health and safety, and discourage unnecessary red tape.

The result should be that health and safety is treated as an essential part of a project's development and not as an afterthought. This ensures that the responsibility lies firmly with all individuals involved at every stage of the construction process.

These regulations require the appointment of a Construction Design and Management Coordinator, whose job it is to advise the client on health and safety issues during the design and planning phases of construction work. Their role includes the following:

- advising the client in the selection of competent designers and contractors
- helping to identify the information that will be needed by designers and contractors

CDM Coordinator

> **KEY POINT**
> The Health and Safety Executive (HSE) is the government body in the United Kingdom responsible for the encouragement, regulation and enforcement of workplace health, safety and welfare regulations and government legislation.

- coordinating the health and safety arrangements during planning and design work
- making sure that the Health and Safety Executive (HSE) is notified of the project (unless the client is a domestic one)
- advising on whether the initial construction phase plan is suitable
- preparing a health and safety file, which should include information for the client to make sure future cleaning, maintenance and alterations can be carried out safely

A summary of the duties of each party and how they are applied is given in the table below (courtesy of the HSE).

	All construction projects	**Additional duties for notifiable projects**
Clients (excluding domestic clients)	• Check competence and resources of all appointees. • Ensure that there are suitable management arrangements for the project welfare facilities. • Allow sufficient time and resources for all stages. • Provide pre-construction information to designers and contractors.	• Appoint CDM Coordinator*. • Appoint principal contractor*. • Make sure that the construction phase does not start unless there are suitable welfare facilities and a construction phase plan is in place. • Provide information relating to the health and safety file to the CDM Coordinator. • Retain and provide access to the health and safety file. (*There must be a CDM Coordinator and principal contractor until the end of the construction phase.)
CDM Coordinators		• Advise and assist the client with his/her duties. • Notify HSE. • Coordinate health and safety aspects of design work and cooperate with others involved with the project. • Facilitate good communication between client, designers and contractors. • Liaise with principal contractor regarding ongoing design. • Identify, collect and pass on pre-construction information. • Prepare/update health and safety file.
Designers	• Eliminate hazards and reduce risks during design. • Provide information about remaining risks.	• Check client is aware of duties and CDM Coordinator has been appointed. • Provide any information needed for the health and safety file.

Principal contractors		- Plan, manage and monitor construction phase in liaison with contractor.
- Prepare, develop and implement a written plan and site rules (initial plan completed before the construction phase begins).
- Give contractors relevant parts of the plan.
- Make sure suitable welfare facilities are provided from the start and maintained throughout the construction phase.
- Check competence of all appointees.
- Ensure that all workers have site inductions and any further information and training needed for the work.
- Consult with the workers.
- Liaise with CDM Coordinator regarding ongoing design.
- Secure the site.
Contractors	- Plan, manage and monitor own work and that of workers.
- Check competence of all their appointees and workers.
- Train own employees.
- Provide information to their workers.
- Comply with the specific requirements in Part 4 of the Regulations.
- Ensure that there are adequate welfare facilities for their workers.	- Check client is aware of duties and that a CDM Coordinator has been appointed and HSE notified before starting work.
- Cooperate with principal contractor in planning and managing work, including reasonable directions and site rules.
- Provide details to the principal contractor of any contractor whom he or she engages in connection with carrying out the work.
- Provide any information needed for the health and safety file.
- Inform principal contractor of problems with the plan.
- Inform principal contractor of reportable accidents, diseases and dangerous occurrences.
Workers/everyone	- Check own competence.
- Cooperate with others and coordinate work so as to ensure the health and safety of construction workers and others who may be affected by the work.
- Report obvious risks.	

Mandatory helmet symbol

COSHH stamp

COSHH data sheet

A3 THE PERSONAL PROTECTIVE EQUIPMENT AT WORK REGULATIONS 1992

Employers have basic duties concerning the provision and use of personal protective equipment (PPE) at work wherever there are risks to health and safety that cannot be adequately controlled in other ways.

PPE is defined in these regulations as all equipment that is intended to be worn or held by a person at work and that protects them against one or more risks to their health or safety. An example of this would be safety helmets, gloves, eye protection, high visibility clothing, safety footwear and safety harnesses, but there are many others. Hearing protection and respiratory (breathing) protective equipment provided for most work situations are not covered by the PPE Regulations because other regulations are in force, which deal specifically with these areas. However, these items need to be compatible with any other PPE provided.

The Regulations require that PPE is:

- properly assessed before use to ensure that it is suitable
- maintained and stored correctly
- provided with training and instructions on how to use it safely
- used correctly by employees.

All employers must provide PPE free of charge, whether it is returnable or disposable. (This also applies to agency workers not in the employer's full employment.) PPE must also be provided to members of the public who are at risk, for example site visitors. If PPE is provided, it must be used, and any lost or damaged equipment reported to the employer.

A4 THE CONTROL OF SUBSTANCES HAZARDOUS TO HEALTH (COSHH) REGULATIONS 2002

The Control of Substances Hazardous to Health Regulations, known as COSHH, are intended to protect people from illness caused by exposure to hazardous substances. The Regulations require employers to do the following:

- Assess the risks to health and safety.
- Decide what precautions are needed to prevent ill health.
- Prevent or control exposure.
- Make sure that the control measures are used and maintained.
- Monitor exposure and carry out health checks if needed.
- Make sure that all employees are properly informed, trained and supervised.

To comply with COSHH, the HSE recommends that employers follow the eight steps outlined on the following page.

Step one	Assess the risks	Assess the risks to health from hazardous substances used in or created by your workplace activities.
Step two	Decide what precautions are needed	Your employer must not carry out work which could expose you to hazardous substances without first considering the risks and the necessary precautions.
Step three	Prevent or adequately control exposure	Your employer must prevent you from being exposed to hazardous substances. Where preventing exposure is not reasonably practicable, then your employer must adequately control it.
Step four	Ensure that control measures are used and maintained	Your employer must ensure that control measures are used and maintained properly and that safety procedures are followed.
Step five	Monitor the exposure	Your employer should monitor the exposure of employees to hazardous substances, if necessary.
Step six	Carry out health surveillance	Your employer must carry out appropriate health surveillance where the risk assessment has shown this is necessary or where COSHH sets specific requirements.
Step seven	Prepare plans and procedures to deal with accidents, incidents and emergencies	Your employer must prepare plans and procedures to deal with incidents and emergencies involving hazardous substances, where necessary.
Step eight	Ensure employees are properly informed, trained and supervised	Your employer should provide you with suitable and sufficient information, instruction and training.

Under the COSHH Regulations, hazardous substances are defined as:

- chemicals – classified under 'Chemicals Regulations' and identifiable by orange hazard warning symbols on the container (care should be taken with unmarked containers)
- any substance that has been assigned a workplace exposure limit
- dusts in concentrations in the air greater than 10mg/m^3 for inhaled dust, or 4mg/m^3 of respirable dust
- biological agents such as bacteria, viruses, fungi and parasites
- asphyxiants such as carbon dioxide and nitrogen
- carcinogens such as radon gas or tobacco smoke.

Hazardous substances can enter the body by:

- breathing in vapours, gases, dusts and fumes
- eating or drinking substances or foods contaminated by hazardous substances

> **KEY POINT**
>
> There are many forms of hazardous substance, for which manufacturers and suppliers produce COSHH data sheets. The data sheet is an invaluable source of safety information and is designed to make you aware of the known hazards associated with a material or substance, advise you of safe handling procedures, and recommend the most effective response to accidents.

- contact with the skin or absorption into the body through the skin causing harm to internal organs or via cuts or wounds
- contact with the eyes by fumes, vapours, liquids and dusts.

A5 REPORTING OF INJURIES, DISEASES AND DANGEROUS OCCURRENCES REGULATIONS 1995 (RIDDOR)

The Reporting of Injuries, Diseases and Dangerous Occurrences Regulations (RIDDOR) applies to all work activities. The Regulations place a legal duty on your employer, the self-employed and people in control of work premises to report certain work-related accidents, diseases and dangerous occurrences by the quickest means possible (see box, left). RIDDOR applies to all work activities but not all incidents are reportable. Those that must be reported are:

- deaths
- major injuries
- 'over-three-day' injuries – where an employee or self-employed person is away from work or unable to perform their normal work duties for more than three consecutive days due to injury
- injuries to members of the public or people not at work, where they are taken from the scene of an accident to hospital
- certain work-related diseases
- dangerous occurrences – where something happens that does not result in an injury, but could have done.

'Gas Safe' registered gas fitters must also report dangerous gas fittings they find, and gas conveyors/suppliers must report certain flammable gas incidents.

> **KEY POINT**
> RIDDOR contacts:
> - Report online at www.hse.gov.uk/riddor/index.htm
> - Report by phone – 0845 300 99 23 (only deaths and major injuries can be reported by phone).

A6 THE ELECTRICITY AT WORK REGULATIONS 1989

The Electricity at Work Regulations place legal responsibilities on employers and employees to ensure that fixed electrical equipment and portable appliances are tested and maintained, and that regular inspections are carried out to ensure that they are safe to use. Verifiable evidence of this is required in the form of:

- documented inspection and testing records such as Portable Appliance Test (PAT) records and test certificates
- evidence that training has been carried out
- electrical authorisations
- the control of work activities
- competent persons.

The Regulations ensure that precautions are taken to avoid death or personal injury from electricity during work activities. The main requirements are as follows:

Electrical testing

- Make sure that all persons working on or near electrical equipment are competent.
- Maintain electrical systems in safe condition.
- Carry out electrical work safely.
- Ensure that equipment is suitable and safe to use in terms of:
  - strength and capability
  - use in adverse or hazardous environments (for example, weather, dirt, dust, gases, mechanical hazards and flammable atmospheres).
- Ensure effective insulation of conductors in a system.
- Ensure effective earthing of the system.
- Ensure that if work is carried out to the earthing system that involves breaking the flow of current, other precautions are taken to maintain the earth continuity.
- Ensure that all components of the electrical system are suitable and safe for use.
- Protect against system overload.
- Provide suitable means for cutting off the supply of electrical current to any electrical equipment and effective isolation of electrical equipment.
- Work should not be carried out on or near a live conductor unless absolutely essential and suitable precautions are taken to prevent injury.
- Ensure adequate working space, access and lighting to all electrical equipment where work is undertaken.

A7 THE WORK AT HEIGHT REGULATIONS 2005

The Work at Height Regulations apply to all work at height where there is a risk of a fall that may cause personal injury. These regulations place duties on employers, the self-employed and any person that controls the work of others such as managers, supervisors or building owners who may use contractors to work at height. As part of the Regulations, duty holders must ensure that:

- all work at height is properly planned and organised
- those people working at height are competent
- the risks from working at height are assessed and the correct work equipment is selected and used
- equipment for working at height is regularly inspected and properly maintained.

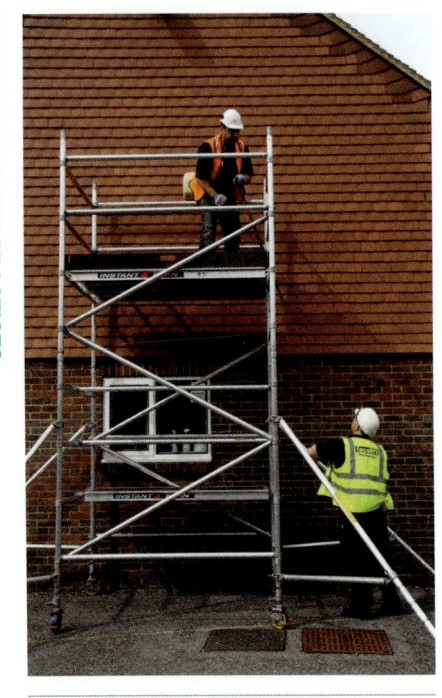

Duty holders must also:

- ensure that working at height is avoided where possible
- use work equipment or other measures to prevent falls where working at height is unavoidable
- where they cannot eliminate the risk of a fall, use work equipment or other measures to reduce the distance of the fall.

The Regulations also include requirements for:

- existing places of work and means of access for working at height
- collective fall prevention equipment such as guardrails and working platforms
- collective fall arresters such as nets and airbags
- personal fall protection such as harnesses and work restraints
- ladders.

Safe working at height

A8 MANUAL HANDLING OPERATIONS REGULATIONS 2005

The Manual Handling Operations Regulations apply to a wide range of manual handling activities including lifting, lowering, pushing, pulling and carrying. In the Regulations, loads are described as being either inanimate – for example a box or trolley, or animate – a person or animal.

The Regulations require employers to:

- avoid hazardous manual handling operations so far as is reasonably practicable
- assess any hazardous manual handling operations that cannot be avoided
- reduce the risk of injury so far as is reasonably practicable, including automating or mechanising the lifting process as much as possible.

Employees have a duty to make full and proper use of any system of work provided by their employer to reduce risks of manual handling injuries.

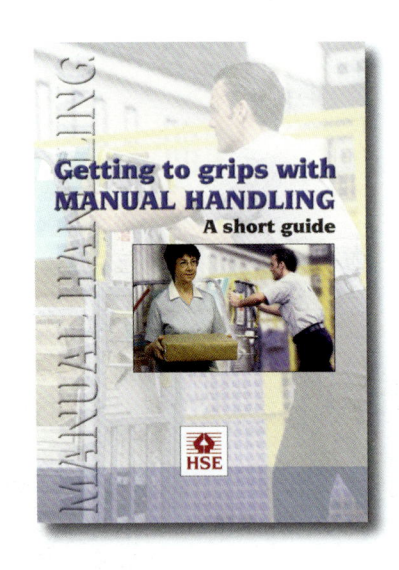

Manual handling regulations booklet

A9 THE SAFETY SIGNS AND SIGNALS REGULATIONS 1996

The Safety Signs and Signals Regulations require employers to provide specific safety signs whenever and wherever there is a risk that cannot been avoided or controlled in other ways. This includes the use of road traffic signs within workplaces to control road traffic movements. The Regulations also place a duty on employers to keep the safety signs in good condition and explain unfamiliar signs to their employees, giving instructions on what they need to do when they see a safety sign.

KEY POINT

The Safety Signs and Signals Regulations apply to all places of work, but do not include signs and labels used in connection with the supply of substances, products and equipment or the transport of dangerous goods.

The Regulations also cover other methods of conveying health and safety information, including the use of illuminated signs, hand and audible signals such as fire alarms, spoken communication and the marking of pipework containing dangerous substances. These are in addition to the traditional safety signs such as prohibition and warning signs. Fire safety signs are also covered.

A10 THE CONTROL OF LEAD AT WORK REGULATIONS 2002

The Control of Lead at Work Regulations apply to all work that exposes any person to lead in any form whereby the lead may be ingested, inhaled or absorbed into the body. This is relevant to plumbers as the lead may be absorbed through the skin when it is being handled or through the fumes breathed in when lead welding.

The Regulations state that the employer must assess the nature and extent of the exposure to lead so that the measures of control will be adequate based on that assessment. Where there is 'significant' exposure to lead, all the Regulations will apply, but below this level only some of the Regulations will apply.

The basic measure to protect employees from absorbing lead is the prevention of the escape of lead dust, fume or vapour into the workplace. Personal hygiene is important in controlling lead absorption and the provision and use of adequate washing facilities and personal protective equipment (PPE) is a basic requirement. Food and drink should not be consumed in any place that may be contaminated by lead. Adequate alternative arrangements should be made. Employees should be given sufficient information and training regarding hazards, precautions and duties under the Regulations.

Working with lead, and the symptoms of lead poisoning will be covered in detail later in this unit.

A11 THE CONTROL OF ASBESTOS REGULATIONS 2006

The Control of Asbestos Regulations prohibits the importation, supply and use of all forms of asbestos. The Regulations follow on from the ban introduced in 1985 for blue and brown asbestos and in 1999 for white asbestos. The ban on the second-hand use of asbestos products such as asbestos cement sheets and asbestos boards and tiles also remains in place.

The ban applies to new use of asbestos. If existing asbestos-containing materials are in good condition, they may be left in place providing that their condition is monitored and managed to ensure they are not disturbed.

Asbestos will be covered further later on in this unit.

General mandatory sign

KEY POINT

An Approved Code of Practice (ACoP), 'Control of Lead at Work', is available and should be used in conjunction with the Regulations.

Asbestos shown in poor condition

A12 BUILDING SERVICES SPECIFIC LEGISLATION

The term 'building services' is used to describe those activities not connected with the construction of the building but related to the services that are installed within the building as it is constructed. The services in a building are:

- water
- gas
- electricity
- heating and ventilation
- telecommunications.

The building services industry has specific legislation to ensure the health and safety of the general public.

The Water Supply (Water Fittings) Regulations 1999

These relate to the supply of drinking water, specifically targeting the prevention of contamination, waste, undue consumption, misuse and erroneous metering. They will be covered in more detail in Chapter 006.

The Gas Safety (Installation and Use) Regulations 1998

These cover the installation, maintenance and use of gas and gas appliances, aimed at preventing carbon monoxide (CO) poisoning, fires and explosions. The Regulations state that all Gas Engineers must be registered with the Gas Safe Register to prove their competency, and that it is the responsibility of landlords to ensure their tenants' pipework and appliances are safe to use.

The 17th Edition IEE Regulations (BS 7671)

These are the national standards to which all wiring should now conform. Any person involved in the design, installation, inspection and testing of electrical installations must have a sound knowledge of the document.

KEY POINT
The IEE Regulations are produced by the Institute of Engineering and Technology (IET), the industry body that covers electrical installation. The 17th Edition contains many major changes that align it with other similar European documents.

A13 THE LEGAL STATUS OF HEALTH AND SAFETY PUBLICATIONS

Health and safety publications can be divided into two distinct groups – mandatory (must be followed by law) and advisory (guidance, which is recommend but not legally enforceable).

Mandatory publications can be:

- Acts of parliament – these create a new law or change an existing one. Their implementation is the responsibility of a specific government department. In the case of health and safety acts, this is the Health and Safety Committee.

- Regulations – these are rules, procedures and administrative codes set by authorities or governmental agencies to achieve a particular objective. They are legally enforceable and must be followed to avoid prosecution.

Advisory publications can be:

- Approved Codes of Practice (ACoP) – these are documents giving practical guidance on complying with the Regulations. Although it is not an offence not to comply with an ACoP, in the case of health and safety ACoPs, proof that their advice has been ignored could be seen as evidence of guilt if an employer or employee faces criminal prosecution under health and safety law. Following an ACoP is considered good practice.

- Guidance notes – these are produced by the Health and Safety Executive (HSE) to help people interpret and understand what is required by law, and to comply with it. They also give technical advice. The course of action set out in guidance notes is not compulsory, but if the guidance is followed, it is usually enough to comply with the law.

ACoP front cover

A14 WHO ENFORCES THE HEALTH AND SAFETY REGULATIONS?

Health and safety law is enforced by the Health and Safety Executive (HSE) and the Local Authority. These bodies work in partnership under the Health and Safety Executive/Local Authorities Enforcement Liaison Committee (HELA). Both employ Health and Safety Inspectors, whose job it is to ensure that the laws are adhered to.

The role of the Health and Safety Inspectors

Heath and Safety Inspectors have the legal right to enter a workplace without giving notice, although notice may be given where the Inspector considers it appropriate. On a normal inspection visit, the Inspector would look at the place of work, work activities and the management of health and safety, and check that the employer is complying with health and safety law. The Inspector may offer guidance and advice or talk to employees, take photographs and samples, serve improvement notices or take action if a risk to health and safety is perceived.

If a breach of health and safety law is found, the Inspector will decide what action to take. The action will depend on the severity of the breach. The Inspector should provide employees or their representatives with information relating to the breach and any necessary action.

There are several ways in which an Inspector may take enforcement action to deal with a breach of the Regulations:

Informal action
Where the breach of the law is comparatively small, the Inspector will advise the duty holder what action to take to conform with the requirements of the law. If asked, this can be given in writing.

Health and Safety Inspector

Improvement notice
More severe breaches will receive a direct order to take specific action to comply with the law. The Inspector will discuss the improvement notice with the duty holder and resolve points of difference before serving it. The notice will say what has to be done, why and by when. The time period to take the corrective action will be a minimum of 21 days, to allow the duty holder time to appeal to an Industrial Tribunal.

Prohibition notice
Where an activity involves a risk of serious personal injury, the Inspector may issue a prohibition notice forbidding the activity either immediately or after a specified time period. This notice will not be lifted and work will not be allowed to resume until corrective action has been taken.

Prosecution
In some cases, prosecution may be deemed necessary. Failure to comply with an improvement or prohibition notice, or a court remedy order, carries a fine of up to £20,000 or 6 months' imprisonment, or both. Unlimited fines and in some cases imprisonment may be given by higher courts.

B GENERAL AND PERSONAL CONSTRUCTION SITE SAFETY

We will now look at construction site safety from a general and personal point of view. We will examine general site hazards and how we can either help or hinder our own health and safety and that of those around us.

B1 ACCIDENT PREVENTION AND REPORTING

An accident is an unexpected or unplanned event that could result in personal injury, damage or sometimes death. When an accident occurs, there are always reasons for it and if there's a reason then there is usually blame.

Accidents do not just happen – they are caused. Finding out what causes accidents is the first step towards preventing them in the future. Accident prevention is something that everyone can practise. It means being able to recognise dangerous situations and take steps to remove the danger, and it is the responsibility of everyone engaged in any way on a construction site.

The causes of accidents
Hazards on site can be divided into three specific groups:

1 general site and work area cleanliness, which can lead to trips, slips and falls

2 equipment and PPE that is inadequate for the job, not provided (in the case of PPE) or defective

3 Personal conduct:

- incorrect manual handling methods
- incorrect methods of working at heights, in trenches and on excavations
- not taking enough care and attention in dangerous environments
- using equipment or carrying out activities without appropriate training
- taking risks.

Some examples of things that can lead to accidents in the workplace are:

- poor storage of materials
- poor weather conditions
- electrical faults
- PPE or clothing not used or worn
- inadequate lighting/heating or noise
- inadequate training and supervision
- defective tools and equipment
- excessive haste or taking shortcuts in order to get the job done
- lack of preparation and failure to comply with instructions and rules of safety
- lack of concentration due to distraction or lack of interest in the job
- unsafe methods of handling and lifting
- failure to use guards provided
- working under the influence of drugs or alcohol.

If safe working practices are followed, accidents in the workplace may be prevented.

SUGGESTED ACTIVITY...
Do any of the things you normally see and do at work have the potential to be a source of danger? Write a list of potential accidents and against each one write down an action you could take to reduce the risk of it happening.

KEY POINT
A hazard is anything that may cause harm, such as chemicals, electricity, gas or working from ladders. The risk is the chance, no matter how high or low, that somebody could be harmed by these and other hazards, together with an indication of how serious the harm could be.

Risk assessments

A risk assessment is a detailed examination of any situation to assess whether enough steps have been taken to manage the risk to individuals. Your employer is legally required to assess the risks in the workplace and to implement measures to control those risks. The law does not expect an employer to eliminate all risk, but they are expected to take steps to ensure health and safety as far as is reasonably practicable.

There are five basic steps to risk assessment (as recommended by the HSE):

Step one	Identify the hazards	Work out how people could be harmed. Do this by: • walking around the site • asking employees what they think • visiting the HSE website for practical guidance • contacting trade associations for advice • checking manufacturers' instructions and COSHH data sheets.
Step two	Decide who might be harmed and how	Identify the group(s) of people at risk.
Step three	Evaluate the risks and decide on precaution	• Can I get rid of the hazard altogether? • If not, how can I control the risks so that harm is unlikely?
Step four	Record your findings and implement them	• A proper check was made. • You asked who might be affected. • You dealt with all the significant hazards, taking into account the number of people who could be involved. • The precautions are reasonable, and the remaining risk is low. • You involved your staff or their representatives in the process.
Step five	Review your assessment and update if necessary	Review the risk assessments every year: • Have more employees joined the company? • Has new machinery or equipment been installed? • Have any fellow workers spotted any problems? • Has anything been learned from accidents or near misses?

Risk assessment form

SmartScreen Unit 001 handout 3

SmartScreen Unit 001 worksheet 1

Method statements

A method statement is usually completed after the risk assessment. This document outlines the way in which a worker should complete a task or process. Included in the method statement is an outline of the hazards involved. It should also include a step-by-step guide on how the work may be completed in a safe manner.

A client may request a method statement as part of a tender process. This allows the client to get an idea of how the company operates. A method statement may also be known as a 'safe system of work'.

Permits to work

When work has been identified as being 'high risk', strict health and safety controls are required. A permit to work is a document produced by those authorising the work and those carrying it out that gives authorisation for named persons to carry out specific work within a nominated time frame. It describes the work and how it will be carried out (more detail is given in the method statement). It also lists the precautions that are required to complete the work safely, based on a written risk assessment.

Work affecting the public and their health and safety

It is not only construction workers that suffer accidents as a result of construction work. Members of the public are killed or injured each year. Accidents can often occur when people are walking close to where buildings are being constructed, refurbished or demolished. Work near to where the general public have access must be planned and executed correctly, taking into account people with pushchairs, people with disabilities, and older people.

The best way to protect the public from the dangers of construction sites is to restrict access:

- Erect a 2m high perimeter fence. If parts of it need to be taken down for access, make sure that it is put back at the end of the day.
- Lock the site gates and any windows and doors at night.
- If work is being done in an occupied property, clear responsibilities need to be established with the occupier for maintaining the fencing.
- If the work is near a school or residential area, enlist the help of the head teacher or the residents' association to discourage children and young people from entering the site.
- Cover trenches, excavations and scaffolds, and remove all ladders.
- Store materials so that there is no risk of them toppling over.
- Lock away hazardous substances.
- Protect passers-by from falling objects from scaffolds by the use of toe boards, brick guards and netting.
- Use plastic sheeting to retain dust, drips and splashes.
- Tie down or remove loose materials from scaffolds.
- Ensure that warning and danger signs are posted on and around the scaffold.
- Initiate other security methods, such as using security guards.

For more information on restricting access to the public, see the HSE website: www.hse.gov.uk/construction/safetytopics/index.htm

Accident reporting

Every accident must be recorded. An accident report book or record forms should be on every site or place of work, usually with the Site Manager, or whoever is in charge of the site or workshop. Accidents where persons require hospital treatment must be recorded at the place of work, even if no treatment was given there. Serious injuries are reportable under RIDDOR (see page 10). It is important that you report any accident that you are involved in to your supervisor as soon as possible.

There is no set place to keep an accident report book, but it needs to be kept in an accessible place and employers must make employees aware of where the book is kept. Often it is kept where first aid is available.

All accidents have to be entered in the accident book and the following information must be recorded:

- name, address and occupation of the injured person
- signature of the person making the entry, address and occupation – must then be dated
- when and where the accident happened
- brief description of the accident, cause and what injury occurred
- if the accident is of such a nature that it has be reported to the HSE.

All accidents that cause death or major injury to an employee or member of the public must be reported to the Health and Safety Executive (HSE) or the Local Authority Administrator for Health and Safety. A major injury is defined as certain fractures, amputations, loss of sight or anything that requires hospital treatment for more than 24 hours.

All accidents in the workplace, whether fatal or otherwise, are investigated. Those involved in the investigation may include:

- the employer
- an investigator from an insurance company, acting on behalf of the employer or employee
- a safety representative, usually from a trade union
- a Health and Safety Inspector from the local authority or the HSE.

B2 SAFETY SIGNS

Safety signs are used on construction sites where risks have not been avoided by other means. Employers are required to provide and maintain safety signs, and workers need to be trained in the recognition of safety signs and symbols. To ensure that the correct number and type of safety signs have been used, an employer must carry out a number of simple tasks:

SmartScreen Unit 001 handout 10

Accident report book

SmartScreen Unit 001 handout 4

- conduct a risk assessment
- ensure fire equipment and emergency exits are clearly indicated
- use signs to prohibit entry into dangerous areas
- make sure that mandatory requirements, such as wearing PPE, are clearly shown
- clearly indicate all first-aid areas and equipment
- use signs to show prohibited behaviour, such as 'no smoking'.

The signs used must communicate their message clearly and effectively. Safety signs must comply with EC Safety Signs Directive (92/58/EEC), the purpose of which is to encourage the standardisation of safety signs throughout the member states of the European Union. Safety signs are divided into six separate groups:

SmartScreen Unit 001 interactive activity 1

Category	Description	Example
Prohibition	**Colour:** A red circular band with a diagonal cross bar on a white background; the symbol within the circle is black **Purpose:** To indicate that a certain behaviour is prohibited	
Hazard	**Colour:** A yellow triangle with a black border and black symbol **Purpose:** To warn of any type of hazard	
Mandatory	**Colour:** A blue circle with a white symbol **Purpose:** To indicate that a specific course of action must be taken	
Fire equipment	**Colour:** A red rectangle or square with a white symbol **Purpose:** To describe the location of firefighting equipment	
Safe condition	**Colour:** A green rectangle or square with a white symbol or text **Purpose:** To provide information about safe conditions, such as emergency exit routes	
Warning	**Colour:** An orange rectangle or square with black edges and a black symbol **Purpose:** To make aware of possible danger	

Occasionally, a sign may be seen that is a mixture of different types of signs. These are known as combination signs.

B3 WORKING WITH HAZARDOUS SUBSTANCES

Section 7 of the Health & Safety at Work Act states that the employer must prevent or control their employees' exposure to hazardous substances.

In most cases, hazardous substances can be divided into six main categories:

Category	Description	Examples
Toxic	Poisons and dangerous substances that have the ability to cause death if ingested, inhaled or absorbed into the body.	Cyanide, asbestos, lead
Harmful	Harmful substances could be in any form – liquid, solid (dust particles) or gas.	Fluxes, solvents, cleaning fluids, chemicals, dust
Corrosive	Substances that have the ability to cause severe burns to exposed parts of the body.	Hydrochloric acid, sulphuric acid, caustic soda
Irritant	Can cause irritation of the skin, eyes, nose and throat.	Fibreglass roof insulation, some paints, solvents and sealants
Oxidising	Induces materials to burn fiercely by adding oxygen to a fire.	Oxygen from welding bottles
Extremely flammable	Has the potential to burn fiercely if the substance is either exposed to a source of ignition or subjected to temperatures close to its flash point so that it spontaneously combusts.	Petrol, LPG, acetylene gas, solvent weld adhesives and cleaning agents

Chemicals

There are many chemicals that may be found on construction sites. These include:

- asbestos
- lead
- fluxes
- cadmium (found in plastics such as PVCu)
- carbon monoxide (from use of blowtorches, welding, generators, gas heaters etc)
- welding fumes (from welding metals such as steel pipes)
- flux fumes from soldering copper tubes and fittings
- spray paints
- cutting oil mists (cutting and threading low carbon steel tubes)
- solvents (these have many uses on construction sites such as cleaning agents)
- jointing compounds.

The effects on your health from exposure to chemicals can range from mild to very severe. In some cases (for example with asbestos) it may be years before the effects are felt.

Working with lead

As part of your job as a plumber, you may be asked to work with lead. This may be in replacing a lead pipe or installing sheet lead weatherings and roof work. Lead is a highly toxic metal that can enter the body through any of the following methods:

- absorption: touching and handling lead without the use of barrier cream
- ingestion: not observing personal hygiene by not washing your hands before eating and drinking after handling lead
- inhalation: breathing lead fumes when lead welding or soldering with leaded solder.

Lead work

Lead is a very powerful neurotoxin that damages the central nervous system and leads to brain and blood disorders. Lead oxide in the form of a white powder from the corrosion of lead is particularly dangerous. The symptoms of lead poisoning are:

- headaches
- tiredness
- irritability
- constipation
- nausea
- stomach pains
- anaemia
- loss of weight.

Continued uncontrolled exposure could cause more serious symptoms such as:

- kidney damage
- nerve and brain damage
- infertility.

What you and your employer must do to protect your health at work when working with lead

If you are exposed to lead or lead compounds (eg lead oxide, dust, fume or vapour from lead welding or smelting) while you are at work, your employer must do the following:

- Assess the risk to your health to decide whether or not your exposure is 'significant' and what precautions are needed to protect you.

- Put in place systems of work, such as fume and dust extraction, to prevent or control your exposure to lead, and to keep equipment in good working order.

- Provide washing and changing facilities, and places free from lead contamination where you can eat and drink.

- Inform you about the risks to your health from working with lead, and the precautions you should take.

- Train you to use any control measures and protective equipment correctly.

- Provide you with protective clothing and arrange for that clothing to be laundered.

- Measure the amount of lead in the air that you are exposed to and tell you the results. If your exposure to lead cannot be kept below a certain level then your employer must issue you with respiratory protective equipment.

- Arrange to measure the level of lead in your body. This is done by a simple blood test administered by a doctor at your place of work. You must be told the results of your tests.

There are ways you can help yourself too:

- Make sure that you have all the information and training you need to work safely with lead, including what to do in an emergency, such as a sudden uncontrolled release of lead dust or fume into the atmosphere.
- Use all the equipment provided by your employer and follow its instructions for use.
- Follow good work practices, keeping your immediate work area as clean and tidy as possible and taking care not to take home any PPE such as overalls or protective footwear.
- Wear any necessary PPE clothing and respiratory protection.
- Report any damaged or defective equipment to your employer.
- Only eat and drink in designated areas that are free from lead contamination.
- Practise a high standard of personal hygiene, washing your hands, face and nails regularly and showering before leaving the site when necessary.
- Do not miss medical appointments with the doctor where you work.

(Taken from HSE recommendations.)

Working with fluxes

Flux is a paste compound that helps solder to adhere to copper tubes and copper-based fittings. The term we use for this process is 'wetting'. There are two basic types of flux used today in the plumbing industry – traditional and self-cleaning.

Traditional fluxes

Traditional grease-based fluxes often contain a chemical called 'rosin' (also known as 'colophony') or zinc chloride. Rosin is a natural, solid resin-type material obtained from pine trees; when heated, it forms acidic particles that can irritate the breathing. This could lead to occupational asthma. Zinc chloride is corrosive and could cause skin irritation, burns and eye damage if it gets in the eye. Take care when using this kind of flux. It is recommended that you check COSHH data sheets for further information regarding these products.

Self-cleaning fluxes

This type of flux is also known as 'active' flux because of its aggressive nature. Most are based on zinc chloride or hydrochloric acid, both of which can cause burns and severe skin irritation, and so careful handling and use is very important. Other self-cleaning fluxes may use natural enzymes as cleaning agents but these are also known to irritate the skin. Again, it is recommended that you check COSHH data sheets for further specific health and safety information regarding these products.

All flux should be handled with care. Use a brush to apply the paste and always wash your hands thoroughly after use.

A typical self-cleaning flux

Working with solvents

A variety of solvents with differing degrees of toxicity are used in construction. They are in paints, adhesives, epoxy resins and other products.

Generally, exposure to excessive amounts of solvent vapours is greater when solvents are handled in enclosed or confined spaces. Care should be taken when using solvent adhesives to solvent weld PVCu pipes and fittings in confined spaces. Solvents can:

- irritate your eyes, nose or throat
- make you dizzy, sleepy, give you a headache or cause you to pass out
- affect your judgment or coordination
- cause internal damage to your body
- dry out or irritate your skin.

When working with solvents, follow the basic instructions as listed below:

- Avoid contact with the skin.
- Avoid contact with the eyes.
- Only use in an open, well-ventilated space.
- Keep away from naked flames because solvents are flammable.
- Store in a well-ventilated, secure area.

Working with asbestos

Asbestos is one of the most dangerous materials that you will come across during your work as a plumber. The HSE estimates that on average, eight joiners, six electricians and four plumbers die every week from an asbestos-related disease. Therefore, it is vital that you know what you should do if you encounter asbestos.

What is asbestos?

Asbestos is a naturally occurring fibrous material that has been used as a building material since the end of the 1940s. It is often mixed with other materials, such as cement, so it is hard to know when you are working with it. Asbestos is not used in any new-builds, but if you work in a building built before the year 2000, it is likely that asbestos has been used during its construction in one form or another.

In the past, it was used extensively for the following plumbing-specific applications:

- flue pipes
- gutters and rainwater pipes
- soil and vent pipes
- pipe insulation (both sprayed on and applied as a paste and wrapped in linen)
- boiler gaskets and fire-proof ropes
- cold water cisterns.

Some domestic uses of asbestos

It may also be found in:

- Artex
- roof and ceiling tiles
- soffit boards
- plaster coatings
- floor tiles and coverings
- asbestos sheeting and corrugated roofing.

There are three main types of asbestos:

1 Chrysotile (white asbestos) – a white curly fibre. Chrysotile accounts for 90 per cent of asbestos in products and is a member of the serpentine group. It is a magnesium silicate.

2 Amosite (brown or grey asbestos) – straight amosite fibres that belong in the amphibole group, and contain iron and magnesium.

3 Crocidolite (blue asbestos) – another member of the amphibole group. Crocidolite takes the form of blue, straight fibres. It is a sodium iron magnesium silicate.

Other forms of asbestos include:

- anthophyllite
- tremolite
- actinolite.

What are the hazards with asbestos?

The presence of asbestos alone does not necessarily constitute a health risk. Providing the fibres are intact and are not disturbed the risk is relatively low. However, once the fibres are loose and enter the atmosphere the risk increases dramatically. The asbestos is inhaled into the lungs, which causes certain types of lung diseases:

1 **asbestosis** – a process of widespread scarring of the lungs

2 **disease of the lining of the lungs (the pleura)** – has a variety of signs and symptoms and is the result of inflammation and the hardening (calcification) and/or thickening of the lining tissue

3 **mesothelioma** – a rare form of lung cancer

> **KEY POINT**
>
> For further information and resources about asbestos visit:
> www.hse.gov.uk/asbestos/

Dealing with asbestos

Asbestos-containing materials should have been identified before work begins but there is always the risk that some may be hidden on site and is not found until work has started. If you think you have found asbestos, stop work at once and alert people that asbestos may be present. Asbestos is a difficult substance to identify, so it is better to assume that a material contains asbestos until proven otherwise. Do not return to the site until it has been deemed safe to do so. The following advice is based on the recommendations of the HSE.

Don't start work if:

- you are not sure if there is asbestos where you are working
- the asbestos materials are sprayed coatings, board or insulation and lagging on pipes and boilers (only licensed contractors should work on these)
- you have not been trained on non-licensed asbestos work – basic awareness is not enough.

You should only continue if:

- the work has been properly planned and the right precautions are in place and you have the correct equipment
- the materials are asbestos cement, textured coatings and certain other materials which do not need a licence
- you have had training in asbestos work and know how to work with it safely.

If you work with asbestos:

- use hand tools and not power tools
- keep materials damp, but not too wet
- wear a properly fitted, suitable mask (eg Disposable FFP3 type) – an ordinary dust mask will not be effective
- don't smoke, eat or drink in the work area
- double bag asbestos waste and label the bags properly
- clean up as you go and use a special (class H) vacuum cleaner, not a brush
- after work, wipe down your overalls with a damp cloth or wear disposable overalls (type S)
- always remove overalls before removing your mask
- do not take overalls home to wash
- wear boots without laces or use disposable boot covers
- put disposable clothing items in asbestos waste bags and dispose of them properly
- do not carry asbestos into your car or home.

Remember: If you are in any doubt, seek expert advice.

Licensed asbestos removal companies

Asbestos removal requires a licence for all asbestos contamination situations where the risk of airborne asbestos particles is high. The Health and Safety Executive Asbestos Licensing Unit issues the appropriate documentation. To be granted a licence, a company will have to demonstrate the necessary skills, competency, expertise, knowledge and experience of work with asbestos, together with excellent health and safety management systems.

Licences, which act as a permit to work, are issued for a fixed time period, after which they have to be renewed. At this time, the recorded performance of the company through HSE and Local Authority Inspectors will be taken into account.

Asbestos disposal

There are three ways to dispose of asbestos and asbestos-containing materials (ACM):

1 The safest way is to hire a specialist asbestos removal company.

2 Less safe is to dismantle the asbestos material yourself, taking the correct precautions with regard to health and safety, and hire a licensed asbestos waste company to dispose of the waste.

3 The least safe way of these three options is to transport it yourself to a site licensed by the Environment Agency. The asbestos will require double wrapping in strong plastic bags and to be clearly marked as asbestos waste. The site will usually make a charge for this service. Before you arrive at the site you will need to telephone them first to advise them of the type, quantity and intended time of arrival of the asbestos you wish to dispose of.

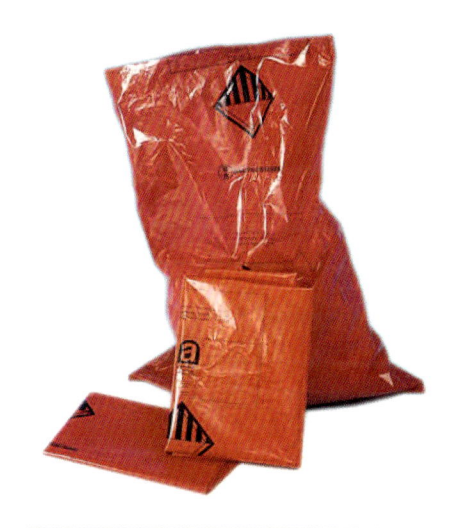

Most licensed sites will only accept certain types and quantities of ACM. Usually these are:

- asbestos produced by the householder from domestic properties
- cement-bonded asbestos sheeting, pipes, gutters or flues in pieces of 150mm or less
- asbestos sheeting, which is in pieces of 150mm or less
- a maximum of six small bags.

Asbestos disposal sacks

SmartScreen Unit 001 worksheet 3

B4 PERSONAL PROTECTIVE EQUIPMENT

Personal protective equipment (PPE) is designed to protect against workplace hazards. See page 8 of this chapter for detail about health and safety law in relation to PPE. Your employer is obliged by law to provide the following:

- suitable protective clothing for working in the rain, snow, sleet etc
- eye protection or eye shields for dust, sparks or flying objects
- respirators to avoid breathing dangerous dust and fumes

- shelter accommodation for use when sheltering from bad weather
- storage accommodation for protective clothing and equipment when not in use
- ear defenders where noise levels cannot be reduced below 80dB(A)
- adequate protective clothing when exposed to high levels of lead, lead dust or fumes or paint.

Safety helmets

While on site there is a danger of materials or objects falling into excavations or from scaffolds, as well as the danger that you may hit your head on protruding objects.

Always wear your personal safety helmet, which you will have to adjust to fit your head. Do not add paint or stickers to your helmet, as it may reduce its effectiveness.

Safety helmets are designed to protect the head of a wearer against falling objects by resisting the penetration and reducing the shock absorption by the head and body. A safety helmet meeting BS EN 397:

- may be used in temperatures as low as −30°C and as high as 150°C
- has electrical resistance up to 440V
- has resistance against molten metal, marked as MM
- is resistant against side squeeze, marked LD, for lateral deformation
- should be replaced once a year or if the hat has been struck by an object.

A typical safety helmet

Safety footwear

You need to protect your feet against various hazards, including damp, cold, sharp objects, uneven ground and crushing. Flimsy footwear and ordinary trainers will not give the protection required. A good pair of boots with steel toecaps (EN20345 – 200 joules) and steel midsole for underneath protection is a mandatory requirement on construction sites.

A typical safety boot

Overalls and work wear

There are numerous types of clothing produced to wear over your normal clothes for protection from dust, dirt and grime. Some have protective kneepads provision, which is especially useful for plumbers, and are designed to last longer. Plumbers should always consider flame retardant work wear where possible.

High visibility jackets and vests are now a mandatory requirement for all construction site workers. The usual colours are fluorescent yellow or orange.

Plumbers' trousers

Eye protection

An injury to the eye could prevent you from working, and could even cause blindness, so it is extremely important to protect your eyes while working. There are around 1000 eye injuries every working day – wearing eye protection could have prevented the majority of these.

Goggles, visors, spectacles, face screens and fixed shields are all forms of eye protection. It is important to wear the appropriate type for the work you are carrying out. Signs must also be used to indicate where there is a risk of anyone sustaining an eye injury.

Types of hazards that can cause eye injuries

Some of the hazards and risks encountered in the workplace that may cause eye injuries are:

- using hammers and chisels
- handling or coming in contact with corrosive or irritant substances such as acids and alkalis
- the use of gas or vapour under pressure
- molten metals
- instruments that emit light or lasers
- abrasive wheels
- chipped or broken tools
- work involving welding or soldering
- threading steel pipe.

Impact resistant goggles

All eye protection should be CE Approved to the relevant European Standards including EN 166 and EN 172. Eye protection is a requirement by law under regulation 4 of the Personal Protective Equipment At Work Regulations 1992 when working in a hazardous area.

In the event of an eye injury:

- no medication is to be applied to the eye
- the eye involved should be washed with clean, cold water if needed, and covered with clean, dry material (if possible, cover the unaffected eye as well to reduce eye movement)
- immediate medical attention should be sought
- a thorough ophthalmic examination should be carried out within 24 hours.

Respirators (respiratory protective equipment)

Dust and fumes are a known hazard to health, especially when inhaled over long periods.

The greatest problem on site and in the workshop is the dust from common substances such as wood, cement, stone, silica and plastics. Cutting and grinding of these materials can often produce great amounts of dust, which can cause breathing problems such as asthma and emphysema. In general the dust is too fine to be seen with the naked eye but problems and symptoms can appear in later years.

Fumes from solvents, paints and adhesives can also cause serious health problems especially if used in confined or unventilated spaces.

By law, employers must make provision for the protection of employees from dust and fumes, and also persons not employed, who may be at risk. As well as providing respiratory protective equipment (RPE), suitable signs must be displayed where there is a chance of anyone coming into contact with dust and fumes from hazardous substances. It is the responsibility of the employer to carry out a risk assessment to determine when RPE is required and what type is appropriate to control the exposure to the hazardous material.

Selecting the correct respirator

A competent person must carry out selection of the correct RPE. The choice will depend upon:

- the nature of the hazard and material
- the amount of dust present
- the period of exposure
- the weather conditions if working outdoors
- if the respirator is suitable for the user, field of vision, communication etc.

There are many types of respiratory protective equipment (RPE) available, including:

- disposable face masks
- half dust respirators
- high efficiency dust respirators
- ventilator visor or helmet respirators
- compressed air line breathing apparatus
- self-contained breathing apparatus.

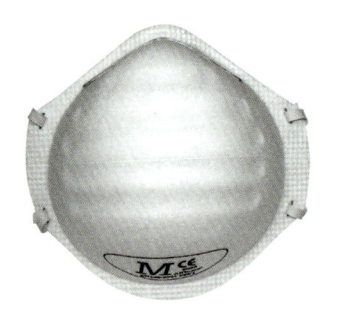

Disposable dust mask

Gloves

Your hands are vulnerable to a wide range of hazards such as cuts, blows, chemical attack and temperature extremes, depending on the type of work being carried out, so it is important to wear the correct hand protection. The various different classifications of hand protection are outlined below.

EN 388

These are gloves designed to protect the hands against mechanical risks associated with the handling of rough or sharp objects that could cut or graze.

EN 388 gloves for mechanical risks

EN 407
These are gloves designed to protect the hands against thermal hazards. Heat can be convected, conducted or radiated, or it may be the flame itself. Cold can be anything from cold water to freezing pipe gases.

EN 374
These are gloves designed to protect the hands against chemicals and microorganisms. Any substance that would irritate, inflame or burn the skin is classed as a chemical hazard. Some substances can cause the skin to become sensitive over a period of time while others have an immediate, painful effect. This type of glove gives protection against chemical splashes and protection against microorganism hazards. They are often recommended specifically by the Control of Substances Hazardous to Health Regulations 2002.

EN 374 gloves for chemical risks

EN 12477
This is the standard for protective gauntlets for welders.

EN 421
These are gloves designed to protect against ionising radiation and radioactive contamination.

Hearing protection
If noise levels reach 80 decibels, employers must carry out a risk assessment and provide information and training to employees. There is an upper noise limit of 87 decibels (taking into account hearing protection) above which workers should never be exposed. The British Standards for ear protection are:

- ear defenders BS EN 3521:2002
- earplugs BS EN 3522:2002
- ear defenders on safety helmets BS EN 3523:2002
- level-dependent ear defenders BS EN 3524:2001
- active noise reduction ear defenders BS EN 3525:2002
- ear defenders with electrical audio input BS EN 3526:2002
- level-dependent earplugs BS EN 3527:2002.

The type of hearing protection you use will depend on the work you are doing. For very noisy situations or long duration work, ear defenders would be the best solution as they offer greater protection than earplugs.

Ear defenders

B5 MANUAL HANDLING

Manual handling operations are an important part of the construction industry but are one of the biggest causes of back problems and time off work. Here, we will look at safe manual handling techniques, including:

- how to avoid manual handling injuries by using correct lifting methods
- how to assess your own lifting capability
- how to decide whether a manual handling activity is safe
- how to safely lift a load, transport it and put it down
- ways of reducing the load
- ways of avoiding manual handling.

As already mentioned, the Manual Handling Operations Regulations 1992 control manual handling and lifting. They require employers to reduce the risks from manual handling, and for employees to adopt the safe working practices as set by the employer.

Here are some points for you to consider before attempting any lifting or handling operation:

- Be aware of your own strength and limitations.
- Decide if it is a one-person operation or if you require help.
- Always use mechanical equipment or aids if available, and ensure that you are trained in their use.
- Be sure of the weight of an item before lifting.
- Wear gloves to protect your hands.
- Wear safety boots to protect your feet.
- Make sure the surrounding area is clear and safe to carry out lifting and movement.

SmartScreen Unit 001 handout 7

Lifting and handling techniques

To avoid injury, the following principles should be followed, as recommended by the HSE:

Kinetic lifting

- Think before lifting/handling.
- Plan the lift.
- Can handling aids be used?
- Where are you moving the load to? Will you need help with the load?
- Don't lift or handle more than can be easily lifted. If the load is too heavy seek advice or get help.

KEY POINT

Often manual handling and lifting can cause immediate pain and injury. This type of injury is called an **acute** injury. Sometimes the result of an injury can take weeks or months, or even years to develop. These types of injuries are called **chronic** injuries.

- Remove obstructions in your way.
- For a long lift, consider resting the load midway on a table or bench to change your grip.

One-person lift

STEP 1 – Assess the load and pathway before you attempt to move it.

STEP 2 – Crouch down with your back straight, and take hold of the load. You should be as close to the load as possible at the beginning of the lift, so you don't have to reach too far.

STEP 3 – Tilt the load away from you to assess the weight and get a good grip.

STEP 4 – Lift smoothly, keeping your back as straight as possible, and bringing the load in close towards your body as soon as you can.

STEP 5 – Once upright, keep the load close to the body and get your balance before moving forwards. Keep your shoulders level and facing in the same direction as your hips. Turn by moving your feet rather than twisting the body.

STEP 6 – Move forwards smoothly and carefully, looking ahead and watching where you are going.

STEP 7 – Place the load down carefully and securely, and straighten up. Then carefully make any adjustments to get it into the right position.

The two-person lift

Awkwardly shaped and very heavy objects should be moved or carried only with the help of other workmates. Appoint a team leader and obey his or her instructions. Try to pick someone of the same height and size so that the effort of each person is the same.

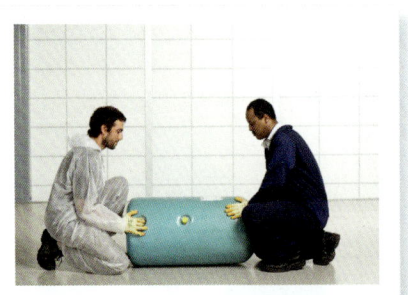

STEP 1 – Crouch down either side of the object, and take hold.

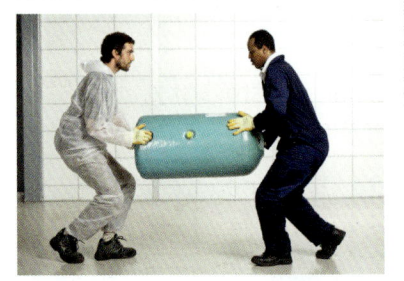

STEP 2 – Count down and lift together, keeping your eyes on your lifting partner and your backs in the correct position.

SmartScreen Unit 001 worksheet 4

A pallet truck

A sack truck

Mechanical lifting gear

There are numerous items of small lifting equipment available to assist with handling materials on site and in the workshop. These range from small brick lifts, slings, barrows and dumpers through to mechanical forklift trucks. Only use these if you are qualified to do so.

- A pallet truck can be used on hard areas for moving heavy loads.
- Barrows are the most common form of equipment for moving materials on site.
- A sack truck can be used for moving bagged materials, heavy boilers and other heavy pieces of plumbing material.

Most large construction sites will have a hired crane of some description, whether it is fixed or a mobile. These are sometimes the only method of getting heavy equipment and appliances to where we need them. Only trained personnel may operate these.

Care should be taken if cranes are on site and you should be aware of where the lifting hook is when you are walking to and from different areas of the site: the area it covers should be off-limits to all non-essential personnel.

B6   FIRST-AID PROVISION IN THE WORKPLACE

People at work can suffer injuries or fall ill at any time. The most important thing is that they receive immediate and appropriate attention. First aid covers the arrangements that should be made to ensure that this happens. It can prevent minor injuries becoming major incidents and can often save lives.

What the law requires

Health and safety regulations require employers to provide adequate and appropriate equipment, facilities and personnel to enable first aid to be given if an employee suffers an accident or injury or falls ill at work. While different working environments have different needs, the minimum first-aid provision in any workplace or construction site should include:

- a suitably stocked and maintained first-aid box. The HSE advises that it should include at least:
    - 20 wrapped sterile adhesive dressings in assorted sizes
    - two sterile eye pads
    - four individually wrapped triangular bandages
    - six safety pins
    - six medium-sized and two large-sized individually wrapped sterile unmedicated wound dressings
    - a pair of disposable gloves
- an appointed person to take charge of first-aid arrangements
- around-the-clock quick access to the first-aid equipment
- a trained first-aider at all times during working hours.

What is an appointed person?

An appointed person is someone your employer chooses to:

- take charge when someone is injured or falls ill, including calling an ambulance if required
- keep stock of the first-aid box and replenish supplies
- be available at all times that people are working on site.

What is a first-aider?

A first-aider is someone who has undergone a recognised first-aid training course, such as those given by Association of First Aiders (AoFA) and recognised by the HSE. The first-aider must hold a current first aid at work certificate.

Employers' first-aid responsibilities

Your employer is required by law to make an assessment of significant risks in your workplace and assess the risks of potential injury and ill health. If a significant number of risks exist, more than one first-aider may be needed. Your employer also needs to assess whether there any specific risks, such as working with hazardous substances, dangerous tools or machinery etc, which could necessitate specific training for first-aiders or extra first-aid equipment. If there are different parts of the workplace that present different degrees of risk, your employer will need to make sure that each area has the relevant provisions.

Your employer may need to review the accident record book to find out the types of injuries and how often they are occurring. This may influence the number of first-aid boxes and their exact location.

First-aid kit

First-aid sign

If your workplace or site is spread out over different floors and buildings, adequate provision must be made for all locations. If any employees travel or work alone, your employer should consider issuing a personal first-aid kit, and providing training on how to use it.

For shift work or out of hours working, your employer needs to ensure that there are enough first-aiders to cover all hours of operation.

There are no legal responsibilities for guests and site visitors but it is good practice to include them in first-aid provision.

Your employer has to inform all employees of the first-aid arrangements by putting up notices telling staff who the first-aiders are and where they can be found as well as where the nearest first-aid box is kept. It is also good practice to make provision here for people who have reading difficulties or whose first language is not English.

Dealing with minor injuries at work

As a plumber you are likely to experience minor injuries from time to time. Here, we will look at the following minor injuries:

- minor cuts
- minor burns
- objects in the eye
- exposure to fumes.

These tips are only for minor injuries. You should seek expert medical attention if you think the wound is more serious or the following circumstances are present:

- a wound will not stop bleeding
- the injury is to the eye or ear
- a wound was caused by a rusty or dirty object
- a cut is deep or wide
- the person's last tetanus injection was more than 10 years ago
- a burn is larger than the palm of your hand or is situated on the neck, face, groin, foot, back of the hand
- signs of infection such as the redness of the skin or fever are present
- the person has lost consciousness.

Cuts

Minor cuts will need treatment to prevent dirt getting into them, as this could cause infection.

The area around the cut should be cleaned thoroughly with soap and warm water. If it is bleeding, apply direct pressure to stem the flow of blood. The extent of bleeding will depend on where the cut is and how deep it is. It is a good idea to wear protective gloves when dealing with cuts that are bleeding. The edges can be held together using butterfly bandages, and applying an antiseptic cream will help to reduce the chance of infection. A bandage or a sticking plaster can then cover the wound. Care should be taken when using plasters as some people suffer reactions to the adhesive.

Burns

Burns need to be treated immediately. Firstly, cool the area with cold running water or by submersing in a clean bucket of clean, cold water. Keep the burn in the water for at least 10 minutes, as this is the single most effective way of stopping the pain. Remove anything that could cause constriction (eg watches or jewellery) before the area starts to swell.

Once the burn has cooled sufficiently, it should be gently washed with clean water and covered with a sterile burns sheet or other suitable non-fluffy material. If no other materials are available, cling film or a clean plastic bag could be used. Do not apply any antiseptic cream or ointments as these have the effect of sealing the heat inside the burn resulting in a more intense pain. Do not pierce or pop any blisters that develop as this could result in the burn becoming infected.

Depending on the severity of the burn, the person should be accompanied to the nearest accident and emergency (A & E) hospital or to a doctor.

Objects in the eye

Objects in the eye can be painful, and could potentially damage the eye. Loose objects such as dust can float on the white of the eye. These can usually be rinsed off. However, you must never touch anything that penetrates the eyeball or rests on the coloured part of the eye (the pupil and iris) because this may permanently damage the eye. Faced with this situation, the person should seek immediate medical attention. The signs to look for are:

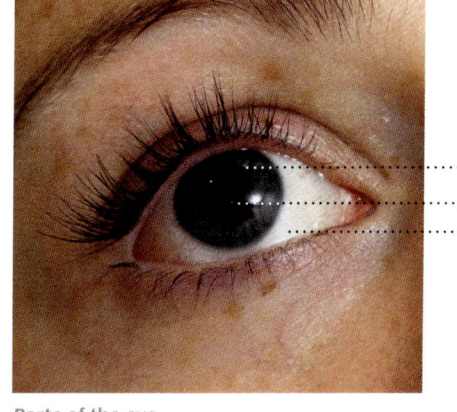

Parts of the eye

- blurred vision
- pain or discomfort
- redness or watering of the eye
- eyelids screwed up in a spasm.

The aim of any treatment you give is to avoid permanent damage. Bearing this in mind, carry out the following examination if you are with someone who thinks they have a foreign body in the eye:

- Sit the person down facing the light.

- Stand behind the person and very gently part the eyelids with a finger and thumb.

- Make sure that you examine every part of the eye by getting the person to look up, then down, then to the left, then to the right.

If you spot an object in the eye:

- Wash it out with clean cold water from a glass or fresh running water from the tap. Tilt the person's head toward the injured eye and place a towel or pad on the shoulder. Pour water from the bridge of the nose so that the water runs across the eye to flush the object out.
- If this doesn't work, lift the object off with a damp corner of a clean tissue or swab (only do this if the object is on the white of the eye and not the coloured part).
- If this still doesn't work, seek medical advice.

Exposure to fumes

You must be very careful when dealing with a person who is suffering as a result of exposure to fumes as you need to ensure that the fumes do not also overcome you. You will have to consider:

- the nature of the fumes (What are they? Where have they come from? Can they be stopped?)
- whether the area is sufficiently ventilated
- whether you can get the person out without becoming overcome by the fumes.

If the person is unconscious, getting them out of the area and into fresh air is absolutely vital. The following should only be carried out if you can minimise your own risk:

- Immediately carry or drag the person to fresh air.
- If the person is not breathing, start cardiopulmonary resuscitation (CPR) immediately and continue it until the person is breathing or help arrives.
- Send someone for help as quickly as possible.

Dealing with serious injury at work

In this section we will examine the best way of dealing with those injuries that are more serious, such as:

- fractures and breaks
- unconsciousness
- electric shock.

Fractures and breaks

A fracture is a break or a crack in the bone. There are two types of fracture:

1. A **simple fracture** where the skin is intact and there is no wound present. There may be a swelling around the area of the fracture.

2. A **compound fracture** where the bone causes a wound or the breaking of the skin. The bone may or may not be visible with this kind of injury.

It is not always obvious that the bone is fractured, but if you are in any doubt, always assume that it is. There are signs to help you identify a fracture, and a few rules to ensure that the injured person is comfortable until the emergency services arrive:

- Talk to the person and ask them questions. (eg did they hear a snap at the time of injury?)
- Look for an open wound that may indicate a hidden fracture.
- Check for pain by gently feeling along the area. The injured person should be able to tell you where the pain is. In a few cases there may be no pain associated with the fracture and the person may be able to move the injured limb, but in most cases the person will be in great discomfort and any movement will cause very severe pain.
- Examine the area for cuts and wounds, and feel for swelling or deformities. You can check for deformity of the limb by comparing it with the opposite side of the body.
- Lightly squeeze the person's fingers and toes to ensure no spinal injury or nerve damage has occurred.
- Seek medical help immediately and remain with the patient until this arrives.

Unconsciousness

A person can faint or fall unconscious for many reasons:

- after strenuous work or exercise
- shock or emotional upset
- excessive heat
- the side effects of drugs and/or medication
- a blow to the head
- a fit or seizure.

Fainting involves the loss of blood to the brain, which leads to dizziness, nausea, cold sweats and a partial or complete loss of consciousness. Someone who has fainted is usually only unconscious for a short time, and they will make a full recovery in a matter of minutes. The real danger here is not the period of unconsciousness, but the damage that can be from the resulting fall.

More serious unconsciousness comes from a blow to the head (called concussion), a fit or a seizure. In these cases, recovery can take much longer and may have underlying health implications later, so you should seek advice immediately.

If someone faints or falls unconscious, you should carry out the following steps:

1. Try to break his or her fall.

2. Check to see whether the person is breathing. Look, listen and feel for breathing for no more than ten seconds. Is the chest rising and falling? Can you feel their breath against your cheek?

3. Ensure the airway is clear. To do this, place one hand on the casualty's forehead and gently tilt the head backwards, then lift the chin using only two fingers.

4. If he or she is not breathing, start resuscitation procedures (see page 43). For the safety and wellbeing of the casualty, it is highly recommended that this is done by a trained first-aider or someone who has completed a CPR course.

5. If the casualty is breathing, loosen clothing that might restrict the flow of blood (such as neck ties or shirt buttons) and place him or her in the recovery position (see page 44).

6. Once a person regains consciousness after a fainting episode, you may find it helps to lay the casualty on his or her back and raise the legs to encourage blood to flow to the brain.

Dealing with electric shock

Electricity is one of the most dangerous elements that we have to deal with. You can't see it or smell it – but if you touch it, it could kill you. You may think that a shock of 1000 Volts would be more deadly than 100 volts but this is not necessarily the case. People have been electrocuted (killed by electricity) by appliances using ordinary household supplies of 230 volts alternating current (AC) and by electrical apparatus in industry using as little as 42 volts direct current (DC).

The real measure of a shock's intensity lies in the amount of current (measured in amperes) that is forced through the body, and not the voltage. Any electrical device used on a house wiring circuit can, under certain conditions, transmit a fatal current. While any amount of current over 10 milliamps (0.01 amps) is capable of producing painful to severe shock, currents between 100 and 200mA (0.1 to 0.2 amps) are lethal.

It is vital to know how to deal with a person who has direct contact with a live electricity power source, how to isolate them from the power supply and how to administer life-saving CPR.

- If you see someone who is in direct contact with electrical current, they need immediate help. The victim may be unable to move because of muscle spasms, or they may be unconscious. Helping such a person is very dangerous. If you touch the victim, you may get caught by the current yourself and become a second casualty.

- Try to turn off and unplug the appliance or, better still, turn off the power at the electrical consumer unit (fuse box). If you cannot turn off the power, get a long piece of wood (a broom handle will do) or any non-conducting material, and try to break the contact between the victim and the electricity.

Volt
Unit of electrical potential.

Amp (and milliamp)
Unit of electrical current, the measurement of ampere.

- Do not move the victim if there is any suspicion of neck or spinal injuries unless there is an immediate danger. Keep him or her lying down and check for a pulse and breathing.

- If the victim is not breathing, call for help and apply mouth-to-mouth resuscitation. If the victim has no pulse, begin cardiopulmonary resuscitation (CPR).

- Once a pulse and breathing have been established, cover the victim with a blanket to maintain body heat, keep the victim's head low, and get medical attention. Stay with the victim until help arrives.

Cardiopulmonary resuscitation (CPR)

CPR is a manual method of maintaining a heartbeat and air supply to a person who is unconscious, has stopped breathing and has no pulse. The aim is to keep blood pumping around the body and maintain a supply of oxygen to the brain and other vital organs so that brain damage does not occur. There are many instances where a person may need cardiopulmonary resuscitation (CPR), for example exposure to fumes or an electric shock.

How do I perform CPR?

The first thing to do in all cases is to send for urgent medical help. If you need to check for a pulse, put your fingers in the groove between the windpipe and the muscles in the side of the neck, and press backwards. If there is no pulse or any signs of breathing or movement, proceed with the following method of CPR:

1. Place the victim on their back on a firm surface and kneel next to their chest.

2. Remove, open or cut any excess clothes. CPR should be performed close to the patient's chest and not through thick garments.

3. Place your hands directly above the sternum (breast bone), one on top of the other, two fingers' width above the point where the lower ribs meet. Only the heel of the hand should touch the chest.

4. Shift your weight forward on your knees until your shoulders are directly over your hands.

5. Keeping your elbows locked straight, repeatedly press down and then release. You must depress the chest of an average adult approximately 5–6 cm with each compression, releasing completely after each compression.

6. Compress the chest about 100–120 times every minute. To get the right speed and rhythm, count out loud. Try to compress and release for equal periods of time.

7. Give the victim two rescue breaths following each set of 30 compressions. To give a rescue breath, take a normal breath, seal your lips around the victim's mouth and blow air in steadily for about one second.

8 Repeat this cycle until the victim breathes, coughs or shows any sign of movement or returned circulation, or medical assistance arrives.

Note: It is highly recommended that CPR is carried out by trained persons. If you are alone with a casualty, and especially if you are untrained, continue chest compressions while calling for assistance.

The recovery position

This is the best position for a casualty who is unconscious but still breathing. Putting someone in the recovery position will ensure the airway remains clear and open. It also enables any vomit or fluid to flow away from the airway to avoid choking.

STEP 1 – With the casualty lying on their back, place the arm nearest you at a right angle.

STEP 2 – Move the other arm across the body so the back of their hand is against their cheek. Get hold of the knee furthest from you and pull up until the foot is flat on the floor.

STEP 3 – Pull the knee towards you turning the casualty on their side, keeping the person's hand pressed against their cheek. Position their top leg at a right angle.

STEP 4 – Make sure that the airway remains open by tilting the head back and lifting the chin. Check that the casualty is breathing.

STEP 5 – Monitor the casualty's condition until help arrives and do not leave them unattended for more than three minutes.

All accidents on site should be properly recorded in the company's accident book and, if necessary, reported. See the section on reporting accidents earlier in this chapter (page 20).

Raising the alarm in an emergency, and the role of the emergency services

In an emergency, time is of the essence. The quicker the emergency services arrive at the scene, the greater the chance that lives will be saved. If calling for help, ensure that you take the following steps to help the emergency services get to you as quickly as possible:

1. Dial 999 and ask for the service that you require: police, fire or ambulance.

2. Once you are connected, speak clearly to the operator. Tell them the nature of the incident, the location and the possible entry points to your workplace or site.

3. Send work colleagues to wait at all the entrances and to assist the emergency services to get straight to the incident once they've arrived. If the site is large, have a chain of people to direct them.

4. On no account leave the injured person. Stay with them and let the emergency services come to you.

5. Stay at the scene until you are not needed. Ask if the injured person should be accompanied to the hospital and, if necessary, go with them.

6. Ask someone to advise the injured person's next of kin without alarming them unduly.

Fire evacuation procedures

If you discover or are informed of a fire or other emergency, sound the alarm immediately. Safe evacuation is an absolute priority. After the alarm has been sounded, notify the fire service of the exact location of the incident.

Once the alarm has been sounded, staff may attempt to deal with the fire, but only if the fire is small and it is safe to tackle it. Never put yourself at risk.

On hearing the alarm:

- all operatives working on the site must respond
- a named person will summon the emergency services by dialling 999
- evacuate to the designated assembly point(s).

C WORKING SAFELY WITH EQUIPMENT ON SITE

During your work in the building services industry you will come into contact with many types of specialist equipment, some of it directly related to your job and some of it not. This part of the chapter covers how to work safely with or around the main types of equipment you will find on site.

C1 SAFETY WITH ELECTRICITY

To comply with the Electricity at Work Regulations (EAW), employers are required to maintain their electrical systems in a safe condition.

According to the Health and Safety Executive (HSE), periodic inspections and testing should be completed as part of this maintenance. Over 1000 electrical accidents and incidents at work are reported to HSE every year and around 30 people die from their injuries. The HSE reports that many deaths and injuries arise from:

- the use of poorly maintained electrical equipment
- work near overhead power lines
- contact with underground power cables during excavation work
- work on or near 230 volt domestic electricity supplies
- fires started by poor electrical installations and faulty electrical appliances.

Electricity supply

The supply of electricity to homes and construction sites is normally provided by either:

- a public supply from a local electricity company
- a site generator (where the use of the public supply is not practicable or is uneconomic).

The supply of electricity to a construction site

To maintain site safety, the supply of electricity to a construction site or workshop should always be distributed by means of a reduced voltage system. This system ensures that the correct voltage is supplied to where it is required:

- Woodworking machines in a workshop require a 400V 3-phase supply.
- Site office lighting requires a voltage of 230V single phase supply.
- Site portable power tools and site lighting require a 110V single phase supply.

Each site voltage has its own colour coding as shown in the table below.

AC operating voltage	Voltage colour coding	Use
25V	Violet	Damp conditions
50V	White	Damp conditions
110V	Yellow	General site voltage
230V	Blue	Domestic and site offices
400V	Red	Fixed machinery

The reduced voltage system must comply with the Electricity at Work Regulations 1989, and the distribution units, sockets and plug adapters should comply with BS 4363:1998 (Specification for distribution assemblies for reduced low voltage electricity supplies for construction and building sites).

To avoid plugs designed for one voltage being connected to sockets of another voltage, there are different positions for the connecting pins in the plugs and sockets as illustrated below:

110V 1 phase – yellow   230V 1 phase – blue   400V 3 phase – red

> **KEY POINT**
> The terms 'single phase' and '3 phase' refer to the fact that there are either one or three live conductors, 'phase' meaning live.

The voltage used on construction sites for site lighting and portable power tools is 110V, colour coded yellow. A 110V, single phase supply is much safer than 230V and so the risk of serious injury from an electric shock is much reduced. The 230V supply (colour coded blue) for general site use is not allowed unless it is through a residual current device (RCD), which disconnects the supply immediately in the event of a fault or shock condition occurring.

Electrical hazards on construction sites and in domestic properties

Electrical hazards occur due to:

- faulty installations
- faulty electrical equipment
- electrical equipment being misused
- cables that are trailing, or buried/hidden, or too close to pipework
- inadequate fuse or over-current protection
- the overloading of electrical sockets and outlets
- maintenance neglect
- use of electrical equipment in wet or damp conditions.

Electric shock is a major hazard. The severity of the shock will depend on the level of current and the duration of the contact:

- At low levels of current (about 1 milliamp) the effect may be only an unpleasant tingle, but enough to cause loss of balance or a fall.

- At medium levels of current (about 10 milliamps) the shock can cause muscular tension or cramp so that anything grasped is hard to release.

- At high levels of current (about 50 milliamps and above) for a period of 1 second, the shock can cause fibrillation of the heart, which can be lethal.

- Electric shock also causes burning of the skin at the points of contact.

Electric shocks are caused by contact between a live conductor and earth. An electric current will always attempt to earth itself, therefore if anything comes between the flow of current and earth, the current will pass through it depending upon its resistance to the flow of current. The human body, because it contains 70 per cent water, is a very good conductor of electricity that offers very little resistance to the flow of electric current.

There are some materials that are poor conductors and will therefore offer greater resistance to the flow of electric current. Some of these materials, such as PVC, are used to shield the electricity and are called insulators.

Electric cable consists of a copper wire (an excellent conductor) and an outer cover or sheath (PVC), which is an excellent insulator. The result is a safe electric cable that can be used as an electrical supply for tools and equipment.

Unfortunately, electric shock is not the only problem. Because electricity can produce great amounts of heat it can ignite the material, causing a fire or explosion.

Electrical installations in the workplace and domestic properties

All electrical installations should comply with BS 7671 and be maintained to prevent danger. The HSE recommends that this includes an appropriate system of visual inspection (looking for visible signs of damage or faults), which will need to be reinforced by thorough testing of the system as necessary.

Formal visual Inspections and tests

During formal electrical inspections, the system will be checked and tested to ensure that:

- the polarity (live and neutral) of the system is correct
- all the fuses, MCBs (miniature circuit breakers) and RCDs (residual current device) are correct and working
- all the cables and cores are effectively terminated
- the equipment is suitable for its environment.

Working in domestic properties

When you are working in domestic properties, there are things you can do to help prevent electrical hazards and accidents from occurring:

- Be aware of any concealed cables in solid and stud walls. Use a cable finder and check the wall before using drills and chisels.
- Do not install pipework too close to electrical cables. Heating pipework can cause the cable to overheat and faulty cables can arc across to the pipe causing a potential electric shock hazard. Pipework

KEY POINT

BS 7671 is the British Standard for the requirements for electrical installations. This is the national standard in the United Kingdom for low voltage electrical installations. It is also used as a national standard by Mauritius, St Lucia, and several other countries that base their wiring regulations on BS 7671.

must be a minimum of 25mm away from electrical cables and 150mm from electrical apparatus.

- Take care when lifting or replacing floorboards as there may be cables underneath.

- Do not overload sockets and outlets with too many appliance connections as this can cause the system to overheat, sometimes with disastrous consequences. As a general rule, one socket = one plug, unless a recognised, independently fused multi-socket is used.

- Look out for damaged cables, sockets and fittings. Report any problems to the customer or your supervisor.

Portable power tool safety

All portable power tools, such as drills, jigsaws, circular saws and angle grinders, should be of the double insulated type. The symbol for double insulated tools is shown below.

Power tools must be subjected to the following safety tests:

- User checks – should be performed before use.

- Formal visual inspection – to be scheduled in accordance with your maintenance schedule and health and safety policy.

- Combined inspection and test (Portable Appliance Testing) – to be carried out by a competent person, usually an external contractor.

Portable Appliance Testing

Portable Appliance Testing (PAT) is the inspection and testing of in-service electrical equipment. This was introduced to enable companies and organisations to comply with the Electricity at Work Regulations.

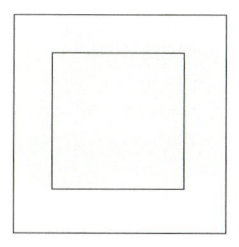

Double insulated tool symbol

To meet these regulations, it is necessary to have in place a programme of inspection and electrical safety testing of portable appliances. Records should be kept of all inspections and tests made and should be kept up to date at all times. PAT testing helps to ensure:

- earlier recognition of potentially serious equipment faults, such as poor earthing
- discovery of inappropriate electrical supply
- discovery of incorrect fuses being used
- monitoring of any misuse of portable equipment
- an increased awareness of hazards linked to electricity.

All of the following 110V equipment being used on a construction site should be given a formal visual inspection on a monthly basis, and combined inspection and testing every three months:

- stationary equipment
- IT equipment
- movable equipment
- portable equipment
- hand-held equipment.

User checks

Before using a portable appliance, you must check that:

- there is a recent PAT label on the equipment
- there are no overheating, or burn marks, on the plug, cable, sockets or equipment
- there are no bare wires or conductors visible
- the cable covering is undamaged and free from cuts or abrasions
- the cable is the correct length
- the cable is out of the way and does not present a trip hazard
- the plug is in good condition (with no cracks, and the pins unbent)
- there are no taped or other non-standard joints in the cable
- the outer covering of the cable is where it should be (with no coloured wires visible)
- the outer casing of the equipment is not damaged or loose
- 'trip-out' devices (RCD adaptors) are working effectively.

When using portable electrical power tools:

- you must wear or use PPE where required, or clothing that is appropriate for the work being carried out
- always ensure that the tool is switched off before connecting it to a power supply
- have the tool checked by an electrician if the cord feels more than comfortably warm or is sparking
- disconnect the power supply before making adjustments or changing accessories such as blades or drill bits

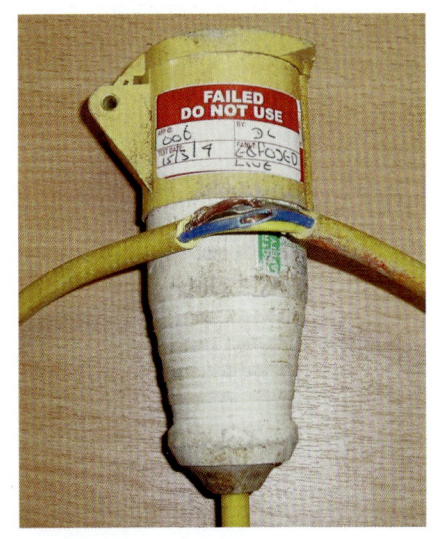

Bare conductors showing on a 110V extension cable

- remove any wrenches, spanners and adjusting tools before switching on the tool
- inspect the cord for any damage, eg fraying
- if the tool is defective, clearly label it with an 'out of service' notice, and replace it immediately
- keep power cords clear of tools to prevent cutting into the cord
- use clamps or a vice to hold and support the piece being worked on to allow you to use both hands for better control of the tool and to help prevent injuries if a tool jams or binds in a workpiece
- use only approved extension cords that have the proper wire size (gauge) for the length of cord and power requirements of the tool you are using to prevent the cord from overheating
- fully unwind any extension cable being used because a coiled extension cable is likely to overheat, which could cause a fire
- for outdoor work, use appropriate extension cords, marked 'W-A' or 'W'
- hang power cords up over aisles or work areas to prevent them becoming trip hazards
- grip the plug rather than the cord when unplugging a tool – pulling the cord causes wear and may adversely affect the wiring to the plug and cause electrical shock to the operator
- keep the work area free from all clutter and debris, as this could become a trip or slip hazard
- keep cords away from heat, water, oil, sharp edges and moving parts as these can damage the insulation and cause a shock
- ensure that cutting tools such as drill bits and blades are kept sharp, clean and well maintained
- store tools in a dry, secure location when they are not being used.

An overheated extension cable that has started to melt

Battery-powered cordless tools

In recent years, the use of battery-powered cordless tools such as drills and jigsaws has become widespread for both construction site and domestic use. Voltages tend to range from 9V to 36V. Cordless tools offer many benefits over those that are mains powered:

- Often the tools are smaller and lighter giving greater flexibility of use.
- There are no extension cables to cause trip hazards.
- There is much lower risk of electric shock.

The disadvantages of using cordless tools are:

- Most are not as powerful as their mains counterparts.
- The power packs require constant recharging, tend to wear out quickly and are costly to replace.
- There is risk of an electric shock from the battery charger.

Cordless tools are still subject to health and safety inspection and testing with regard to:

- PAT testing of the battery charger
- disposal of spent battery packs in line with Local Authority guidelines as they contain nickel-cadmium and should not be disposed of in domestic waste
- leaking batteries – some contain acid so you must not allow a leaking battery to make contact with your skin
- spent battery packs – these must not be burnt as they are liable to explode
- storage restrictions – temperatures guidelines exist for most cordless power tools.

Safe isolation procedures for electrical supplies

Plumbers often need to work on electrical supplies for repair or replacement of equipment such as electric showers and immersion heaters. The correct isolation of electrical supplies and systems is vital if accidents are going to be avoided. To work on electrical installations, you must have proven your competency by gaining certification of Part P of the Building Regulations (BS 7671).

In domestic properties, the type of electricity supply is 230V, single phase. The Electricity at Work Regulations require that live work is not undertaken unless it is impracticable to work on the circuit when it is dead.

All electrical circuits must be properly switched off, isolated and whenever possible be locked in the 'off' position. You must then prove that the circuit is dead by the use of an approved voltage indicator, usually a multimeter or a safety voltage indicator. Volt sticks and neon screwdrivers are not suitable for this purpose.

To safely isolate an electrical supply, you must carry out the following steps:

1 Identify the circuit or the equipment you wish to work on.

2 Make sure that it is convenient to isolate the supply.

3 Isolate the supply at the consumer unit by switching off the miniature circuit breaker (MCB), residual current device (RCD) or removing the fuse.

4 Using an approved voltage indicator:

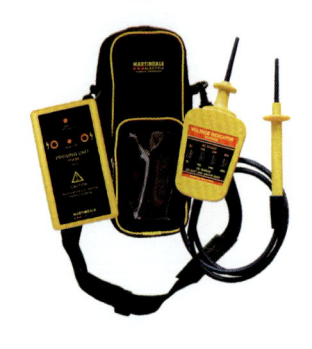

Safety voltage indicator

a firstly, check the indicator is working on a known live supply by testing live/neutral, live/earth, neutral/earth

b then use the indicator to check that the circuit you wish to work on is dead

c then recheck that the indicator is still working on the known live supply once again.

5. Lock off the isolator (RCD, MCB) using an approved lock, or keep the fuse you have removed in a safe place. To be absolutely sure that no one can put the fuse back in, the safest place is in your pocket.

6. Place a notice or sign at the consumer unit advising that the circuit is off and must not be turned back on.

Temporary continuity bonding

Temporary continuity bonding involves the use of two crocodile clips joined by 10mm^2 earth cable. This is called a temporary continuity bonding clip.

All gas, water and central heating copper pipework should be bonded to the main electrical equipotential bonding system. In other words, copper pipework must be earthed. When we cut into a copper pipe, we are disconnecting all pipework from the earth system that occurs after the cut. If a fault to earth already exists, then all the pipework after the cut could become live.

By carrying out temporary continuity bonding before removing or replacing metal pipework, we are providing a continuous earth for the pipework in order to prevent an electric shock in the event of any electrical fault. Once the connection has been made, the bonding clips can be safely removed.

MCB safety lock

Equipotential bonding
A system where all metal fixtures in a domestic property such as hot and cold water pipes, central heating pipes and gas pipes, radiators, stainless steel sinks, pressed-steel enamelled washbasins, steel and cast iron baths are connected together through earth bonding so that they are at the same potential voltage everywhere.

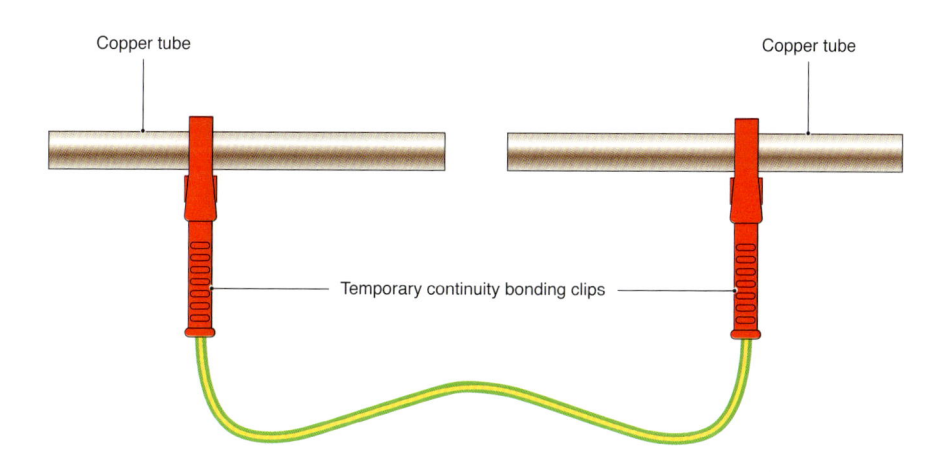

The use of temporary continuity bonding clips

> **KEY POINT**
>
> The thread direction refers to the direction in which fittings (such as hoses and regulators) screw on to their bottles. Fuel gases have left-hand threads so that their fittings cannot be mistakenly used with right-hand thread oxygen cylinders, and vice versa.

C2 SAFETY WITH GAS HEATING EQUIPMENT

Part of a plumber's work involves the use of bottled gases, both of the flammable and non-flammable type. Using bottled gas of any type can be dangerous and requires special consideration. The four main types of gases you may come across in your work are shown in the table below, along with the cylinder colour and the thread direction.

Bottled gas	Cylinder colour	Thread direction
Propane (C_3H_8) – a highly flammable liquid petroleum gas (LPG) that is used for soldering processes. It is heavier than air, which makes it especially dangerous when working in trenches and confined spaces as any leaks would collect at low level. Propane has a distinctive smell, which is like rotten eggs.	Signal red	Left hand
MAPP (methylacetylene-propadiene propane) gas – also used for soldering processes, but has a much hotter flame than propane. Usually only supplied in small cylinders for plumbing work, MAPP has a distinctive garlic-like smell.	Yellow	Left hand
Acetylene (C_2H_2) – used in conjunction with oxygen when undertaking welding and brazing processes. Plumbers usually only use oxyacetylene sets when lead welding. Acetylene is a colourless, odourless gas. When contaminated with impurities it has a garlic-like smell. Acetylene burns with a sooty flame that produces lots of carbon when used without oxygen. It is lighter than air.	Maroon	Left hand
Oxygen (O_2) – bottled liquid oxygen is a very powerful oxidising agent: organic materials will burn rapidly in the presence of oxygen. Used in conjunction with acetylene, oxygen hardens the flame, increasing the temperature. Although oxygen itself is not flammable, it can induce other materials to combust fiercely. Never use oxyacetylene near jointing compounds or grease because oxygen reacts violently in their presence.	Black	Right hand

Many companies operate a written permit system when using fuel gases. This is known as 'hot work'. The permit details the type of work to be done, how and when it is to be carried out and the precautions to be taken.

Gas equipment should not be used unless you have received adequate training in:

- safe use of the equipment
- the precautions to be taken
- use of the correct type of fire extinguishers
- the means of escape, raising the fire alarm and calling the fire brigade.

The safe storage and handling of bottled gases is crucial. Listed below are the points to consider when storing or handling bottled gas:

- Always keep full cylinders separate from empty ones.
- Oxygen cylinders should be stored at least three metres away from cylinders containing acetylene or LPG, or separated by a wall.
- All gas cylinders must be stored away from combustible materials, sources of ignition, and any corrosive, toxic or oxidant material.
- Gas cylinders should preferably be kept on a hard surface (not soft ground) in a secure, open-air compound. The enclosures must be properly labelled.
- Acetylene and LPG cylinders should always be kept upright during use, storage and transport, even if they are empty. Any vertically stacked cylinders should be secured to prevent falling.
- Oxygen cylinders can be stacked horizontally, but no more than four cylinders high and wedged to prevent rolling.
- Cylinders should be shielded from direct sunlight or other heat sources to avoid excessive internal pressure build-up. This is because pressure build-up could lead to a gas leakage or, in extreme cases, bursting of the cylinder.
- Never lift oxyacetylene or LPG bottles by their control valves.
- Gas cylinders must be treated with care and not subjected to shocks or falls.
- When they are transported around a site, cylinders should be secured upright to avoid any violent contact that could weaken the cylinder walls.
- Cylinders should only be transported on purpose-designed trolleys of the correct size. Three-wheeled trolleys are safer than two-wheeled.
- Trolleys for transporting cylinders should be manufactured to BS 2718.

Equipment used with oxy/fuel gases

As well as the cylinders themselves, there are several other pieces of equipment that we need before we can start using our oxyacetylene bottle set. The main components of oxy/fuel gas equipment are listed on the following page.

A **flashback arrester** to protect cylinders from flashbacks and backfires. Flashback arresters (also called flame traps) must be fitted into both oxygen and acetylene gas lines to prevent a flashback flame from reaching the regulators.

A **control valve** to shut off or isolate the gas supply, usually the cylinder valve situated at the top of the cylinder. It has a square key to open and close the valve. As a general rule when using oxyacetylene, both the oxygen and acetylene bottles should have their own key, which should be left on the bottle during the welding process so that the bottle can be isolated quickly in an emergency.

Flexible hoses to convey the gases from the cylinders to the blowpipe. Hoses between the torch and the gas regulators should be colour-coded – red for acetylene and blue for oxygen. Fittings on the oxygen hose have right-hand threads (non-flammable gas), while those on the acetylene hose have left-hand threads (flammable gas).

A **pressure regulator** fitted to the outlet valve of the gas cylinder, used to reduce and control gas pressure. Most modern regulators work with a two-stage system – the initial stage dispenses the gas at a set rate from the storage cylinder, and the second stage handles the pressure reduction. On a two-stage system, the device has two pressure gauges. One gauge tells how much gas is remaining in the cylinder, and the other tells the pressure of the gas being released.

Cylinders of oxygen and fuel gas (usually acetylene, although propane is also used).

Non-return valves to prevent oxygen reverse flow into the fuel line and fuel flow into the oxygen line. The valves can be used to prevent conditions leading to flashback, but should always be used in conjunction with flashback arresters.

A **blowpipe** or other burner device where the fuel gas is mixed with oxygen and ignited.

Oxyacetalene set

KEY POINT
A flashback is where the flame burns in the torch body, accompanied by a high-pitched whistling sound. It will occur when flame speed exceeds gas flow rate so that the flame can pass back through the mixing chamber into the hoses. Most likely causes are: incorrect gas pressures giving too low a gas velocity, hose leaks or loose connections.

Oxyacetylene equipment safety checks
Before using welding equipment, it is wise to check the condition and operation of your equipment. As well as normal equipment and workplace safety checks, there are specific procedures for oxyacetylene. You should check that:

- flashback arresters and non-return valves are present and in good condition in both oxygen and acetylene lines
- the hoses are the correct colours, with no sign of wear or damage, and that they are as short as possible and not taped together
- the regulators are the correct type for the gas being used

- a bottle key is in each bottle
- the bottles are securely fastened by chains to the bottle trolley and that the trolley is in good condition
- there are no physical signs of damage to the bottles or valve assembly.

It is recommended that oxyacetylene equipment is checked at least annually. Regulators should be taken out of service after five years. Flashback arresters should be checked regularly in line with the manufacturer's instructions, and with some types, it may be necessary to replace if flashback has occurred.

Assembling and purging the oxyacetylene equipment

After your initial checks, when assembling the gauges, hoses and blowpipe, follow the advice listed below:

1. Make sure that each regulator is the correct type for the cylinder it is to be attached to.

2. Open the oxygen valve assembly briefly before attaching the oxygen regulator. This is to eliminate the potential for a dust explosion. **Never** open the acetylene control valve to 'blow-out' as this could cause a fire.

3. Inspect the regulator and cylinder valve for the presence of any oils or grease. If present, **do not use**.

4. Make sure that the adjusting screw on the regulator has not been damaged.

5. Wipe the connection seats with a clean cloth.

6. Connect the gauges to the cylinders using the correct coloured hoses (see above). Tighten them with the correct sized, open-ended spanner. Take care not to damage the brass threads.

7. Inspect the torch. Check that inlet connections are in good condition for a tight connection. Check for obvious physical damage to the torch. Check that the threads are satisfactory on the head of the torch to correctly tighten in the tip.

8. Make sure that the acetylene regulator is turned off by turning the regulator handle anticlockwise out a few turns, then turn the gas valve on top of the cylinder on. Only turn the control valve one turn of the wrist. This allows the bottle to be turned off quickly in an emergency. Never allow acetylene gas pressure to exceed 15 PSI. At higher pressures, acetylene becomes unstable and may ignite spontaneously or explode.

9. After turning on the acetylene cylinder control valve, open the regulator valve by turning the handle clockwise. This should be done very slowly, while watching the low-pressure gauge. Open only until the pressure indicated is 0.14 bar (2 PSI).

10. Open the gas valve on the blowpipe handle until you hear gas escaping. This is to purge the air from the acetylene hose. Then watch the low-pressure gauge to see if the pressure remains steady during flow, to ensure that you have the regulator set correctly.

11. Close the acetylene valve on the torch.

12. Check for leaks by using suitable non-greasy leak detection fluid. Never use an open flame to check for leaks.

13. Turn the oxygen regulator pressure off by turning the regulator handle a few turns anticlockwise then proceed with the steps in the list below to adjust the oxygen pressure.

To adjust oxygen pressure:

1. Open the oxygen cylinder control valve all the way to open.

2. Open the regulator valve slowly, watching the low pressure gauge as you do so, until the pressure reads 0.14 bar (2 PSI).

3. Open the oxygen valve on the blowpipe to allow the atmosphere to vent out of the hose until the hose is purged, about three to five seconds for an eight-metre hose.

4. Close the blowpipe valve.

5. Check for leaks by using suitable non-greasy leak detection fluid or soapy water.

What to do in the event of leakage

Never use a leaking cylinder. If you smell gas and detect a leak:

- Close the cylinder valve.
- Clearly label the bottle as 'leaking'.
- Remove the cylinder to an outdoor location and post 'no smoking' and 'keep clear' signs.
- Call the gas supplier to collect the cylinder as soon as possible.

Safe lighting and extinguishing procedures for oxyacetylene equipment

To light oxyacetylene equipment, follow the steps listed below:

1. Open the acetylene blowpipe valve a quarter of a turn and light the acetylene with a friction-type lighter. **Never** light the oxyacetylene torch with a mixed gas.

2. Adjust the acetylene flame to the desired velocity.

> **KEY POINT**
> Make sure that you purge both acetylene and oxygen lines (hoses) prior to igniting the torch. Failure to do this can cause serious injury to personnel, and damage to the equipment.

3 For lead welding, open the oxygen blowpipe valve and adjust to a neutral flame (equal amounts of acetylene and oxygen).

4 For brazing or bronze welding, open the oxygen blowpipe valve and adjust to a slightly oxidising flame (slightly more oxygen than acetylene).

To extinguish oxyacetylene equipment, follow these steps:

1 Close the acetylene blowpipe valve first, then close oxygen blowpipe valve.

2 Turn off both acetylene and oxygen control valves on the cylinders.

3 Turn the acetylene regulator handle anticlockwise until it is loose.

4 Open the acetylene blowpipe valve to release the pressure from the regulator.

5 Close the acetylene blowpipe valve.

6 Turn the oxygen regulator handle anticlockwise until it is loose.

7 Open oxygen blowpipe valve to release the pressure from the regulator.

8 Close the oxygen blowpipe valve.

Safe use of liquid petroleum gas (LPG)

Liquid petroleum gas (LPG) is the generic name for the family of carbon-based flammable gases that are found in coal and oil deposits deep below the surface of the earth. They include:

- methane
- ethane
- butane
- propane.

Of these, generally only butane and propane are commercially available as bottled LPG. Plumbers regularly use propane when soldering copper tubes and fittings.

The main risks associated with using LPG are those of fire/explosion, carbon monoxide poisoning, asphyxiation and extreme cold. Below is a list of facts about LPG that are essential to know if planning to work with this gas.

- LPG (propane or butane) is a colourless liquid, which easily evaporates into a gas when exposed to the outside air. One litre of liquid propane creates 250 litres of gas.

- It has no smell. A distinctive odour is added to help detect leaks.

- It can burn or explode when it is mixed with air in the correct ratio and if it comes into contact with a source of ignition.

- It is heavier than air, so it tends to sink towards the ground. It can flow for long distances along the ground, and can collect in drains, gullies, cellars and trenches.

Commercial propane cylinder

KEY POINT

Propane turns from its liquid state to a gas (ie it boils) at −42°C, while butane boils at −4°C. Propane can therefore be used when the temperature is much colder, a distinct advantage when working on construction sites in winter.

- LPG is supplied in pressurised cylinders to keep it liquefied. The cylinders are strong and not easily damaged but the control valve at the top can be vulnerable to impact.
- Leaks can occur from valves and pipe connections, mostly as a gas.
- If the gas is drawn from the cylinder too quickly, the control valve is likely to freeze.
- LPG liquid can cause cold burns if it comes into contact with the skin.
- LPG equipment should be used in a well-ventilated space to prevent the build-up of carbon dioxide (CO_2). Take particular care when using in a confined dry space, such as a loft.
- When connecting hoses and blowtorches, always check for leaks with a suitable leak detection fluid.
- Always turn the cylinder off at the control valve when it is not in use.

LPG regulators, hoses and blowtorches

Most blowtorches that we use today require a regulator to control the amount of gas that flows from the cylinder and a hose that connects from the regulator to the blowtorch. The regulator should have an adjustable pressure setting control. High-pressure hoses are usually coloured orange and are manufactured to BS 3212.

There are many different types of blowtorch available. Most have a range of interchangeable aeration nozzles of differing sizes so that the right nozzle can be chosen for the type of work being carried out. Some blowtorches connect straight onto a small propane or MAPP 400g gas cylinder.

A modern plumber's blowtorch

C3 FIRE SAFETY

An important part of learning and understanding fuel gases is having an awareness of what they produce as an end result – fire. Fire is one of the most destructive elements that exist, but it is something that plumbers rely upon in their work, for example when soldering and welding.

Combustion

Combustion is a chemical reaction in which a substance reacts violently with oxygen to produce heat and light. The substance is known as a fuel and can be a solid such as wood, a liquid such as petrol or a gas such as acetylene. The oxygen is known as an oxidiser or an oxidising agent within this process. To create combustion or fire, we need a third element in the form of heat or an ignition source. These three elements of fuel, oxygen and heat combine into what is known as the fire triangle. All three of the elements need to be in place for combustion to happen. Take any one of them away and combustion will not take place.

Fire triangle

If we remove the fuel then combustion will not occur because there is nothing for the fire to consume. Fuel can be removed naturally as the fire consumes it, mechanically by physically removing it, or chemically by rendering the fuel incombustible.

If we remove the oxygen, the fire will extinguish itself because the fuel has nothing to react with. There are several ways that we can 'suffocate' a fire: by the use of foam, powder or carbon dioxide.

Without a source of heat, fire can neither start nor continue. If we douse a wood fire with water, the water turns to steam. This effectively removes the heat from the fire as the heat is transferred from the wood to the water.

Understanding these processes is the basis for all firefighting techniques and the fire extinguishers we use.

Classification of fire and fire extinguishers

There are six classes of fire, each one using a different source of fuel. Each class of fire requires a different type of fire extinguisher, although some extinguishers can be used on more than one class of fire. The classes of fire are:

Class A	Solids such as paper, wood, plastic
Class B	Flammable liquids such as paraffin, petrol, oil
Class C	Flammable gases such as propane, butane, methane
Class D	Metals such as aluminium, magnesium, titanium
Class E	Fires involving electrical apparatus
Class F	Cooking oil and fat

There are four classes of fire extinguisher. Each fire extinguisher is coloured red but has a different coloured panel on it to show its content.

SmartScreen Unit 001 handout 9

Types of fire extinguisher

On the next page is a table explaining where each of the different types of fire extinguisher can be used.

Class A	Class B	Class C	Class D	Class E	Class F
Water					Special wet chemical fire extinguisher
Foam	Foam				
Powder	Powder	Powder	Powder	Powder	
	CO_2	CO_2	CO_2	CO_2	

SmartScreen Unit 001 worksheet 5

Fighting small fires

Small fires can be brought under control easily and safely with the use of a portable extinguisher. Before you tackle the fire, make sure that everyone else has left the area and the fire service has been notified. Follow the steps in the list below when fighting a small fire:

1. Stand approximately 3 metres away from the fire with the extinguisher, keeping your exit route at your back so that the fire cannot trap you.

2. Remove the safety pin, enabling the extinguisher for use.

3. Aiming low, point the hose at the base of the fire and squeeze the lever above the handle to release the substance.

4. Move the hose from side to side to ensure you cover all of the flames.

5. Continue this until the flames have been extinguished, repeating the process if the fire reignites.

You should have the extinguished fire inspected by the fire service, even if you are sure the fire is out.

Fire safety in the plumbing industry

The use of soldering and welding equipment presents plumbers with the potential to cause fires in homes, factories and commercial properties. You should take the following precautions to eliminate the fire risk from your everyday work as much as possible:

- Always carry a dry powder or CO_2 fire extinguisher with you when soldering or welding.

- Always use heat-proof mats when soldering next to wall coverings and skirting boards.

- Move furniture and carpets away from the soldering area.

- Never point your blowtorch directly at combustible materials.

- When soldering joints under a suspended floor, check to make sure there is nothing that could catch fire before you solder.

- Never replace floorboards etc after soldering activities until you are sure that there is nothing smouldering underneath the floor.

- When lead welding on a flat roof, damp off the substrate before welding begins.

C4 WORKING AT HEIGHT

Most of the work carried out in the construction industry is above ground level. Some of this work can be done at a normal working height of up to 1.5 metres without the use of steps or ladders, but there will be some occasions when you will be asked to work at heights above this.

There are various types of equipment that can be used when you are required to work at height. These are:

- step ladders
- ladders
- roof ladders
- trestle scaffolds
- tower scaffolds (mobile and fixed)
- tubular scaffolds (fixed)
- mobile elevated working platforms and mobile mini tower scaffolds.

Each of these types of equipment is designed for a specific purpose and use and should not be misused.

SmartScreen Unit 001 handout 8

Ladders

In this section, we will look at various different types of ladder, their uses and the safety precautions that must be taken when using them.

Ladders are generally used for access to and egress from the working platform or scaffold but may be used for some light work at high levels. Ladders that are manufactured and supplied in the United Kingdom and the European Union are constructed to the same standards and must be classified correctly:

- timber ladders manufactured to BS 1129:1990 (British)
- aluminium ladders manufactured to BS 2037:1994 (British)
- timber and aluminium ladders manufactured to EN 131:1993 (European).

The standards detail the dimensions, markings and testing requirements for ladders, which include deflection, torsion, rigidity, straightness, loading and performance. These standards apply to all portable ladder types, including stepladders, platform steps and extension ladders. They do not apply to special single-use ladders and fixed access ladders, eg pole ladders, loft ladders and static roof access ladders.

A modern stepladder

There are three main classifications of ladder. For each class of ladder, there is a safe working load that it is designed to support, which includes the weight of a single person plus their equipment. This is known as the 'maximum static load'. The table on the next page shows the three main classes of ladders and their details.

Classification	Duty rating	Max static vertical load	Application	Colour identification
Class 1	130kg	175kg	Industrial – These have the highest rating in terms of strength and quality. They are used in heavy-duty industrial applications and environments.	(blue)
Class EN 131	115kg	150kg	Commercial – This is a Europe-wide classification, which replaces the old British Class II Ladder Standard. In the UK this classification is known as BS EN 131 and is most suitable for light trade work and heavy-duty DIY use.	**Either** (green) / Or (yellow)
Class 3	95kg	125kg	Domestic – These are not suitable for use within any commercial or trade environment. They are for light domestic use only.	(red)

Ladders should be colour coded to provide a visual sign of their classification (see table above). Colour identification can often be found on the rubber feet of ladders and steps and/or the user instructions and warning labels on the stiles.

Pole ladders
This type of ladder is generally made of timber. The stiles are cut from one tree trunk sliced down the middle. This ensures strength and durability. Pole ladders are used on fixed ladder installations for access to scaffolds and can be up to 12 metres in length. Some pole ladders have wire reinforcement to provide extra strength. They have a Class I rating and will safely support a maximum load of 175kg.

Single-section ladders
These are usually made to Class I standard from lightweight aluminium or timber. Timber standing ladders are made from Douglas fir, redwood, white-wood or hemlock. Lengths up to 10 metres are available. These ladders are often called 'standing ladders'.

Multi-section ladders
These are often called extension ladders. They consist of either two or three sections, which can be slid apart to give the required height, and are available in various lengths of 2.5 to 3.5 metres (when closed). A two-section ladder can give a length of up to about 8 metres and should be suitable for most two-storey properties. Three-section ladders can give lengths up to about 10 metres.

On smaller ladders, the ladder may be extended by hand and secured with stay locks that rest on a selected rung. On larger ladders, the sections are extended by means of a rope loop and pulley system running down the side of the ladder.

A timber pole ladder

A double-extension ladder

A roof ladder

Multi-section ladders can be made of timber, aluminium and glass reinforced plastic (GRP).

Roof ladders
This type of ladder should always be used when working on a sloped roof. It should always be accessed from a scaffold, not a ladder. The roof ladder has two wheels at the upper end, which allows it to be pushed up the roof without damaging the slates or tiles. On the other side to the wheels, the ladder is formed into a hook, which fits over the top ridge of the roof and stops the ladder from slipping down the roof.

Crawl boards
Crawl boards are used for working on fragile roofs. They help to spread the weight across the roof to lower the risk of the roof giving way. They are used for access only and are not intended for carrying tools or materials and should be used with extreme care.

Stepladders
These are normally used for inside work but can be used outside if the ground is firm and stable. They can be made from timber, aluminium or glass reinforced plastic (GRP) and come in a range of sizes and heights. Timber stepladders are vulnerable to damage and can warp or twist, whereas aluminium and GRP stepladders are lighter, stronger and rot-proof.

Stepladders consist of a set of stiles supporting flat steps, which are spaced at around 250mm. When the steps are opened, a locking bar ensures that the steps are at the correct working angle and this prevents the steps from collapsing.

Raising and lowering ladders

Ladders should be erected with the sections in the closed position. Never try to raise an extended ladder. Extension ladders should be raised one section at a time and should be at the correct height before the ladders are used. Two people are required to raise and lower ladders. Follow the steps below for raising ladders.

STEP 1 – Lay the ladder flat, one person at the foot and other at the head. The ladder foot should be at the correct distance from the wall for the final angle.

STEP 2 – The person at the foot of the ladder should keep one foot on the bottom rung to prevent the ladder slipping. He should hold the stiles to steady the ladder as it is lifted, and reach forward to receive the ladder.

STEP 3 – The person at the other end lifts the ladder above his head. He should lift the ladder by moving hand over hand, walking towards the foot of the ladder, raising it as he goes. This is continued until the ladder is in the upright position.

STEP 4 – The correct safety angle for the erected ladder is 75°, or a ratio of 4 metres upwards to 1 metre outwards.

To lower the ladder from the upright position, the above process is reversed. One person may raise lighter ladders alone, provided the bottom is placed against a firm stop before the lifting procedure is started.

Tying ladders

Ladders must have a firm and level base on which to stand. They must be securely fastened, preferably at the top, but if this is not possible then at the bottom is acceptable. If neither way is possible, a person must 'foot' the ladder, holding both stiles and paying attention at all times. This will prevent the base from slipping outwards or the ladder from falling sideways.

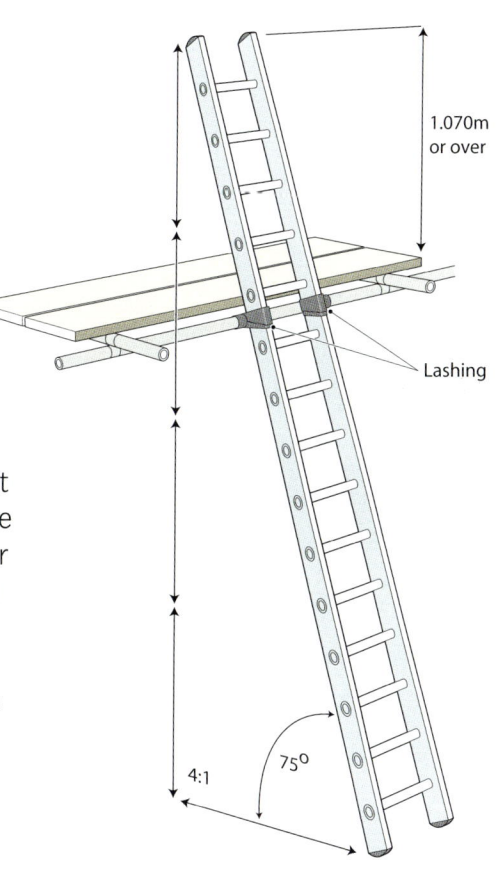

Footing a ladder
Standing with one foot on the bottom rung, the other firmly on the ground.

A correctly erected and lashed ladder

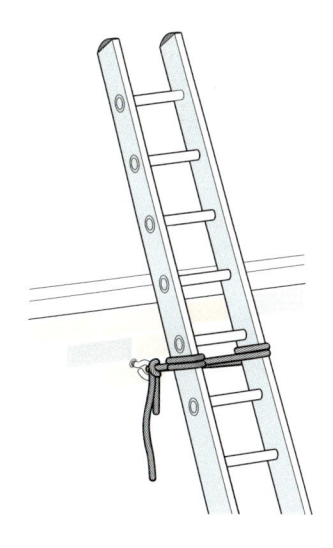

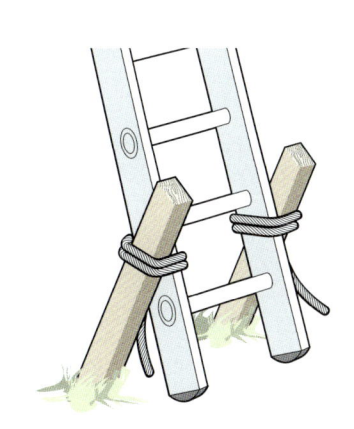

Tying and lashing ladders

Lifting and carrying ladders

When moving ladders more then a few metres, they should be lowered and carried by two people, one at either end. To move ladders over short distances, rest them on the shoulder then find the correct balance and angle before lifting vertically by grasping the rung just below normal arm's length.

A single-person ladder lift over a short distance

STEP 1 – With one person at either end, crouch down and prepare to lift the ladder.

STEP 2 – Lift and carry the ladder supported on one shoulder, braced by the nearest hand.

A two-person ladder lift

Storing ladders

- Ladders should be stored in a covered, well-ventilated area, which is protected from the weather and away from dampness or heat.

- Ladders must not be stored by leaning against a wall or building as they may fall.

- Never hang a ladder vertically from a rung as this could cause the rung to warp.

- Do not store ladders where a child might be able to gain access to them.

- Store the ladder horizontally on a rack or wall bracket. Always support the lower stile at 1-metre intervals.
- Keep wooden ladders clear of the ground to avoid contact with water or dampness.

Safety checks

Ladders, roof ladders, crawl boards and stepladders should not be used if:

- they have broken, weak or repaired stiles or rungs
- they have faulty ropes, guide brackets, latching hooks or pulley wheels
- they are painted (as paint can hide defects on wooden ladders)
- they have missing safety feet.

If any defects are found, a 'do not use' notice must be placed on the equipment and the defects reported to your supervisor.

Safe working with ladders

Consider the following safety guidance when working with all types of ladder:

- Ladders must extend five rungs (1 metre) above the working platform.
- Never stand on the top platform of a stepladder. You are at a safe working height when your knees touch the top platform.
- Never erect a ladder or stepladder on an uneven or fragile surface or soft, loose ground.
- Never stand a ladder or stepladder on a box or other unsteady base to gain extra height. Always use a ladder that is long enough to reach the job.
- Never climb ladders if the rungs are slippery, icy or greasy.
- Never attempt to carry too much equipment up a ladder.
- Never overreach when working from a ladder. Always move the ladder if you cannot reach something easily.
- Never stand side on to work – always face the job.
- Take care when erecting ladders to avoid overhead obstructions such as electric cables.
- Always ensure that the ladder is erected to the correct angle and is securely fastened.

- Ladders must be lowered and stored in a safe place overnight. If this is not possible, a scaffold board at least 2 metres long should be firmly fixed to the rungs to prevent unauthorised access.
- Always use ladder equipment in accordance with the manufacturer's instructions.

Scaffolds

Scaffolds are a much safer way of working at height, but extreme care must still be taken. You need to be aware of your surroundings at all times and take care with tools and equipment.

Tower scaffolds

There are two types of tower scaffold:

- static
- mobile.

Both kinds of tower scaffold can either be constructed at the place where it is needed or made by a manufacturer with standard sections that fit together. You must hold the relevant certification and be registered to erect any type of fixed scaffolding.

Static tower scaffolds are constructed from regular tubular scaffolding components. Access to the tower must be by a fixed ladder

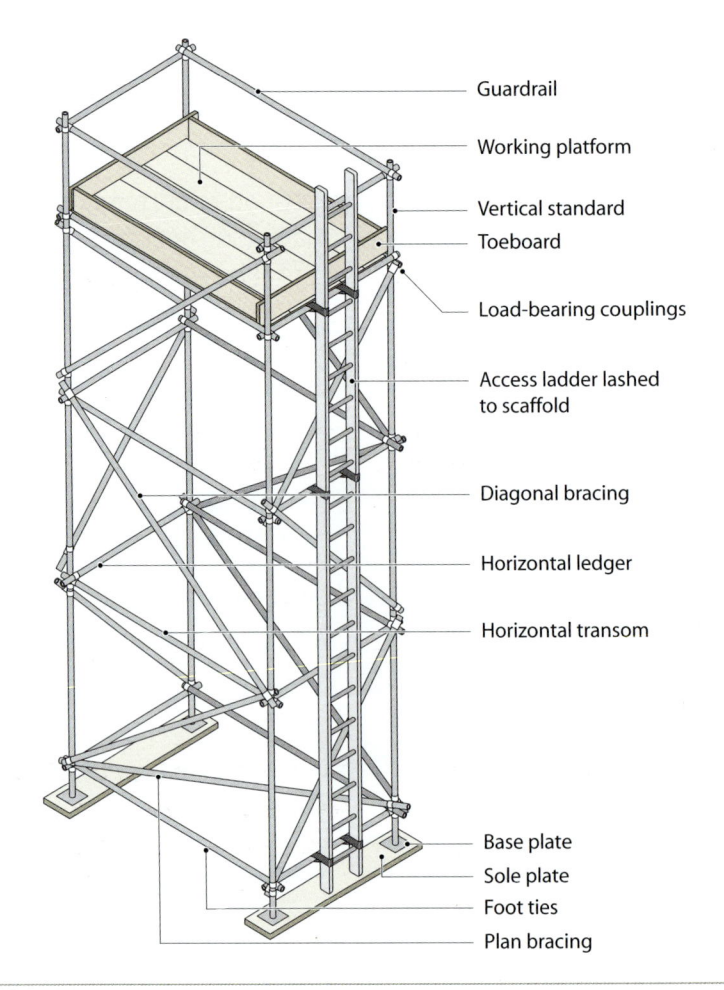

A correctly erected static tower scaffold

from either inside or outside of the tower. Care must be taken when leaning ladders on the outside of scaffolds to ensure that the stability of the tower is not affected.

Scaffold towers should be designed to carry a load of 150kg/m² spread over the whole working platform plus its own weight. A special design will be required for any tower scaffold where extra loadings from materials on the working platform are required or where wind may occur.

Mobile tower scaffolds are used mainly for light work of a short duration, such as installing boilers and flues at high level. They are usually manufactured from lightweight aluminium and are easily dismantled and erected. This type of scaffold should only be used where the ground is sufficiently firm and level. All wheels on mobile scaffolds must be lockable and kept locked when the scaffold is being used. Tower scaffolds should never be moved with persons still on the working platform.

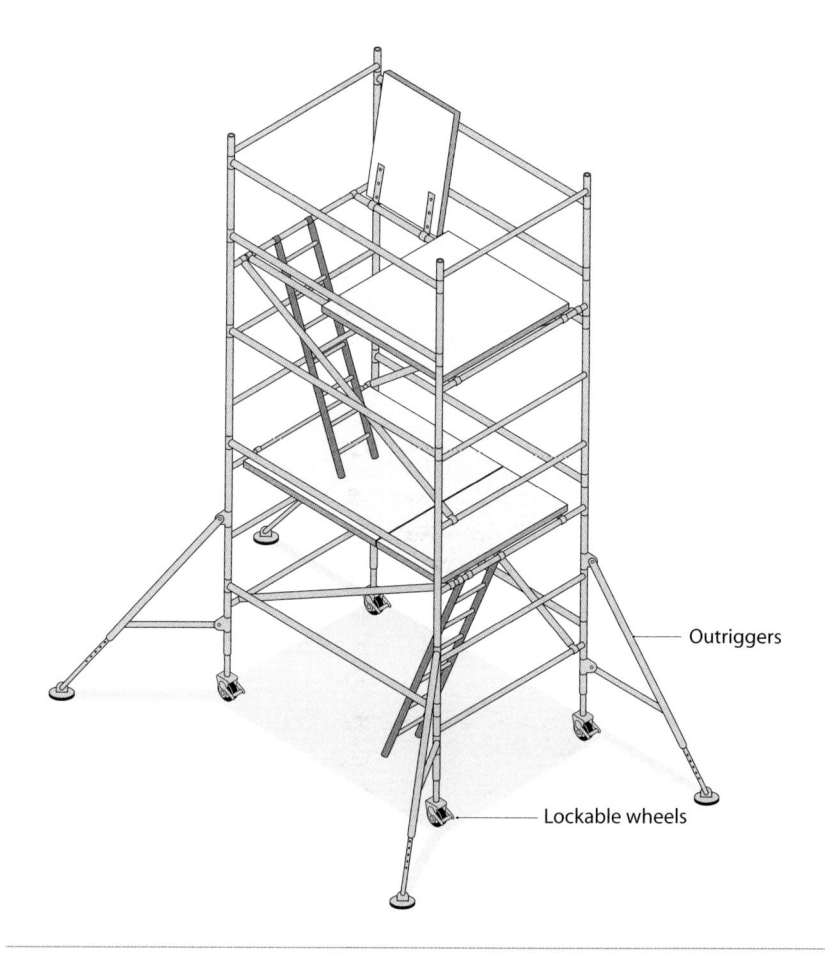

A mobile tower scaffold

The working height of a tower scaffold may be calculated by taking the measurement of the shortest side and multiplying it by 3. For example, if the tower scaffold has the dimensions of 4 metres long by 2 metres wide, then 2 × 3 = 6 metres.

If extra working height is needed then the base measurement can be increased by the use of outriggers. These are special attachments that connect to the bottom of the tower at the corners, giving a greater overall base measurement. Outriggers also help to give greater stability to a scaffold tower

Tubular scaffolds

There are two types of tubular scaffold:

- independent
- putlog.

An **independent scaffold** is completely independent from the building. It must still, however, be tied to the building for stability. The main applications for this scaffold are:

- access to stonework and brickwork on masonry buildings
- access to solid or reinforced concrete structures
- maintenance and repair work.

The independent scaffold consists of two rows of vertical uprights, which are called standards. These are joined together by horizontal scaffold tubes, called ledgers, that the scaffold boards rest on. These are joined by diagonal strengthening tubes, known as transoms. Strength is achieved by triangulation with cross bracing at every lift to ensure a rigid construction. The ground should be firm and level, and base plates should be used under every standard, with wooden sole plates supporting the base plates to spread the load.

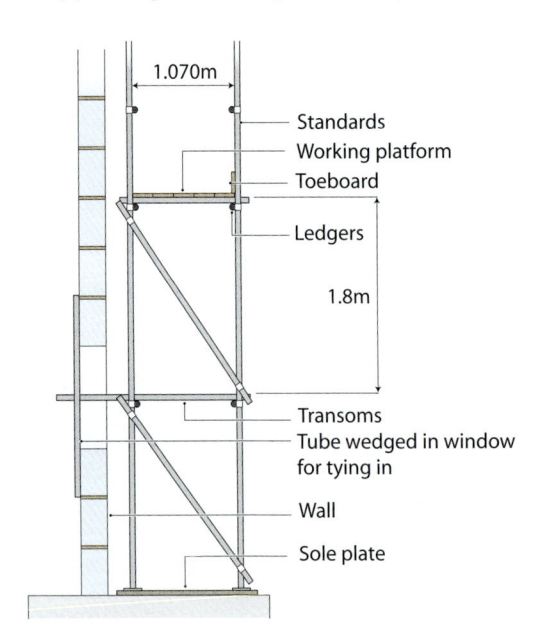

An independent scaffold – side view

A **putlog scaffold** is also known as the dependent or bricklayer's scaffold. It is visually similar to the independent scaffold but has one major difference: there is only one row of standards, which means that the scaffold is reliant on the building for support. The inner row of standards is replaced by the brickwork. The rest of the scaffold functions in the same way as the independent scaffold.

The putlog scaffold can be erected to existing brickwork, but is usually used with new building work. Working platforms are supported by the 'putlog' tubes by allowing the flat end (known as the spade) of the putlog to rest flat on the brickwork. Putlogs should never be removed or the scaffold will be in danger of collapse. Putlog scaffolds should be tied to the building at 4-metre intervals vertically and 6-metre intervals horizontally.

Access to and from a tubular scaffold is usually gained by using a ladder. To safely allow this, a suitable gap should be left in the handrail and toe board so that workers on site can access the working platforms. The ladder should be secured both at the top and bottom and extend at least 1 metre (five rungs) above the working platform.

The final 'step-off' rung from the ladder to the working platform should ideally be just above the level of the platform to prevent tripping. The gap left between ladder and guardrail should not exceed 500mm.

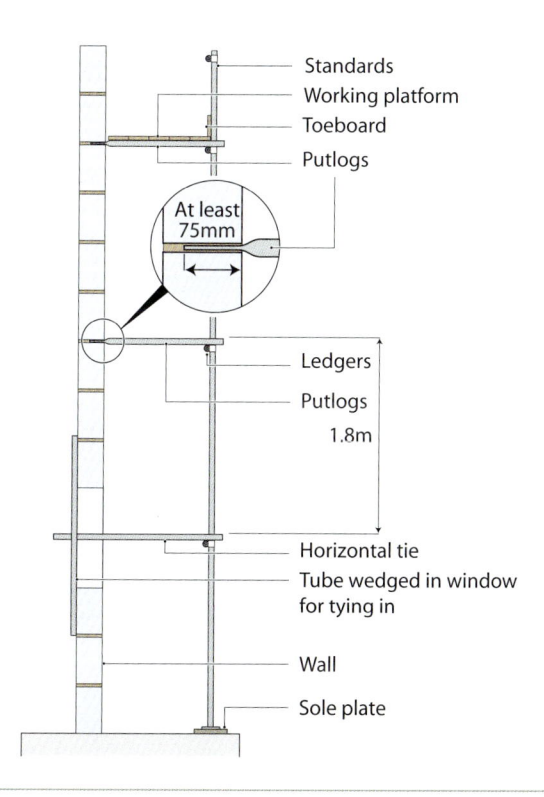

A putlog scaffold – side view

Guardrails and toe boards

Keeping platforms clear of all loose and unused materials should reduce the risk of causing injury from falling materials. Working platforms more than 2 metres high must have guardrails, brick guards and toe boards installed as these provide a method of preventing materials and tools from being kicked off the edges of working platforms.

Toe board
A board placed around a platform or on a sloping roof to prevent personnel or materials from falling.

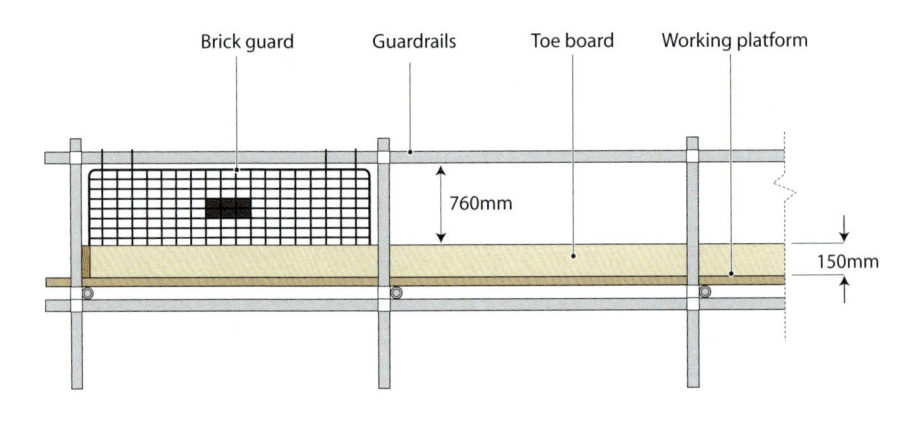

Guardrails and toe boards

Working platforms

The working platform is the area of the scaffold where the work activities are carried out. Certain rules must be followed to ensure safety:

- Materials must be placed as evenly as possible to spread the load.
- Working platforms must be kept free from hazards including ice, snow and trip hazards.
- Gaps between the scaffold boards should be kept as small as possible.
- 50mm-thick scaffold boards must be at least 150mm wide.
- 32mm-thick and 38mm-thick scaffold boards must be at least 200mm wide.
- Boards must be placed evenly on their supports.
- Boards must not project more than four times their thickness beyond their end support, and no less than 50mm.
- Boards must be bound with a steel strap at each end to prevent the board splitting.
- Split or damaged boards must be replaced.

Mobile elevating work platforms

Mobile elevating work platforms (MEWPs) include cherry pickers, scissor lifts and vehicle-mounted booms. MEWPs provide a way of working at height without the need for a scaffold. They are designed to allow the worker to access the work quickly and easily. They may be used either indoors or outdoors.

When working with MEWPs, employers should:

- choose the correct MEWP for the task
- identify and manage the risks involved with working from MEWPs through correct risk assessment and correct training.

MEWP in use

Training in the use of MEWPs is essential to prevent serious injury or death. The following list outlines safety checks that should be carried out before use:

- Gate enclosures and securing chains must be in the closed position before raising the lift.
- The MEWP must be prevented from making movement after it has been placed into the final work position by switching off the control panel.
- The maximum operating weight including personnel, equipment, supplies and tools must not be exceeded.
- Personnel must not work on MEWPs when high winds, storms, ice or snow are present or expected.
- MEWPs are not cranes and must not be used as such.
- The MEWP must not travel with personnel in the basket while it is raised, unless the equipment is specifically designed for this activity and the correct regulations are observed.
- MEWPs must only be used on stable, flat and structurally sound flooring or ground.
- Where other moving vehicles are present, the work area must be marked with clear warning signs, roped off areas or other effective means of traffic control.
- Loose objects, materials and tools must not be allowed to gather on the floor of the MEWP.

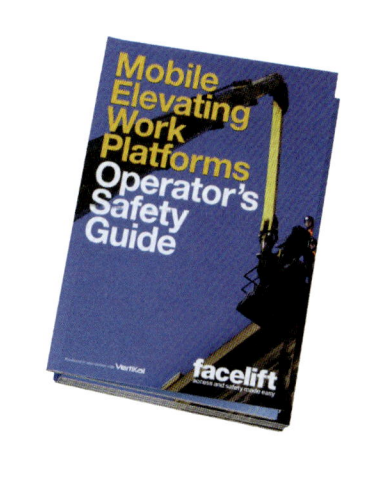

MEWP safety guide

C5 WORKING IN TRENCHES AND EXCAVATIONS

The need for plumbers to work in trenches and excavations is limited as most of this type of work is done by others on construction sites, such as groundworkers or the service providers such as the gas company or the water authority. However, there is still a need to understand the planning and working practices of working in trenches and excavations.

Every year construction workers are killed and injured when the excavations and trenches they are working in suddenly collapse. Deaths have occurred in both shallow and deep excavations, so it is important that any excavation work is properly planned, managed, supervised and carried out to prevent accidents.

Many types of ground are, to some extent, self-supporting, but this should not be relied upon when working in a trench. It is vital that precautions are taken to ensure that excavations are adequately supported.

The maximum depth that a trench or excavation can be dug without support is 1.2m. On the average person, 1.2m would be around waist height. The significance here is that the chest would be above ground level and so breathing would not be restricted in the event of trench collapse.

Beyond this depth, the trench sides should be either:

- battered – a method by which the sides of the trench are sloped away from the trench bottom (the angle of the slope would be decided by the type of ground, but usually 45° is considered adequate)
- benched – the sides of the trench are cut into steps away from the trench side
- supported – using a proprietary trench support system.

The general requirements for safe trench and excavation design are as follows:

- The ladder used to gain access should be secured in position to the trench supports and in long trenches access should be spaced at regular intervals.
- The spoil from the trench should be at least a metre away from the edge of the trench to prevent trench collapse. One square metre of earth can weigh as much as a tonne and the added weight against the weak edge of the trench could cause collapse or earth slide.
- The edge of the trench must have a 2-metre high barrier placed around it at least 1 metre away from the edge to stop people from falling into the trench. It must also have a toe board to stop tools and materials from being accidentally kicked in.
- Vehicle stops must be used to prevent vehicles and plant getting too near to the edge and to stop a build-up in the trench of poisonous carbon monoxide fumes.
- The use of propane gas is prohibited as this gas is heavier than air and any leak could gather at the trench floor.
- Trenches and excavations must have a secure ladder (or several if the trench is long) for quick emergency evacuation.
- Warning notices and signs should be placed at regular intervals along the trench length.

Trench support sheets and braces

Proprietary trench support systems

In order to support the walls of an excavation and prevent trench collapse, a preliminary trench is dug and its walls shored up by means of a trench box or trench shield placed inside the trench. A series of piles are driven into the soil below the trench box or trench shield as the excavation is made deeper

Trench safety

There are many things that you have to be aware of to maintain your own personal safety when working in trenches:

- Always wear the correct PPE. Arguably the most important piece of

PPE is the high visibility (hi-viz) jacket or vest, followed by your hard hat.

- Never work in an unsupported trench deeper than 1.2m and never work ahead of the trench supports.
- Be aware of where the access points and ladders are. This could be vital in an emergency situation.
- Be aware of plant and vehicles approaching the trench.

C6 WORKING IN CONFINED SPACES

A confined space is an area that is enclosed where there is a risk of death or serious injury from hazardous substances or dangerous conditions, such as lack of oxygen or being overcome by fumes.

During plumbing installations and maintenance, you may be required to work in:

- tanks and cisterns
- trenches
- sewers
- drains
- flues
- ductwork
- unventilated or poorly ventilated rooms
- small roof spaces and under floors.

All of these constitute confined spaces and precautions need to be put in place to ensure your health and safety.

The risks of working in confined spaces

Every year, a number of people are killed or seriously injured working in confined spaces in the construction industry, from those involving complex plant to unventilated or poorly ventilated rooms.

Those killed include not only people working in the confined space but also those who try to help them without the proper training and equipment. Dangers occur because of:

- lack of oxygen
- poisonous gas, fumes or vapour
- liquids and solids suddenly filling the space
- fire and explosions
- residues left behind, which may give off fumes, vapour or gas
- hot working conditions.

Legal duties and obligations

The Management of Health and Safety at Work Regulations 1999 require that a suitable assessment of the risks for all work activities is carried out so that decisions can be made on what measures are necessary for safety.

For work in confined spaces this means identifying the hazards present, assessing the risks and determining what precautions to take. In most cases, the assessment will include consideration of:

- the task
- the working environment
- tools and materials
- the suitability of those carrying out the task
- arrangements for emergency rescue.

CONCLUSION

It is no coincidence that this chapter is the longest in the book, such is the importance of health and safety in the modern construction industry. We, as plumbers and apprentices, have a duty of care towards ourselves, those that we work with and those whom we come into contact with. The ultimate responsibility of how we behave, how we work and how we respond to accidents and incidents rests with us. By taking notice of health and safety and following the rules that are in place to safeguard us, we too can reduce the likelihood of accidents and, ultimately, save lives. The key message that we must always remember is: health and safety is everyone's responsibility.

001 TEST YOUR KNOWLEDGE

SmartScreen Unit 001 revision sample questions

1. Which Act of Parliament is the key piece of health and safety legislation in the UK?

2. Under Section 2 of the HASAWA, what is the duty of every employer?

3. Under the Construction (Design and Management) Regulations, everyone involved in a construction project should be working towards three main aims. What are these?

4. Who is responsible for providing PPE?

5. What does COSHH stand for?

6. What is RIDDOR?

7. Give two examples of building services specific legislation.

8. Which UK government body enforces health and safety legislation?

9. Identify the following safety sign types:

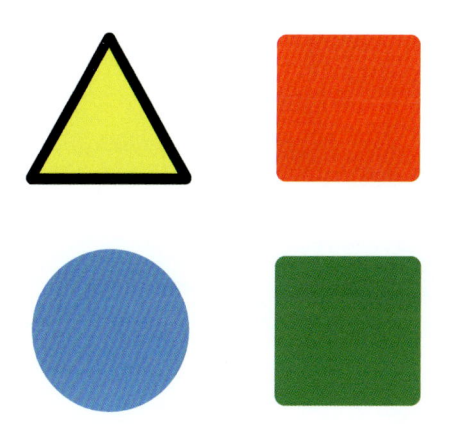

10. What are the three ways in which lead can enter the body?

11. What are chrysotile, amosite and crocidolite?

12. Which four types of injuries are classified as minor?

13. What do the initials CPR stand for?

14. Identify the following electrical voltages from their colour coding:

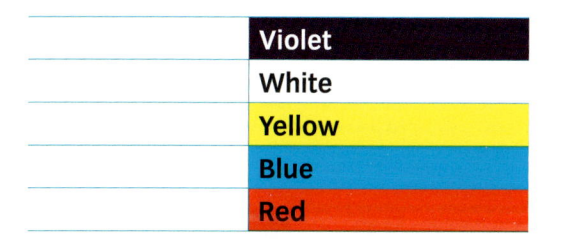

	Violet
	White
	Yellow
	Blue
	Red

15. How often should stationary electrical equipment on site be PAT tested?

16. What is the purpose of temporary continuity bonding?

17. Identify the bottle colour of the following common gas bottles:
    - Propane
    - Acetylene
    - Oxygen

18. True or false? LPG is heavier than air.

19. Identify the following fire types:

Class A	
Class B	
Class C	
Class D	
Class E	
Class F	

20. What is the correct angle that a ladder should be erected to?

21. At what depth would a trench require support before work could be undertaken?

002
UNDERSTAND HOW TO COMMUNICATE WITH OTHERS WITHIN BUILDING SERVICES ENGINEERING

IN THIS CHAPTER, YOU WILL COVER:

A The construction industry management hierarchy
- A1 The structure of the site management team
- A2 The key roles of the site management team
- A3 The building contractor and employees
- A4 Members of the on-site team
- A5 The inspectors

B Documentation and sources of information in the building services industry
- B1 The relationship between you and your employer
- B2 Knowledge and understanding of your job
- B3 Knowledge and understanding of on-site documentation
- B4 Communication between the company and the customer
- B5 Company policies and procedures

C Formal and informal methods of communication
- C1 Methods of communication at work
- C2 Effective communication strategies
- C3 Conflicts in the workplace
- C4 The effects of poor communication at work

Test your knowledge

There are many working relationships that exist within the construction industry: between the company and its customers/suppliers, between a manager and the workforce and between the different trades on site. How well these relationships function can play a significant role in the overall effectiveness of how any job or construction site is run with regard to completion times, handover and customer satisfaction.

There are some important factors that can determine whether these relationships are healthy and productive or whether they will have a harmful effect on the workplace. Good relationships are built on cooperation, where each individual is working towards the achievement of common objectives.

In this chapter, we will look at the various roles of the people within the construction industry hierarchy, from the Architect to the tradespeople and apprentices on the site, as well as investigating the roles of site visitors who can influence what we do and the way we work. We will also learn about the various systems of information and communication available to us to ensure the smooth running of site operations.

A   THE CONSTRUCTION INDUSTRY MANAGEMENT HIERARCHY

The construction of any building is a complex process that requires a group of professionals, known as the construction team, to work together to produce what the client has requested. In this first section of the chapter, we will take a closer look at the construction team. We will consider the role that each individual has in the overall construction project and their responsibilities within the management structure.

A1   THE STRUCTURE OF THE SITE MANAGEMENT TEAM

Within each construction project, there is a site management team. This usually follows a recognised structure by which the team operates and communicates. This is illustrated in the line diagram below.

It is important that all members know their roles and responsibilities within the management structure to ensure the smooth running of the project and that any problems are dealt with as quickly as possible.

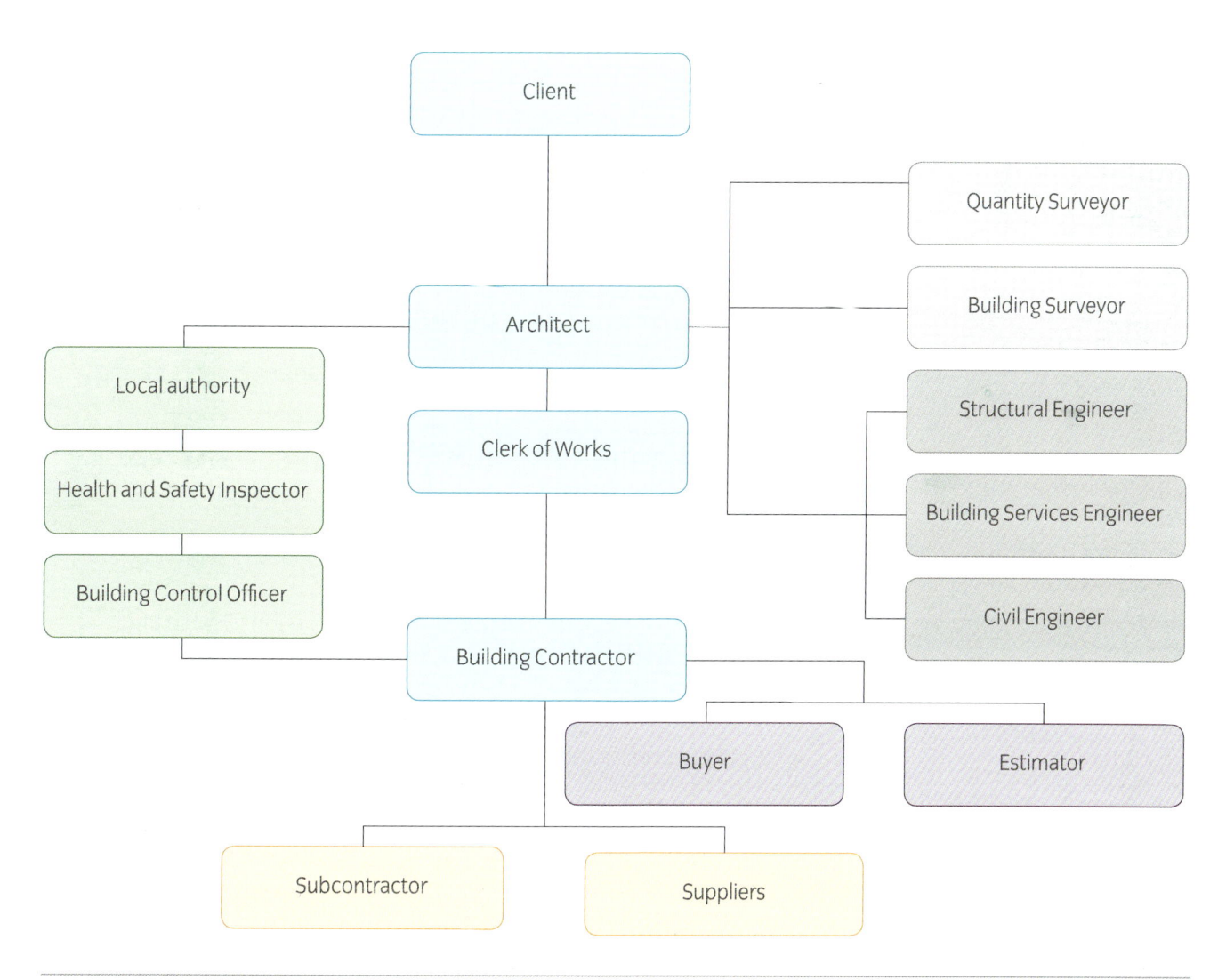

Structure of the site management team

81

SmartScreen Unit 002 handout 2

A2 THE KEY ROLES OF THE SITE MANAGEMENT TEAM

The management of construction projects requires a good comprehension of modern management systems, as well as expert knowledge of the design and construction process. Construction projects have a specific set of objectives, which must be completed within a given time frame and on budget to a specific set of rules and regulations.

The management of any large construction site usually falls into two tiers:

- those that only visit the site occasionally, usually senior management
- those that are permanently site-based.

In this section we will look at the first tier.

The client

Clients are arguably the most important part of the project because they are the reason for the construction of the building. The client, either directly or indirectly, employs everyone else who has connection with the construction project, and finances the whole project. The client can vary from a single individual to a large consortium or organisation.

Under the Construction (Design and Management) Regulations 2007 (see Chapter 001, pages 5–7), clients (with the exception of domestic clients who intend to live in the completed building) have responsibilities with regard to the health and safety of all those people directly or indirectly employed as part of the project. On all projects clients will need to:

- ensure the competence of all team members and ensure that they are adequately resourced and appointed early on in the project
- ensure that there are suitable management arrangements for the project welfare facilities
- allow sufficient time and resources at every stage of the project from concept to completion
- provide pre-construction information to designers and contractors so that regulations can be followed.

Where projects are notifiable under the Construction (Design and Management) Regulations 2007 (CDM), clients must also:

- appoint a CDM coordinator
- appoint a principal contractor
- make sure that construction work does not start unless a construction phase plan is in place and there are adequate welfare facilities on site
- provide information relating to the health and safety file to the CDM coordinator

KEY POINT

Projects are notifiable under the Construction (Design and Management) Regulations 2007 (CDM) when they are to last more than 30 days or involve 500 person days of construction work.

- keep the health and safety file and provide access to it if required.

The Architect

The Architect (or designer) is considered to be the leader of the management team. It is their responsibility to convert the client's requirements into a building design and working drawings, seek planning permissions, advise the client on materials and generally supervise all aspects of the construction work until handover to the client.

The Architect, like the client, has direct responsibilities under the Construction (Design and Management) Regulations 2007 and these were discussed in Chapter 001 (see page 6).

> **KEY POINT**
>
> The Architect must be registered with the Architects Registration Board (ARB) whose duties and functions are defined by the Architects Act 1997. It was established to regulate the architects' profession in the UK. Many architects are also members of the Royal Institution of British Architects (RIBA).

The Surveyor (Building Surveyor)

The role of the Surveyor (also known as the Building Surveyor) is to position the building on the land plot and to ensure that the Building Regulations are followed during the planning and construction phases of new buildings and extensions, and conversions to existing properties. He or she resolves problems arising from the Building Regulations and other relevant legislation. The Surveyor will also make site visits at different stages of construction to ensure that the building process is being properly carried out.

> **KEY POINT**
>
> The Building Regulations set standards for the design and construction of buildings, primarily to ensure the safety and health of people in or around those buildings, but also for energy conservation and access to buildings. They are divided into Documents or Parts named after letters of the alphabet, such as Document L Conservation of Fuel and Power and Document H Building Drainage.

The Quantity Surveyor

The Quantity Surveyor (QS) is an accountant who advises as to how the building can be constructed within the client's finances. The QS also measures the amount of labour and materials needed to complete the building from looking at the Architect's drawings. These details are then collated into a document called the Bill of Quantities, which is used by building contractors to produce an estimate.

As work progresses, the QS will produce measurements and variations of the work carried out to date so that the main contractor can receive interim payments. At the end of the contract, the QS will also prepare the final account to be presented to the client. In addition to these duties, the QS may also advise the Architect on the cost of any variations to the original contract or any additional work completed.

Specialist engineers

These are hired as part of the Architect's team to assist in the design of the building with regard to their specialist fields. There are three main types of specialist engineer:

The Civil Engineer

The Civil Engineer is the designer of the roads into and out of the building, along with any bridges, tunnels etc that may be required. He or she may also be involved in the design of drainage and water requirements to the building or complex.

The Structural Engineer

The Structural Engineer is someone who works closely with the Architect to find the most efficient method of constructing the project. The engineer calculates the loads, taking into account wind and rain, and the weight of the building itself. The frame and foundations can then be designed to support those loads.

The Building Services Engineer

This person designs the internal services within the building, such as heating and ventilation, hot and cold water supplies, air conditioning and drainage. Building Services Engineers are responsible for ensuring the cost-effective, environmentally sound and sustainable design and maintenance of engineering services in buildings.

Their areas of responsibility include all equipment and materials involved with heating, lighting, ventilation, air conditioning, electrical distribution, water supply, sanitation, public health, fire protection, safety systems, lifts, escalators, facade engineering and acoustics. With the current emphasis on sustainability, this role is at the cutting edge of designing, developing and managing new technologies that integrate into existing systems and services.

> **KEY POINT**
>
> Although their role increasingly demands a multidisciplinary approach, Building Services Engineers tend to specialise in one of the following areas:
> - electrical engineering
> - mechanical engineering
> - public health

The Clerk of Works

The Clerk of Works (CoW) may also be referred to as the Project Manager. Appointed by the Architect, the CoW is the Architect's representative on site. He or she ensures that the building is constructed in accordance with the drawings while maintaining quality at all times. This includes checking the standard of the work and the quality of the materials. The CoW will make regular reports back to the Architect as work progresses and he or she will also keep a diary in case of any disputes, make any necessary notes on the weather, and note any stoppages.

On large sites, the CoW will be a resident member of the management team, while on smaller sites he or she will visit periodically.

Clerk of Works

The Local Authority

The Local Authority has the overall responsibility for ensuring that all works carried out conform to the requirements of the relevant planning and the Building Regulations. It may also show interest in site health and safety in collaboration with the Health and Safety Executive. The Local Authority employs two different people to carry out roles.

The Planning Officer is responsible for processing planning applications, listed building consent applications, conservation area consent applications and advising on planning issues.

The Building Control Officer is responsible for ensuring that regulations on public health, safety, energy conservation and disabled access are met, working to the Building Regulations. A Building Control Officer's job involves:

- checking plans and details of new constructions and alterations to existing buildings
- regular inspections of work in progress to ensure that the construction work is in accordance with the Building Regulations
- management of buildings and structures identified as being in a dangerous condition
- management of the demolition of derelict buildings
- management of improved access to buildings for people with disabilities
- guidance and advice on all types of buildings and construction problems outside of Building Regulations control.

> **SUGGESTED ACTIVITY...**
> Who is the Local Authority in your area? Check out its website and see what services it offers to the construction industry.

A3 THE BUILDING CONTRACTOR AND EMPLOYEES

In this section, we will examine the role of the Building Contractor and the members of the team directly employed by him or her.

The Building Contractor

The Building Contractor will enter into a contract with the client to carry out the work in accordance with the drawings, the Bill of Quantities and the specification. Every contractor develops his or her own methods of pricing and tendering for the work and, depending on the size of the job, this will determine the company's staff requirements.

The Building Contractor will employ specialists within the construction industry to undertake certain key roles. These include the following roles.

The Estimator

Breaks the Bill of Quantities down into labour, materials and plant, and applies a set payment rate for each one. This represents the amount it will cost the Contractor to complete each stage of the project. Added to this will be a set percentage for overheads and profit.

Overheads
Costs that include those of the site office and site/administration staff salaries.

The Buyer

Sources and purchases all the materials needed. He or she will obtain quotes for the materials in the quantities required, together with delivery times and quality assurances.

The Planning Engineer

Is responsible for the pre-contract planning and identifying the most economic and efficient way to use labour, plant and materials.

The Plant Manager

Is responsible for all the items of mechanical plant used by the Building Contractor (either owned by the Contractor, or hired from a company) to carry out a specific task. The Plant Manager is also responsible for maintenance and repair and the training of plant operators.

The Safety Officer is responsible for carrying out safety inspections, investigations of accidents and safety training and inductions on the site. He or she must also complete safety records. This person is accountable to the senior management for all aspects of on-site health and safety.

The Contracts Manager supervises the creation and management of planning and building operations contracts, liaising with head office staff and site agents as needed.

A4   MEMBERS OF THE ON-SITE TEAM

So far we have looked at the roles and responsibilities of the site management team. In this section, we consider the on-site workers who report to the site management.

Subcontractors

Subcontractors enter into a contract with the main Building Contractor for a specific or specialised part of the contract, such as plumbing, heating and ventilation, air conditioning, electrical installation, plastering, bricklaying or joinery/carpentry. The contract may be for labour only, with the Building Contractor purchasing the materials. Alternatively, it may be on a supply-and-fix basis, where the Subcontractor purchases the materials as well as carrying out the work. The Architect may specify a nominated Subcontractor in the initial contract, who would, with the client's permission, be used.

The Site Supervisor

Also known as the Construction Manager, the Site Supervisor is the Building Contractor's main representative on site. He or she is responsible for the general day-to-day running of the site. This can include preparing budgets, hiring team members, handling deliveries and overseeing construction duties.

The Trade Supervisor

Each of the different trades on site will have their own supervisor. The Trade Supervisor will be responsible for the overall running of their company's contract on the site. Their tasks include:

- determining work requirements and allocation of duties to the operatives under their direct control
- consulting with other managers to coordinate activities with other trades
- maintaining attendance records and rosters
- explaining and enforcing regulations
- overseeing the work of the workforce and suggesting improvements and changes

- holding discussions with workers to resolve grievances
- performing the tasks of their trade.

The on-site trades

No construction site can function without the on-site trades. Working to the Architect's drawings, it is the trades that build the Architect's vision. The trades can be divided into two main groups – craft operatives and building operatives.

Craft operatives

Craft operatives are skilled craftsmen who perform specialist tasks. Their roles include the following:

- **Bricklayers** – Construct the building to the Architect's specifications using a range of building materials including brick, block and stone.

- **Carpenters/joiners** – Provide a vital function on site during the initial building phase, fitting door and window frames, floor joists and roof trusses. During the second phase, they fix internal doors, skirting boards, architraves etc.

- **Plumbers** – Perform three key functions on domestic construction sites: installing hot and cold water supplies, central heating and gas installations. On large construction sites, the plumber's work will be restricted to hot and cold water supplies only. In some cases, specialist companies will perform the gas and heating installations.

- **Electricians** – Install and test all electrical installations on site, including power, lighting, fire and smoke alarms and security systems, usually running the cables in trunking or conduits for neatness.

- **Heating and ventilation/air conditioning engineers** – A very specialist trade, especially where the installation of air conditioning is concerned. Their work mainly involves the installation of large-diameter pipework for heating systems and air conditioning ductwork.

- **Gas fitters** – Install natural gas and LPG lines in domestic properties and in commercial or industrial buildings. On some sites they may also install large appliances and pipelines.

- **Plasterers** – Responsible for wall and ceiling finishing, dry lining and external rendering, if required, using a mixture of both modern and traditional techniques.

- **Painter and decorators** – Responsible for wall and ceiling finishing, including painting skirting boards and architraves and any specialist decorating such as murals, frescos etc.

- **Tilers** – Responsible for internal and external tiling of walls and floors and specialist tiling, such as swimming pools and wet rooms.

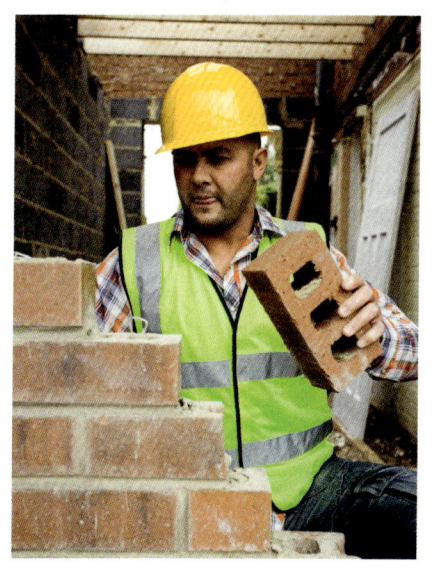

On-site trades

> **KEY POINT**
>
> Craft operatives such as plumbers, electricians, joiners and bricklayers, have served a formal apprenticeship. This usually takes around four years to become fully qualified, with a formal City & Guilds (or equivalent) competency qualification being achieved.
>
> Specialist building operatives are often trained 'in-house' by the company that employs them, or they may have undergone formal training courses. These operatives quite often do not serve a formal apprenticeship.

Building operatives

Building operatives are labourers who carry out practical tasks. These include:

- **General building operatives and groundworkers** – Usually mix concrete, lay drains, offload materials and generally assist the craft operatives.

- **Specialist building operatives** – Include scaffolders, glaziers, suspended ceiling installers.

The following flow diagram illustrates the structure and roles of the Building Contractor and their employees.

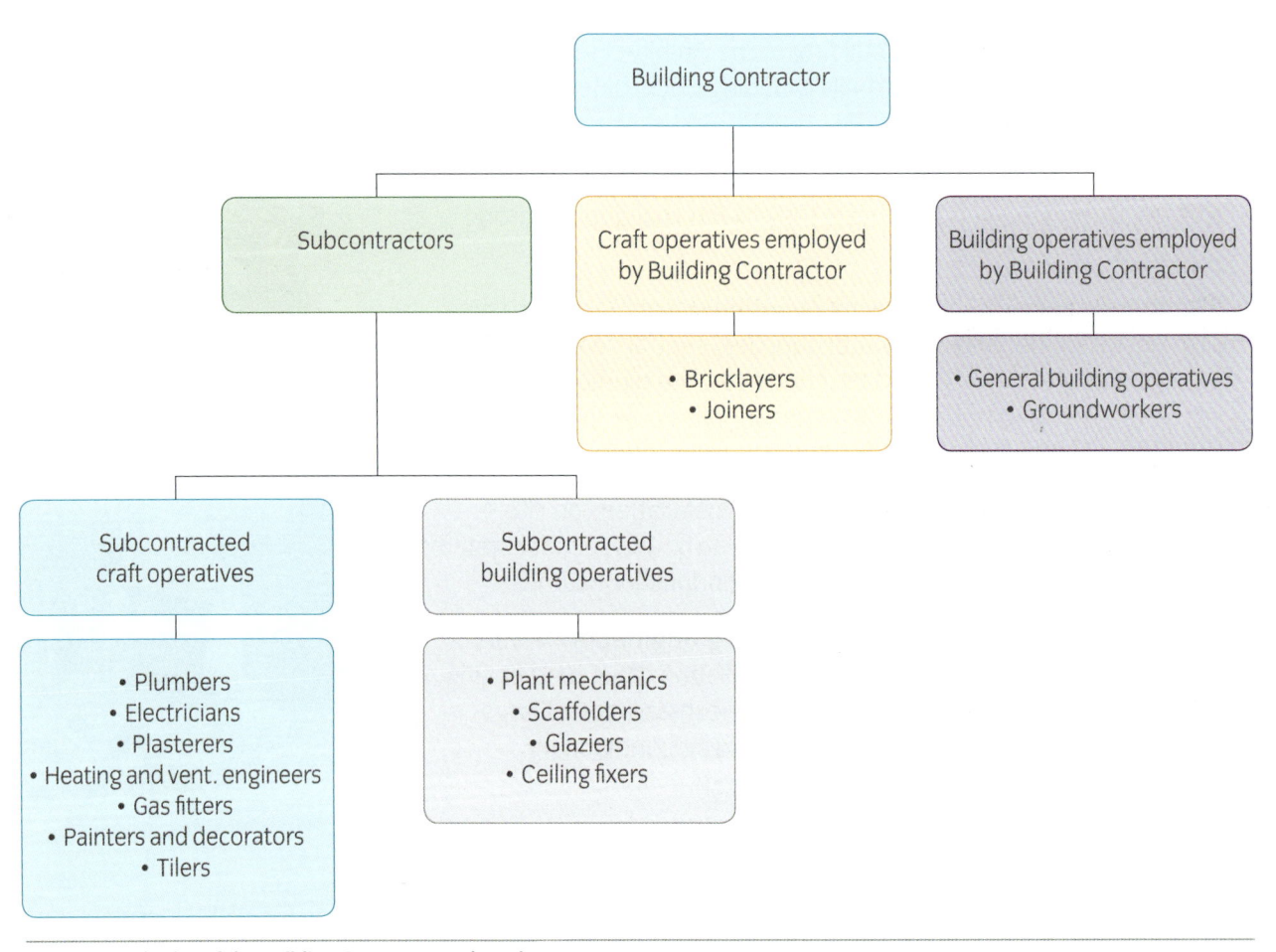

Structure and roles of the Building Contractor and employees

A5 THE INSPECTORS

There are other outside visitors to the construction site whose sole focus is health and safety. These are the inspectors. Their role is to check that everyone involved in a building project is complying with the rules and regulations to ensure that the structure, the people who work in it and on it, and the services that the eventual occupiers will use, are safe and without risk. There are four types of inspectors:

- The Health and Safety Inspector
- The Building Control Inspector

Trades working together on site

- The Water Inspector
- The Electrical Services Inspector

The Health and Safety Inspector
The Health and Safety Inspector usually works for the Health and Safety Executive, but can also be employed by the Local Authority. It is this inspector's duty to ensure that all health and safety law is fully implemented by the Building Contractor. Their role was covered in detail in Chapter 001, pages 15–16.

The Building Control Inspector
The Building Control Inspector (now more generally known as the Building Control Surveyor) works for the Local Authority and makes sure that each of the Building Regulations documents are observed in the planning and construction stages of new buildings.

Building Control Surveyors need to know the Building Regulations and how to interpret them accurately as they have the power to reject plans that fail to meet the Regulations. They may also have to use their professional judgement and skill to offer advice on acceptable solutions to meet statutory requirements, should any problems arise. They will make site visits at different stages of construction to ensure that all construction work is being properly carried out.

The Water Inspector
Water Inspectors are employed by the local water undertaker. The key objective of the Water Inspector's role is to reduce the risk of contamination of the public water supply from backflow of any fluid by providing advice and guidance on regulation compliance in new and existing premises. The Water Inspector enforces the Water Regulations by inspecting a range of plumbing installations.

The Water Inspector carries out hands-on inspections:

- in a percentage of new domestic premises
- in all new non-domestic premises/connections
- based on potential risk in existing premises.

KEY POINT
The Health and Safety Inspector is also known as the Factory Inspector.

SmartScreen Unit 002 handout 4

KEY POINT
For more information on the water undertaker, see Chapter 006.

He or she will also carry out reactive inspections following:

- requests to inspect due to water quality problems
- requests from customers for advice and resolution of plumbing problems with old or new systems.

Electrical Services Inspector

Electrical inspections must be made on all new electrical installations. However, it is very important that they are made on commercial/industrial properties. These are undertaken by the local electrical supply company but, because these companies are now privately owned, they usually employ private subcontractors to inspect the installations and issue test certificates on their behalf. The fees for these services are paid for by the customer.

B DOCUMENTATION AND SOURCES OF INFORMATION IN THE BUILDING SERVICES INDUSTRY

The relationships we have at work will vary. For example, the relationship we have with our employer will be different from the relationship we develop with work colleagues and customers. In this section, we will look at the sources of information and strategies you will need to adopt to make each of these relationships as effective as possible.

B1 THE RELATIONSHIP BETWEEN YOU AND YOUR EMPLOYER

One of the key points about running a successful business is the relationship between the employer and the employee. Businesses are successful partly because the management and staff work together, are motivated and engage in useful dialogue.

In the past, pay and working conditions varied from employee to employee, and the employer had the power to 'hire and fire' as they saw fit. However, these days employers and employees are actively encouraged to engage in discussions about matters across the whole spectrum of a business, including their respective rights.

There are many forms of legislation that your employer (and you, the employee) must be aware of and follow with regard to employment. The main acts are summarised below.

The Equality Act 2010

The Equality Act came into force in October 2010. This act provides a single legal framework with clear, simplified law that will be more effective at tackling disadvantage and discrimination. It was implemented by the Equality and Human Rights Commission (EHRC). The EHRC was formed on 1 October 2007, and has responsibility for the promotion and enforcement of equality and non-discrimination laws in England, Scotland and Wales.

The Employment Relations Act 2003

This act provides a range of contract rights for employees, including:

- entitlement to an itemised pay statement
- entitlement to the National Minimum Wage
- terms and conditions of work set out in writing
- protection against unfair dismissal at work
- the right to have a written contract of employment.

The Sex Discrimination Act 1975

This act protects employees against discrimination on the grounds of gender, for example:

- in job advertisements
- in selection of employees for jobs
- in promoting employees
- in offering training and career development opportunities.

The Equal Pay Act 1970

The Equal Pay Act sets out that women and men should receive the same pay for doing the same type of work or ranked as being of the same value.

The Race Relations Act 1976

This act makes discrimination on the grounds of race illegal in the same way as the Sex Discrimination Act (see above).

The Disability Discrimination Act 1995

This act applies to companies who employ over 20 people – these companies are required to accommodate the needs of disabled people with regards accessibility and rights.

The Data Protection Act 2018

The Data Protection Act gives everyone the right to know what information is held about them, and sets out rules to make sure that this information is handled properly.

The Freedom Of Information Act 2000

This act gives you the right to ask any public body for all the information they have on any subject you choose. Unless there's a good reason, they have to provide it within a month. You can also ask for all the personal information they hold about you.

B2 KNOWLEDGE AND UNDERSTANDING OF YOUR JOB

To be a successful plumber, it is important that you know your job well. This partly comes from knowing where and how to access information about all aspects of the work, and to use the information so that you can install the systems correctly and without risk to the customer. There are three main types of information that you will need to access, which are outlined on the next page.

KEY POINT

To find out more about employer and employee rights, visit:

www.direct.gov.uk/theory/theory–employment-legislation–261.php

KEY POINT

From 1 October 2010, the Equality Act replaced most of the Disability Discrimination Act (DDA). However, the Disability Equality Duty in the DDA continues to apply.

Regulations

Plumbing is one of the most regulated trades within building services engineering. Failure to comply with regulation often results in prosecution. Regulations in the plumbing industry include:

- Water Supply (Water Fittings) Regulations
- Gas (Installation and Use) Regulations
- Building Regulations.

These are the main regulations that workers in the plumbing industry must comply with, and these will be discussed in later chapters of this book.

British Standards and approved codes of practice

These provide guidance on interpreting and following regulations. The British Standards are not enforceable, but they set out a series of recommendations so that the minimum standard to comply with the regulations can be achieved. By adhering to the recommendations within the British Standards, the regulations will be seen to be satisfied. Often the regulations and the British Standards will make reference to one another and it may even be the case that the regulations make reference to more than one British Standard.

However important the regulations and the British Standards are, they are not our primary source of information when installing equipment and appliances. Manufacturers' guidance overrides both of these.

Manufacturers' guidance

Manufacturers' installation, servicing/maintenance and user instructions are the most important documents you will have access to when installing, servicing and maintaining equipment and appliances. They tell us in basic installation language what we must do for correct and safe operation of their equipment. This guidance must be followed, otherwise:

- the terms of the warranty will be void
- the installation may be dangerous
- we may inadvertently be breaking the regulations.

In some instances, it may seem that the instructions contradict the regulations or the British Standards. This is because regulations are reviewed only periodically, whereas manufacturers are moving forward all the time with new, more efficient products, so their information may be more up to date. In these cases, follow a simple but effective rule: the manufacturers' guidance must be followed at all times, even if it contradicts the regulations and British Standards.

B3 KNOWLEDGE AND UNDERSTANDING OF ON-SITE DOCUMENTATION

No construction site can function without certain documents and a certain amount of day-to-day paperwork. Each of these documents has an important function.

Time sheet
A time sheet is completed by each employee (or sometimes by the trade foreman) on a weekly basis. Details of hours worked and a description of the jobs carried out is included. Time sheets are used by employers to calculate wages and provide information for planning future estimates.

Delivery note
Also known as a delivery advice note, this is a document that lists the type and amount of materials that are delivered to site. It should be checked against the actual materials delivered and only signed if the materials on the note and the materials delivered are the same. A copy should be retained for administration purposes.

SmartScreen Unit 002 handout 5

Job specification
A job specification is a description of the installation that is being quoted for, complete with the types of materials and appliances that the installation contains. Occasionally, it may specify the manufacturer or British Standard of the materials the installation is to use.

Working drawings
These are the plans, elevations and details needed by the contractor (along with the job specifications) so that an estimate can be obtained and the building can be constructed. These need to show all dimensions and to be accurately scaled.

KEY POINT
Working drawings are sometimes known as building services drawings.

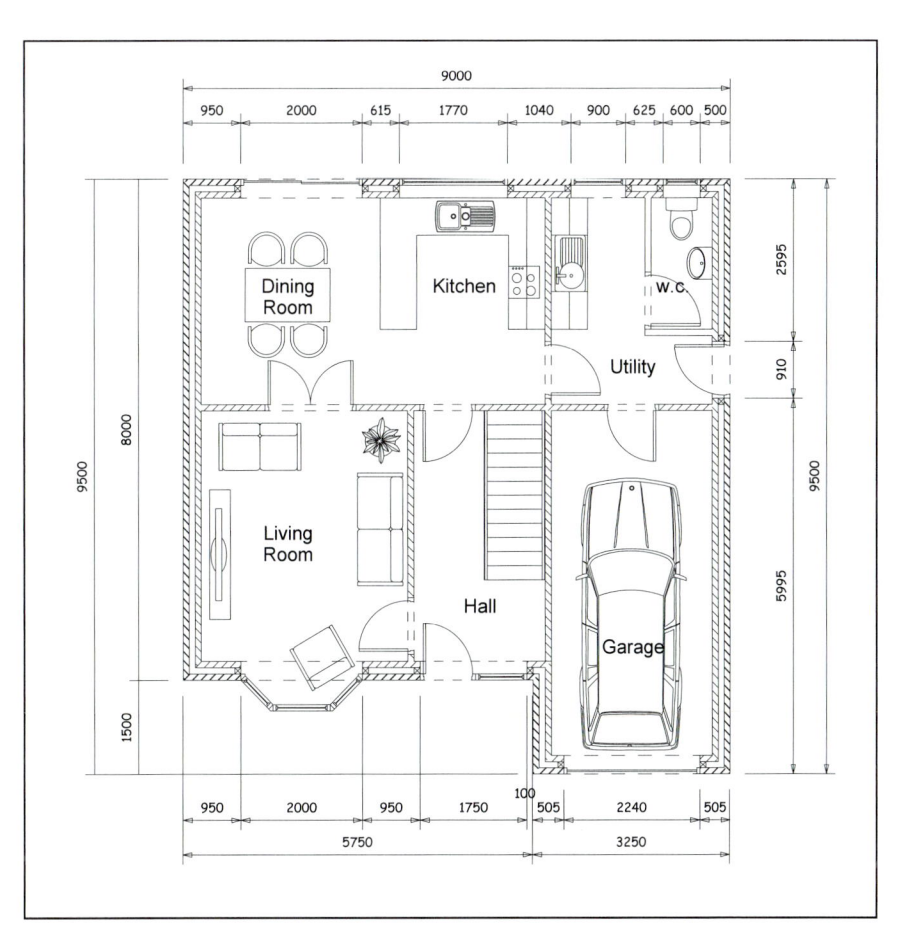

Working drawings

002 COMMUNICATING WITH OTHERS

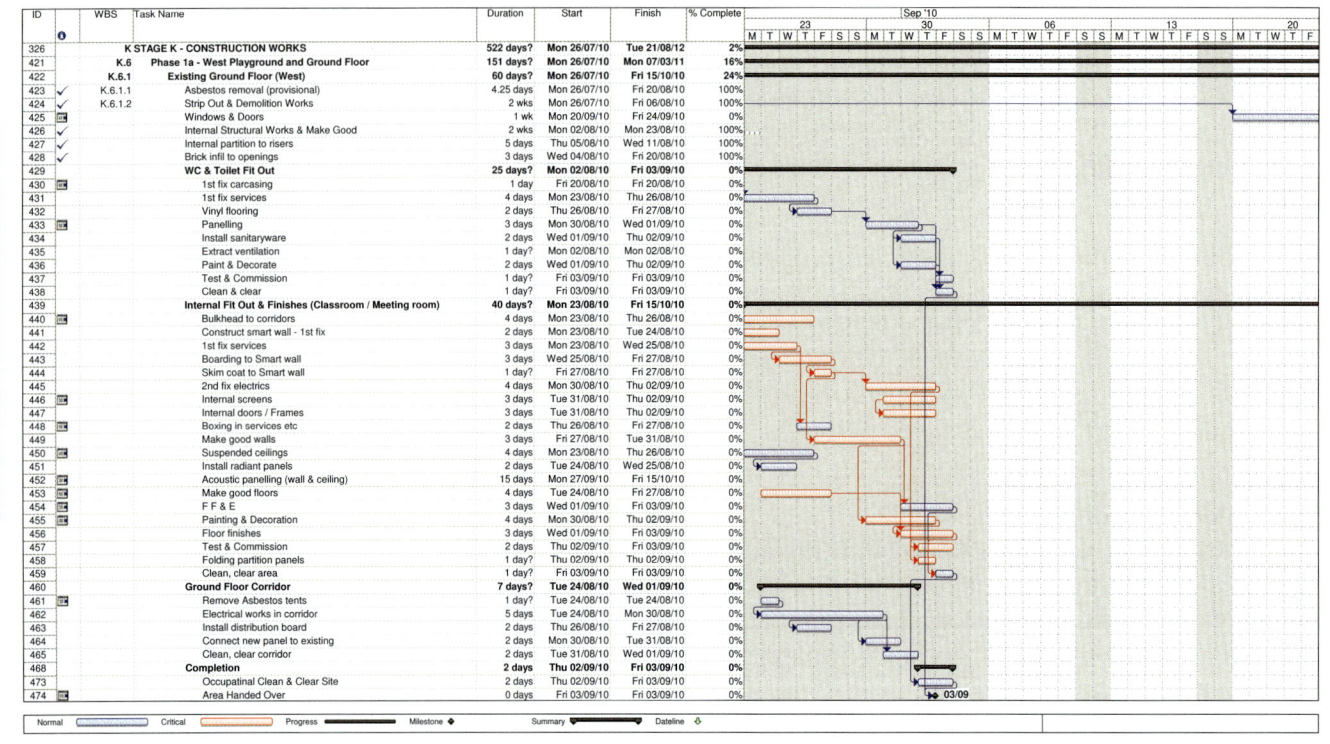

Gantt chart

KEY POINT
Another name for a work programme is a Gantt chart. The chart is named after Henry Gantt, who designed the chart around 1910–1915.

SmartScreen Unit 002 worksheet 4

Work programme
The work programme has a number of purposes:

- establish dates for work to start and finish
- illustrate the required labour and plant during the contract
- show the order of operations
- provide information for monitoring work progress.

Policy documents

Health and safety file
This is a document held by the client. In it, health and safety information is recorded and kept for use at the end of a construction project. It is a legal requirement of the CDM Regulations 2007. The type of information contained in the health and safety file is designed to help those in positions of responsibility to identify key health and safety risks that may be encountered on site, and to provide operating and maintenance manuals for the building and any equipment installed.

Customer care policy
Also known as a customer service document/charter. Good customer care makes for loyal customers, and customers are a good source of positive advertising. A customer care policy sets out the standard of service the company aims to provide the customer with respect to:

- the response times to enquiries and call-outs
- reliability and punctuality
- courtesy shown by operatives towards the customer
- recording, monitoring and publishing compliments and complaints

- consultation with customers on a regular basis
- periodic review of customer services and possible areas of improvement.

Environmental policy

This document is a written statement of intent to manage human activities to prevent, reduce or remove any harmful effects on the environment and the earth's natural resources. It also aims to ensure that any changes that are made to the environment by humans do not have harmful effects on the population or wildlife. Environmental statements often make several commitments:

- lower pollution and waste
- use energy and resources efficiently
- minimise the environmental impact on natural habitats and bio-diversity of new developments
- minimise the environmental impact of raw material extraction.

An environmental policy is implemented through Environmental Management Systems (EMS). Writing an environmental policy is currently voluntary in the UK, and the structure and content are not yet regulated under UK legislation.

> **SUGGESTED ACTIVITY...**
> Does the company you work for have an environmental policy? What does it contain? How does it affect the way that you work as a plumber? Find out what measures your company is taking to protect the environment.

B4   COMMUNICATION BETWEEN THE COMPANY AND THE CUSTOMER

Communication between the company and the customer takes place at every stage of the contract – from the initial contact, to customer care at the contract completion. Written communication can take the form of the following:

Quotations and estimates

Both of these documents show prices for the work to be completed. A quotation is a fixed price and cannot vary. An estimate is not a fixed price and can go up or down if the estimate was not accurate or the work was completed ahead of schedule. Most contractors opt for estimates because of this flexibility.

Invoice/statement

This is the document issued at the end of any contract as a demand for final payment. Invoices and statements can be from the supplier to the contractor for payment for materials supplied, or from the contractor to the customer for services rendered. Usually a period of time is allowed for the payment to be made.

Statutory cancellation rights

Legislation gives the customer the legal right to cancel contracts after signing, provided work has not already started. There is usually no penalty for cancellation if the cancellation is confirmed in writing within a specific time frame. Most cancellation periods start when the customer receives notification of their right to cancel, which should be at least seven days before work commences.

Handover information

At the end of any contract, the customer must be given certain information. For large contracts, this includes the health and safety file mentioned above. For small domestic contracts, a file should be made that contains any manufacturers' information, installation, servicing and user instructions, the appliance warranty information, contact numbers of key personnel within the company, and a letter of thanks for their custom. During the handover process, the customer should be shown where all control valves are and how to use any appliances and controls that have been installed.

B5 COMPANY POLICIES AND PROCEDURES

Company working policies and procedures highlight what is important for the company image and link this to its daily operations. Well-written policies and procedures allow employees to understand their roles and responsibilities and allow management to guide operations without needing to constantly intervene.

Companies may have policies and procedures relating to the following:

- **Behaviour** – Organisations demand certain behaviour and professionalism towards their customers and management. Customers demand a certain respect, efficiency and professional attitude towards the work and themselves.
- **Timekeeping** – Customers do not expect and will not tolerate lateness unless it is unavoidable. If lateness cannot be helped, then the customer should be informed at the earliest opportunity.
- **Dress code** – A company uniform or dress code presents a positive, professional image that the customer comes to recognise. Many companies and organisations have a set company uniform, which employees are expected to keep in reasonable order. Some companies have a laundry policy where uniforms and company workwear is cleaned free of charge.
- **Contract of employment** – A contract of employment is a mutual agreement between the employer and the employee, which is the basis of the employment relationship. A contract is made when an offer of employment is accepted.

Plumber presenting a professional image

SmartScreen Unit 002 worksheet 2

Limits to personal authority

As with most trades, plumbing follows a set pattern with regard to the roles and responsibilities of the qualified operatives. Each member of the team will have certain expectations placed on them by the management of the company or organisation. The higher the qualifications of the employee, the more responsibility will be given. The following are examples of the level of responsibility that different roles will generally be given:

Apprentice Plumber

Apprentice Plumbers have very little responsibility with regard to plumbing installations. They will be under constant supervision from the plumber they work with, but as they gain experience they may work on simple installations and maintenance tasks. They will, however, have responsibility for maintaining the company image with regard to time-keeping, appearance and customer care. Their main task is learning their trade to the best of their ability.

NVQ Level 2 Qualified Domestic Plumber

Domestic Plumbers qualified up to NVQ Diploma Level 2 are able to install 'non-complex' hot and cold water systems as well as domestic sanitation pipework and basic central heating pipework. They work under their own initiative with some supervision. They may also have some responsibility for improving business products and services.

NVQ Level 3 Qualified Domestic Plumber

Domestic Plumbers qualified up to NVQ Diploma Level 3 have much more responsibility than those at Level 2. At Level 3 they will be gas qualified and may be included on the company gas safe register. They are capable of running their own jobs, taking responsibility for domestic hot and cold water and domestic heating installations, and working on their own initiative without supervision. They will be able to undertake unvented hot water installations and work to the Water, Gas and Building Regulations. They may also have responsibility for improving business products and services and initiating some basic system design.

Plumbing Supervisor

Plumbing Supervisors will have many years of experience. They will be capable of design and installation across a broad spectrum of systems and have knowledge of the regulations and British Standards. They will have good managerial and organisational skills and hold at least a Level 3 qualification in plumbing and sometimes a Level 5 qualification, such as an HNC in building services engineering. They will also have responsibility for improving business products and services and have overall responsibility for the operatives and installations under their supervision.

C FORMAL AND INFORMAL METHODS OF COMMUNICATION

A company cannot function properly without proper methods of communication, whether it is a formal letter, email, memo, fax or verbal instructions. Formal and informal communications take place in the workplace every day.

C1 METHODS OF COMMUNICATION AT WORK

There are a number of ways that companies communicate with customers, staff and suppliers, and other companies.

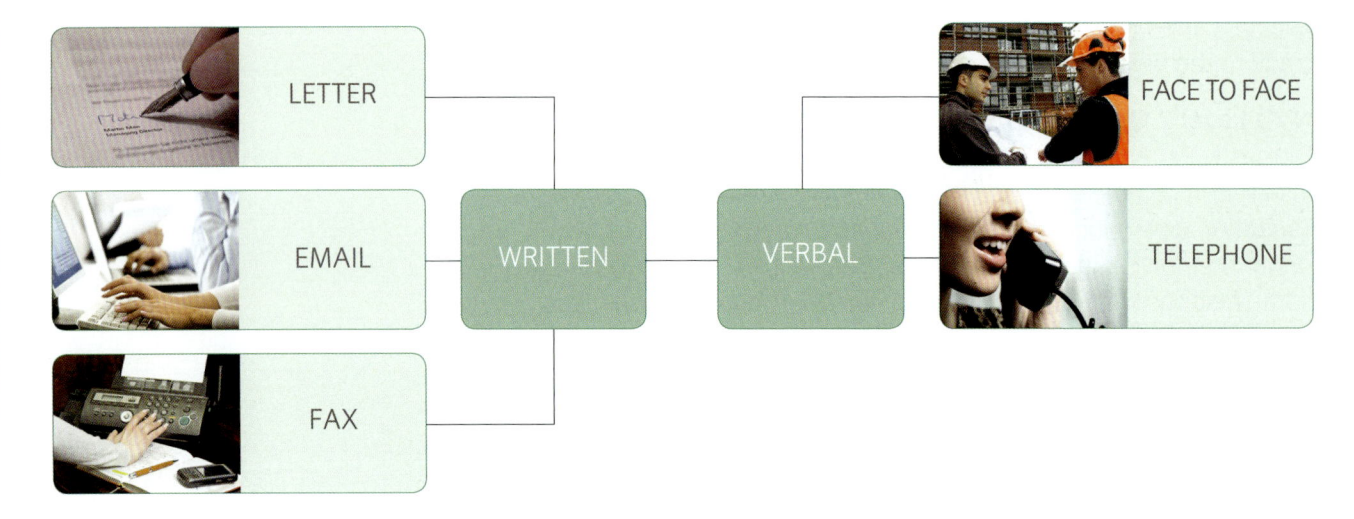

LETTER

EMAIL — WRITTEN — VERBAL — FACE TO FACE

FAX   TELEPHONE

 SmartScreen Unit 002 worksheet 1

Written communication

Letters
The letter is an official method of communication and may be easier to understand than verbal communication. Good written communication can help towards the success of any company by portraying a professional image and building goodwill. Official company business should always be in written form, usually on company headed paper and with a clear layout. The content of the letter must be well written, using formal English, correct grammar and should be divided into logical paragraphs. Examples of types of business letters are sales letters, information letters, general enquiries or problem-solving letters.

Emails
The email has emerged as a hugely popular form of communication because of the speed at which it is transferred to the recipient. As with letters, an email should be well written and laid out, using correct grammar and spelling to convey professionalism, whether the recipient is a client, customer or colleague.

Faxes
Faxes are used mainly for conveying documents such as orders, invoices, statements and contracts where the recipient may wish to see an authorising signature. Again, the basic rules apply with regard to layout, grammar and content. Remember to always use a cover page that is appropriate for your company. This is an external communication that reflects the business and company image.

Verbal communication
The spoken word is, more often than not, our main method of communication. In order to present a professional image and communicate effectively, you must consider what you are saying, your tone of voice, your body language and the response of your listener. Good verbal communication involves listening carefully as well as speaking. Verbal communication should always be backed up with written confirmation to prevent confusion.

C2 EFFECTIVE COMMUNICATION STRATEGIES

The general rule of good, effective communication is that you should think beforehand about the kind of information you will need to give and what information you will expect to receive. You should always make sure that your language, tone and body language are appropriate to what you are saying and who you are saying it to. Good communication is crucial if you are to carry out your job safely and efficiently and you may need to adapt your communication skills to deal effectively with some individuals and groups. The principle behind effective communication is making sure that both parties completely understand each other. You may need to take into account the following factors:

Physical disabilities

Communicating with a customer, colleague or site visitor with a hearing impairment may mean that you need to:

- use written notes or drawings to reinforce verbal information
- use appropriate written information (such as a sales leaflet, manufacturers' literature or guides) to make sure that both you and the other person understand what is being referred to
- if available and appropriate, use other means of technology (telephone amplifiers etc) to help communication.

Communicating with a customer, colleague or site visitor with a visual impairment may mean that you need to:

- give more verbal detail than you would usually use
- describe any diagrams or visual aids that you are using
- keep the other person informed of his or her surroundings, eg who is present, who has left the room, etc.

Special learning needs

Communicating with a customer, colleague or site visitor with special learning needs may mean that you need to:

- if possible, make sure a responsible person is present to hear any important information
- keep information short and relevant, and avoid using too much technical information or jargon
- consider using visual aids and diagrams to back up information.

Language differences

Construction sites are often multicultural places, so you may be dealing with colleagues, clients and site visitors for whom English is not their first language, or who speak with a different accent or dialect.

SmartScreen Unit 002 handout 3

SmartScreen Unit 002 interactive activities 2 and 3

Accent refers to the way in which people pronounce their words. Dialect is different. A person's dialect is a combination of the way they pronounce words, the vocabulary they use and the grammatical structures they use. A person's accent and dialect are often a result of where in the country they live or were brought up, but other factors such as social class and gender may also play a part.

Communicating with a customer, colleague or site visitor whose first language is not English may mean that you need to:

- speak clearly and avoid using slang words
- use short sentences and simple words
- ask questions to confirm that you have been understood
- use diagrams and visual aids to back up verbal information
- use an interpreter, if possible, or ask if a family member can act as an interpreter.

Communicating with a customer, colleague or site visitor whose accent is different from your own may mean that you need to:

- use the correct terminology in work situations – avoid using local slang that may not be understood
- ask questions to confirm that you have been understood
- refer to product catalogues or manufacturers' literature to make sure that you are both talking about the same thing.

In all these cases, you should always show respect for the other person in the way in which you communicate. Keeping your body language open and engaged with good eye contact will help you to judge how the information is being received.

C3 CONFLICTS IN THE WORKPLACE

When people work together in groups, there will be occasions when individuals disagree and conflicts occur. Whether these disagreements become full-blown feuds or instead fuel creative problem-solving is, in large part, up to the person in charge. Conflicts can occur for many reasons, such as unfair working conditions or pay structures, language or cultural differences, unacceptable behaviour (racism, sexism or bullying) or simply a clash of personalities.

It is important to deal with workplace conflicts quickly and effectively because if they are left unresolved they could affect morale, motivation and productivity, and potentially cause stress and even serious accidents. Conflicts may occur between:

- employer and employee – may need union involvement or some form of mediation
- two or more employees – may need employer intervention
- customer and employer – may need intervention by a professional body

---

SmartScreen Unit 002 interactive activity 4

KEY POINT
Different names for tools, systems etc from around the UK include: handibender vs scissor bender, lump hammer vs club hammer, troffins vs guttering, tin snips vs shears.

SmartScreen Unit 002 handout 6

KEY POINT
Reaching an agreement through discussion is known as negotiation.

- customer and employee – may need employer intervention.

Dealing with workplace conflicts

There are several ways that employers may deal with disagreements. The steps that should be taken are as follows:

1. Identify the problem. Make sure that everyone involved knows exactly what the issue is, and why he or she is arguing. Talking through the problem helps everyone to understand that there is a problem, and what the issues are.

2. Allow every person involved to clarify his or her perspective and opinions about the problem. Everyone should have an opportunity to express his or her opinion. A time limit may even be established for each person to state his or her case. All participants should feel safe and supported.

3. Identify and clarify the ideal end result from each person's point of view.

4. Work out what can reasonably be done to achieve each person's objectives.

5. Find an area of compromise to see if there is some part of the issue on which everyone agrees. If not, try to identify long-term goals that mean something to all parties.

Informal counselling is one method that helps managers and supervisors to address and manage conflict in the workplace. This may be in the form of:

- meetings
- negotiation/mediation sessions
- other dispute-resolving methods.

It is important that employees know that there is someone to go to if a conflict develops, or if they witness unacceptable behaviour. Employees should first discuss the problem with their immediate supervisor.

In the plumbing industry, more serious workplace conflicts can usually be resolved by the Joint Industry Board (JIB). In cases where the matter cannot be resolved, then mediation or union involvement may be required. This may involve the use of an impartial outside mediator, for example from the Advisory, Conciliation and Arbitration Service (ACAS). Mediation is usually the first step. Further options available include conciliation, where the aim is to mutually agree a solution, or arbitration, an alternative to an industrial tribunal available for certain types of dispute. In both cases, the decisions that come out are legally binding on both parties.

> **KEY POINT**
>
> Conciliation is an alternative dispute resolution process whereby the parties to a dispute agree to use the services of a conciliator, who then meets with the parties separately in an attempt to resolve their differences. Collective conciliation is when a group of employees is involved, and individual conciliation when it is only one employee involved in the dispute.

SmartScreen Unit 002 worksheet 3

> **KEY POINT**
>
> The Joint Industry Board for Plumbing and Mechanical Engineering Services is the registration body for the plumbing industry in England and Wales. It is associated to the Construction Skills Certificate Scheme (CSCS) registration schemes. The JIB-PMES comprises the Association of Plumbing and Heating Contractors (APHC) and Construction Confederation, representing plumbing employers and the Amicus trade union, representing employees.
>
> For further details see the website at:
> http://www.jib-pmes.org.uk/

C4 THE EFFECTS OF POOR COMMUNICATION AT WORK

The effects of poor communication can be extremely harmful to both a business and its personnel. If poor communication exists, goals will not be

> **KEY POINT**
>
> For more information and advice on ways of resolving disputes and avoiding conflict, visit the ACAS website:
> www.acas.org.uk.

achieved and this could develop into problems within the company. It can lead to demotivation of the workforce, and the business will not function as a cohesive unit. The effects are obviously negative. Employees may:

- become mistrustful of management and, often, each other
- argue and reject their manager's opinions and input
- file more grievances related to performance issues
- not keep their manager informed and avoid talking to management
- do their best to hide their deficiencies or performance problems
- refuse to take responsibility.

Poor communication in the workplace can disrupt the organisation and cause strained employee relations and lower productivity. This can often result in:

- time lost as instructions may be misunderstood and jobs may have to be repeated
- frustration developing, as people are not sure of what to do or how to do a task
- materials being wasted
- people feeling left out if communication is not open and effective
- messages being misinterpreted or misunderstood, causing bad feelings
- people's safety being put at risk.

All of these problems will eventually filter down to existing and potential customers, and when that happens, customer confidence will disappear. This could, in severe cases, even lead to a collapse of the company.

CONCLUSION

During this chapter, we have looked at the varied personnel of the construction industry, from the client through to the building contractor and the workers on site. We have seen how the relationships between the trades are interwoven, with everyone working towards a common objective – a successful, quality building and a happy client.

Effective working relationships are crucial if the construction process is to be successful, but they are also often fragile and it is important to know that help is at hand if these relationships, for whatever reason, break down.

The interaction between the many members of the construction team and the smoothness of the construction process is the most visible testament that effective working relationships at all levels of construction management, tradesperson and labourer are just that – effective and working!

002 TEST YOUR KNOWLEDGE

SmartScreen Unit 002 revision sample questions

1. Who is considered to be the leader of the management team?

2. Who is the most important member of any construction project?

3. Who advises on how the building can be constructed within the client's budget?

4. Which member of the management team is appointed by the Architect to oversee works on site the Architect's behalf?

5. Name the three types of specialist engineers that may work with the Architect to assist in the design of a building.

6. Who would install large diameter low carbon steel heating pipework on large construction sites?

7. Who will visit a construction site to ensure that the building is being constructed correctly?

8. Which act was implemented by the Equality and Human Rights Commission (EHRC)?

9. Which piece of legislation gives you the right to know what information is held about you by a public body?

10. Which document gives details of hours worked by an employee?

11. Which is regarded as a more formal method of communication – a letter or an email?

12. Who should ideally be present if you are communicating on site with someone who has special learning needs?

13. When communicating with a person whose first language is not English, what can we use to help as a back-up to verbal information?

14. If a conflict develops with a work colleague, who should be the first person that you inform?

15. Informal counselling is one method that helps managers and supervisors to address and manage conflict in the workplace. What form may this take?

003
UNDERSTAND HOW TO APPLY ENVIRONMENTAL PROTECTION METHODS WITHIN BUILDING SERVICES ENGINEERING

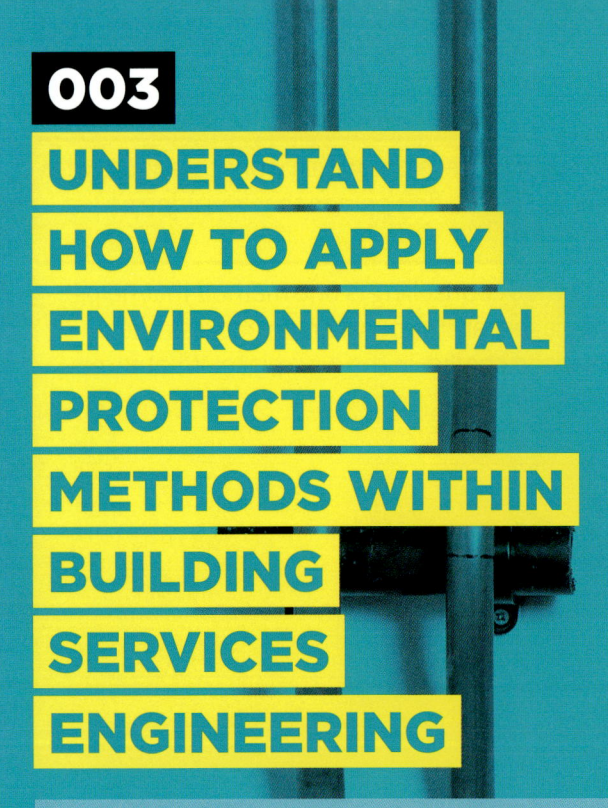

It is important to be aware of environmental issues when carrying out your work. This includes how you use energy, resources and materials, how you dispose of our waste, and how that impacts on the environment around you, both locally and nationally.

In this chapter, we will investigate alternative methods of generating and using energy and how best to limit the effects of fossil fuels on our environment (in particular, their contribution to climate change). We will look at current energy conservation legislation and examine ways to limit wastage of materials and resources. We will also explore new methods of water conservation that will influence both the way we, as plumbers, work and the systems we install. Energy efficiency and resource conservation should be key considerations in your plumbing work, which could result in cost savings to your customers.

IN THIS CHAPTER, YOU WILL COVER:

A Energy conservation legislation
- **A1** General energy conservation legislation
- **A2** Construction energy conservation legislation
- **A3** Building services energy conservation guides

B Sources of energy
- **B1** High carbon
- **B2** Low carbon
- **B3** Zero carbon

C Basic operating principles of green plumbing technologies
- **C1** Solar thermal systems
- **C2** Biomass solid fuel
- **C3** Heat pumps (water, air and ground source)
- **C4** Combined heat and power (CHP)
- **C5** Combined cooling, heat and power (CCHP)
- **C6** Wind turbines
- **C7** Photovoltaic power
- **C8** Energy saving and conservation organisations
- **C9** Energy rating tables (energy performance)

D Energy efficiency and the customer
- **D1** Design, installation and handover

E Caring for the environment – a plumber's responsibilities
- **E1** Adopting a green attitude to plumbing activities
- **E2** Disposing of waste materials
- **E3** Waste management legislation
- **E4** The benefits of good waste management practice
- **E5** Methods of safely disposing of waste materials
- **E6** Licensed waste management companies
- **E7** Waste carriers' licence
- **E8** Recycling
- **E9** Recycling processes
- **E10** Disposing of hazardous materials

F Water conservation and protection
- **F1** Statutory legislation for water wastage and contamination – the Water Supply (Water Fittings) Regulations 1999 and the Water Act 2003
- **F2** Water efficiency calculations for new dwellings
- **F3** Methods for reducing water consumption in the home
- **F4** Rainwater harvesting
- **F5** Grey water recycling

Test your knowledge

A ENERGY CONSERVATION LEGISLATION

In this first section of the chapter we will be investigating the various pieces of energy conservation legislation and how they impact on the construction industry and, more specifically, the building services industry, of which plumbing is a part.

A1 GENERAL ENERGY CONSERVATION LEGISLATION

In 1997, a conference was held in Kyoto, Japan, with the sole aim of reducing the world's greenhouse gas (GHG) emissions. The result of this was the Kyoto Protocol. This is an international agreement, which has set targets for 37 industrialised countries and the European Union to reduce GHG emissions by around 5 per cent between 2008 and 2012.

The Kyoto Protocol places more responsibility on developed nations, such as the UK and the EU. It recognises that developed countries are mostly responsible for the current high levels of GHG emissions in the atmosphere through more than 150 years of industrial activity. The result of the Kyoto Protocol in the UK was the Climate Change Act 2008.

The Climate Change Act 2008

The Climate Change Act came into force in the UK in November 2008. It sets a target for the UK to reduce carbon emissions to 80 per cent below 1990 levels by 2050. It includes a provisional reduction target of 34 per cent by 2020 (with a possible further increase to a 42 per cent cut by international agreement) and establishes the idea of carbon budgets.

The Committee on Climate Change (CCC) acts as an advisor to the UK Government on setting these targets and budgets and the measures in order to achieve them.

EU Directive on the Energy Performance of Buildings

The Directive on Energy Performance of Buildings (2002/91/EC) is the main legislative document in the European Union that aims to improve energy performance in buildings (a key and cost-effective way to reduce CO_2 emissions and energy costs). The Directive insists that all member states must apply minimum energy performance ratings on new and existing buildings, ensure that certification of energy performance of buildings is conducted and that regular inspections on boilers and air conditioning systems is completed to ensure that they remain energy efficient.

See http://www.ec.europa.eu/energy/efficiency/buildings/buildings_en.htm for information on the Directive. The Directive was updated in 2010, and the new documentation can be accessed at http://www.diag.org.uk.

KEY POINT

The six major greenhouse gases covered in the Kyoto Protocol's targets are: carbon dioxide, methane, nitrous oxide, sulphur hexafluoride, hydrofluorocarbons (HFCs) and perfluorocarbons (PFCs).

KEY POINT

The Climate Change Act 2008 is the first legally binding long-term framework for reducing emissions to come into force in any country.

KEY POINT

For more information on the Climate Change Act, visit the Carbon Trust website: http://www.carbontrust.co.uk

The Energy Performance of Buildings (Certificates and Inspections) (England and Wales) Regulations 2007

These Regulations put into practice articles 7, 8 and 9 of the EU Energy Performance of Buildings Directive, which states the requirements for clients and landlords to produce energy performance certificates when buildings are constructed, rented out or sold. All buildings will eventually require an energy performance certificate (EPC) and in the case of public buildings the EPC will have to be on prominent display.

The regulations can be accessed at: http://www.communities.gov.uk/documents/planningandbuilding/pdf/322911.pdf.

Installing loft insulation

The Code for Sustainable Homes

The Code for Sustainable Homes is a national standard to guide industry in the design and construction of sustainable homes. It has been developed to enable a change in sustainable building practice for new homes, which should lead to continuous development and innovation in sustainable home building techniques.

The code has been written by the Government, working closely with the Building Research Establishment (BRE) and Construction Industry Research and Information Association (CIRIA). The Code complements the system of energy performance certificates for new homes introduced in 2008 under the European Energy Performance of Buildings Directive, and builds on the most recent changes to Building Regulations in England and Wales.

Home Energy Conservation Act (HECA) 1995

The Home Energy Conservation Act 1995 (HECA) places obligations on Local Authorities to make plans to increase domestic energy efficiency in their area by 30 per cent over 10–15 years. The UK Government has estimated that a 30 per cent reduction in UK domestic energy use is possible through energy conservation and increased energy efficiency. The objectives of the Act are to reduce emissions of carbon dioxide (CO_2) to help combat climate change, while also reducing fuel poverty by lowering the cost of fuel bills.

This legislation is due to be reviewed to measure its effectiveness and then updated or replaced.

A2 CONSTRUCTION ENERGY CONSERVATION LEGISLATION

When a building is designed, the Architect must have an understanding and knowledge of building structure and the property's uses of materials so that the building complies with the Building Regulations. An awareness of energy conservation, and knowledge of recyclable materials and sustainable sources of energy is essential.

In many cases, the Architect will employ building services engineers to design energy efficient systems for hot water supply, heating and air conditioning.

In the Building Regulations, there is specific legislation for energy conservation for both new buildings and the renovation of existing ones. For information on the Building Regulations, see: http://www.direct.gov.uk/en/HomeAndCommunity/Planning/BuildingRegulations.

Building Regulations Part F: Ventilation

Although not directly an energy efficiency document, Part F of the Building Regulations deals with indoor air quality. Part L (see below) of the Building Regulations requires that architects ensure that buildings are airtight to retain heat and save energy, and includes ventilation as a method of controlling indoor temperatures. However, this 'airtight' approach can create poor air quality inside the buildings if they aren't ventilated properly, and this can lead to sick building syndrome (SBS). If too much ventilation is applied it can lead to excessive building heat loss also. That is why Part F is important. This document defines ventilation as the removal of stale air from the building and replacing it with fresh outside air. The main aim is to:

- remove airborne pollutants and odours
- control humidity
- provide outside air for breathing.

Sick building syndrome
A combination of ailments associated with an individual's place of work or residence.

Part F also encourages energy-efficient controls as an important part of ventilation systems. A popular method with heating, ventilation and air conditioning (HVAC) designers is to use ventilation systems with heat recovery capability. Heat recovery systems reduce energy costs by extracting stale air from the building and then removing and reusing the useful heat or cooling energy to either warm or cool the incoming fresh air. This can save up to 30 per cent on initial costs of the heating and cooling equipment, as well as lowering energy costs. Choosing efficient types of fan motor and using energy-saving control devices will also help.

Building Regulations Part L: Conservation of Fuel and Power: 2010 (Part J in Scotland and Part F in Northern Ireland)

Part L of the Building Regulations* controls the insulation values of building elements, the heating efficiency of boilers, the insulation and controls for heating appliances and systems together with hot water storage, lighting efficiency and air permeability of the structure. It also sets out the requirements for SAP (Standard Assessment Procedure) calculations and carbon emission targets for dwellings. It is divided into four parts:

- Part L1a – New domestic properties
- Part L1b – Existing domestic properties

* Note that this part of the Building Regulations is Part L in England and Wales, but Part J in Scotland and Part F in Northern Ireland.

KEY POINT
The installation of ventilation systems is usually completed by air conditioning engineers or heating and ventilation engineers.

> **KEY POINT**
>
> Further information on Building Regulations may be found at:
> http://ecologicenergyratings.co.uk/Building-Regulations.php

- Part L2a – New buildings other than domestic properties
- Part L2b – Existing buildings other than domestic properties

We will look at Parts L1a and L1b only, as these deal with domestic properties.

The Updated Building Regulations Part L1a/L1b 2010

These new regulations use the Government's Standard Assessment Procedure for Energy Rating of Dwellings (SAP 2009) as the only method to demonstrate compliance with the Building Regulations and to provide energy ratings for dwellings.

The latest version of Part L, which came into force on 1 October 2010, seeks to further reduce fossil fuel consumption and the emissions of carbon dioxide from a building by reducing the amount of energy needed to heat, cool and ventilate the building, supply hot water and provide it with electric lighting. This is measured by the building's carbon dioxide emission rate or 'carbon footprint'. Part L describes how this can be achieved, including:

- reducing the heating demand by adequately insulating the building fabric, using suitable windows and doors to reduce air leakage
- limiting the demand for building cooling by insulating pipework and cylinders, and reducing solar heat gain
- providing energy-efficient controls to heating, hot water, ventilation, cooling and lighting
- ensuring that systems are properly designed, installed and commissioned, and that instructions are given so that they can be operated efficiently and effectively.

From October 2010, the target is to reduce carbon dioxide emissions from a dwelling to 25 per cent lower than the level set by the 2006 Approved Documents Part L1A and L1B.

Part L requirement

L1A and L1B states the following:
Reasonable provision shall be made for the conservation of fuel and power by:

1 Limiting heat gains and losses:

   a through windows, doors and roofs and other parts of the building fabric; and,

   b from pipework, ducts and hot water storage vessels used for hot water services, space heating and space cooling;

2 Designing, installing and commissioning energy-efficient building services with effective and energy-efficient controls; and,

3 Providing the owner with sufficient information about the building, the building services and their maintenance requirements so that

- SmartScreen Unit 003 handout 1
- SmartScreen Unit 003 worksheet 1

the building can be operated using no more fuel and power than is absolutely necessary.

Since 1 October 2010, all boilers fitted will have to be Band A or SEDBUK 2005 90 per cent. The main changes to Part L are:

- SEDBUK ratings will disappear and be replaced by minimum percentage efficiency figures.
- Gas boiler efficiency will be listed on either SEDBUK 2005 or SEDBUK 2009 lists.
- Only gas boilers with an efficiency of 90 per cent or above (SEDBUK 2005, previously known as SEDBUK Band A) can be installed. If using SEDBUK 2009 data, the minimum efficiency is 88 per cent.

Changes to SEDBUK

According to the Domestic Building Service Compliance Guide that accompanies Part L, the terms Band A, Band B etc were dropped as of 1 October 2010 to avoid confusion when the forthcoming Energy-using Products directive (EuP) is introduced. EuP will measure the energy efficiency of systems and will provide a label similar to the current EU energy labels currently seen on washing machines, refrigerators and other domestic appliances.

The new SEDBUK 2009 ratings will take account of summer and winter seasonal efficiency, predominantly from the heating of hot water, rather than being calculated at just one point through the year. This will reduce the overall efficiency by around 2 per cent. This will mean that a current SEDBUK Band A boiler will either be SEDBUK 2005 90 per cent or SEDBUK 2009 88 per cent, depending on what the manufacturer's information states, in line with the Standard Assessment Procedure (SAP). If the efficiency is not stated as SEDBUK 2009, it will be assumed to be SEDBUK 2005.

A key point here is that all gas boilers must now be SEDBUK 2005 90 per cent or SEDBUK 2009 88 per cent.

Oil boilers must have a minimum efficiency of 85 per cent. There are also now minimum efficiencies for solid fuel boilers and individual heating appliances and a requirement for:

- cleansing and water treatment for wet central heating systems
- commissioning certificate to confirm completion
- provision of a suitable set of operating and maintenance instructions
- minimum standard for hot water storage cylinders
- minimum standard for heating and hot water controls.

Controlled services, elements and fittings are subject to Building Control inspection, and a Building Notice will normally have to be provided. An alternative to this is for the work to be self-certified by a qualified, competent person who has registered with a self-certification body.

KEY POINT

What does boiler efficiency mean in monetary terms? If a gas boiler is 88 per cent efficient, for every £1 spent by households on their fuel bill, 88 pence will produce usable heat. The remaining 12 pence is lost in heat wastage out of the flue.

Vaillant boiler

> **KEY POINT**
> Registration bodies are Gas Safe for gas boilers, OFTEC for oil boilers and HETAS for solid fuel installations.

> SmartScreen Unit 003 interactive activity 1

> **KEY POINT**
> CHeSS is updated periodically to reflect changes in Building Regulations and other government advice. To find the latest specifications, search on the website of the Energy Saving Trust:
> www.energysavingtrust.org.uk.

A3 BUILDING SERVICES ENERGY CONSERVATION GUIDES

The building services industry, and especially plumbing and heating, produces a number of guides to help the installer comply with energy conservation legislation. They should be used in conjunction with the Building Regulations and relevant British Standards.

Domestic Building Services Compliance Guide 2010

The Building Regulations make reference to the Domestic Building Services Compliance (DBSC) Guide 2010. DBSC Guide lays down rules for minimum boiler energy efficiency requirements. This has been developed by the Government in cooperation with the heating industry, and covers all aspects of the installation and design of domestic heating. This includes gas, oil, solid fuel, hot water, solar thermal, underfloor heating, combined heat and power and heat pumps (these will be explained later in this chapter), as well as detailed information on the insulation requirements for central heating pipework. This will help ensure heating and hot water systems comply with the requirements of the Building Regulations Part L with regard to energy efficiency.

Central Heating System Specifications (CHeSS) 2008 CE51

CHeSS was produced by the Building Research Energy Conservation Support Unit (BRECSU) to create a set of common standards for energy efficiency, which domestic heating installers and manufacturers should work towards. It gives recommendations for good practice and best practice and for the energy efficiency of domestic wet central heating systems. Customers and clients should use these specifications to ensure that their heating installations meet current good or best practice, and installers should also use them to design and quote for systems.

The Guide to the Condensing Boiler Installation Assessment Procedure for Dwellings

The Guide to the Condensing Boiler Installation Assessment Procedure for Dwellings is a government publication that lays down revised guidance for the efficiency of hot water central heating gas and oil fuelled boilers installed in new and existing domestic dwellings.

The guide looks at the installation of condensing boilers and covers:

- possible installation difficulties
- outline of the assessment procedure
- the purpose of the assessment procedure
- how to carry out the assessment
- explaining flue terminal siting and dealing with extended flue lengths
- boiler location
- connection of condensate drain.

B SOURCES OF ENERGY

The Earth's energy resources are changing. Many nations are still heavily reliant on fossil fuels for heating, hot water and power generation, but the carbon dioxide that these fuels produce are hugely damaging to the Earth's climate. As the world's natural resources of coal, oil and gas run out, we have to find other, cleaner sources of energy.

B1   HIGH CARBON

Fossil fuels are formed by anaerobic decomposition of buried, dead carbon-based plants. The age of the plants and the resulting fossil fuel is millions of years, occasionally exceeding 650 million years. Plants absorb energy from the sun through photosynthesis. When they die, this stored energy is usually released in the decaying process, but under anaerobic decomposition conditions, it remains trapped. All fossil fuels produce carbon dioxide (CO_2) and water vapour when they are burnt. During the combustion process, the fuel will release as much CO_2 as the fuel has taken in during its lifetime. For example, if the fuel being burnt was a 600-year-old oak tree that died millions of years ago, then the fuel will release 600 years' worth of CO_2 regardless of whether the fuel is gas, oil or solid fuel.

Fossil fuels have a high carbon content. They produce vast quantities of CO_2, one of the most damaging of all known greenhouse gases (GHGs), and are used extensively for both domestic and industrial heating and for power generation. There are three basic types of fossil fuels used as energy sources in building services engineering, all of which are classed as non-renewable sources of energy.

- solid fuel – coal, coke and peat
- fuel oil
- gas – natural gas and liquid petroleum gas (LPG).

KEY POINT

Fossil fuels are also known as hydrocarbons. These fall into three categories: light, medium and heavy.

SUGGESTED ACTIVITY...

How environmentally friendly is your home? Think about the types of fuel you use at home and see if there are ways in which you can use less fuel. Could controls be added to the central heating to make it more efficient? Are you using energy-saving light bulbs? List the places where improvements could be made.

Solid fuel

Coal is created from the remains of plants that lived and died between 100 and 400 million years ago, when large areas of the earth were covered with huge swamps and forest bogs. Coal is classified into four main types, depending on the amount of carbon, oxygen and hydrogen present. Anthracite, which has the highest carbon content, is the most commonly used for domestic heating. The higher the carbon content, the more energy the coal contains. Coal is still used for central heating boilers, both domestic and industrial, and for steam and electricity generation.

UK coal mine

Coke is produced by heating coal in coke ovens to around 1000°C. During this process, the coal gives off methane gas and coal tar, both of which can be cleaned and reused. Coke burns clearly and without a flame and gives out a lot of heat but it has to be mixed with coal, as it will not burn by itself. Coke is a smokeless fuel that is valued in industry because it has a calorific (heat) value higher than any form of natural coal. It is widely used in steel making and in certain chemical processes but can also be used in many domestic boilers and room heaters.

Peat is an organic material that forms over hundreds of thousands of years from the decay of plant material in the absence of oxygen, in boggy, water-logged ground. This encourages the growth of moss, which forms the basis of the peat. Peat is a poor quality fossil fuel that is easily cut and dried. It has a high carbon content, but still much lower than that of coal, with large amounts of ash produced during combustion. It is used in many domestic fires, room heaters and peat burning stoves.

Fuel oil is a liquid by-product of crude oil, produced during petroleum refining. There are two main categories under which it is classified. One is distillate oils, such as diesel fuel, and the others are residual oils, which includes heating kerosene.

Around 95 per cent of boilers burning fuel oil in domestic properties use kerosene, which is also known generically as C2 grade 28 second viscosity oil. This is the preferred oil fuel grade for domestic heating, due to its clean combustion. Modern oil central heating boilers only require a single annual service, if being used with an atomising pressure jet burner. It is the only oil grade that can be used with balanced or low level flues.

Kerosene is very good in cold weather, and remains fluid beyond −40°C, although it does tend to thicken slightly during extremely cold weather.

A domestic kerosene oil tank for oil-fired heating

Kerosene is a high carbon fuel and is clear or very pale yellow in colour. Newer boilers have a label inside the casing with information on nozzle size and pump pressure that show that the boiler has been set up to use kerosene. It may also reference the British Standard for kerosene, BS 2869 grade C2.

Gas (natural gas and liquid petroleum gas – LPG) is a light hydrocarbon fuel found naturally wherever oil or coal has formed. Despite its association with global warming, it is one of the cleanest, safest and most useful of all energy sources, being used for a variety of industrial and domestic applications such electricity generation, heating and cooking. All gas, whether natural or liquid petroleum gas (LPG) produces carbon dioxide, nitrogen and water vapour as byproducts of the combustion process.

Typical composition of natural gas		
Methane	CH_4	70–90%
Ethane	C_2H_6	0–20%
Propane	C_3H_8	0–20%
Butane	C_4H_{10}	0–20%
Carbon dioxide	CO_2	0–8%
Oxygen	O_2	0–0.2%
Nitrogen	N_2	0–5%
Hydrogen sulphate	H_2S	0–5%

- **Natural gas** is a mixture of gases found naturally in coal and oil deposits. Because of this, it has reasonably high carbon content. It consists of over 90 per cent methane, but also contains other flammable gases such as ethane, propane and butane. It is a major source of fuel for domestic and industrial use and electricity generation. It is found in many parts of the world such as Russia, the USA, Saudi Arabia and the North Sea. Before natural gas is introduced to the gas pipe network, it must be rigorously processed and cleaned to remove the waxy oil deposit it contains. This deposit is called naptha. Natural gas has no smell, so a distinctive 'rotten eggs' type smell, called mercaptan, is added. It has a specific gravity of 0.6–0.7, which means it is lighter than air.

- **Liquid petroleum gas (LPG)** is generally processed from petroleum refining, or it can be extracted from natural gas. It is stored in liquid form by compressing it. This reduces the gas's volume by 274 times. In other words, one litre of liquid makes 274 litres of gas. There are two types of LPG commercially available, shown in the table on the following page.

Gas	Chemical symbol	Specific gravity	Boiling point	Characteristics and use
Propane	C_3H_8	1.5	−45°c	The most widely available of all LPG gases. It is used for cooking, heating equipment including boilers and fires, cars and many industrial applications. Propane is 1.5 times heavier than air. Propane is available in a range of bottle sizes and can also be stored in bulk for domestic and commercial use.
Butane	C_4H_{10}	2.0	−4°c	Butane has a slightly higher calorific value than propane but its use is limited to camping and portable equipment because of its relatively high boiling point. Butane is available in a range of bottle sizes. It is twice as heavy as air.

As with natural gas, LPG has no smell, so a 'stenching agent' is added. The calorific value of LPG is about 2.5 times higher than that of mains gas so more heat is produced from the same volume of gas.

Calorific value

Calorific value (CV) is a measure of heating power and is dependent upon the structure of the fuel. The CV refers to the amount of energy released when a known volume of fuel is completely combusted under specific conditions.

Calorific values of common solid fuels, liquid fuels and gases

The basic calorific value for solid and liquid fuels is the gross calorific value at constant volume. For gases, it is the gross calorific value at constant pressure. Solid and liquid fuels are measured in mega joules per kilogram (MJ/kg) and gases are measured in mega joules per cubic etre (MJ/m^3).

The calorific values of different solid and liquid fuels and gases are shown in the following table:

Solid and liquid fuels	Gross calorific value (MJ/Kg)
Anthracite (4% water)	36
Coal tar fuels	36–41
General-purpose coal (5–10% water)	32–42
High-volatile coking coal (4% water)	35
Peat (20% water)	16
Kerosene	47
Gaseous fuels at 15°C	**Gross calorific value (MJ/m³)**
Commercial butane	118
Commercial propane	94
Natural gas	39

Electricity from non-renewable sources

Large amounts of coal, oil and gas are consumed in UK power stations. The fuel is burnt to boil water, which in turn produces steam. The steam drives large turbines, which then generate electricity. The electric current is transported to houses and factories by large copper power cables.

A coal-fired power station in the UK

> **KEY POINT**
>
> Sulphur Dioxide (SO_2) and rainwater (H_2O) combine to make a very weak solution of Sulphuric Acid (H_2SO_4), a major contributor to acid rain.

There are 18 UK coal-fired power stations in the UK and these only achieve an average of 36–39 per cent efficiency. Some UK power stations have carbon dioxide (CO_2) and sulphur dioxide (SO_2) scrubbing, which involves the separation of these harmful gases from the flue gas emissions after the combustion process to help reduce the overall effects of greenhouse gas emissions on the environment.

B2 LOW CARBON

Low carbon fuels can be classified as those made from renewable sources. In this part of the chapter, we will look at some of the more common low carbon fuels.

Solar thermal
Solar thermal technology uses the heat from the sun to generate domestic hot water supply. This offsets the water heating demand from other sources, such as electricity or gas.

Biomass
The term biomass can be used to describe many different types of solid and liquid fuels. It is defined as any plant or animal matter used directly as a fuel or that has been converted into other fuel types before combustion. When used as a heating fuel, it is generally solid, for example wood pellets, vegetal waste (including wood waste and crops used for energy production), animal materials/wastes and other solid biomass.

Hydrogen fuel cells
In a hydrogen fuel cell, the conversion of chemicals takes place and this process produces electricity cleanly and efficiently.

Unlike a battery, which stores the chemicals inside it and 'dies' when all the chemicals have been converted into electrical energy, a fuel cell never dies and as long as the flow of chemical (in this case hydrogen and oxygen) is maintained, the fuel cell continues producing electricity indefinitely. A by-product of this process is the production of water used with much success by NASA as a power supply and drinking water supply on the space shuttle.

Heat pumps
A heat pump is an electrical device with reversible heating and cooling capability. It extracts heat from one medium at a low temperature (the source of heat) and transfers it to another at a high temperature (called the heat sink), cooling the first and warming the second. They work in the same way as a refrigerator, moving heat from one place to another. Heat pumps can provide space heating, cooling, water heating and air heat recovery. There are several different types:

- ground source heat pumps
- air source heat pumps
- water source heat pumps
- geothermal heat pumps.

Combined heat and power (CHP)
Combined heat and power (CHP) is a plant where electricity is generated and the excess heat generated is used for heating. It is used primarily for district heating systems but micro-CHP has also been developed for domestic properties.

Combined cooling, heat and power (CCHP)
Very similar to CHP, combined cooling, heat and power uses the excess heat from electricity generation to achieve additional building heating or cooling.

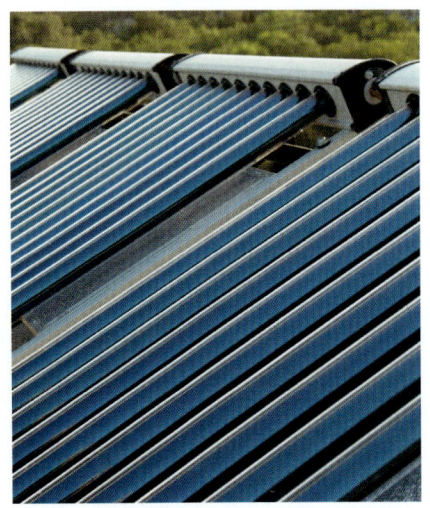

Solar hot water collector

Biomass chips and pellets

Installation of ground source heat pump pipework

B3 ZERO CARBON

When the net carbon dioxide emissions from all the fuel used is zero, this is called 'zero carbon fuel'. This type of fuel is classified as 'renewable energy'. There several types of this kind of energy.

Electricity generated by wind power

This is the world's fastest growing renewable energy. Wind is a meteorological event that is caused by the differences in air pressure under the influence of the sun. High pressure moves towards low pressure – the greater the difference in pressure, the greater the power of the wind. This air movement is used to drive large blades, which are connected to electricity generating turbines. These in turn generate electricity. To be efficient as a method of power generation, the wind needs to be constant at 12mph and above.

Wind farms are becoming more and more common and can be sited at both on-land and off-shore locations around the United Kingdom.

Electricity generated by tidal movements

The kinetic movement of sea tides that flow towards and away from land can be used to drive underwater turbines, in much the same way as wind drives wind turbines. The tides move vast amounts of water twice a day, and it is anticipated that tidal electricity generation could provide up to 20 per cent of the UK's energy needs.

There are eight locations around the UK that have the potential to have tidal power stations built. These include the estuaries of the river Severn, Humber, Avon, Solway and the Dee.

Hydroelectric power

Hydroelectric power is when the gravity movement of large amounts of water drives turbines, which in turn generate electricity. There are many hydroelectricity generators in the UK, most of which are located in Scotland. However, the largest hydropower station in the UK is located at Dinorwig in North Wales, which has a power output of 1728 megawatts (MW) of electricity.

Solar photovoltaic

Solar photovoltaic (PV) panels, or 'arrays', produce direct current (DC) electricity from the power of solar radiation. These can be used to run domestic lighting and appliances, or to charge batteries. They do not need direct sunlight and can generate electricity even on a cloudy day.

C BASIC OPERATING PRINCIPLES OF GREEN PLUMBING TECHNOLOGIES

Green technology is playing an increasingly important role within the plumbing industry as new installation methods and equipment become available. In this section, we will take a brief look at some of the new equipment available and the technology behind it.

> **KEY POINT**
>
> The amount of CO_2 saved by using wind power could be 10 billion tonnes by 2020 and 34 billion tonnes by 2030.

SmartScreen Unit 003 handout 2

SmartScreen Unit 003 worksheet 2

SmartScreen Unit 003 interactive activity 2

C1 SOLAR THERMAL SYSTEMS

A solar thermal water heating system uses roof-mounted solar collectors. These are positioned to face south in order to capture the heat generated by the sun. The solar collector can be either a series of vacuum tubes, known as an evacuated tube collector, or a flat panel, known as a flat plate collector. Both of these are filled with a transfer fluid (usually a mixture of water and antifreeze).

On average during the summer months, 1m^2 of solar panel will deliver around 1kW of energy; therefore 1m^2 is needed for every occupant of a dwelling. The minimum recommended area per dwelling is 2.5m^2. This will supply about 80 per cent of the hot water demand during the summer (eight months of the year) and around 20 per cent over the winter season (four months of the year). This is an average of 60 per cent over the whole year. A conventional gas or oil boiler, or an electric immersion heater, will be required for the remaining 40 per cent, or in case the solar thermal system should fail.

The components of a solar thermal hot water system

Solar hot water systems require certain components (some of them specialised) to enable the system to work effectively. These are shown in the diagram below:

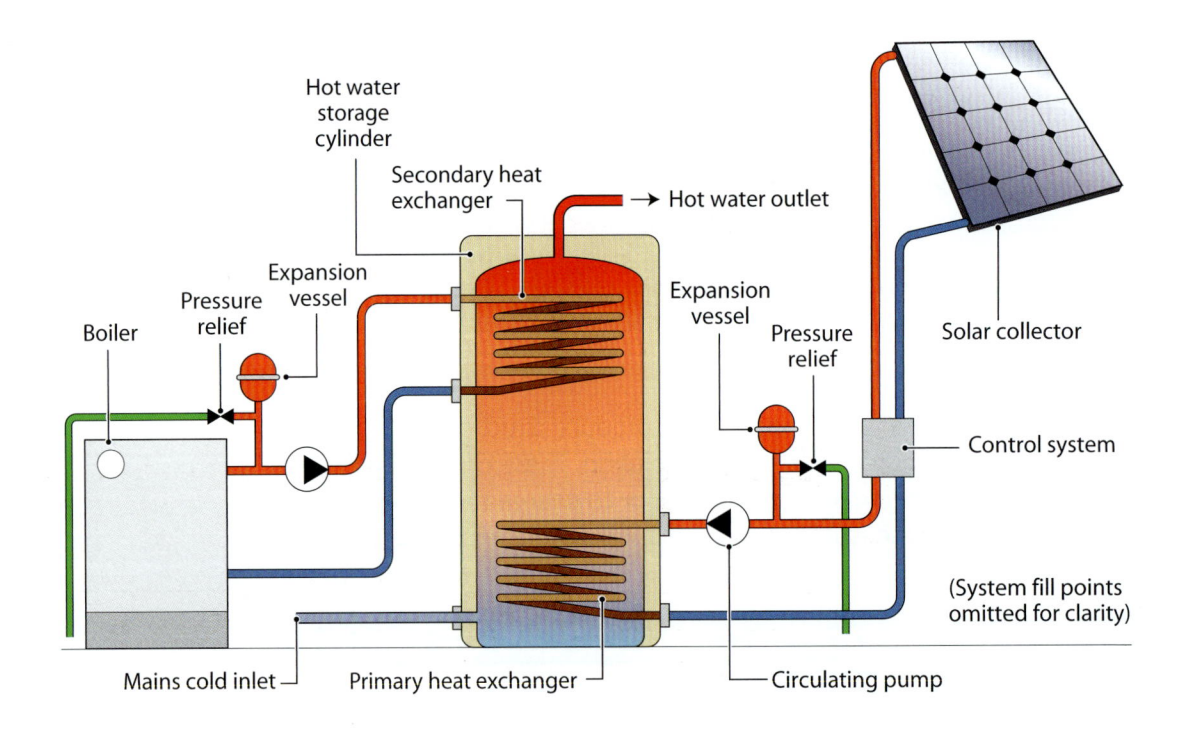

Working principles of solar hot water

How solar thermal panels work

1 The sun heats the fluid in the solar collector.

2 When the thermostat senses that the panel is 6°C above the temperature inside the hot water storage cylinder, the circulation pump will start to run.

3 The heated fluid is then pumped from the solar collector to the heat exchanger coil in the hot water storage cylinder.

4 Here, the heated fluid gives off its heat into the cylinder of stored water before returning back to the collector to be reheated. This process continues until the hot water storage cylinder is at the required temperature.

ACTIVITY

This activity is a group experiment to build a small solar collector. For this experiment, you will need:

- 1 × old, small radiator, painted matt black and mounted on a board at 45°
- 1 × small galvanised steel cistern (not plastic!)
- 6m × 15mm copper tubing
- 1 × 15mm tee
- 2 × 15mm compression tank connectors
- 2 × 15mm × ½ BSP male adapters

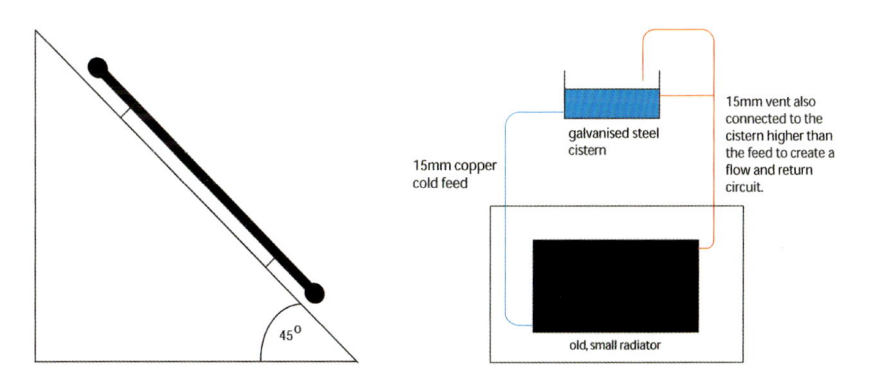

Once built, face the assembly south on a sunny day. Fill with cold water and take a temperature reading.

Take temperature readings every 30 minutes for the next 4 hours. Plot a graph of the results. Provided the weather remains sunny, you should see a steady increase in temperature over the time period.

Warning! The water both in the cistern and the radiator is likely to get **very** hot.

C2 BIOMASS SOLID FUEL

Biomass boilers fall into two separate categories:

1 **Batch fuelled** – these are usually small appliances of below 50kW output, fuelled by logs, 'lump wood' or pellets. They can be either stoves or room heaters that directly heat the room in which they are installed, or hot water/heating boilers.

2 Continuously fired – With a continuously fired appliance, the fuel is added directly to the combustion air in the correct proportion of fuel-to-air mix. This gives the desired heat output for the appliance. Combustion air is regulated to match the heat output. Generally, continuously fired units produce lower emissions than batch fuelled. The fuels for these appliances include wood chips and pellets.

How do modern biomass boilers work?

The fuel (wood chips or pellets) is fed from the fuel store to the boiler automatically by a worm screw (sometimes called an auger screw) feeding system. This is a large screw-shaped blade that is constantly turning. As it turns, it automatically moves the fuel towards the fire bed. The fuel then burns in the combustion chamber, where a regulated flow of combustion air ensures that it burns cleanly and efficiently. The hot gases, from the combustion of the fuel, heat the water in the heat exchanger, which is then circulated to the central heating system and hot water storage cylinder.

Biomass boilers are technically solid fuel appliances but, unlike coal-fired solid fuel boilers, they are very controllable. Heating and hot water controls ensure a high degree of adjustment that allows the user the same degree of temperature control as modern condensing gas and oil boilers.

1 Combustion chamber
2 Back of boiler / base of flue connection
3 Fire pot
4 Gas / oil back burner connection
5 Heat exchangers
6 Auger
7 Fan
8 Chain drive
9 Hopper
10 Auger motor
11 Drop box

A typical biomass boiler

C3   HEAT PUMPS (WATER, AIR AND GROUND SOURCE)

A heat pump is basically a machine that warms or cools a building by moving heat from a low temperature reservoir to another reservoir at a higher temperature. An example of this would be a household refrigerator that creates heat in the process of making the refrigerator cold. The process is known as the vapour compression refrigeration cycle, and involves compressing a gas (called the refrigerant) with a

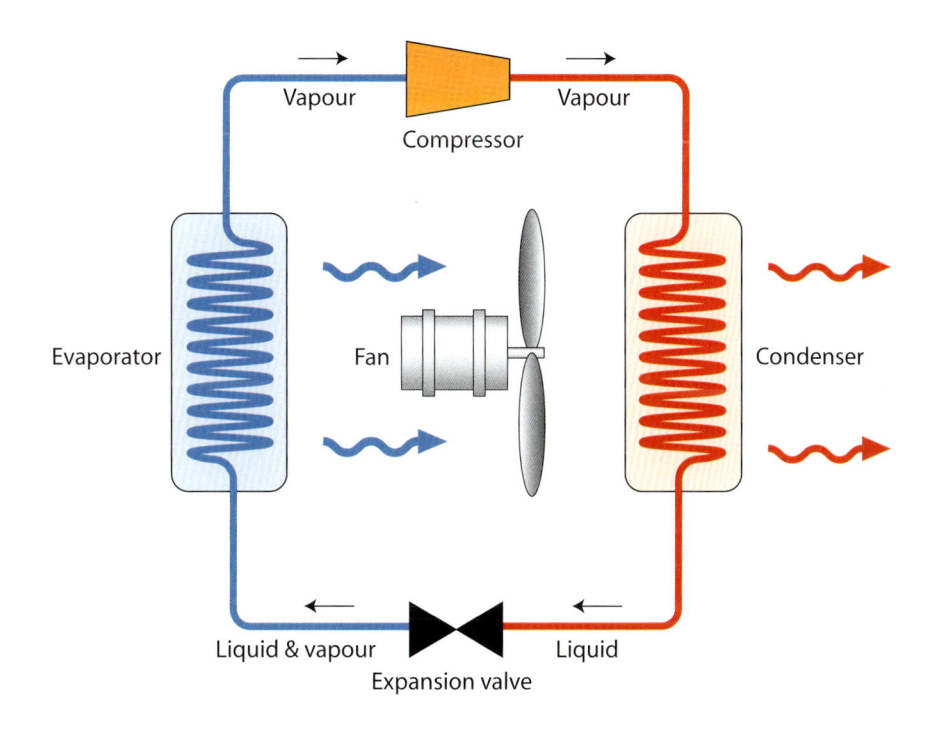

The principles of heat pumps

compressor until it becomes a liquid. This process generates useful heat that can be used to warm a building. When the pressure is released through an expansion valve, very cold temperatures are generated, which can be used for cooling a building. This process is reversible. There are several different types of heat pumps:

- **Air to air** – used mostly in commercial buildings as reverse cycle heat pumps that can be used for both heating and cooling.

- **Air to water** – used to heat swimming pools, and for providing domestic hot water and space heating for dwellings.

- **Water to air** – can use wells or boreholes, but can also be installed with many units connected together on a common closed water loop to transfer energy from hot points of a building to cold parts.

- **Ground to air** – using the constant ground temperatures to provide the heat source with warm air delivered to the building.

- **Ground to water** – same as above but used with underfloor heating systems, radiators or wall heaters.

C4   COMBINED HEAT AND POWER (CHP)

During electricity generation, a lot of excess heat in the form of steam is produced. This is usually wasted. Combined heat and power (CHP) – also known as co-generation – is a process where electricity is produced and the surplus heat is used for district heating systems or industrial processes.

> **KEY POINT**
> On average, large centralised power stations throw away two-thirds of the energy they generate. CHP utilises this wasted power to generate useful heat for many homes from one central boiler plant. This is known as district heating.

The environmental benefits of CHP are significant. Eighteen per cent of the EU's total gross electricity generation is predicted to be by CHP in 2010, with many properties and businesses benefitting from the heat energy it produces. This equates to a reduction of 150 megatonnes of CO_2 when compared with 1994 emissions (4 per cent).

CHP system types

The CHP process can be based on a variety of fuels, including fossil fuels, biomass, solar, geothermal and nuclear. The heat production can be from steam turbines, gas turbines and combustion engines.

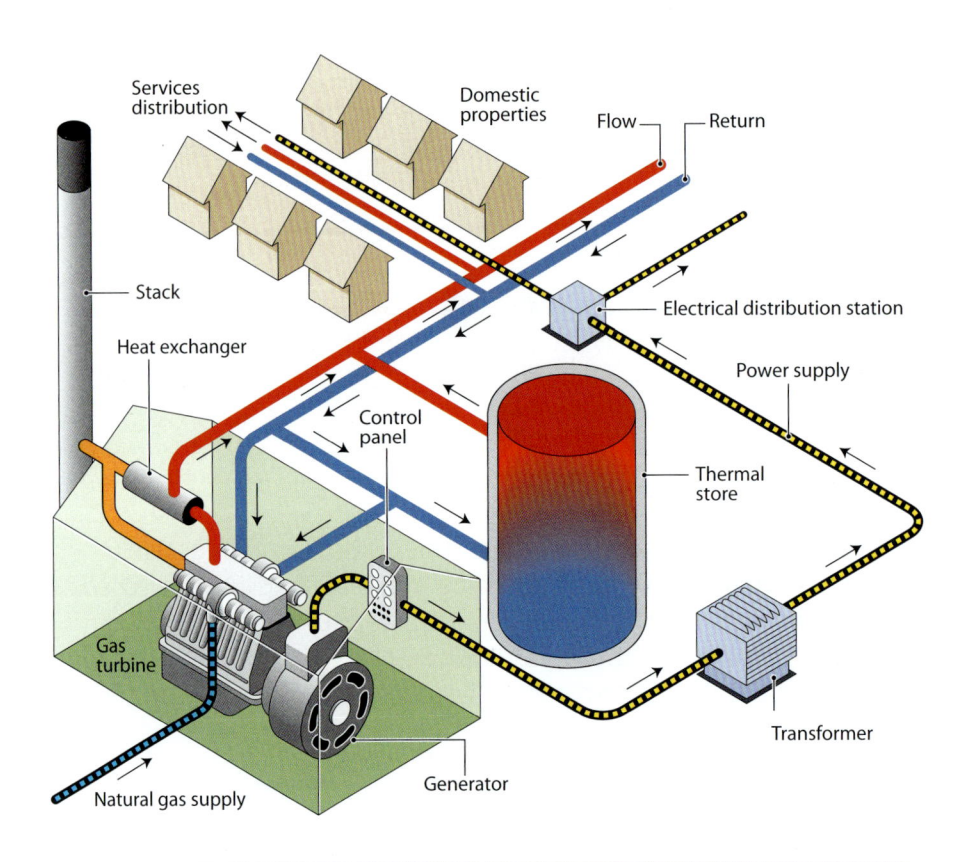

The principles of combined heat and power

CHP for district heating usually supplies heat produced centrally at one or more locations. This is supplied to an unrestricted number of both industrial and residential customers, using either steam or high temperature hot water. There are several variations:

- **Backpressure power plant** – this is the simplest form of CHP, where both electricity and heating are produced in a steam turbine, generated by a steam boiler using coal, oil or gas type fossil fuels.

- **Extraction condensing power plant** – this is used mainly for electricity generation only, though some steam may be extracted from the turbine for heating purposes.

- **Gas turbine heat recovery power plant** – heat is generated from the hot flue gases in the turbine; the most common fuels are natural gas and oil.
- **Reciprocating engine power plant** – these units use a reciprocating engine (usually diesel fuelled) coupled to a heat recovery boiler. This combination generates steam for electricity generation and heat production.

C5   COMBINED COOLING, HEAT AND POWER (CCHP)

Similar to CHP, combined cooling, heat and power also incorporates an absorption chiller into a 'tri-generation' scheme (as opposed to co-generation for CHP) to provide district cooling as well as district heating.

All of the processes that are present in the vapour compression cycle (see heat pumps on pages 120–21) are also present in the vapour absorption refrigeration process. In the absorption process, the refrigerant is usually a mixture of water, lithium bromide or ammonia.

Compression of the refrigerant takes place in the thermal compressor powered by steam (or other heat source). The refrigerant is condensed in the condenser and is evaporated in the evaporator. The refrigerant generates the cooling effect in the evaporator when the pressure created by the condenser is released. Excess heat is released to the atmosphere via the condenser.

Unlike the vapour compression cycle, two different devices called the absorber and the generator carry out the process of suction and compression. The major difference between the two systems is the method of energy input into the system. The vapour absorption process uses steam or hot water, which can be provided by natural gas, oil or solid fuel.

C6   WIND TURBINES

As we learnt in section B3 (page 117), electricity generated by wind power is a type of zero carbon fuel, as it is the natural kinetic energy of the wind that is used to drive a bladed turbine and generate electricity. The turbine blades are designed to be aerodynamic and, like an aircraft wing, have a curved cross-section. The wind blowing over the blades causes them to 'lift' and rotate. The greater the velocity of the wind, the faster the blades turn, generating more electricity.

The blades drive a shaft, which is connected to a generator. It is this generator that generates the electricity. This is then transmitted through the national grid to homes, shops, offices and factories.

Types of wind turbines

Modern wind turbines fall into two groups:

- **horizontal axis** – usually two or three blades facing into the wind
- **vertical axis** – named the Darrieus wind turbine after its inventor, but also often called the 'eggbeater' because of its shape.

Outputs range from 100kW for domestic use to several megawatts. Larger turbines are grouped together to form wind farms, providing bulk power to the national grid. Small wind turbines are often used in conjunction with diesel engines and photovoltaic panels. These are termed 'hybrid wind systems' and are mostly used where connection to the national grid proves difficult.

Most turbines generate electricity at wind speeds of 8–16 mph but will continue to work up to 55 mph. At speeds above this, the turbines are designed to shut down as the high wind could damage the generator.

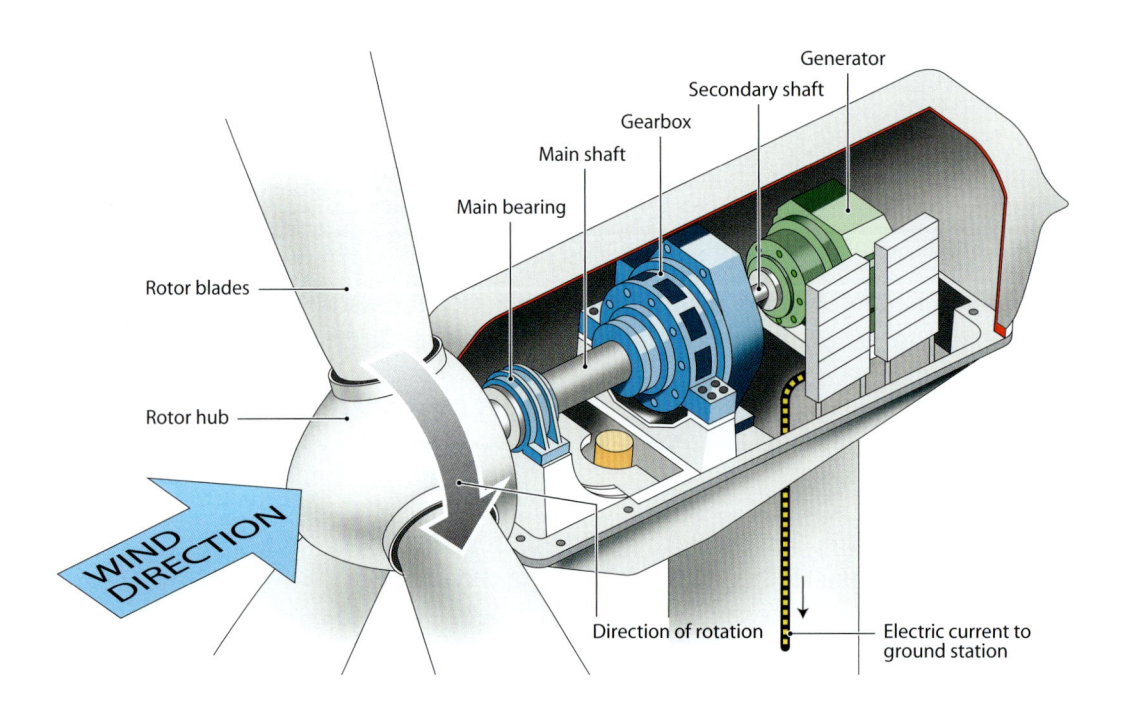

The working principles of a wind turbine

C7 PHOTOVOLTAIC POWER

On a sunny summer's day, around 1000 watts per m^2 are shining on the surface of the UK. Solar photovoltaic (PV) panels, or 'arrays', harness this free power and convert it into usable electricity.

Each solar panel is made from one or, often, two layers of a semi-conducting material, usually silicon. When the light strikes the panel, some of the light is absorbed by the silicon, which knocks electrons loose, allowing them to move freely. The free moving electrons flow in one direction under the influence of the electric field and this creates the current. By placing conductors at the top and bottom of the panel, the current can be drawn off and used. The current, together with the voltage, defines the wattage (or power) that the panel can produce. The stronger the light, then the more electricity is produced.

Photovoltaic power is classified as a green, renewable energy source, which produces no CO_2 at all. PV panels installed on a domestic property can save around a tonne of CO_2 per year, amounting to over 25 tonnes during its life span.

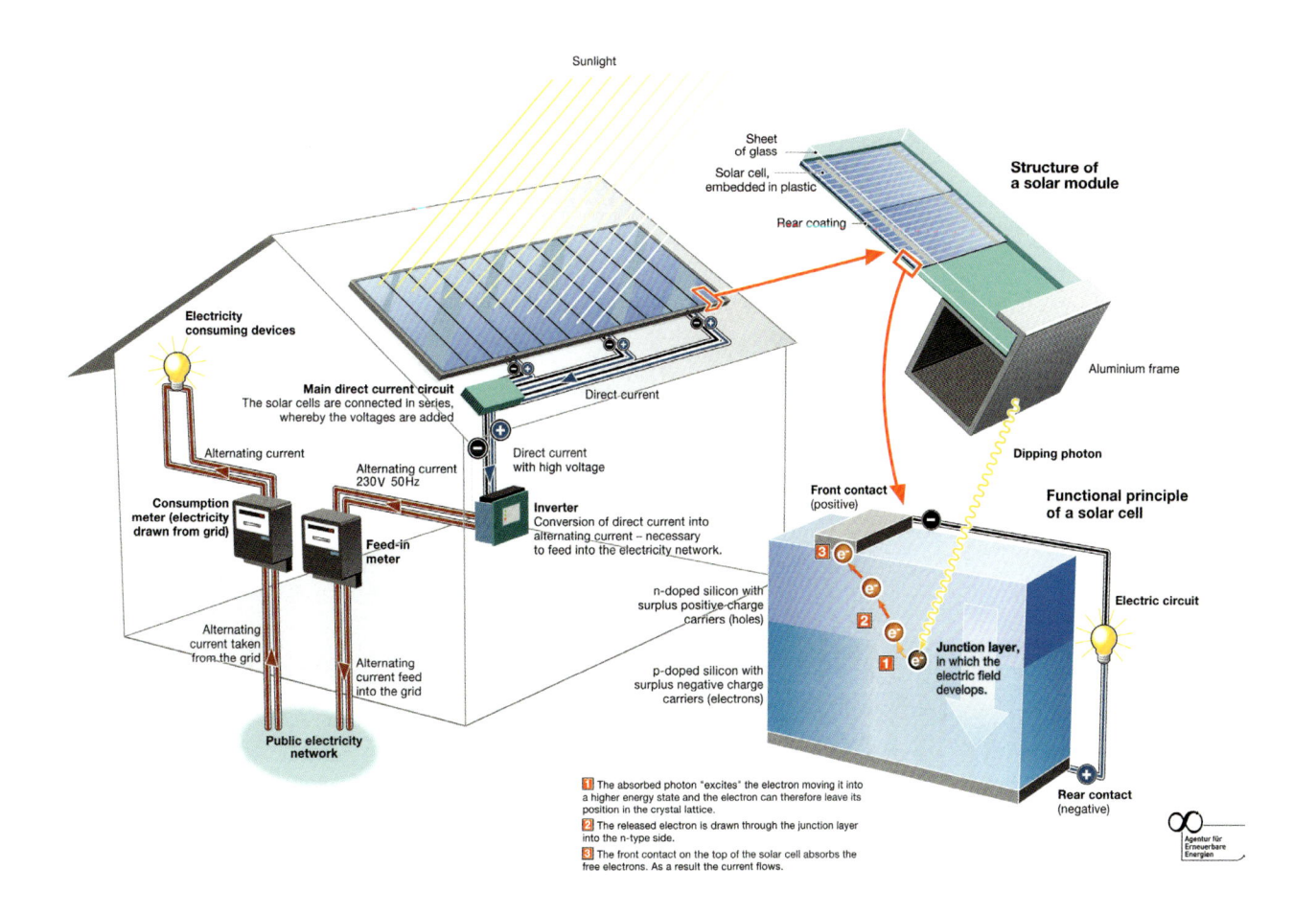

The principles of photovoltaics

C8 ENERGY SAVING AND CONSERVATION ORGANISATIONS

Energy conservation simply means increased efforts to reduce energy consumption through efficient energy use and increased use of green energy technology.

There are many organisations and government agencies to help with energy conservation and energy efficiency, and a selection of them are listed below.

> **The Energy Saving Trust** (EST) is recognised as an industry leader in helping to reduce CO_2 emissions. It is an independent, non-profit organisation, which was set up after the 1992 Rio 'Earth Summit'. It is jointly funded by the British Government and the private sector and is a recognised authority on alternative energy sources, offering advice to both industry and the public. The organisation's objectives are to help fight climate change by promoting the sustainable use of energy and energy conservation, and to cut carbon dioxide emissions in the United Kingdom. Find out more at:
> http://www.energysavingtrust.org.uk/

The Carbon Trust is an independent, non-profit organisation set up by the UK Government with support from businesses to encourage and promote the development of low carbon technologies. Its key role is to support British business in reducing carbon emissions through funding, support of technological innovation and by encouraging more efficient working practices. Find out more at: http://www.carbontrust.co.uk/

The Department of Energy and Climate Change (DECC) is the UK Government department responsible for sustainable energy deployment and for coordinating the country's response to climate change. DECC was established in 2008. Find out more at: http://www.decc.gov.uk/

The National Energy Foundation (NEF) is an independent British charity, established to encourage more sustainable use and generation of energy. Find out more at: http://www.nef.org.uk/

C9 ENERGY RATING TABLES (ENERGY PERFORMANCE)

Around 27 per cent of the UK's carbon dioxide emissions come from domestic properties. Energy Performance Certificates (EPCs) give information on how to make buildings more energy efficient and reduce carbon dioxide emissions. All dwellings, whether rented, bought or sold, need to have an EPC. They contain information regarding:

- the building's energy usage and carbon dioxide emissions
- recommendations to reduce energy use and carbon dioxide emissions.

EPCs include ratings that make comparisons between the current energy efficiency and CO_2 ratings against those that the building could potentially achieve if improvements were made.

The rating measures the energy efficiency and CO_2 emissions on a scale from A to G, A being the most efficient and G the least efficient. All buildings are measured using the same calculations and grade D is the most common achieved to date.

The recommendations

EPCs provide a detailed report advising on what could be done to help reduce the amount of energy the building uses. The report details:

- suggested improvements such as cavity insulation
- possible savings if the recommendations are carried out
- the energy efficiency and CO_2 emissions of the building after improvements have been made.

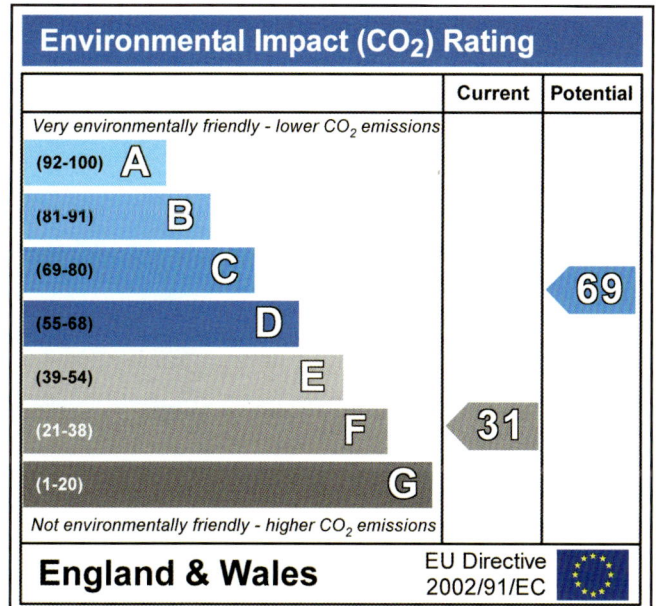

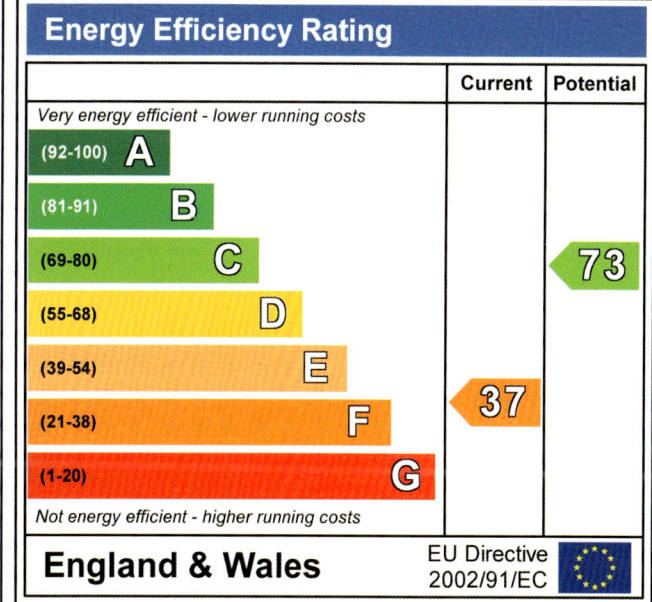

Energy performance tables

Which buildings require an EPC?

An EPC is required when a building is constructed, rented or sold, and this includes all commercial buildings. If it has walls, a roof and uses energy to maintain an 'indoor climate', in other words if it has heating, air conditioning or mechanical ventilation, then an EPC is required. The building can be whole, part of a larger building but with its own heating, such as a flat or apartment, or a building containing separate units. Public buildings such as colleges and libraries are required to display their EPC prominently.

The following buildings are exempt from having an EPC:

- places of worship
- temporary buildings
- industrial sites, workshops and non-residential agricultural buildings that use very little energy
- stand-alone buildings smaller than 50 m² that are not providing living accommodation.

Energy ratings for electrical appliances

Environmentally friendly electrical appliances carry energy efficiency rating information, much in the same way as central heating boilers and buildings. This is a standard that has been developed by the Energy Saving Trust in conjunction with industry and the UK Government that appears on all new electrical appliances, or 'white goods'.

An appliance that carries the Energy Saving Recommended certification mark can be guaranteed to surpass the standard energy efficiency regulations. The certification mark is used in conjunction with the European Union Energy Label for laundry and refrigeration appliances. The label is a useful indicator of how efficient the product is in various categories.

An overall rating is denoted a score by letter, with A being the most efficient (for fridges this is now A++) and G being the least efficient rating available. There are also individual ratings based on other appliances. For example, with washing machines, the appliance is rated on kWh per cycle, washing performance, spin drying performance, load capacity and noise. Each of these can help you make your purchase based on your specific requirements.

D ENERGY EFFICIENCY AND THE CUSTOMER

An energy efficient system begins at the design stage. How we approach the design of, say, a central heating system or a cold water system will ultimately have a direct effect on the efficiency of the system after installation. There are set stages that each design must go through to ensure that the maximum efficiency possible is realised. All too often, a good design is ruined by poor materials, wasteful/mediocre installation techniques and incorrect commissioning procedures. Poor handover instructions and feedback to the customer merely worsens the problem.

D1 DESIGN, INSTALLATION AND HANDOVER

To ensure that a system is working to its full potential, it must be:

- designed to a group of set criteria in line with the regulations in place
- sized to give the correct flow rates or heat outputs in line with the design criteria
- installed using recognised materials, in line with good practice while observing the regulations and British Standards/European Standards (BS/EN BS/CE)
- commissioned to give optimum performance, which will involve:
  - filling each system, ie cold water, hot water and heating, and flushing in accordance with the regulations
  - soundness testing to ensure that any water leaks are dealt with
  - setting and adjusting the shut off levels of any float operated valves
  - checking flow rates at terminal outlets, taps etc, to ensure that they meet the design specification
  - checking the temperatures at all hot terminal outlets, taps etc

Plumber customer care

- running central heating systems to maximum working temperature and checking flow and return temperatures, radiator temperatures, operation of controls, thermostats and time clocks/programmers
- checking the CO_2 emissions of space heating and hot water appliances
- benchmarking the systems in line with manufacturers' recommendations.

- handed over to the customer, ensuring that the customer is fully aware of its operating procedures and characteristics.

Customer care

A crucial part of any installation is the handover to the customer. A requirement of the Building Regulations Document L1a/b is that all building owners should be fully briefed as to how they can ensure that their building service and maintenance requirements use no more fuel than is reasonably necessary.

At the end of the installation and commissioning stage, it is good practice to provide the customer with a file that can be referred to over the life of the system. The file should contain the following:

- the contact details of key members of staff of the company
- emergency contact telephone number(s) in case of breakdown or a burst pipe
- the gas emergency telephone number (for any gas appliance)
- all of the manufacturers' installation and maintenance instructions for the appliances and controls
- the benchmarking certificate.

The customer should be given careful instruction on the operation of the system, including:

- the appliance and all controls and thermostats
- how to make adjustments to the timing and temperature controls
- what routine maintenance is needed to ensure that the system efficiency is maintained through the service life of the system and its components
- the need for annual routine boiler and system maintenance.

It is also a good idea to walk the customer through the emergency controls in the property, such as stop taps, gas emergency control valve, electric consumer unit, etc.

It is very important to give the customer the opportunity to ask questions.

SmartScreen Unit 003 handout 3

SmartScreen Unit 003 worksheet 3

SmartScreen Unit 003 interactive activity 3

SUGGESTED ACTIVITY...
A customer is considering a move to solar hot water. In groups, discuss the benefits and pitfalls of solar hot water with regards to price, installation costs and short-term outlay compared to long-term pay back times.

E CARING FOR THE ENVIRONMENT – A PLUMBER'S RESPONSIBILITIES

E1 ADOPTING A GREEN ATTITUDE TO PLUMBING ACTIVITIES

There are many ways in which plumbers can help the environment. Adopting a 'green' attitude and working practices helps to reduce CO_2 emissions and material wastage, which could contribute to saving the environment. It could also save the customer money and increase the profits of the company. Here are a few tips:

- Wherever possible, use compression type fittings. Although the initial cost is higher than capillary fittings, they do not require the use of a blowtorch and soldering equipment and so do not contribute directly to global warming through CO_2 emissions. The time taken to install compression fittings is also less than capillary fittings, which saves on installation costs.

- Introduce a waste reduction programme in line with the Government's Waste and Resources Action Programme (WRAP).

- Use off-site fabrication techniques for some parts of large or complicated installations.

- Plan work activities to minimise waste of materials. Exact measurement and cutting and bending of copper and low carbon steel tubes to precise dimensions helps to eliminate wastage of materials. Accurate measurement of materials also assists in precise estimation of jobs.

- Avoid over-purchasing of materials.

- Develop accurate store-keeping techniques on larger sites for booking materials in and out to minimise waste, loss and theft.

- Keep fragile materials such as WC pans and wash basins in a locked store and stack them safely to avoid damage and breakages.

- Avoid theft by keeping all materials in a locked store.

E2 DISPOSING OF WASTE MATERIALS

Construction in the UK produces vast quantities of waste every year. The Strategy for Sustainable Construction quotes the following statistics for construction waste in England:

- Construction projects in England use around 400 million tonnes of materials every year.

- Around 100 million tonnes of waste is produced.

- An estimated 25 million tonnes of construction design and engineering inert waste is disposed of in landfill sites every year.

The UK Government, via the Waste Strategy and the Sustainable Construction policy, proposes a reduction of 50 per cent in the levels of construction, demolition and excavation waste compared to levels in 2008. They propose that this is achieved by 2012, and that it can be achieved by the 'three Rs' rule:

1 **Reducing** – minimising the amount of waste produced.

2 **Reusing** – reusing items as many times as possible.

3 **Recycling** – recycling what you can and disposing of what is left in a responsible way, and as a last resort.

By following these rules the construction industry can:

- generate income
- reduce costs by buying less material and maximising skip space
- comply with the Regulations
- reduce accidents on site by ensuring a tidy site
- reduce CO_2 emissions
- help to save natural resources.

E3   WASTE MANAGEMENT LEGISLATION

The Site Waste Management Plans Regulations 2008 took effect on 6 April 2008. Much like the CDM Regulations (see Chapter 001, pages 5–7), these are relevant to the construction industry, including the client, planners, developers, contractors and those buying materials for construction projects in England. Under the Regulations, site waste management plans (SWMPs) are compulsory in England for every construction project with an estimated cost greater than £300,000. For projects costing more that £500,000, SWMPs with much greater detail are required. Either the client or the main contractor has the overall responsibility for the SWMP. Under the Regulations, construction management must:

- estimate the amount and type of waste such as inert, non-hazardous or hazardous, which is expected to be produced. The Environment Agency must be notified that hazardous waste is being produced before it is removed

- show how they intend to improve efficiency in using materials

- set out how they intend to reuse, recycle and lawfully dispose of such waste, which helps compliance with the waste duty of care and reducing fly-tipping and other unlawful waste disposal methods

- update the SWMP during the project, recording all receipts and references to waste transactions with authorised waste removers or receivers, eg waste transfer notes, the identity of the remover, whether the site to which the waste is being taken is licensed or exempt

- measure actual performance of the SWMP against estimates.

Site waste management programmes apply to all construction work including:

- preparatory work such as demolition and excavation
- civil engineering and engineering projects
- projects involving maintenance, alteration and decoration of existing structures
- the installation, maintenance or removal of related services such as electrical, gas, water, sewage and telecommunications.

To comply with the SWMP Regulations, construction companies must:

- complete waste transfer notes before any waste leaves the site
- ensure that all waste carriers have a valid waste carriers' registration certificate
- ensure that all wastes are disposed of at a correctly licensed site
- complete notification for hazardous waste to the Department for Environment, Food and Rural Affairs (DEFRA).

E4 THE BENEFITS OF GOOD WASTE MANAGEMENT PRACTICE

The benefits to the environment of good waste management are obvious, such as less environmental pollution, less unusable land. But for the construction industry itself, the benefits of helping the environment can also be financial. Financial benefits include:

- lower disposal costs, eg reduced hiring skips, landfill tax and gate fees
- avoidance of waste transportation costs
- greater reuse/recycling of materials on site, saving on raw materials purchased
- lower levels of waste materials.

Construction companies should consider the following when aiming to save money and benefit the environment at the same time.

Buying and storing materials

- Order the right amount of materials.
- Arrange for 'just in time' deliveries to reduce storage and loss of materials.
- Consider the source of materials. Recycled can be cheaper.
- Think about the packaging used for materials delivered to the site – can this be recycled?
- Make sure that deliveries are rejected if damaged or incomplete.

- Make sure that storage areas are safe, secure and weatherproof if necessary.
- Store liquids away from drains to avoid pollution from spillages.

Activities on site
- Make sure that the options for the use of reclaimed and recycled construction materials are considered.
- Recycle spoil, demolition materials and surplus construction material to avoid the need to transport materials.
- Keep the site tidy to reduce loss of materials and waste.

Training and awareness
- Promote good practice awareness as part of health and safety induction/training for workers on site.

Segregating waste
- Segregate different types of waste using different skips. There should be skips for wood, inert and mixed materials, with a special skip for waste metals, which may generate some income.
- Or use a licensed waste management company to deal with waste. They may be able to recover recyclable materials from mixed skips.

Segregated skips

E5 METHODS OF SAFELY DISPOSING OF WASTE MATERIALS

The construction industry produces vast quantities of waste materials each year. Most of this waste can be classified into three distinct groups as follows:

Hazardous	Non-hazardous	Inert
• Batteries • Asbestos • Oil and oily rags • Waste electrical equipment • Solvent-based glues and adhesives • Paints • Sealants • Chemicals • Lead	• Timber • Plastics/packaging • Paper/cardboard • Empty tins/tubes • Water-based glues and paints • Scrap metal (not including lead)	• Bricks • Hard concrete • Asphalt • Glass • Ceramics/tiles • Sand and gravel

Research has identified the true cost of the average builder's skip of waste as being:

Skip hire charge:	£110
Labour to fill skip:	£160
Original cost of material in skip	£1095
True cost of the waste:	£1365

Each phase of a construction process will produce different needs in terms of identifying wastage. By reducing the volume of waste and then finding the best methods of collection and recycling, it helps to keep costs down and the construction site a safer environment in which to work.

Construction companies should use licensed waste disposal companies to remove the waste from construction sites in line with their site waste management plan (SWMP). The proposals for the disposal of waste should be submitted at the planning application stage of the project and should detail the following:

- waste produced/recycled/reclaimed
- steps to minimise waste and maximise the use of recycled materials
- management of waste on site and waste leaving the site
- relevant evidence of waste carrier/waste transfer in the form of a waste transfer note.

E6 LICENSED WASTE MANAGEMENT COMPANIES

A licensed waste disposal contractor will conduct a site survey with the customer to assess the potential for segregating the different types of waste generated. This often includes options for waste compaction, on-site segregation and container services.

KEY POINT

If you have waste after completing a customer's job, it is **your** responsibility to remove or have it removed and **not** that of the customer.

Technical advisers will often assist site management to develop site waste management plans (SWMPs), setting up a fully auditable recycling certificate to verify their waste recycling performance. This helps to measure key performance indicators (KPI), and has several key advantages:

- tailored waste system to match project commitments on waste recycling
- 'Duty of Care' documentation and Pre-Treatment Declaration
- management report on recycling performance.

E7 WASTE CARRIERS' LICENCE

It is the law that companies and the self-employed must register with the Environment Agency if they transport or arrange the disposal or recovery of commercial, industrial or household waste. This is known as controlled waste. This means that a plumber who is removing waste materials and scrap metals from a completed job in a company van will need a waste carrier's licence.

Licensed waste carrier company

However, a licence is not needed if:

- the waste is owned by the person transporting it (unless it is building or demolition waste)
- it is being transported in or out of the country only by sea or air
- it is exempt waste, which comprises only animal by-products, only mines or quarry waste, or only agricultural waste.

E8 RECYCLING

Much of the waste materials from construction sites can be recycled or put to other uses. Waste such as concrete, asphalt, dry walling, wood and metals all have a recyclable value, and finding a way to recycle them is a must for those companies who wish to 'go green'. In addition, as we have already seen, these materials can generate much-needed income. There are many ways to recycle, including the following:

- Isolate and separate the different types of waste into colour-coded skips. Using this method reduces the time spent by waste management companies in separating the different forms of waste.
- Ask suppliers to deliver goods to site with less packaging and ask if excess materials can be returned for reuse.
- Use local salvage companies for wood items and reinforced steel joists (RSJs). These can be found using the internet searching under 'Construction Materials Salvage'.

Sorting construction waste

- Think about which waste materials will be mostly generated, such as top soil, hardcore, timber etc, and concentrate on recycling those.

- Keep recycling skips away from those for general rubbish. This avoids confusion as to what can go in where.

- Education can play a large role in getting the workforce in a green, recycling way of thinking. Using posters to highlight recycling and waste reduction can have a positive effect on the operatives on site.

Plumbers and recycling

As with everyone within the building services industry, plumbers have a duty of care to 'know' their waste. It is **illegal** to do the following:

- Mix hazardous waste, such as asbestos and lead, with non-hazardous waste.

- Give waste materials to anyone who is not a licensed waste carrier or waste management company – it is the plumber's responsibility to check, so ask to see their licence.

- Dump waste at the side of the road or in unauthorised places (known as fly-tipping, and carries a maximum fine of £50,000 or up to five years' imprisonment).

- Burn waste such as polystyrene and plastics, as these give off very toxic fumes including cyanide gas.

Plumbing contractors must:

- make sure that a waste ticket is produced for every waste collection

- keep a record of all transfers of waste to a waste carrier for a period of two years

- use a hazardous waste ticket when recycling hazardous substances such as asbestos and lead (must be kept for three years)

- notify the site supervisor immediately if they come across any hazardous materials or waste in the course of their work.

Dos	Don'ts
Reduce the amount of waste produced by storing materials carefully and using accurate installation techniques.	Throw broken power tools into the bin when renewing or replacing. Instead, dispose of them at a recycling centre.
Reuse materials instead of purchasing new ones, such as copper tube that is in good condition (providing it is not for hot and cold water supplies).	Pour anything except clean water down road drains (this is illegal, as road drains feed directly into water courses).
Recycle as much as possible – for help and advice, visit: www.wrap.org.uk	Pour anything hazardous or anything that could cause a blockage down household drains.
Design pipe runs to use as few fittings as possible and save resources by fitting the minimum size of boiler, cistern and radiators for the job.	

KEY POINT

Be prepared to offer a full consultancy service regarding green technology and how it can save the customer money. Either carry a portfolio of leaflets and brochures around with you, or show the information on screen via a laptop or email PDFs/web links. These could be sent along with your quotation.

Plumbers can be excellent advisors on green installations, advising customers on which methods and systems can save money and the environment. This advice may include any of the following, as appropriate:

- water saving devices such as dual flush, low water content WCs, grey water recycling systems and rainwater harvesting
- solar power hot water heating systems, heat pumps etc
- good standards of insulating pipework, water cisterns and cylinders
- low water use shower heads, which can save up to 70 per cent of water against normal shower heads
- full and proper instruction on how to get the best out of heating systems and their controls.

E9 RECYCLING PROCESSES

When a material is recyclable, such as cardboard and plastics, they usually carry the international recycling symbol shown below. In general, recyclable materials can be divided into four main groups:

- metals
- plastics
- cardboard
- wood

Metals

Metal for recycling is divided into two groups. These are ferrous (metals that contain iron) and non-ferrous. Metals should be separated into their respective groups before being collected by the licensed waste remover.

003 ENVIRONMENTAL PROTECTION METHODS

137

Once they have been collected, they are smelted into ingots and sold to manufacturers. Most metals can be recycled indefinitely and do not lose any of their properties during the recycling process. All metal has a monitory recycling value.

Ferrous metals include:

- iron
- steel.

Non-ferrous metals include:

- copper
- brass
- bronze
- aluminium
- lead
- nickel
- zinc.

Plastics

There are around 50 individual groups of plastics, with many hundreds of varieties. All plastic is recyclable. To make sorting and grading of plastics for recycling easier, the American Society of Plastics Industry has developed a standard marking code to help identify and sort the main types of plastic. These types and their uses within the plumbing industry are:

KEY POINT

On no account should plastic be burnt. Plastics give off dangerous amounts of acrid, thick black smoke that contains a deadly poison called cyanide when combusted.

Symbol	Code	Description
2	HDPE	High-density polyethylene – water pipe
3	PVCu	Unplasticised polyvinyl chloride – waste pipes, guttering, soil and vent pipes, drainage
4	MDPE	Medium density polyethylene – blue water pipe
5	PP	Polypropylene – push-fit waste systems, overflow pipe and cold water storage cisterns
6	PS	Polystyrene – protective packaging
7	OTHER	Any other plastics that do not fall into any of the above categories. An example is melamine, which is often used in kitchen and furniture manufacturing.

Mechanical plastics recycling

Mechanical recycling of plastics involves the melting, shredding or granulation of waste plastics. Plastics must be sorted prior to mechanical recycling. In the UK, most sorting for mechanical recycling involves the use of trained staff, who manually separate the plastics by polymer type and colour. New technology is being introduced that will sort the plastics automatically, using techniques such as x-ray fluorescence, infrared and near infrared spectroscopy, electrostatics and flotation. After separation, the plastic is either melted down directly and moulded into a new shape, or melted down after being shredded into flakes and then processed into granules called regranulate.

There is a wide range of products made from recycled plastic, including polyethylene bin liners and carrier bags; PVCu pipes, flooring and window frames; building insulation board; video and compact disc cassette cases; fencing and garden furniture; water butts, garden sheds and composters; seed trays; anoraks and fleeces; fibre filling for sleeping bags and duvets; and a variety of office accessories.

Plastic recycling

Cardboard/paper

The UK produces over 8 million tonnes of cardboard each year, which amounts to about 140 large cardboard boxes for every person in the country. Almost everything that a plumber installs is delivered in a cardboard box, from boilers to pipe fittings. Cardboard is the largest single element of public solid waste worldwide.

Cardboard is made from cellulose fibres that are manufactured from wood pulp. When it is recycled, the cardboard is soaked and agitated to release the fibres, which can then be re-pulped. This process can be repeated up to five times before the fibres disintegrate and become useless.

Many companies employ specialist recycling and waste removal companies to collect their waste, but some of these disregard cardboard collection as uneconomical and much of the cardboard waste ends up at landfill sites.

Most cardboard (without plastic coating and too much ink) will be biodegradable. Cardboard in shredded form can be used for animal bedding and home insulation. It can also be used as a fuel as it releases almost twice as much heat as per kilogram as other sources, does not release toxic fumes into the atmosphere and leaves only ash as its by-product.

Wood

About 50 million cubic metres of wood are used in the UK every year, of which two-thirds is imported. Nearly 50 per cent of this goes into paper manufacturing. Whilst wood is a sustainable resource, recycling

> **KEY POINT**
> Because biodegradable materials are mainly carbon based, they will eventually form harmful methane gas when buried in landfill sites. Methane is known to have greater greenhouse properties than CO_2.

wood still has a positive environmental impact. Wood recycling ensures that the waste does not end up in landfill sites where it produces greenhouse gases because it is biodegradable.

Wood waste from businesses consists mainly of packaging, such as wooden crates, and demolition. It is estimated that UK businesses produce about 1.4 million tonnes of wood waste each year. The reuse of wood should be the first option as this uses the least energy.

If the wood cannot be reused in some way, wood recycling can make the waste into a variety of products such as mulch, chipboard or biomass fuel. Wood recycling means that new, virgin wood is not felled and processed as this uses more energy and water, which are less renewable resources than the wood itself. If 'new' wood is to be purchased, make sure that it is Forest Stewardship Council certified. This ensures that the supply of wood used is sourced from sustainable forests.

Local Authorities may offer a waste collection and wood recycling service.

E10 DISPOSING OF HAZARDOUS MATERIALS

During the course of your work within the building services industry, you may come across hazardous substances that will have to be disposed of safely. In this section of the chapter, we will look at how to dispose of:

- asbestos
- refrigerants
- electrical and electronic equipment
- lead.

Disposing of asbestos

Asbestos was covered in detail in Chapter 001, but here we will look at its safe disposal.

It is very hard to identify asbestos visually, but if you suspect that you have found asbestos or asbestos-containing materials (ACMs), you should follow these guidelines:

- Leave the asbestos alone. Provided that you do not disturb or damage the material, it will remain reasonably safe.
- Never drill, saw or break asbestos materials.
- Always seek professional advice before removing asbestos materials.
- Do not attempt to remove asbestos lagging, spray coatings or insulation board. These materials can only be safely removed by a licensed contractor.
- It may be necessary to take a sample to identify the type of asbestos. Never attempt to do this yourself – employ a suitably trained licensed asbestos contractor to sample or do a survey of the premises.
- If asbestos is found in site waste, for example in a skip, seek advice immediately to ensure that it is removed by a licensed asbestos removal contractor.

- Above all, ensure the health and safety of yourself and others by using licensed asbestos removal companies to dispose of **any** asbestos.

What to do if you are concerned about asbestos

The Health and Safety Executive (HSE) provides information and advice on the use and removal of asbestos. Contact the HSE infoline on 0845 345 0055 or contact them online at:
https://www.hse.gov.uk/feedback.htm

Disposing of refrigerants

If you have to dispose of an old refrigerator, freezer or chiller cabinet, or dismantle a cold store room, you should make sure that it is done safely to prevent accidents or harm to the environment.

What is the law?

Refrigerants are chemicals that boil at ultra low temperatures. When a refrigerant turns from a liquid to a gas, it creates a dramatic cooling effect (the vapour compression refrigeration cycle, see pages 120–21). This process is used in freezers, refrigerators, air conditioning cooler batteries and cold stores. The problem is that many of these chemicals are toxic and some contain harmful chlorofluorocarbons (CFCs) and hydrochlorofluorocarbons (HCFCs) that deplete the Earth's ozone layer. These are known as ozone depleting substances (ODS).

Since 1 October 2000, the law has stated that it is the duty of care of the owner of the equipment to make sure that all ODSs are removed at Environment Agency approved facilities before the equipment is disposed of.

The 'Duty of Care' is a law that says you must take all reasonable steps to keep waste safe. If you give waste to someone else, you must be sure that they are authorised to take it and can transport, recycle and dispose of it safely.

How to identify refrigeration equipment that may contain ozone depleting substances

The age of small domestic refrigeration equipment should give a general idea as to whether or not refrigeration equipment may contain ODSs:

- Before 1994, almost all refrigeration appliances used CFC refrigerant (CFC R12).

- After 1994, these were generally replaced with HFC as refrigerant (R134a).

- Modern refrigerators and freezers are manufactured using HFC (R134a) or hydrocarbon (HC600a) refrigerants.

Refrigerators and freezers are marked with an 'appliance rating plate'. This is a metal plate or label that is found either on the back of the appliance or inside it. The plate contains important information about the appliance, such as the model and serial number. In almost all cases it will also state what refrigerant was used in the appliance.

> **KEY POINT**
>
> Do not incinerate asbestos. Asbestos has a tendency to explode when exposed to extreme heat, releasing harmful fibres into the atmosphere.

> **KEY POINT**
> The hole in the Earth's ozone layer that appeared because of CFC usage has begun to repair itself as a direct result of the banning of materials that contain HCFCs and CFCs, such as refrigerants.

How to dispose of refrigeration equipment that contains ozone depleting substances

1 Consult the local waste collection authority (district, borough or unitary council). The Local Authority is not obliged to accept it from a business, but may be able to offer some advice on how to dispose of the equipment locally.

2 Consult a specialist refrigeration disposal company.

3 Refrigeration units may be refurbished and resold anywhere within the EU.

4 Always remember that fly-tipping is illegal. Dumped refrigeration equipment can pose a hazard to small children, pets and livestock, who may become trapped inside.

5 For more information, see the Department of the Environment, Food and Rural Affairs (DEFRA) website at:
http://www.defra.gov.uk/environment/waste/topics/hazwaste/fridges/index.htm

Disposing of lead

As we discovered in Chapter 001, lead is a highly toxic metal. Much new lead is extracted from ores dug from underground mines. Processing lead ore involves the mining, crushing, filtering, roasting and finally smelting the ore to extract the virgin metal, which is a highly energy intensive process. It is especially environmentally important, therefore, to recycle lead wherever possible.

If lead is recycled, it should be handled carefully and disposed of at a licensed waste contractor. Here the scrap lead is smelted into ingots and resold to industry. Lead recovered from scrap requires far less energy than producing new lead from ore.

Disposing of electrical and electronic equipment

When old and obsolete electrical and electronic equipment is not recycled, raw materials have to be processed to make new ones. This takes significant amounts of energy and fuel and causes environmental damage due to the manufacturing process.

At the beginning of the twenty-first century an estimated six million tonnes of electrical and electronic equipment was lost because of non-recycling. This amounted to the loss of much recyclable material, including:

- 2.4 million tonnes of ferrous metal
- 1.2 million tonnes of plastics
- 652,000 tonnes of copper
- 336,000 tonnes of aluminium
- 336,000 tonnes of glass.

Waste electrical goods

The production of these raw materials and the products made from them results in environmental damage from mining, transport, water and energy usage. According to a recent study by the United Nations, the manufacture of a new PC uses 240kg of fossil fuel, 22kg of chemicals and around 1500 litres of water. This is very similar to the quantities used in making a small car!

The disposal of electronic and electrical appliances in landfill sites or through incineration creates a number of environmental problems because of the toxic nature of many of the substances, including arsenic, bromine, cadmium, halogenated flame retardant, hydro-chlorofluorocarbons (HCFCs), lead, mercury and printed circuit boards.

When electrical and electronic goods are disposed of, it is important to ensure that they are recycled safely to avoid dangerous chemicals and substances from damaging the environment. The symbol shown to the right appears on many electrically operated items and batteries and means that they should not be placed in the normal household waste.

Disposing of batteries

Many tools we use today are battery powered. Batteries have a finite life, which means that, even if they can be recharged a number of times, they will eventually need replacing. There are many different types of batteries, of which the main categories you will come across are:

- Wet-cell – Often used in industry or to power vehicles. Some types are rechargeable, such as lead-acid car batteries.

- Dry-cell non-rechargeable – General-purpose disposable household batteries. Types include zinc carbon and zinc chloride batteries (used for low-drainage appliances such as remote controls, radios, torches or clocks), alkaline manganese batteries (suitable for similar uses, but longer lasting and less likely to leak), and primary button cells of various types (often used for watches, hearing aids and calculators).

- Dry-cell rechargeable – General-purpose rechargeable household batteries. As well as the uses listed for dry-cell non-rechargeable, rechargeable batteries also include nickel cadmium, nickel metal hydride and lithium-ion batteries, which are frequently used for power tools, laptop computers and mobile phones.

By using rechargeable batteries, the number of batteries to be disposed of is reduced. However, it should be remembered that 80 per cent of rechargeable batteries contain nickel cadmium, a known human cancer-causing material, and need to be disposed of safely.

From January 2010, manufacturers of batteries were legally required to provide collection and recycling facilities for their disposal under new European Union regulations.

Many large supermarkets that sell batteries now have collection bins for old, used batteries. Town halls, libraries and even schools may also set up collection points. Look out for the 'Be Positive' signs in shop windows and in stores to find out where these collection points are. Many local councils already collect batteries as part of their recycling collection service, or provide special used battery bins at the local waste and recycling centres. It is a good idea to get in touch with your local council to find out what battery recycling choices are available in your area.

Disposing of batteries

KEY POINT

For more details about battery recycling, and the new laws, visit the website of the Department for Environment, Food and Rural Affairs (DEFRA) at www.defra.gov.uk/.

F WATER CONSERVATION AND PROTECTION

Sustainable water management is essential to safeguard the water environment and future supplies to ensure future demand may be met. Although water is a renewable resource, there is not an infinite supply and there are limits to freshwater availability.

Throughout the UK, the demand for water is increasing and predictions show that this trend is likely to continue for the next 25 years. Climate change may also affect demand further.

Cutting water consumption is one of the easiest ways to save water. Adopting a systematic approach to reducing water use could cut it by up to 30 per cent. But cutting water consumption is just half of the problem. Contamination is also a huge risk to the water supply.

Contamination from industry and farming, floods, broken drains, old lead pipework and poor plumbing installations add further to the pressure on the nation's water supply. In this section, we will look at the environmental impact of water wastage and contamination. Also we will examine how to calculate water consumption for new dwellings and methods of reducing water consumption in all buildings.

F1 STATUTORY LEGISLATION FOR WATER WASTAGE AND CONTAMINATION – THE WATER SUPPLY (WATER FITTINGS) REGULATIONS 1999 AND THE WATER ACT 2003

The Water Supply (Water Fittings) Regulations 1999 are designed to safeguard our water supply from contamination, wastage, misuse, undue consumption and erroneous metering by strictly controlling the way we install and use the water and its systems of pipework. Although they will be covered in more detail in Chapter 006, here we will cover only the key points around how the regulations contribute towards safeguarding the supply of clean, wholesome water by the water undertaker.

1. **Contamination of water** – Contamination of the water supply can take many different forms, from back siphonage and backflow into the water main through faulty or poor plumbing installations, to contamination of the water source from chemical spills, floods and other environmental disasters. Once contamination gets into the water source it is often difficult to find the cause. Plumbers have a legal duty under the Regulations to ensure that the water supplied by the water undertaker does not become contaminated as a result of their work or during the course of their work, and to ensure that any water returned to the water source during the commissioning stage is not contaminated.

Often, the chemicals that plumbers use during the course of their work have the potential to contaminate the water source, ie a river or stream. When large, industrial and commercial plumbing installations are disinfected with chlorine solution, for example, it is important that the water is disposed of in a safe and legal manner. Simply draining the heavily chlorinated water into a water supplier's drainage system is not acceptable, and permission should be gained first by contacting the local water undertaker. This will ensure they are aware of any potentially harmful concentrations of chlorine that can be detected and dealt with accordingly.

Other ways in which plumbers can inadvertently cause contamination are:

- incorrect use of oils and jointing compounds such as penetrating oils and linseed-oil-based compounds
- incorrect use of solvents and sealants
- incorrect use of leaded solders and non-potable fluxes
- incorrect flushing and commissioning procedures
- incorrect use of plumbing fittings, such as lead locks, which actively cause galvanic action leading to heavy contamination from lead pipes on domestic properties.

2 **Wastage of water** – Over 2000 litres of water every month are wasted in homes and much more from street leaks caused by burst water mains etc. Every litre of water that is lost through leakage will almost certainly never by recovered and so it is important that leakage is reduced as much as possible. Water wastage occurs through:

- dripping taps – this can waste as much as 5000 litres of water per year and also has the effect of costing more in fuel if the water being wasted is hot water
- running overflows from float-operated valves in cisterns
- leaking service pipes
- water main leaks in the road from the water undertakers supply, which can account for as much as 30 per cent of all water usage in some areas of the UK because of old and defective water mains
- burst pipes that go unreported or unrepaired
- over-use of water in the home. The average usage of water in the UK is 150 litres of water per person per day! This could be cut by 15 per cent through simple water-saving techniques and water-efficient taps and fittings.

F2 WATER EFFICIENCY CALCULATIONS FOR NEW DWELLINGS

On 6 April 2010, Part G (regulation 17.K) of the Building Regulations included for the first time the need for a water efficiency calculation to be carried out for every new dwelling. The calculation is required to show the potential consumption of wholesome water per person per day. The calculation is based on fixtures and fittings in a dwelling and any grey water or rainwater harvesting systems installed. Each dwelling is required to have a maximum consumption of 125 litres per person per

Every drop wasted is a drop less to drink

day, which is equivalent to the BRE's (formerly the Building Research Establishment) Code for Sustainable Homes level 1 standard (see page 106). The table below shows the sustainable home's maximum consumptions.

Performance target	Maximum consumption of potable water (litres/person/day)
17.K Compliance	125
Code for Sustainable Homes (level 1/2)	120
Code for Sustainable Homes (level 3/4)	105
Code for Sustainable Homes (level 5/6)	80

The water efficiency calculator on the following page can be used to calculate the daily consumption for a new dwelling. This ensures that the correct amount of water storage and usage can be accurately assessed before the installation begins. Manufacturers' literature should be consulted for accurate flow rate data for taps, showers, float-operated valves etc. Insert the relevant figures into the table, and then add or multiply the columns together as instructed. For full guidance on water efficiency calculations, visit the web address below, from which this table is sourced.

http://www.planningportal.gov.uk/uploads/br/water_efficiency_calculator.pdf

The water calculator for new dwellings

Installation type	Unit of measure	Capacity/ flow rate (1)	Use factor (2)	Fixed use (litres/ person/ day) (3)	Litres/ person/day = [(1)×(2)]+(3) (4)
WC (single flush)	Flush volume (litres)		4.42	0.00	
WC (dual flush)	Full flush volume (litres)		1.46	0.00	
	Part flush volume (litres)		2.96	0.00	
WC (multiple fittings)	Average effective flushing volume (litres)		4.42	0.00	
Taps (excluding kitchen/utility taps)	Flow rate (litres/minute)		1.58	1.58	
Bath (where shower also present)	Capacity to overflow (litres)		0.11	0.00	
Shower (where bath also present)	Flow rate (litres/minute)		4.37	0.00	
Bath only	Capacity to overflow (litres)		0.50	0.00	
Shower only	Flow rate (litres/minute)		5.60	0.00	
Kitchen/utility room sink taps	Flow rate (litres/minute)		0.44	10.36	
Washing machine	Litres/kg dry load		2.1	0.00	
Dish washer	Litres/place setting		3.6	0.00	
Waste disposal unit	Litres/use	If present = 1 If absent = 0	3.08	0.00	
Water softener	Litres/person/day		1.00	0.00	
	(5)	Total calculated use (litres/person/day) = (sum column 4)			
	(6)	Contribution from grey water (litres/person/day)			
	(7)	Contribution from rainwater (litres/person/day)			
	(8)	Normalisation factor			
	(9)	Total water consumption (Code for Sustainable Homes) = [(5) − (6) − (7)] × (8) (litres/person/day)			
	(10)	External water use			
	(11)	Total water consumption (building regulation 17.K) = (9) + (10) litres per person per day			

003 ENVIRONMENTAL PROTECTION METHODS

SmartScreen Unit 003 interactive activity 4

F3 METHODS FOR REDUCING WATER CONSUMPTION IN THE HOME

To achieve the BRE recommendation of 125 litres per person per day, generally, water consumption needs to be reduced. There are many ways that this can be done:

- using dual-flush WC cisterns
- fitting flow-reducing valves to reduce the flow of water to taps and other terminal fittings
- repairing dripping taps and overflows
- changing existing taps to low water use spray taps
- changing shower heads to low flow type shower heads
- taking more showers instead of baths
- installing rainwater harvesting or grey water recycling systems.

We will look at the last point in more detail.

F4 RAINWATER HARVESTING

Rainwater harvesting has the potential to save a large volume of mains

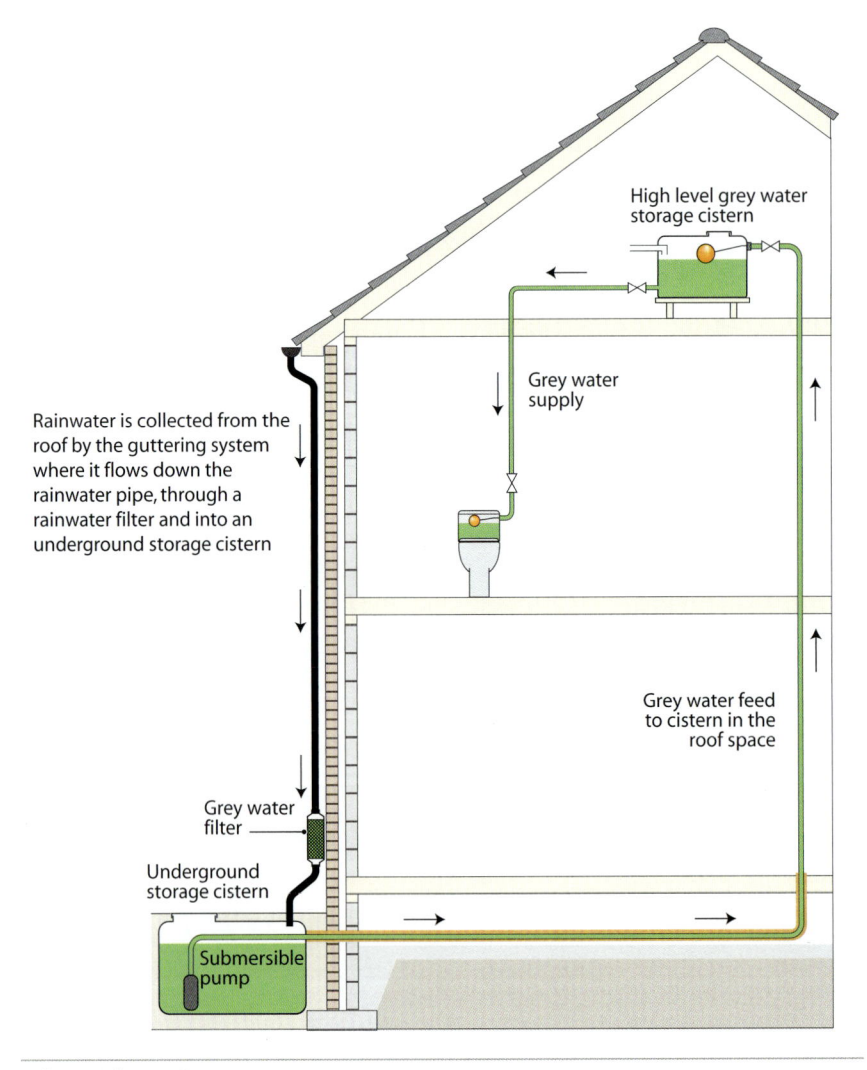

Rainwater harvesting

water and reduce pressure on resources because water that would otherwise be lost can be used to flush toilets, water gardens and feed washing machines.

Rainwater harvesting systems can be installed at domestic or commercial sites and average households can expect to save up to 50 per cent of their water consumption.

Harvesters are usually installed beneath the ground in an underground storage cistern or on the roof of a flat-roofed building, and the water is collected via surface water gullies. A typical four-bedroom house will capture enough water to keep a 5000 litre cistern in use through most of the year.

F5　GREY WATER RECYCLING

Waste water from baths, showers, washing machines, dishwashers and sinks is often referred to as grey water.

About 30 per cent of all water used in the average household is used for WC flushing. The water from baths, showers and wash basins can be collected, cleaned and reused for this purpose.

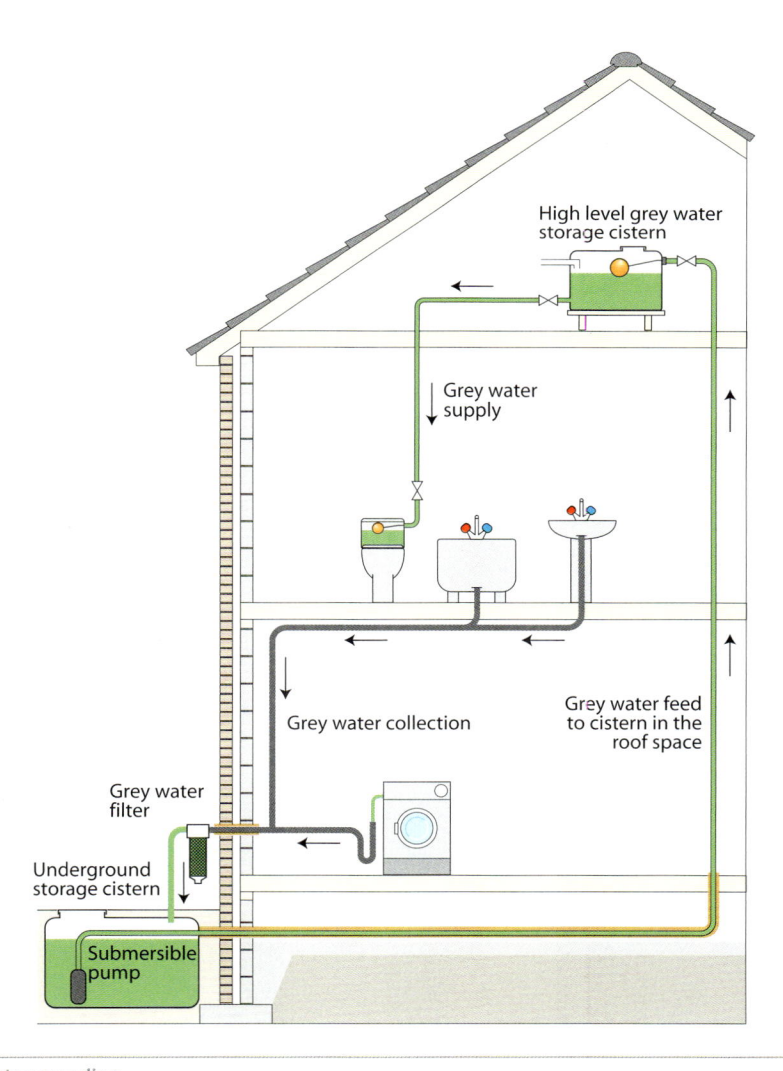

SmartScreen Unit 003 handout 4

SmartScreen Unit 003 worksheet 4

Grey water recycling

Grey water is usually clean enough for use in WCs with only minimal disinfection or microbiological treatment. Problems can arise, however, when the warm grey water is stored because it quickly deteriorates and the bacteria it contains quickly multiply, making the water smell. This can be overcome by filtration and treatment with chemicals.

There must also be a means of protecting the mains water against contamination by backflow from a grey water system in order to comply with the Water Supply (Water Fittings) Regulations 1999 and it must be kept completely separate from all other water supply systems.

CONCLUSION

We have seen as we have worked through this chapter that the plumbing industry is changing. Public opinion and new legislation now dictate that we adopt an environmentally friendly ethos with regard to the installation of the equipment and materials we use. New green technology is becoming more and more prevalent and the option now exists for us to offer these as an alternative to conventional systems. This will mean learning new installation techniques and technical specifications and the adoption of a new 'green' attitude to our trade as we work together to reduce the carbon emissions that do so much harm to our planet. In the coming decades, this will be our challenge.

003 TEST YOUR KNOWLEDGE

SmartScreen Unit 003 revision sample questions

1. Which piece of legislation sets targets for UK carbon emissions?

2. Which Building Regulation document advocates the conservation of fuel and power?

3. Which publication covers all aspects of the installation and design of domestic heating, including gas, oil, solid fuel, hot water, solar thermal, underfloor heating, combined heat and power and heat pumps?

4. Which is the preferred measure of the seasonal efficiency of a boiler installed in typical domestic conditions in the UK?

5. Wind farms are becoming a frequent sight on the UK landscape but what form of energy do the wind turbines they contain generate?

6. Name two fuels that would be classed as low carbon.

7. What is Grade C2 28 second viscosity oil better known as?

8. Which flammable gas is usually present where coal and oil exist?

9. What form of renewable energy harnesses the sun and converts it into usable electricity?

10. Which organisation helps to reduce carbon emissions?

11. What is an EPC and which buildings must have them on display?

12. Which plumbing pipe jointing technique is not considered environmentally friendly?

13. What is mixing hazardous waste with normal waste considered to be?

14. Which dangerous material are you likely to come across that cannot be recycled?

15. What does the following symbol mean?

16. Which Building Regulation Document deals with water efficiency of new dwellings?

17. What is the maximum allowed water usage per person per day?

18. Name the system that collects water from surface water gullies.

19. Name the type of system that reuses bath water to flush WCs.

20. How much of the water used in the home is used for WC flushing?

004 UNDERSTAND HOW TO APPLY SCIENTIFIC PRINCIPLES WITHIN MECHANICAL ENGINEERING SERVICES

Plumbing contains a lot of science. The laws of physics and chemistry are involved in one form or another in almost everything that we do, from the installation of cold water systems and hot water systems to central heating and drainage. It is the science behind these laws that gives us the theory to enable us to design and install these systems correctly and efficiently. In this chapter we will be investigating some of the laws of physics and chemistry that we use in our day-to-day activities.

IN THIS CHAPTER, YOU WILL COVER:

A The SI units of measurement
- A1 SI base units
- A2 Derived SI units

B The properties of materials
- B1 Relative density of solids, liquids and gases
- B2 Principal applications of solid materials
- B3 Principal properties of solid materials
- B4 Oxidisation, corrosion and degradation of solid materials
- B5 Preventing corrosion

C The properties of liquids
- C1 Water
- C2 Refrigerants
- C3 Glycol
- C4 Fuel oils (kerosene)
- C5 Lubricants

D The principal applications of gases
- D1 Types of gases
- D2 Gas laws

E Energy, heat and power
- E1 Temperature
- E2 Measuring temperature
- E3 States of matter
- E4 Sensible and latent heat of liquids and gases
- E5 Methods of heat transfer
- E6 Energy, heat and power calculations

F The principles of force and pressure
- F1 The SI units of force and pressure
- F2 Velocity and acceleration
- F3 Flow rate
- F4 Force
- F5 Pressure
- F6 The relationship between velocity, pressure and flow rate in plumbing systems
- F7 Factors affecting flow rate

G The principles behind simple machines
- G1 Simple machines
- G2 Basic mechanics – moments of a force (torque)
- G3 Centre of gravity
- G4 Action and reaction – Newton's third law of motion
- G5 Equilibrium

H The principles of electricity
- H1 The basic principles of electron flow
- H2 The measurement of electrical flow
- H3 The units of electrical measurement
- H4 The types of electrical current
- H5 Material conductivity and resistance
- H6 Ohm's Law
- H7 Voltage, current and resistance in series and parallel circuits
- H8 The requirements for earthing of electrical circuits

Test your knowledge

A THE SI UNITS OF MEASUREMENT

The SI system of measurement is a universal, unified, self-consistent system of measurement units based on the m/k/s (metre/kilogram/second) system. We will use these measurement units as reference points throughout this chapter. The international system is commonly referred to throughout the world as SI after the initials of Système International Unité. The units can be categorised into two main groups:

1 Base units
2 Derived units.

A1 SI BASE UNITS

Measure of	Base SI unit	Symbol
Length	metre	m
Mass	kilogram	kg
Time	second	s
Electric current	ampere	A
Thermodynamic temperature	kelvin	K

KEY POINT

The derived SI units are combinations of the seven base units by a system of multiplication and division calculations. There are 21 derived units of measurement, some of which have special names and symbols.

A2 DERIVED SI UNITS

Measure of	Base SI unit	Symbol
Area (length × width)	square metre	m^2
Volume (length × width × height)	cubic metre	m^3
Volume of liquid (length × width × height × 1000)	litre	l
Velocity	metre per second	m/s
Acceleration	metre per second squared	m/s^2
Density	kilogram per cubic metre	kg/m^3
Specific volume	cubic metre per kilogram	m^3/kg
Force (mass (kg) × acceleration (m/s^2))	newton (kgm/s^2)	N
Pressure	pascal (N/m^2)	Pa
Energy, work, quantity of heat	joule	J
Power	watt	W
Electric potential	volt	V
Electric resistance	ohm	Ω

B THE PROPERTIES OF MATERIALS

There are many materials that you, as plumbers, will deal with in your working life. Each one will have different characteristics, such as weight, melting point, density and strength. It is important that we know and understand the materials we work with to ensure that the correct material is used for a given application.

B1 RELATIVE DENSITY OF SOLIDS, LIQUIDS AND GASES

Relative density is the ratio of the density of a substance to the density of a standard substance under specific conditions. For liquids and solids, the standard substance is usually distilled water at 4°C. For gases, the standard is usually air at the same temperature and pressure as the substance being measured.

When we talk about a material's relative density, we are basically comparing the mass of that material against water or air. In both cases the water and air have a relative density (or specific gravity) of 1. Below are the relative densities of some common substances we use in the plumbing industry:

KEY POINT
Relative density is also known as 'specific gravity'. This term is usually used when talking about gases.

Solids		
Substance	Relative density	Mass/m^3
Water (1m^3 of water has a mass of 1000kg at 4°C)	1	1000kg
Copper	9	9000kg
Steel	7.48–8.0 (depending on the grade)	7480–8000kg
Lead (milled) (cast)	11.34 11.30	11,340kg 11,300kg
Brass	8.4	8400kg
PVCu	1.35	1350kg
Polypropylene	0.91	910kg
Gas		
Gas	Specific gravity	Mass
Air	1	
Natural gas	0.6–0.7	Lighter than air
Propane	1.5	Heavier than air
Butane	2.0	Heavier than air
Hydrogen	0.069	Lighter than air

B2 PRINCIPAL APPLICATIONS OF SOLID MATERIALS

The solid materials used in the plumbing industry can be classified into three distinct groups:

- those made from metals
- those made from plastics
- those made from ceramics and fireclays.

Metals

Metals are one of the main materials used in the plumbing industry. They can be found in the form of pipes, tubes and fittings, in the manufacture of boilers, radiators and other heating appliances and sundry items such as solder, screws and nails.

Metals can be subdivided into four specific groups:

- **Pure metals** – These are the metals that are derived directly from the ore and contain very little in the way of impurities. Below is a table of the more common metals and the ore from which they are extracted:

Metal	Ore	Main producers	Type	Photograph
Iron	Pyrite Marcasite Hematite Magnetite	England Mexico Brazil Australia	Ferrous	
Copper	Malachite Chalcopyrite Turquoise Azurite	North America Chile Cyprus Canada Germany	Non-ferrous	
Aluminium	Gibbsite Bauxite Cryolite	Brazil Jamaica India Australia Guinea	Non-ferrous	
Lead	Galina Cerussite	England Germany Australia North America	Non-ferrous	
Zinc	Sphalerite Zincite	Australia Canada China Peru North America	Non-ferrous	
Tin	Cassiterite	Malaysia Thailand China Indonesia Bolivia Russia	Non-ferrous	

Corrosion
Any process involving the deterioration or degradation of metal components.

- **Alloys** – An alloy is a mixture of two or more metals. These types of mixed metals are used extensively in the plumbing industry. Alloys used include brass (copper/zinc), bronze (copper/tin), gunmetal (copper/tin/zinc), lead-free solders (nickel/tin or copper/tin) and steel (iron/carbon).
- **Ferrous metals** – Those metals that contain iron such as steel and cast iron. These metals corrode easily because of the formation of ferrous oxide otherwise known as rust.
- **Non-ferrous metals** – These metals do not contain iron. Non-ferrous metals include copper, lead, tin, zinc, aluminium and nickel. Non-ferrous metals do not rust but can corrode over time.

Plastics

Just as plumbers should know their metals, they should also know their plastics if mistakes during installation are to be avoided. There are many different plastics that a plumber uses in his or her day-to-day work for installing hot and cold water supplies, central heating, guttering and rainwater pipes, and above and below ground drainage systems.

There are two basic types of plastics – thermoplastics and thermosetting:

- **Thermoplastics** – A type of plastic made from polymer resins, which becomes liquid when heated and hard when cooled. When frozen, however, a thermoplastic becomes brittle and subject to fracture. These characteristics are reversible and it can be reheated, reshaped and frozen repeatedly. This quality also makes thermoplastics recyclable.

There are many different types of thermoplastics, some of which are used extensively in plumbing systems. Each type varies in crystalline organisation and density. Below is a table of the thermoplastics commonly used in the plumbing industry and what they are used for:

Types of plastic	Uses	Characteristics
PVCu **cuPVC**	Unplasticised polyvinyl chloride. Used extensively for: • cold water mains • cold water installations (chlorinated unplasticised polyvinyl chloride) • solvent-welded and push-fit soil and vent pipes • solvent-welded waste and overflow pipes • underground drainage pipes • gutters and rainwater pipes.	• Not suitable for hot water installations. • Can be solvent welded.
Polyethylene **MDPE** **HDPE**	MDPE (medium density polyethylene) Used for: • underground cold water mains (coloured blue) • cold water storage cisterns • underground gas pipes. HDPE (high density polyethylene) Used for: • underground cold water mains (coloured black).	• Cannot be solvent welded. • Degrades under direct sunlight.

Polypropylene	Used for: • push-fit waste and overflow pipe • cold water storage cisterns.	• Cannot be solvent welded. Slightly greasy to the touch. • Degrades under direct sunlight.
Polybutylene	Used for: • push-fit hot and cold water installations • central heating installations.	• Cannot be solvent welded.
ABS	Acrylonitrile butadiene styrene. Used for: • water supply – potable water for apartments, offices, commercial installations • solvent-welded waste and overflow pipes.	• Can be solvent welded. • Severely degrades under direct sunlight.

- **Thermosetting** – These are rigid plastics, such as polyester and epoxies, which are resistant to higher temperatures than thermoplastics. Once it has set, a thermosetting plastic cannot be remoulded. Its shape is permanent and it does not melt when heated.

SmartScreen Unit 004 handout 2

Ceramics and fireclays

Ceramics and fireclays are mainly used for sanitary appliances and tiles. There are three varieties that plumbers may use widely in their work:

- **Vitreous china** – This is a clay material with an enamelled surface used to manufacture bathroom appliances such as WCs and cisterns, wash hand basins and bidets, as well as soap dishes and other sundry bathroom items. It is made from very watery clay, known as 'slip', which is then spray enamelled and fired in a kiln at high temperature.

- **Fireclay** – This is used primarily for heavy-duty appliances, such as certain types of sinks (Belfast, London, cleaners' and butlers') and shower trays where there is greater risk of damage and a higher temperature of water may need to be used. As with other clays, this clay is highly malleable in raw form. It can be moulded, extruded and shaped by hand. It is also used in the manufacture of building products such as chimney pots.

- **Ceramic tiles** – These have many applications and are used extensively in bathrooms, kitchens, floors and swimming pools. The origin of the tile can be identified by looking at its reverse. This is known as the 'biscuit' of the tile. Tiles made in the UK usually have a white-coloured biscuit, Italian tiles are usually cream in colour, and Turkish and Spanish tiles have a dark red biscuit.

B3 PRINCIPAL PROPERTIES OF SOLID MATERIALS

Solid materials are made up of many molecules. How these molecules are arranged and how they behave under certain conditions will determine their properties. A solid material is assessed by its:

- strength (tensile and compressive)
- hardness

- ductility
- malleability
- conductivity (heat and electricity).

Tensile strength

Broadly speaking, the tensile strength of a material is a measure of how well or badly it reacts to being pulled or stretched. Some materials, such as plastics, will stretch or elongate before breaking. Others, such as metals, will also deform in a similar way but not by as much. Hard materials, such as concrete and brick, will not deform at all but will simply snap.

A tensile strength test is also known as a tension test and is the most fundamental type of mechanical test that can be performed on a material. The tests are simple and relatively inexpensive. By simply pulling on a material under specific conditions, how the material will react to being pulled apart will quickly become apparent. The point at which the material fractures is its tensile strength.

Tensile strength is measured in units of force per unit area. In the SI system, the unit is newton per square metre (N/m² or Pa – pascal).

Compressive strength

Compressive strength is the maximum stress a material can sustain when being crushed. Hard materials, such as concrete or cast iron, will shatter under compressive stress, while others such as plastics and some metals may distort in shape. This is called barrelling.

Compressive strength is calculated by dividing the maximum load by the original cross-sectional area of a specimen in a compression test and is measured in units of force per unit area. In the SI system, the unit is newton per square metre (N/m² or Pa – pascal), as with tensile strength.

Shear strength

Shear strength is the ability to withstand stress caused by a pair of opposing forces acting along parallel lines of action through the material. For example, cutting paper with scissors or ripping a substance apart.

Ductility

Ductility is a mechanical property that describes by how much solid materials can be pulled, pushed, stretched and deformed without breaking. It is often described as the toughness of a material to withstand deformation. In materials science, ductility specifically refers to a material's ability to deform under tensile stress. This is often characterised by the material's ability to be stretched into a wire. Copper is one of the most ductile materials a plumber will use because it is easily bent and softened into various shapes.

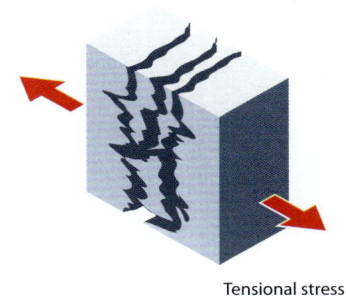

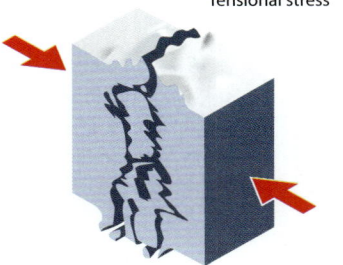

Tensional stress

Compressional stress

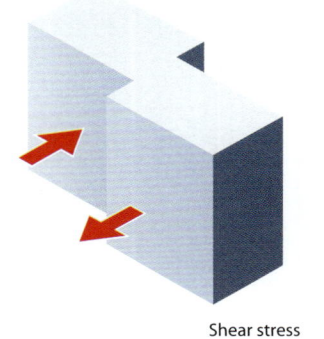

Shear stress

Tensile, compressive and shear stress

Malleability

Malleability can be defined as the property of a material, usually a metal, to be deformed by compressive strength without fracturing. If a metal can be hammered, rolled or pressed into various shapes without cracking or breaking or other detrimental effects it is said to be malleable. This property is essential in sheet metals such as lead that need to be worked into different shapes.

Hardness

Hardness is the property of a material that enables it to resist bending, scratching, abrasion or cutting.

Hardness of minerals can be assessed by reference to the Mohs scale, which ranks the ability of materials to resist scratching by another material. There is a good reason for grouping materials this way. If an unknown material is discovered, one way to find out the nature of the material is by seeing how hard it is.

The Mohs hardness scale starts at 1 for the softest material, and goes up to 10 for the hardest.

Material	Hardness scale
Talc	1
Gypsum	2
Calcite	3
Fluorite	4
Apatite	5
Feldspar	6
Quartz	7
Topaz	8
Corundum	9
Diamond	10

Diamond is the hardest material, which explains why it is used on many cutting edges.

SmartScreen Unit 004 handout 1

SmartScreen Unit 004 worksheet 3

Conductivity

Conductivity is the property that enables a metal to carry heat (thermal conductivity) or electricity (electrical conductivity).

Thermal conductivity

Here, heat is transferred from molecule to molecule through the substance. How fast or how well the heat travels will determine the material's thermal conductivity. For example, metals such as copper transfer the heat quickly and are said to be good conductors of heat, whereas other materials, such as polyurethane, allow the passage of heat only very slowly and so are poor conductors of heat. The inability of polyurethane to allow the passage of heat makes it a very good insulator with the ability to keep heat in. Thermal conductivity is measured in watts per metre kelvin (W/mK).

Electrical conductivity

This is the ability of a material to allow an electrical charge or current to pass through it. Materials that allow an electrical current to flow freely, such as copper and gold, are known as good conductors, whereas those that do not allow the passage of an electrical current, such as wood, ceramics and PVC, are known as insulators.

B4 OXIDISATION, CORROSION AND DEGRADATION OF SOLID MATERIALS

All solid materials will corrode or degrade over time. The amount that materials corrode or degrade will depend upon the material's resistance and the environment in which the material exists.

Oxidisation of metals

Metals are oxidised by the presence of oxygen in air. This process is a form of corrosion. Electrons jump from the metal to the oxygen molecules. The negative oxygen ions that are formed penetrate into the metal, causing the growth of an oxide on the metal's surface. As the oxide layer increases, the rate of electron transfer decreases. Eventually the corrosion stops and the metal becomes passive. However, the oxidisation process may possibly continue if the electrons succeed in entering the metal through cracks, pits or impurities in the metal, or if the oxide layer is dissolved.

Corrosion

Corrosion is the main reason for metals deteriorating. Most metals will corrode on contact with water (and moisture in the air), acids, salts, oils and other solid and liquid chemicals. Metals will also corrode when exposed to some gases, such as acid vapours, ammonia gas and any gas containing sulphur.

Corrosion specifically refers to any process involving the deterioration or degradation of metal components. Most commonly known is the rusting of steel and iron, where the formation of ferrous oxide occurs. The corrosion process is usually electrochemical.

> **KEY POINT**
>
> The electrochemical process involves the passage of a small electrical charge between two metals that are at opposite ends of the electromotive series of metals. The stronger, noble metal is called the cathode and the weaker metal is known as the anode. When these two dissimilar metals are placed in an electrolyte such as water, an electric charge is generated and the anodic metal is 'eaten' away by the cathodic metal. A by-product of this reaction is the generation of hydrogen gas. The process accelerates when heat is present.

When rusting occurs, the metal atoms are exposed to an environment containing water molecules. Here they give up electrons and become positively charged ions.

This effect can occur locally to form a pit or a crack or it can extend across a wide area to produce general corrosion.

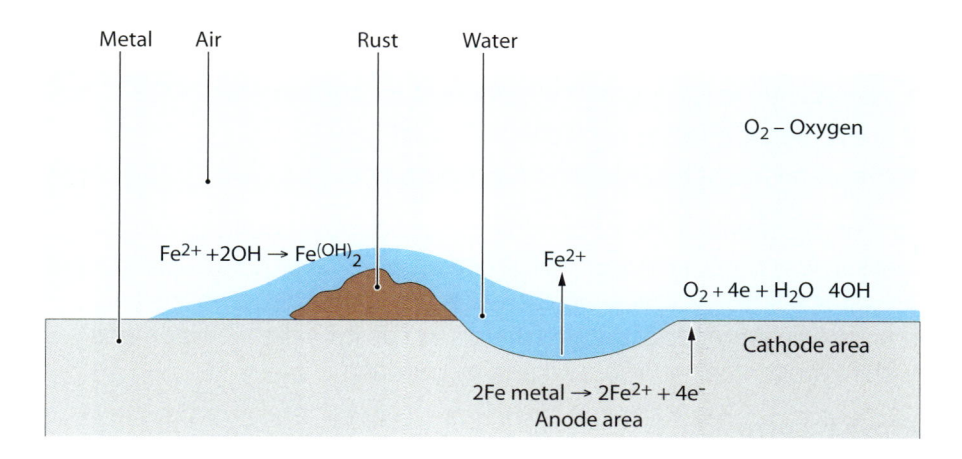

How rust is formed

There are many other forms of metal corrosion that can occur within plumbing and heating systems, including:

- dezincification
- galvanic corrosion
- erosion corrosion
- pitting corrosion.

Dezincification of brass

Brass is an alloy mixture of copper and zinc. Dezincification of brass is a form of selective corrosion, often referred to as de-alloying, that happens when zinc is leached out of the alloy leaving a weakened and brittle porous copper fitting. This commonly occurs in chlorinated tap water or in water that has high levels of oxygen. Signs of dezincification are a white, powdery zinc oxide as a coating on the surface of the fitting, or if the yellow brass turns a shade of red. Selective corrosion can be a problem because the weakening of the fitting leaves it vulnerable to possible failure and eventual leaks.

Dezincification

Electrolyte
A fluid that allows the passage of electrical current, such as water. The more impurities (such as salts and minerals) there are in the fluid, the more effective it is as an electrolyte.

**CATHODIC
(most noble)**

**Copper
Lead
Tin
Nickel
Iron
Chromium
Zinc
Manganese
Aluminium
Magnesium**

**ANODIC
(least noble)**

Electromotive series of metals

Galvanic corrosion

Galvanic corrosion (also called galvanic action, 'dissimilar metal corrosion' and often wrongly termed 'electrolysis') occurs when two dissimilar metals are in contact with each other through the presence of an electrolyte. Metals are graded through the electromotive series (also known as the electrochemical series) of metals (see diagram, left). The further apart the metals are in the series, the stronger the chance of galvanic corrosion.

For galvanic corrosion to occur, three conditions must be true:

1 electrochemically opposed metals must be present
2 these metals must be in electrical contact
3 the metals must be exposed to an electrolyte.

One of the metals is the more noble cathodic metal and the other is the weaker, less noble anodic metal. When an electrolyte is introduced, such as water, a small electrical DC current is generated between the two metals. The stronger of the two metals will destroy the weaker metal with hydrogen being produced as a by-product.

Erosion corrosion

Erosion corrosion occurs in tubes and fittings because of the effects of fast-flowing fluids and gases. The increased turbulence caused by pitting on the internal surfaces of a tube can result in rapidly increasing erosion rates and eventually a leak. Erosion corrosion can also be encouraged by poor workmanship. For example, burrs left at cut tube ends can cause disruption to the smooth water flow and this can cause localised turbulence and high flow velocities, resulting in erosion corrosion.

The effects of erosion corrosion

Pitting corrosion

Pitting corrosion is the localised corrosion of a metal surface and is confined to a point or small area and takes the form of cavities and pits, which may be covered with products of the corrosion. It is generally a form of electrolytic corrosion. Pitting is one of the most damaging forms of corrosion in plumbing, especially in central heating radiators, as it is not easily detected or prevented.

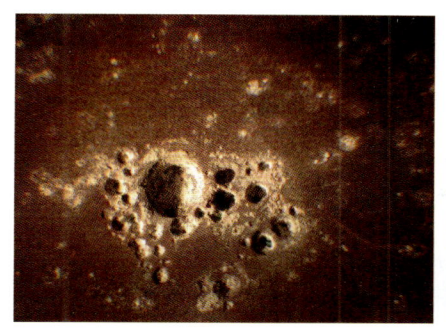

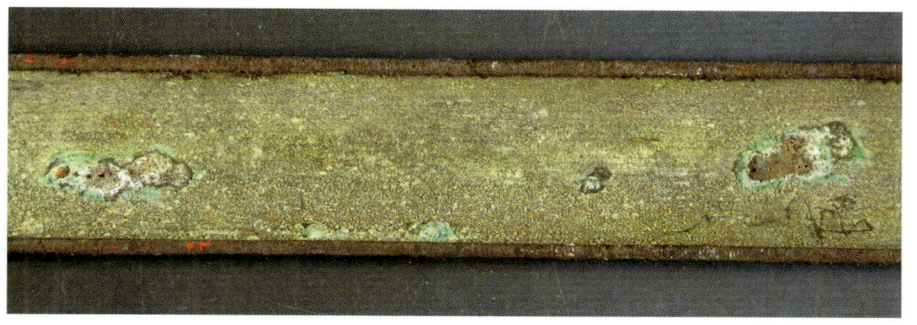

The effects of pitting corrosion

Degradation of plastics

The use of plastics is becoming common in the plumbing industry. Everything from hot and cold water services to central heating and drainage can now be installed in some form of plastic material. However, problems can occur with plastics under certain conditions. Degradation of plastics can occur from a variety of causes, such as:

- heat (thermal degradation)
- light (photo degradation)
- oxygen (oxidative degradation)
- UV (ultraviolet) degradation.

Heat (thermal degradation)

One of the limiting factors when using plastics in high temperature applications is their tendency not only to soften but also to thermally degrade. In some instances, thermal degradation can occur at temperatures much lower than those at which mechanical failure is likely to occur.

All plastics experience some form of degradation during their life. The chemical reactions that occur with thermal degradation lead to both physical and optical changes, such as:

- reduced ductility, making the plastic brittle and easily fractured
- a condition that makes the plastic soft and chalky in texture, usually as a result of exposure to direct sunlight for long periods of time
- colour changes
- cracking.

Light (photo degradation)

This occurs due to the action of light, whether from natural sunlight or electrical fluorescent lighting, and generally causes a yellowing of PVC. It is generally more pronounced on light-coloured PVC, such as PVC window frames, but it can affect all colours.

Oxygen (oxidative degradation)

This is decomposition of the plastic due to the presence of oxygen, which alters the plastic properties. Colour change is often the first sign

> **KEY POINT**
> Photo degradation takes place in direct light, even electric light, whether heat is present or not. Ultraviolet degradation takes place in daylight, whether the sun is present or not. Its effects occur even on cloudy days and as such it is generally down to the climate.

of oxidative degradation, coupled with a change in flow, mechanical and electrical properties of the plastic (even if the colour change is not noticeable). Polypropylene, polyethylene and acrylonitrile butadiene styrene (ABS) are the plastics most severely affected. Polyvinyl chloride, however, is unaffected by oxidative degradation.

UV (ultraviolet) degradation

This is very similar to photo degradation but has much more severe effects. Most plastics are vulnerable to degradation by the effects of direct exposure to the ultraviolet part of the daylight spectrum. Ultraviolet solar radiation is present even on cloudy days. When ultraviolet attack occurs, the colour of the plastic may change and its surface will become brittle and chalky. This can happen over a very short time period and will lead to cracking and eventual failure.

Polypropylene waste pipes and medium density polyethylene water pipes are adversely affected by UV degradation with acrylonitrile butadiene styrene (ABS) pipework and fittings being severely compromised with prolonged exposure to the UV daylight spectrum.

B5 PREVENTING CORROSION

Corrosion is one of the most destructive processes to plumbing and heating systems but there are methods we can employ that prevent and protect from corrosion:

- Galvanisation is one method of protecting steel from rusting by coating it with a thin layer of zinc.

- Greasing and oiling are commonly used to prevent rusting. The grease and oil prevents water and moisture penetration.

- Chrome plating and anodising prevent corrosion of metals by coating the metal, creating a barrier between the metal and the corrosive environment.

- Wet central heating systems can be protected from corrosion by the use of fluid chemical additives known as corrosion inhibitors, which are mixed with the system water.

- Plastics can be protected from the effects of UV light by painting.

- Sacrificial anodes (magnesium rods) placed inside hot water storage cylinders protect the cylinder from electrolytic corrosion.

Anodising
Coating one metal with another by electrolysis to form a protective barrier from corrosion.

C THE PROPERTIES OF LIQUIDS

The plumbing industry is primarily concerned with liquids in one form or another, with water being the most common fluid we deal with. Liquids you may come across in your working life include:

- water
- refrigerants
- glycols and antifreeze

- fuel oils
- lubricants.

C1 WATER

Water is the most abundant compound on earth. It covers seven tenths of the earth's surface and is the key to life on earth. Water has many uses, including hot and cold water supplies and wet central heating systems.

The properties of water

- Water is a colourless, odourless and tasteless liquid. Any taste it does have comes from the minerals that may be dissolved in it, which often explains why water tastes different in different parts of the country.

- Water can exist in all three states of matter: liquid (water), solid (ice) and gas (steam).

- Water has a maximum density of 1000kg per cubic metre (m^3) at 4°C. At this temperature, water is at its densest. When the temperature of water is either raised or lowered from 4°C, water loses density. This peculiar behaviour is known as the 'anomalous expansion' of water. At 100°C, water has a density of 958kg/m^3 and at 0°C, its density is 915kg/m^3. This can also be expressed as a percentage: when heated, water expands by 4 per cent, and when cooled it expands by 10 per cent. When water is turned to steam, it expands by 1600 times, so 1m^3 of water will transform into 1600m^3 of steam!

- The boiling point of water at sea level is 100°C. If the pressure is raised from this, the boiling point increases. At 1 bar pressure, the boiling point of water is 120°C. Similarly, if the pressure is lowered, then the boiling point decreases. At the top of Mount Everest, the boiling point of water is only 69°C.

- Water freezes at 0°C – but again, pressure can affect this. If the pressure increases then the freezing point is lower. Dissolved minerals can also affect the freezing point.

- The relative density of water is 1. This is the measurement that all other solids and liquids are measured against.

- The specific heat capacity of water is 4.19kJ/kgK. Specific heat capacity of a substance is the amount of heat required to raise the temperature of 1kg of the substance by 1°C (or by 1K). In the case of water, it takes 4.19kJ of heat to raise 1kg of water by 1°C.

- Pure water is a poor conductor of electricity – it is the presence of dissolved minerals that makes water a good conductor of electricity. Seawater, for example, is a very good conductor of electricity because of the dissolved salts and minerals it contains.

- Water is a poor conductor of heat, compared to most metals. In fact, water is a better insulator of heat than it is a conductor. That is why it takes so much energy to raise the temperature of water through 1°C (see specific heat capacity).

KEY POINT

The effects of the changes in density of water can benefit water heating by creating heat circulation by convection. We will deal with heat transfer through water on page 177.

Pressure		Boiling point of water
bar	kPa	°C
0	0	100
1	100	120.42
2	200	133.69
3	300	143.75

- Water is known as the 'universal solvent' – almost all substances dissolve in water to a certain extent. Because of this, it is almost impossible to get chemically pure water on earth.
- Water is classified as being hard or soft. The hardness and softness of water is measured by the pH value. This is a measure of the acidity or alkalinity of the water. See the table below:

Type of water	pH value	Base	Notes
Neutral	7	N/A	Neutral water is neither soft nor hard.
Soft	Below 7	Acidic	Water is made soft by the presence of CO_2. It is particularly destructive to plumbing systems containing lead as it can dissolve the lead, making the water contaminated. Because of its lead-dissolving capability, soft water is known as 'plumbo-solvent'. Soft water lathers soap easily.
Temporary hard water	Above 7	Alkali	Temporary hard water contains calcium carbonate ($CaCO_3$), otherwise known as limestone. This kind of water can be softened by boiling but this leaves behind limescale residues, which can block pipes and other plumbing fittings and appliances. When water reaches 65°C, the calcium in the water reforms in a process known as precipitation, causing scaling within plumbing systems. Lathering of soap is difficult.
Permanently hard water	Above 7	Alkali	Permanently hard water contains magnesium and calcium chlorides and sulphates in the solution. It cannot be softened by boiling.

- Water goes through several stages to be turned into steam. At atmospheric pressure (0 bar/0kPa), the boiling point of water is 100°C. To raise the temperature of water from 0 to 100°C takes 419kJ/kg of energy. To turn boiling water at 100°C to steam at 100°C takes a further 2257kJ/kg of energy. At this point, the steam is said to be saturated steam. In other words, it is saturated with heat. The total heat, therefore, to turn water at 0°C to steam at 100°C takes 2676kJ/kg of heat energy (419 + 2257) and it will only remain at 100°C while liquid water is present. When all the water has been evaporated to steam, then its temperature can be increased beyond the 100°C point. This is known as superheated (or dry) steam. Increasing the temperature of steam above 100°C at atmospheric pressure can only take place without the presence of liquid water and in a pure gaseous state unless we raise the pressure of the water or the saturated steam.

Capillary attraction

Capillary attraction is the process where water (or any fluid) can be drawn upwards through small gaps against the action of gravity. The wider the gap, the less capillary attraction takes place. It is of particular interest to plumbers as it has the ability to cause problems within some plumbing systems, such as the following:

- It can cause water to be drawn up underneath tiles and roof weatherings resulting in water leaks inside the building.

- It can initiate water trap seal loss in above-ground drainage systems. In this instance there are two forces at work – capillary attraction and siphonic action.

Conversely, it is also the process plumbers use to make soldered capillary joints on copper tubes and fittings.

Before capillary attraction can take place, two processes need to be present. These are adhesion and cohesion.

Water is fluid because of cohesion. This is the way in which the water molecules 'stick' together to form a mass rather than staying as individuals. This is because water molecules are attracted to other water molecules. The cohesive quality gives water a slight film on its surface, which is known as the surface tension.

Water is also attracted to other materials and so it tends to stick to whatever it comes into contact with. This is known as adhesion. When water is placed in a vessel or a glass, the adhesion qualities of the water give it a slightly curved appearance. This is known as the meniscus and can be convex (outward curve) or concave (inward curve).

Capillary attraction

C2 REFRIGERANTS

Refrigerants are chemicals that are used in both liquid and gas states. All refrigerants boil at extremely low temperatures, well below 0°C.

When a refrigerant gas is compressed, it changes its state to a liquid. During this process a lot of heat and pressure are generated. When the pressure is released quickly, it generates cold. A refrigerant's ability to change its state quickly with such wide temperature changes allows it to be used in refrigeration plants, air conditioning systems and heat pumps. The process is known as the vapour compression refrigeration cycle.

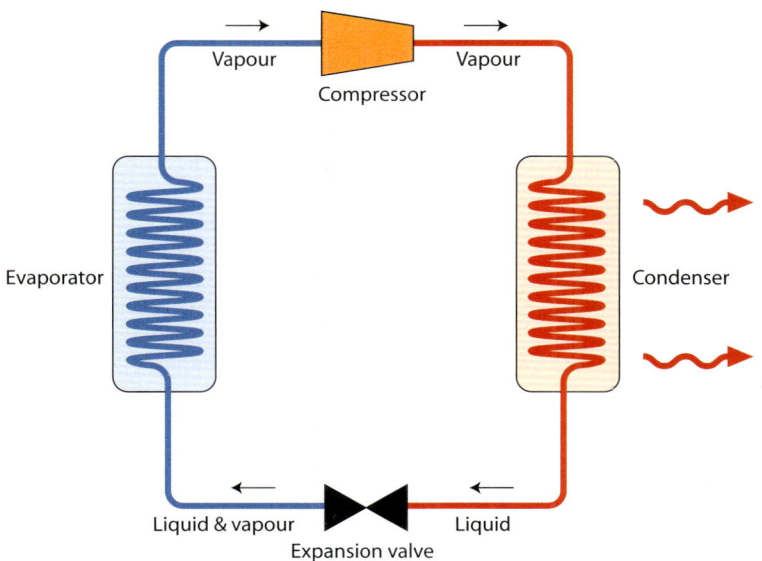

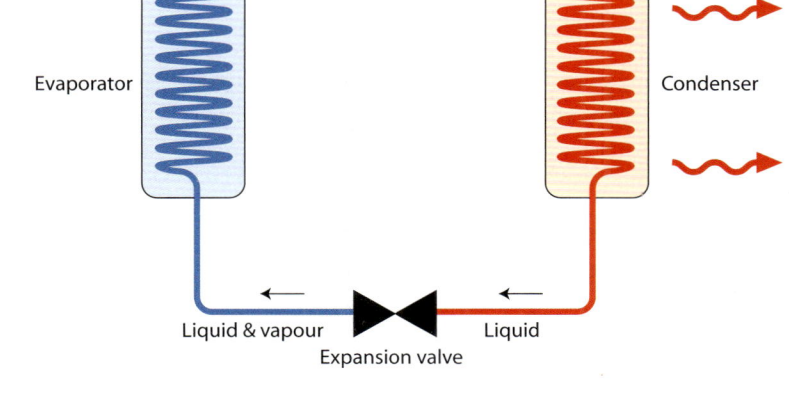

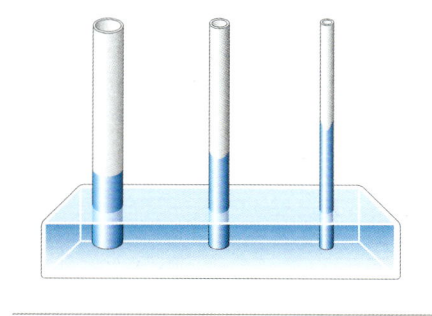

Vapour compression refrigeration cycle

The refrigerant vapour enters the compressor, which compresses it, thus generating heat. The compressed vapour then enters the condenser where the useful heat is removed and the vapour condenses to a liquid refrigerant. From here the liquid refrigerant then passes into the expansion valve where rapid expansion takes place converting the warm liquid into a super-cold vapour/liquid mix. This creates the refrigeration effect. The vapour/liquid mix passes through the evaporator, where final expansion to a vapour takes place. This then enters the compressor for the cycle to begin again.

C3   GLYCOL

Glycol is the chemical solution used as antifreeze in solar hot water systems. It is used for protecting solar panels from freezing during the winter when it is mixed with water in the sealed solar panel circuit. It is available in two forms – propylene glycol and ethylene glycol. Propylene glycol is the preferred chemical for solar panels as ethylene glycol is highly toxic. The antifreeze should be checked regularly as the anti-freezing capability diminishes with time, and the solution can become corrosive with age.

C4   FUEL OILS (KEROSENE)

Kerosene is a fuel oil that is used with most domestic oil-fired boilers (see Chapter 003, page 112). Kerosene is a thin, clear liquid formed from hydrocarbons, and has a density of 0.78–0.81 g/cm^3. It is made from the distillation of petroleum at temperatures between 150°C and 275°C. The flash point of kerosene is between 37 and 65°C and it will spontaneously combust at 220°C. The heat of combustion of kerosene is 43.1 MJ/kg, and its higher heating value is 46.2 MJ/kg.

C5   LUBRICANTS

A lubricant is a substance, often a liquid or a type of grease. It is introduced between two moving surfaces to reduce the friction, thus improving efficiency and reducing wear. There are many types of lubricants in the plumbing industry:

- **silicone grease and spray** – used for general lubrication of plumbing parts for water and drainage systems; also used when jointing push-fit plastic pipe systems to lubricate the rubber seals

- **graphite paste** – used for lubrication of gas taps

- **cutting oils** – used when threading low carbon steel pipe, helping to prevent overheating of the cutting dies

- **penetrating oils** – used to help loosen tight and rusted joints.

D   THE PRINCIPAL APPLICATIONS OF GASES

In this section of the chapter, we will look at the principal uses of gases in the building services industry, together with their properties and the scientific laws that apply to them.

D1   TYPES OF GASES

There are six principal gases in the building services industry. We will look at each of these briefly below.

Air

Air has limited uses within the plumbing industry:

- It can be used as a heating medium in warm air heating systems. Here, a warm air heater, usually fired by gas, warms the air. The warm air is distributed to the property by means of a fan.
- It can be used as a pressure charge in expansion vessels. These are usually installed in sealed heating systems and some unvented hot water storage vessels.
- Air at high pressure can be used to clear blocked drains.

Steam

Once the preferred method of heating, the use of steam has declined over recent years. However, because of new, more efficient system designs, steam is being used as a heating medium for:

- new combined heat and power applications – steam can be used to generate electricity and warm properties in district heating systems
- electricity generation
- hot water production using large hot water calorifiers
- heating systems – the steam is used instead of water in the heat emitters.

Liquid petroleum gas

Liquid petroleum gas (LPG) can be used for heating appliances such as boilers, cookers and fires. It is also used in plumbers' blowtorches for soldering capillary fittings. There are two basic types:

- **butane** – used mainly as a camping gas
- **propane** – the most widely used LPG in the building services industry.

Natural gas

This is the most widely used fuel in the UK. Natural gas has many applications, both domestic and industrial. It is used as a fuel for:

- gas fires
- cookers
- room heaters
- condensing central heating boilers
- water heaters
- electricity generation
- industrial heating and processes.

Carbon dioxide

Carbon dioxide is used as a freezing agent with pipe freezing kits, and also in fire extinguishers.

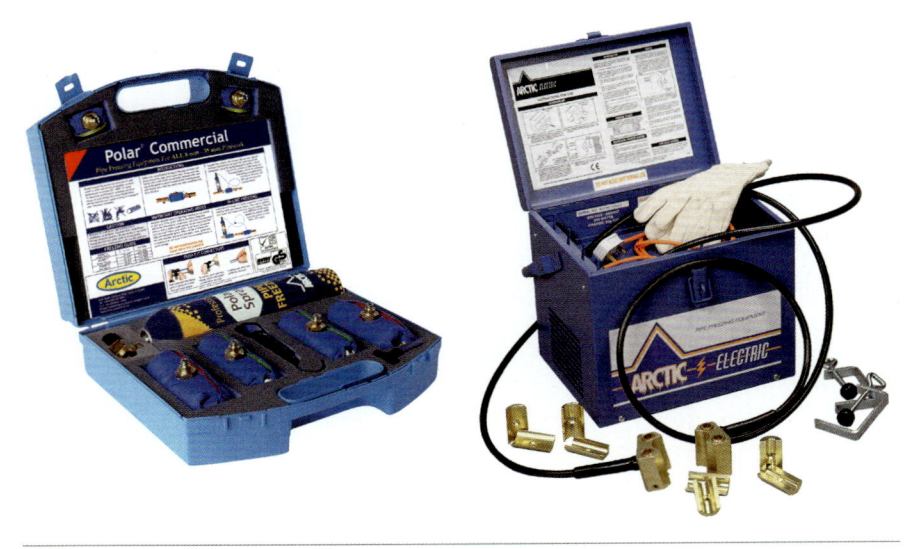

A typical pipe freezing kit and electrical freezing kit

The sixth gas is refrigerant gas. For details on this, see section C2 on page 167.

D2 GAS LAWS

Gases behave very differently from solids and liquids. Unlike solids and liquids, gases have neither a fixed volume nor a fixed shape. They are moulded completely by the container in which they are held. There are three variables by which we measure gases:

- **Pressure** – This is the force that the gas exerts on the walls of its container and is equal on all sides of the container. For example, when a balloon is inflated, the balloon expands because the pressure of air is greater on the inside of the balloon than the outside. The pressure is exerted on all surfaces of the balloon equally and so the balloon inflates evenly. If the sealed end of the balloon is opened, the air will move from the area of high pressure (inside the balloon) to the area of low pressure (outside the balloon) and the balloon will be forced away at speed. Pressure is measured as force per unit area. The standard SI unit for pressure is the pascal (Pa). However, in plumbing it is more likely that pressure will be measured in bar pressure (1 bar = 100kPa) or millibar (1mbar = 100Pa).

- **Volume** – The volume of gas in a given container is affected by temperature and pressure. Pressure is constant if temperature is constant. If temperature is increased, then both the volume and pressure increase.

- **Temperature** – An important property of any gas is its temperature. The temperature of a gas is a measure of the mean kinetic energy of the gas. The gas molecules are in constant random motion (kinetic energy). The higher the temperature, the greater the kinetic energy and the greater the motion. As the temperature falls, the kinetic energy decreases and the motion of the gas molecules diminishes.

Charles's Law

Charles's Law was discovered by Jacques Charles in 1802. It states that the volume of a quantity of gas, held at constant pressure, varies directly with the Kelvin temperature. But what does that mean?

Gases expand when they are heated up and they contract when they are cooled down. In other words, as the temperature of a quantity of gas at constant pressure increases, the volume increases. As the temperature goes down, the volume decreases.

It can be explained with the following analogy:

If a sealed copper pipe was pressurised to, say, 20mbar at room temperature and then placed in direct sunlight where the pipe could warm up, then the pressure inside the pipe would rise. The rise in pressure would be inversely proportional to the rise in temperature. If the pipe was allowed to cool down to room temperature, then it would return to its original pressure. This concept is particularly important when testing gas installation pipework as the pressure in the pipework can rise when subjected to an external temperature, such as sunlight.

The mathematical expression for Charles's Law is shown below:

$V_1/T_1 = V_2/T_2$
where:
V = Volume   T = Temperature

Boyle's Law

Boyle's Law states that the volume of a sample of gas at a given temperature varies inversely with the applied pressure. In other words, if the pressure is doubled, the volume of the gas is halved.

The principle of Boyle's Law can be found in a simple child's balloon. If the balloon is inflated to a set pressure and then squeezed, the pressure inside increases as the space inside the balloon decreases. If the space inside the balloon were halved, then the pressure would double.

Boyle's law can also be expressed as:

Pressure multiplied by volume is constant for a given amount of gas at constant temperature.

To put this in mathematical terms:

$P \times V$ = constant (for a given amount of gas at a fixed temperature).
Since $P \times V = K$, then:
$P_i \times V_i = P_f \times V_f$
where:

V_i = initial volume P_i = initial pressure
V_f = final volume P_f = final pressure
K = constant

The table below illustrates the point:

Sample of gas at constant temperature and varying pressure				
Test	Pressure	Volume	Formula	Calculation
1	100kPa	50cm^3	P × V = K	100 × 50 = 5000
2	50kPa	100cm^3	P × V = K	50 × 100 = 5000
3	200kPa	25cm^3	P × V = K	200 × 25 = 5000
4	400kPa	12.5cm^3	P × V = K	400 × 12.5 = 5000
5	25kPa	200cm^3	P × V = K	2.5 × 2000 = 5000

An example of Boyle's Law can be seen in expansion vessels that use air under pressure to cushion the effects of water as it expands. When the water is heated it expands into the expansion vessel and the pressure rises.

E ENERGY, HEAT AND POWER

The relationship between energy, heat and power is such that it is almost impossible to have one without the other two. Below is a list of units for energy, heat and power:

- **The unit of heat** – The joule is a unit of heat. The amount of heat energy required to raise the temperature of 1g of water from 0°C to 1°C is 4.186 joule (equals 1 calorie).

- **The unit of energy** – Also the joule (see above).

- **The unit of power** – The watt is the SI unit for power. It is equivalent to one joule per second (1J/s), or in electrical units, one volt ampere (1V·A).

- **Specific heat capacity** – The specific heat capacity of a substance is the amount of heat required to change a unit mass of that substance by one degree in temperature. It is measured in kilojoules per kilogram per degree Celsius (kJ/kg/°C).

Heat energy is transferred because of temperature difference. For example, heat passes from a warm body at high temperature to a cold body at low temperature. The transfer of energy as a result of the temperature difference alone is referred to as heat flow. The watt, which is the SI unit of power, can be defined as 1 joule per second (J/s) of heat flow.

In the following part of the chapter, we will investigate the energy/heat/power/temperature relationship and its implications for the building services industry.

E1 TEMPERATURE

Temperature is simply the degree of hotness or coldness of a body or environment and is expressed in terms of units or degrees designated on a standard scale, usually Celsius (centigrade) (°C) or Kelvin (K).

- **Celsius (°C)** – This scale, using increments of 1 degree (1°), is the most widely used by the building services industry. In simple terms, it has a zero point (0°C), which corresponds to the temperature at which water will freeze. When this scale is used, the degree symbol (°) should accompany it, eg 21°C.

- **Kelvin (K)** – This has the same increments as the Celsius scale but has a minimum temperature that corresponds to the point at which all molecular motion would stop. This temperature is often called absolute zero and is equal to −273°C. Therefore:

  −273°C = 0K; or,

  temperature K = temperature °C + 273

The degree symbol (°) is not used when using the Kelvin scale, ie 21K. The two scales (°C and K) are for the most part, interchangeable. The SI unit of temperature is the Kelvin, but when discussing temperature difference, Celsius or Kelvin may be used and since both scales correspond with each other, temperature difference is uniform. In other words, a 1°C temperature difference is equal to 1K temperature difference.

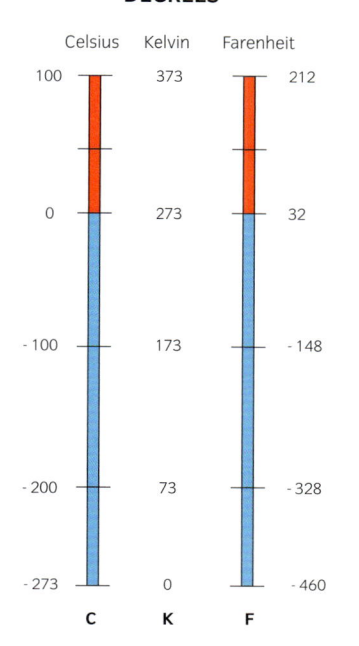

E2 MEASURING TEMPERATURE

Many methods have been developed for measuring temperature. Most rely on measuring some physical property of a working material that varies with temperature.

Glass thermometer

This is one of the most common devices for measuring temperature. It consists of a glass tube filled with mercury or some other liquid. Temperature increases cause the fluid to expand, so the temperature can be determined by measuring the volume of the fluid. These thermometers are usually calibrated so that the temperature can be read by observing the level of the fluid in the thermometer.

Gas thermometer

This type of thermometer measures temperature by the variation in volume or pressure of a gas.

Digital thermometer

These are probably the most common thermometers used in the plumbing industry. Dual digital thermometers can read two temperatures simultaneously, instantly giving the temperature difference between two points. This is essential when benchmarking central heating boilers for reading the temperature of both flow and return pipes.

KEY POINT

Celsius is named after the Swedish astronomer, Anders Celsius (1701–1744). The Kelvin scale is named after the Belfast-born engineer and physicist William Thomson, First Baron Kelvin (1824–1907).

DEGREES

Celsius	Kelvin	Farenheit
100	373	212
0	273	32
-100	173	-148
-200	73	-328
-273	0	-460

Glass thermometer

SmartScreen Unit 004 worksheet 8

Thermocouple thermometer

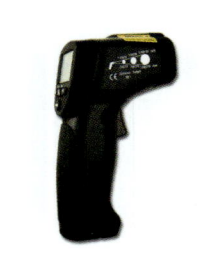

Infrared thermometer

Thermocouple
This device is a connection between two different metals that produces an electrical voltage when subjected to heat. This senses a temperature difference. Thermocouples are a widely used type of temperature sensor for measurement and control when used with digital thermometers. They can also be used to convert heat into electrical power.

Thermistor
Thermistors are resistors that vary with temperature. They are constructed of semiconductor material with a resistivity that is especially sensitive to temperature. When the temperature is measured, the resistance of the thermistor responds in a predictable way.

Infrared thermometer
These use infrared energy to detect temperatures. They detect actual energy levels by the use of an infrared beam and so the thermometer does not actually need to touch the surface for an accurate temperature measurement.

E3 STATES OF MATTER

Everything around us is made up of matter that can exist in three states: solid, liquid and gas. Each of the phase changes is associated with either an increase or decrease in temperature. For example, if heat energy is applied to ice, it melts to form water. And if more heat energy is applied to the water, it reaches its boiling point, where it vaporises, evaporating to steam. The process can also work in reverse. When the heat is given up by the steam, it condenses back to water. Each of these phase changes is given a name:

- ice (solid) to water (liquid) is called melting
- water (liquid) to steam (gas) is called evaporation/vaporisation
- steam (gas) back to a water (liquid) is called condensation
- water (liquid) to ice (solid) is called freezing/solidification
- ice (solid) to steam (gas) is known as sublimation
- steam (gas) to ice (solid) is known as deposition.

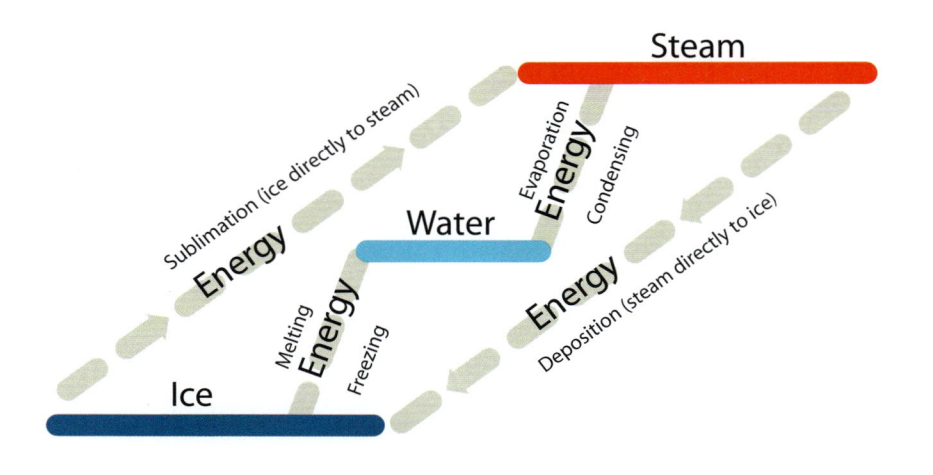

States of matter

E4 SENSIBLE AND LATENT HEAT OF LIQUIDS AND GASES

Sensible heat

When heat is applied to a liquid, its temperature will rise as heat is added without a change of state. The resulting increase in heat is known as 'sensible' heat. This process can be reversed. When heat is removed from the liquid and its temperature decreases, the heat that is removed is also called sensible heat. In other words, sensible heat is any heat that causes a change in temperature without a change of state.

Latent heat

Changes of state, as we have already seen, are the result of a change in energy. Solids can become liquids, liquids can become gases and each change of state is reversible. The heat that causes any change of state is known as latent heat. Latent heat, however, does not affect the substance's temperature. For example, water boils at 100°C. The heat required to raise the water to its boiling point of 100°C is sensible heat. The heat required to keep it boiling at 100°C is latent heat.

Look at the image below, which shows how sensible and latent heat work together.

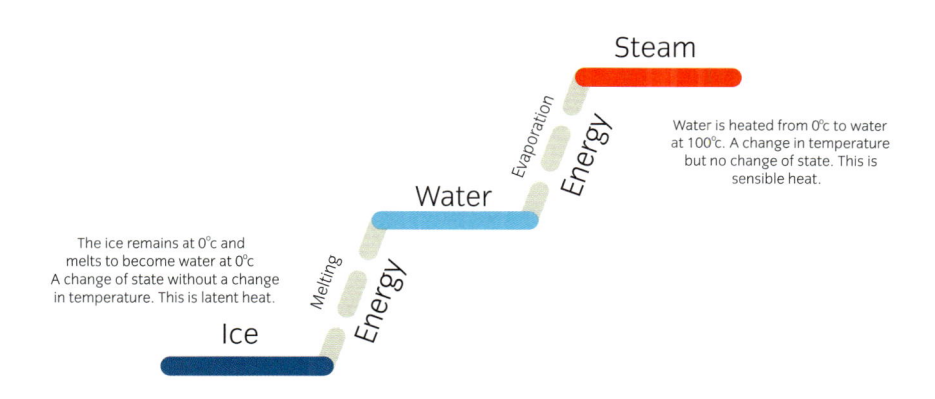

The ice remains at 0°C and melts to become water at 0°C. A change of state without a change in temperature. This is latent heat.

Water is heated from 0°C to water at 100°C. A change in temperature but no change of state. This is sensible heat.

Sensible and latent heat

E5 METHODS OF HEAT TRANSFER

So far we have investigated temperature and heat and how this affects the different states of matter. Now we will consider the methods of heat transfer. There are three methods by which heat can be transferred through a substance or from one substance to another:

1 conduction
2 convection
3 radiation.

Conduction

Conduction happens when heat travels through a substance by the heat being transferred from one molecule to another.

Consider a piece of copper tube. If heat is applied to one end, before long the heat will have travelled through the material so that the effects of the heat will be felt at the other end. This occurs because kinetic energy in the form of heat is being passed from one copper molecule to another very quickly. When the copper is cold, the atoms move very slowly. As heat is applied, these atoms gain speed and collide with the slower, cooler atoms. In this way, some of the kinetic energy is passed through the material, the slow atoms becoming faster and colliding with other slow atoms and so on.

Not all substances, however, transfer heat at the same rate. Some materials such as plastic or wood are very poor at transferring heat, with little or no heat transference occurring at all.

Most metals are very good conductors of heat and because of this they are also very good at conducting electricity. Materials that do not transfer heat well, such as plastic, are known as insulators.

The rate at which a material will transfer heat is known as the coefficient of thermal conductivity, which is measured in W/m/K. It can be found by using the following equation:

$$\text{Thermal conductivity} = \frac{\text{heat} \times \text{distance}}{\text{area} \times \text{temperature difference}}$$

Below is a table of some common substances with their coefficient of thermal conductivity:

Material	Thermal conductivity W/m/K
Silver	406.0
Copper	385.0
Gold	310
Aluminium	205.0
Brass	109.0
Steel	50.2
Lead	34.7
Concrete	0.8
Polyethylene HD	0.5
Wood	0.12–0.04
Polystyrene expanded	0.03

From the table you can see that silver is the best conductor of heat, with copper being a close second.

The poorest conductor of heat is expanded polystyrene, which is an excellent insulator of heat.

Convection

Convection is heat transfer through a fluid substance, such as water or air. Convection occurs because heated fluids (due to their lower density) rise, and cooled fluids fall.

As water or air is heated it expands, which makes it less dense and therefore lighter. If a cooler, denser material is above the warmer layer, the warmer material will rise through the cooler material. The lighter, rising material will release its heat into the surrounding environment, become denser (cooler), and fall because of the effect of gravity. The process then starts over again. In a hot water system, this process is known as gravity circulation.

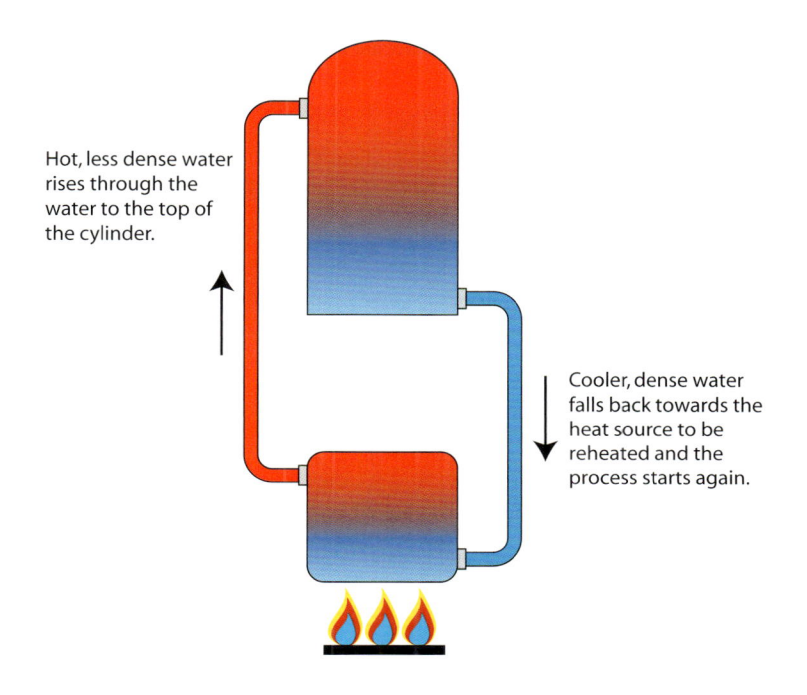

Hot, less dense water rises through the water to the top of the cylinder.

Cooler, dense water falls back towards the heat source to be reheated and the process starts again.

Gravity circulation (or convection)

Modern radiators in central heating systems use two methods of heat transfer with convection being the main heat transfer method. The other is radiation.

Radiation

Radiation heat transfer is thermal radiation from infrared light, whether visible or not, which transfers heat from one body to another without heating the space in between. Like all forms of light, thermal radiation travels in straight lines.

Consider the heat from the sun, which travels millions of miles through the vacuum of space to heat the earth. The heat can be felt from a distance because it travels in waves that are emitted from the heat of the sun. Radiation is the heat transfer method that makes solar hot water collectors in solar hot water systems so effective.

Radiation heat can also be felt from a hot radiator, even though there is no visible heat source or flame. This is because the heat is being radiated as thermal energy.

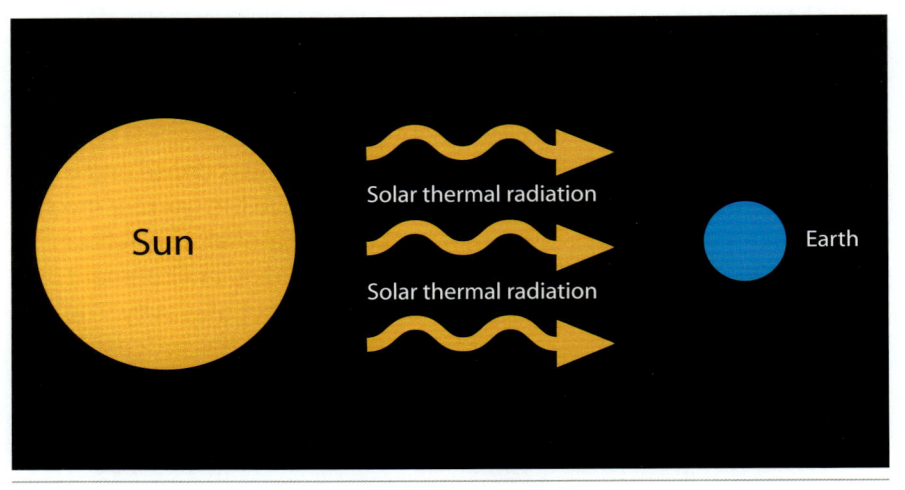

Thermal radiation

E6 ENERGY, HEAT AND POWER CALCULATIONS

In this part of the chapter, we will look at simple energy, heat and power calculations, using information we have previously discovered. To recap, the SI units of measurement of energy, heat and power are:

- energy: the joule (J)
- heat: the joule (J)
- power: the watt (W)
- specific heat capacity: kilojoules per kilogram per degree Celsius (kJ/kg/°C).

Remember: The specific heat capacity of water is 4.186kJ/kg/°C.

Calculations using the specific heat capacity of water

Example 1:
How many kilojoules would it take to heat 100 litres of water from 30°C to 80°C?

> The formula for this is:
>
> L × Δt × SHC of water
>
> where:
>
> L = litres
> Δt = temperature difference
> SHC of water = 4.186

Therefore:

100 × (80 − 30) × 4.186 = **20930 kJ**

> **SUGGESTED ACTIVITY...**
> Using the formula shown in Example 1, calculate how many kilojoules it would take to heat 140 litres of water from 4°C to 65°C.
>
> Answer: 35748 kJ

Example 2:

We can develop this concept further to calculate how many kilowatts it would take to raise the temperature of the 100 litres of water by 50°C. To do this, we need to state a time frame. Let us assume that the 100 litres of water is required in 1 hour. The calculation would then become:

$$\frac{L \times \Delta t \times \text{SHC of water}}{\text{Time (in seconds)}}$$

where:

L = litres
Δt = temperature difference
SHC of water = 4.186
1 hour in seconds = 3600

Therefore:

$$\frac{100 \times (80 - 30) \times 4.186}{3600} = 5.81\text{kW}$$

> **SUGGESTED ACTIVITY…**
> Using the formula shown in Example 2, calculate how many kilowatts it would take to raise the temperature of the 140 litres of water from 4°C to 65°C in two hours.
>
> Answer: 4.96 kW

Example 3:

How many seconds would it take for 20kg of water to be heated by 15°C using a 3kW heating element? Remember: water has a specific heat capacity of 4.186kJ/kg/°C and that 1W = 1J/s.

The formula for this is:

$$\frac{\text{kg} \times t \times \text{SHC}}{\text{kW}}$$

where:

kg = kilograms
t = temperature
kW = kilowatts
SHC = specific heat capacity

Therefore:

$$\frac{20 \times 15 \times 4.186}{3} = 418.6\text{s or } 6.976 \text{ minutes}$$

> **SUGGESTED ACTIVITY…**
> Using the formula shown in Example 3, calculate how many seconds it would take for 42kg of water to be heated by 30°C using a 3kW heating element.
>
> Answer: 1758.12 seconds or 29.30 minutes.

F THE PRINCIPLES OF FORCE AND PRESSURE

In this part of the chapter, we will look at the scientific principles of force and pressure and investigate how they apply to the building services industry.

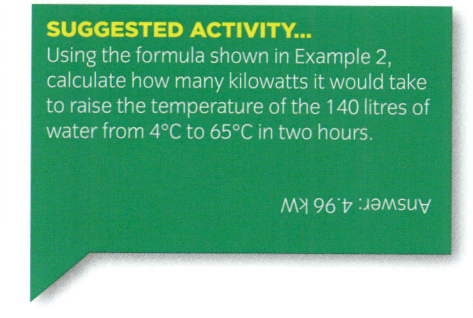

SmartScreen Unit 004 worksheet 4

F1 THE SI UNITS OF FORCE AND PRESSURE

Velocity	metres per second	m/s
Acceleration	metres per second squared	m/s²
Flow rate	metres cubed per second	m³/s
Force	newton (equal to kgm/s²)	N
Pressure, stress	pascal (equal to N/m²)	Pa

F2 VELOCITY AND ACCELERATION

Velocity
Velocity is the measurement of the rate at which an object changes its position. In order to measure it, you need to know both the speed of the object and the direction in which it is travelling. It is measured in metres per second (m/s).

Acceleration
Acceleration is a measure of the rate at which an object increases its velocity. It is measured as a change of velocity over a period of time and as such is directly proportional to force. It will increase and decrease linearly with an increase or decrease in force if the mass remains constant. It is measured in metres per second squared (m/s²).

Acceleration due to gravity
This is the rate of change of velocity of an object due to the gravitational pull of the earth. If gravity is the only force acting on an object, then the object will accelerate at a rate of 9.81 m/s² downwards toward the ground.

F3 FLOW RATE

In plumbing, flow rate is defined as an amount of fluid that flows through a pipe or tube over a given time. It is usually measured in metres cubed per second (m³/s). However, in plumbing systems, flow rate is usually measured in litres per second (l/s).

> To convert from m³/s to l/s, multiply m³/s by 1000:
>
> 1 m³/s × 1000 = 1000 l/s
>
> To convert from l/s to m³/s, multiply l/s by 0.001:
>
> 1000 × 0.001 = 1

Flow rate can also be measured in kilograms per second (kg/s). Since 1 litre of water has a mass of 1 kilogram, then 1 litre per second (l/s) = 1 kilogram per second (kg/s).

> 1 l/s = 1 kg/s.

F4 FORCE

Force is an influence on an object, which, acting alone, will cause the motion of the object to change. If the object at rest is subjected to a force it will start to move. For example, consider water in a pipe connected to a cistern at one end and a tap at the other. When the tap is closed, the water is not moving and so is said to be at rest. When the tap is opened, the force of gravity will move the water out of the tap causing water to flow. It is measured in newtons (N).

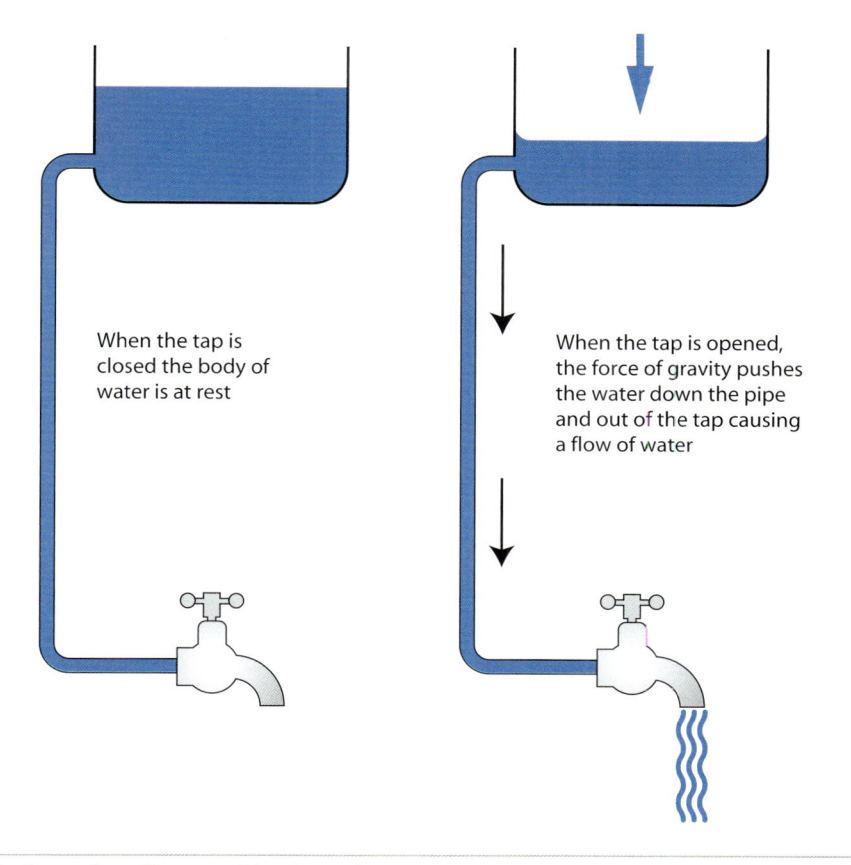

KEY POINT

The unit of the force of gravity is the newton. It is the force required to accelerate a mass of 1kg at 1 metre per second, every second. On earth, that force of acceleration (known as gravitational pull) is 9.81 metres per second per second, or 9.81 m/s². Therefore, if we multiply the mass of an object (in kg) by 9.81, the result is measured in newtons (kgm/s²).

When the tap is closed the body of water is at rest

When the tap is opened, the force of gravity pushes the water down the pipe and out of the tap causing a flow of water

The force of gravity on a cold water system

Calculating force

Consider the cistern in the diagram above. If the cistern contained a mass of water equal to 40kg, then by multiplying the mass by the force of gravity, the force of the cistern acting downwards can be calculated:

40 × 9.81 = 392.4N

SUGGESTED ACTIVITY...

If a cistern in a roof space contains a volume of 100 litres of water and 1 litre = 1kg, what is the force acting on the platform it is standing on?

Answer: 981 newtons

F5 PRESSURE

In physics, pressure is defined as force per unit area. For an object sitting on a surface, the force pressing on the surface is the weight of the object measured in newtons per square metre (N/m²). However, in different orientations it might have a different area in contact with the surface and will therefore exert a different pressure. For example, if a cistern measuring 1m long × 0.5m wide × 0.7m high was placed in a roof space, then what pressure would it exert if:

SmartScreen Unit 004 worksheets 5 and 7

- it was placed on its bottom?
- it was placed on its side?
- it was placed on its end?

Before we can attempt these calculations, we must first find the mass of the cistern in kg. The formula for this is:

Length × width × height = volume in m^3

$1 \times 0.5 \times 0.7 = 0.35$
Volume × 1000 = litres
$0.35 \times 1000 = 350$ litres

Since 1 litre of water has a mass of 1kg

350 litres = 350kg

From earlier calculations we know that to find the force of an object:

kg × gravity = N

$350 \times 9.81 = 3433.5N$

therefore: a cistern measuring 1m × 0.5m × 0.7m has a force of 3433.5N

The formula for finding pressure is:

$$\frac{force}{area} = N/m^2$$

Area of cistern on its bottom = $1 \times 0.5 = 0.5$
$3433.5 \div 0.5 = 6867$ N/m^2 pressure
Area of cistern on its side = $1 \times 0.7 = 0.7$
$3433.5 \div 0.7 = 4905$ N/m^2 pressure
Area of cistern on its end = $0.5 \times 0.7 = 0.35$
$3433.5 \div 0.35 = 9810$ N/m^2 pressure

From these calculations, we can see that the greater the surface area for a given mass, the less force will be exerted by that mass. This is of particular importance when placing large cisterns in roof spaces since a greater surface area that we can rest the cistern on will spread the load of the cistern.

Static pressure of water

The unit of water pressure is the pascal. The pressure exerted by water is due to its mass and is determined by the height of the column of water. For instance, if the pressure exerted by a water main is 300 kilopascals (kPa) it will balance a column of water about 30m high. This pressure is equivalent to a head of water of 30m. Therefore, 10m of head = 100kPa.

Water pressure in plumbing systems is usually measured in bar pressure. 10 m of head = 100 kPa = 1 bar.

> **SUGGESTED ACTIVITY...**
> What is the pressure exerted by a block of lead with a cross sectional area of $4m^2$ with a mass of 4000kg?
>
> Answer: 9810 N/m^2

Static head of water in plumbing systems is measured from the bottom of the water source, ie the cistern, to the outlet as shown on the next page.

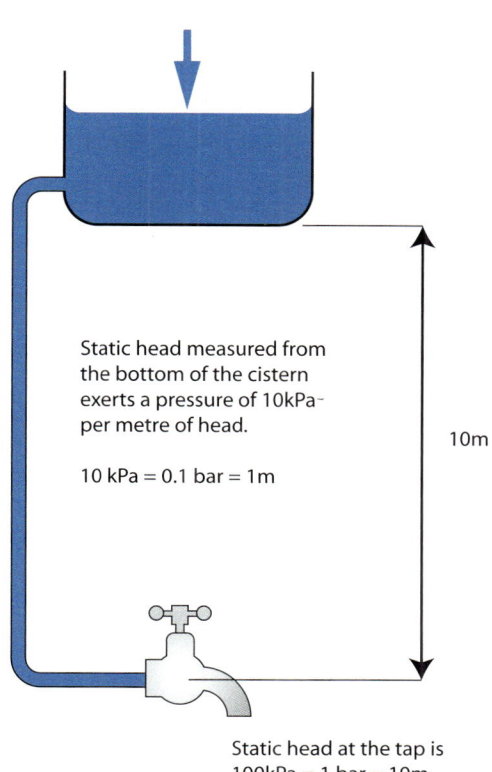

Static head measured from the bottom of the cistern exerts a pressure of 10kPa per metre of head.

10 kPa = 0.1 bar = 1m

10m

Static head at the tap is
100kPa = 1 bar = 10m

The head of pressure on a cold water system

> **SUGGESTED ACTIVITY...**
> If the vertical distance between the bottom of a cold water cistern and the tap is 16m, what is the pressure at the tap in:
> 1 kilopascals?
> 2 bar?
>
> Answers: a. 160kPa, b. 1.6 bar

Below is a table of conversion for common units of head of pressure:

Kilopascals kPa	Bars	Metres head of water	Pounds per square inch psi
10	0.1	1	1.42
20	0.2	2	2.84
30	0.3	3	4.27
40	0.4	4	5.68
50	0.5	5	7.11
100	1	10	14.22
150	1.5	15	21.33
200	2	20	28.44
250	2.5	25	35.55
300	3	30	42.66
350	3.5	35	49.77
400	4	40	56.88
450	4.5	45	63.99
500	5	50	71.10

Dynamic pressure

Also called 'working pressure', dynamic pressure is the pressure of water while it is in motion. In other words, it is the pressure of flowing water. If the pressure of the water is increased, the velocity and flow rate will also increase.

Atmospheric pressure

Atmospheric pressure is the amount of force or pressure exerted by the atmosphere on the earth and the objects located on it. The more pressure there is, the stronger that force will be; at sea level, the atmospheric pressure is 101.325kPa. This is known as 1 atmosphere (atm). Atmospheric pressure decreases with height.

The principle of a siphon (siphonic action) due to atmospheric pressure

The principle of a siphon is to discharge water from a high vessel to a lower vessel using atmospheric pressure and the cohesive properties of water. It is used to good effect in WCs to flush the water from the cistern to the WC pan.

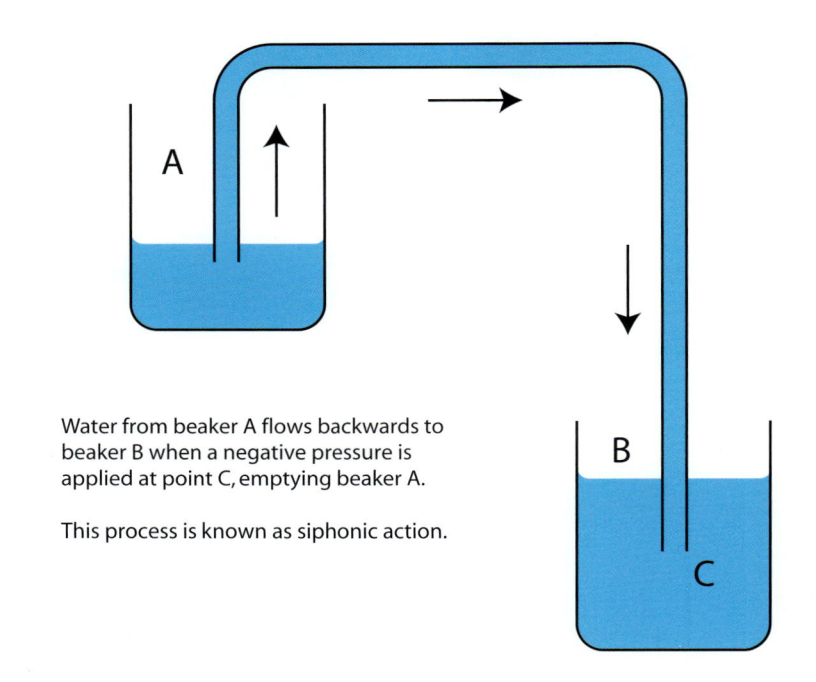

Water from beaker A flows backwards to beaker B when a negative pressure is applied at point C, emptying beaker A.

This process is known as siphonic action.

Siphonic action

The principle of a siphon can be understood with reference to the diagram above. The two beakers are both at atmospheric pressure, but they are at different levels. The pressure at beaker B is greater because it is lower. The outlet from the hose at B must be lower than the inlet of the hose at A for flow to take place. When suction is applied to the end of the hose at B, the water will flow upwards over the top of beaker A where the atmospheric pressure is slightly lower. Here, gravity and the cohesive nature of water will empty the contents of beaker A into beaker B.

F6 THE RELATIONSHIP BETWEEN VELOCITY, PRESSURE AND FLOW RATE IN PLUMBING SYSTEMS

As we have already discovered, if pressure is applied to a pipe full of water, the effect is to increase the velocity and therefore flow rate of the water. The more pressure that is applied, the greater the velocity and the flow rate become.

A similar effect can be seen when a pipe is suddenly reduced in size and this can be seen in a hosepipe. If the end of a flowing hosepipe is suddenly reduced then the speed increases and the water shoots further away, but the pressure and flow rate will be reduced. This is called the Bernoulli Effect, which states that if a pipe is reduced in size, then: ' … an increase in the speed of the fluid occurs simultaneously with a decrease in pressure or a decrease in the fluid's potential energy.'

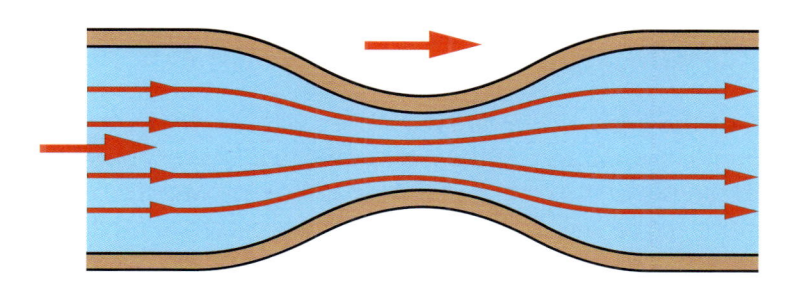

Increased fluid speed, decreased internal pressure

The Bernoulli Effect

Similarly, if the pipe suddenly increases in size, then the velocity of the water will decrease but the pressure will increase slightly. The flow rate remains constant.

F7 FACTORS AFFECTING FLOW RATE

As we have seen, flow rate is unaffected by sudden increases in pipe size but there are elements in plumbing systems that can severely affect the flow rate. Flow rate is affected by:

- **Changes in direction** – Any change in direction of a pipe will offer resistance to the flow of the water. That resistance will, in effect, be an increase in the overall length of the pipe. For example, an elbow installed in the run of copper pipe will offer resistance equivalent to 0.37m of pipe, so if 10 elbows are used then the length of the pipe has, theoretically, increased by 3.7m. Machine-made bends offer slightly less resistance at 0.26m of pipe. This will also vary with the material of the pipe.

Type of fitting	Nominal pipe size * (mm)					
	8	10	12	15	22	28
	Equivalent length (mm)					
Capillary elbow	0.16	0.21	0.28	0.37	0.60	0.83
Compression elbow	0.24	0.33	0.42	0.60	1.00	1.30
Square tee piece	0.27	0.37	0.49	1.00	1.6	2
Swept tee piece	0.22	0.29	0.38	0.60	0.75	1
Manifold connection	0.60	1.00	1.20	n/a	n/a	n/a
Minimum radius (machine) bend	0.12	0.16	0.20	0.26	0.41	0.58
* Copper tubes to BS EN 1057 R250							

- **Size of pipe** – The greatest factor in the flow rate of any system is the size of the pipe itself. The bigger the bore of the pipe then the better the flow rate will be.

- **Pressure** – Pressure increases flow rate. The greater the pressure, the greater the flow rate.

- **Length of the pipe** – Flow rate diminishes with length because of the frictional resistance of the walls of the pipe. Water flows faster down the centre of the pipe than it does at the pipe wall. The nearer the water is to the wall of the pipe, then the greater the frictional resistance and so the slower the water becomes. The frictional resistance of the pipe is slowing the flow rate constantly. The greater the length, the more the frictional resistance, the greater the loss of flow rate. To counter this effect, the pipe size should be increased initially at the start of the pipe run and then reduced as length increases.

- **Frictional resistance of the internal bore of the pipe** – Different materials offer different frictional resistance. Polybutylene pipe, for instance, has the smoothest bore of all common pipe materials, and low carbon steel the roughest. Therefore, the flow rate of low carbon steel for a given size will have much less flow rate than a polybutylene pipe.

- **Constrictions such as valves and taps** – Taps and valves offer a lot of resistance to the flow of water. Some stop taps can increase pipe length by an equivalent of up to 6m per valve.

G THE PRINCIPLES BEHIND SIMPLE MACHINES

Simple machines are those which aid with the lifting and moving of loads that are too heavy to be lifted or moved on their own. There are four main types:

- levers
- wheel and axles
- pulleys
- screws.

These machines give a mechanical advantage (velocity ratio) to human effort, meaning they multiply the force that is put into them. There are two types of mechanical advantage:

1 **Ideal mechanical advantage (IMA)** – Purely theoretical based upon an 'ideal machine', which does not exist.

2 **Actual mechanical advantage (AMA**) – This is the mechanical advantage of a real machine such as a wheelbarrow (lever). Actual mechanical advantage takes into consideration real world factors, such as energy lost because of friction.

G1 SIMPLE MACHINES

Here, we will look at the machines themselves and their possible uses in everyday working life.

Levers

In physics, a lever is a rigid object that can be used with a pivot point or fulcrum to multiply the mechanical force that can be applied to another, heavier object. Levers are examples of mechanical advantage. The calculation for finding out how a lever functions is:

$$\text{Mechanical advantage} = \frac{\text{Load}}{\text{Effort}}$$

There are three classes of lever:

The first class lever is a simple seesaw arrangement where the long arm (force effort) is proportional to the short arm (load). Examples of this are:

- the lever arm of a float-operated valve
- claw hammer
- water pump pliers (double lever).

The second class lever is a variation of the first class lever, where the force and the load are both on the same side of the pivot. Examples of this are:

- wheelbarrow
- crowbar.

With the third class lever the load is at the opposite end to the force. Examples of this are:

- the human arm
- tools, such as a hoe or scythe
- spades and shovels.

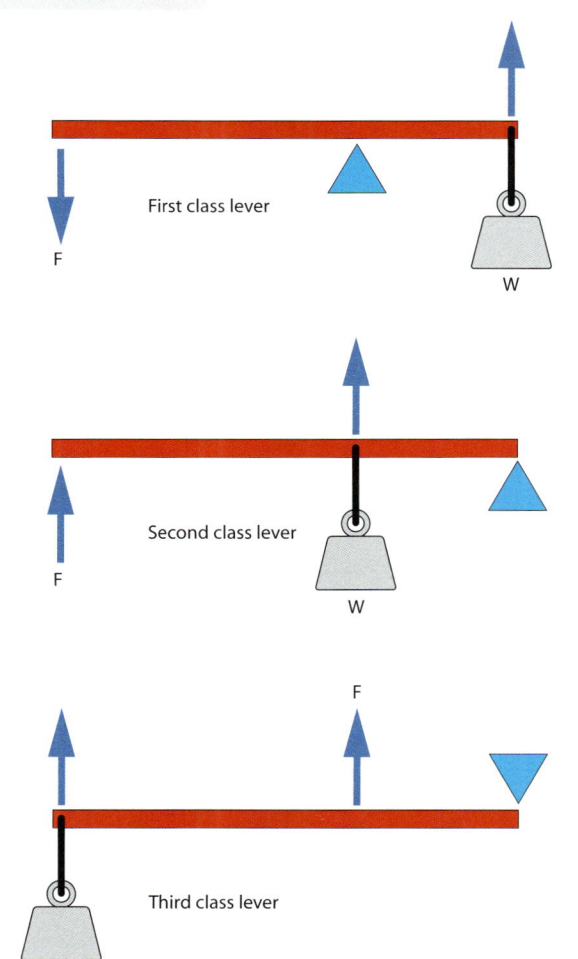

Wheels and axles

A wheel and axle is defined as:

'A mechanical device consisting of a grooved wheel, which is turned by a chord or chain with a rigid axle for winding up the weight.'

The wheel and axle is composed of a wheel, which is larger than the diameter of the axle. Either of these can be used as the effort arm and the resistance arm and this depends where the force is applied. The force is usually applied to the wheel rather than the axle to gain the maximum output. The point where the axle joins the wheel is known as the fulcrum and this acts as the point where the force from the larger wheel is transferred to the smaller axle.

The wheel and axle multiplies the 'torque' during the turning motion.

Both the wheel and the axle have ropes wound around them. The load is lifted by pulling on the rope around the wheel so that the wheel and axle is rotated once, therefore:

$$\text{Mechanical advantage} = \frac{\text{Radius of the wheel}}{\text{Radius of the axle}} = \frac{R}{r}$$

Spanners and screwdrivers use the principle of wheel and axle.

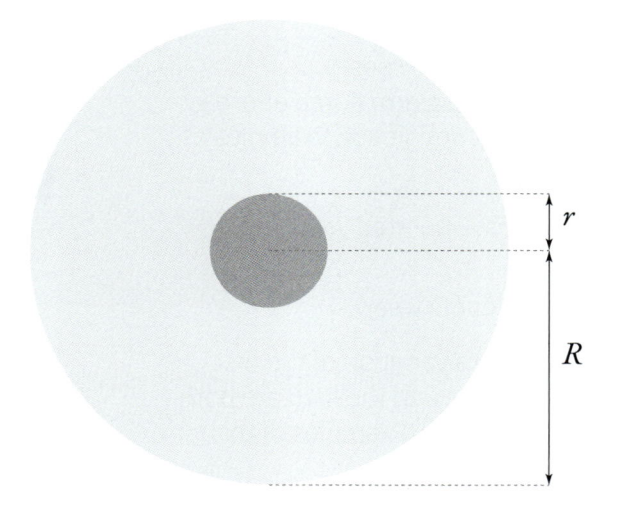

Effort arm
In mechanics, the arm where the force is applied.

Resistance arm
In mechanics, the arm where the load is concentrated.

Torque
The property of force that is exhibited when an object rotates around its axis.

Pulleys

A pulley is a collection of one or more wheels over which a rope or chain is looped to aid lifting heavy objects. Pulleys are simple machines, in other words, they multiply the lifting forces.

A single pulley reverses the direction of the lifting force. When the rope is pulled down the weight lifts up. If a lift of 100kg is needed, an equal force of 100kg must be exerted. A lift of 1m high needs to be pulled downwards 1m.

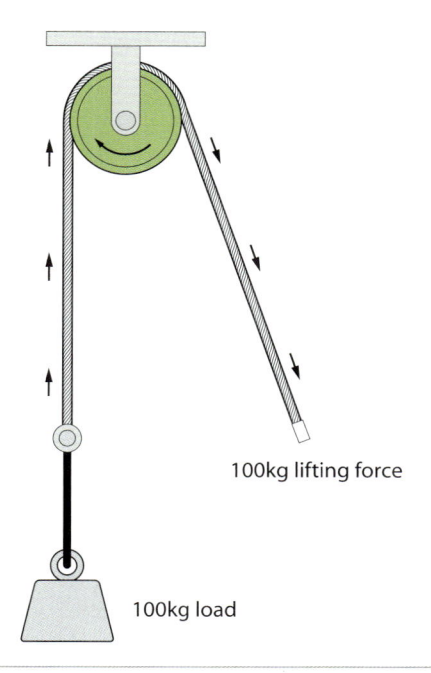

A single pulley system

If more ropes and wheels are added, the effort needed to lift the weight is reduced. If the 100kg weight is supported by two ropes instead of one, the lift effort is halved. This gives a positive mechanical advantage. The bigger the mechanical advantage, the less force is needed.

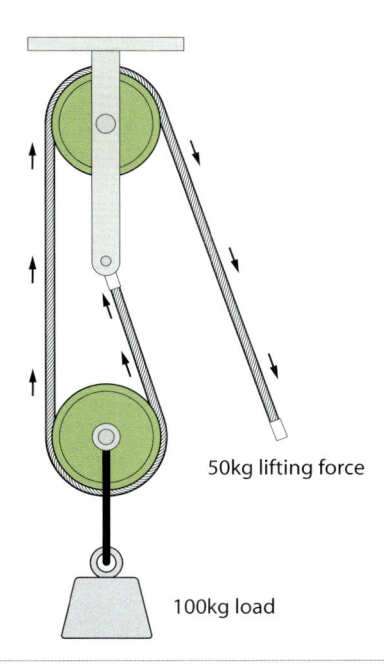

A double pulley system

If four wheels are used and held together by a long rope or chain that loops over them, the 100kg weight is supported by four ropes, which means that each rope is supporting a quarter of the total 100kg weight. This means that only a quarter of the force (25kg) is needed to lift the weight (100kg). This system is known as a block and tackle.

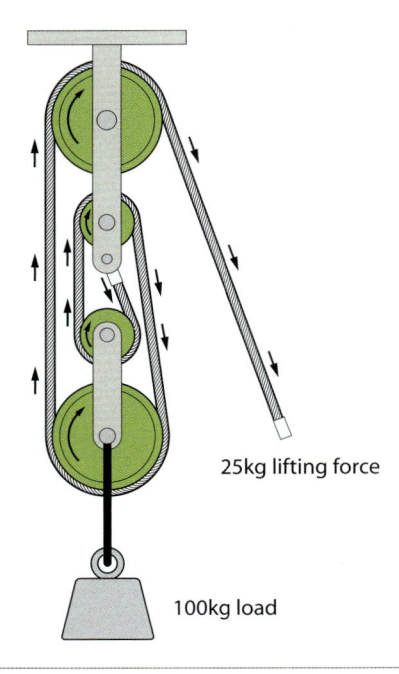

A four-pulley system (block and tackle)

Screws

A screw is a machine that converts rotation into a straight-line motion that can be placed vertically, horizontally or at an angle. It is basically a cylinder or wedge with an incline plane wrapped around it. It was originally designed as a simple water pump (the Archimedes screw), a task for which it is still used today. It can be found in many objects such as screw fixings, bolts and threads on pipes. It can also be seen on drills and auger bits and as a means of moving solid fuel such as coal towards a boiler by its rotary motion.

The following formula is used to calculate the mechanical advantage of a screw:

$$MA = \frac{\pi \times D}{L}$$

where:

MA = mechanical advantage

π = 3.142

D = diameter

L = length

G2 BASIC MECHANICS – MOMENTS OF A FORCE (TORQUE)

In physics, the moment of a force is the measure of the turning effect (or torque) produced by a force acting on a body. It is equal to the applied force and the perpendicular distance from its line of action to the pivot, about which the body turns. The turning force around the pivot is called the moment. Its unit of measurement is the newton-metre.

The moment of a force can be worked out using the formula:

Moment = force applied × perpendicular distance from the pivot

If the magnitude of the force is F and the perpendicular distance is d, then:

Moment = F × d

An example of this would be a spanner turning a bolt. It is much easier to turn the bolt using a long spanner than it is using a short spanner. This is because more torque (turning force) can be applied at the bolt (pivot) for less effort. A long spanner is an example of a force multiplier.

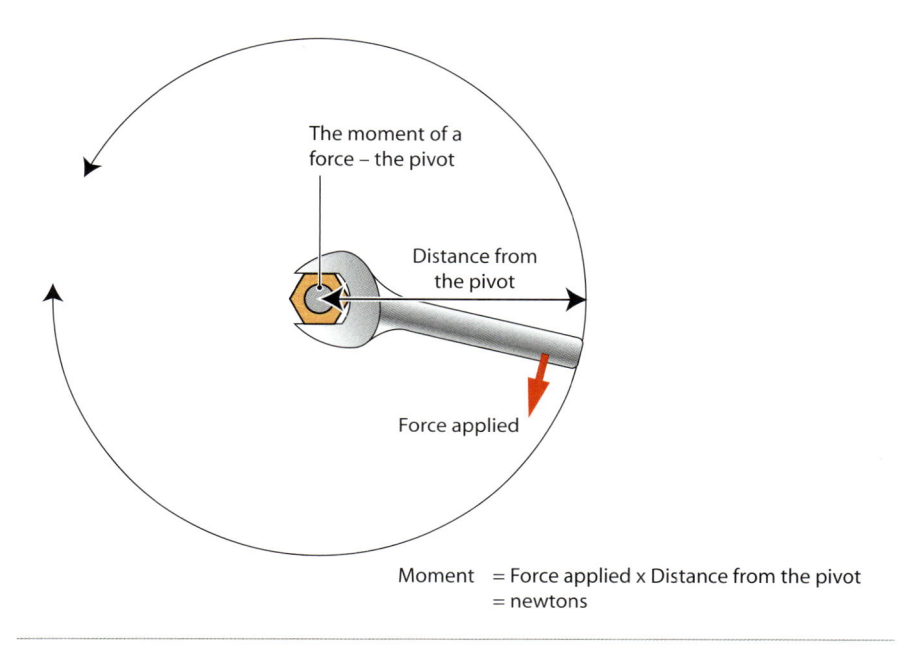

Moment = Force applied x Distance from the pivot
= newtons

The moment of a force

G3 CENTRE OF GRAVITY

In physics, the centre of gravity of an object is the imaginary point where all of the weight of the object is concentrated. This concept is especially important when designing large structures such as multi-storey buildings and bridges, or predicting the gravitational effect on a moving object or body. Another term for it is the centre of mass.

The centre of gravity will vary from object to object. In symmetrically shaped objects it will coincide with the geometric centre.

In irregular (asymmetrical) shaped objects, the centre of gravity may be some distance away from the centre of the object. In hollow objects such as a ball, it may be in free space, away from the object's physical form.

KEY POINT

For many solid objects, the location of the geometric centre follows the object's symmetry. For example, the geometric centre of a cube is the point of intersection of the cube's diagonals.

G4 ACTION AND REACTION – NEWTON'S THIRD LAW OF MOTION

A push or a pull (action) on an object can often result in movement (reaction) when the pull or push is greater than the weight of the object. If both action and reaction are equal then no movement takes place because the object is pushing or pulling against the action with equal force. This is known as contact force and is a result of contact interactions (normal, frictional, tensional and applied forces are all examples of contact forces). Other forces are a result of 'actions-at-a-glance' interactions (gravitational pull, electrical and magnetic). These two types of force have one thing in common: for every force applied there is an equal opposing force and as such is subject to action and reaction. When a person sits on a chair (action), the downward force of the person provokes an upward force in the chair (reaction). The person and the chair have equal force and so equilibrium exists. If the person were to be too heavy for the chair, the chair would collapse (reaction).

This is Newton's third law of motion, which states: 'Every action has an equal but opposite reaction.' This means that for every force that an object is subjected to:

1 there is an opposing force from the object
2 both action and reaction forces are equal
3 forces always come in pairs (points 1 and 2).

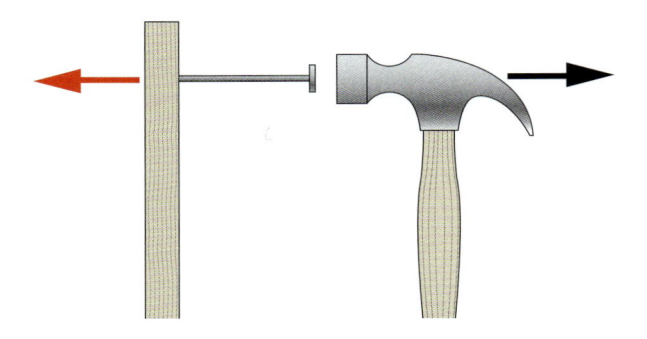

Action and reaction

G5 EQUILIBRIUM

When all the forces acting on a stationary object are balanced, the object is said to be in a state of equilibrium. The forces are balanced when all forces (left, right, front, back, up and down) are the same. In the left-hand image opposite, all forces are 50N and are therefore equal forces in equilibrium.

The same can apply for unequal forces. They too can be in a state of equilibrium, provided left and right forces are equal but not necessarily the same as the equal up and down forces.

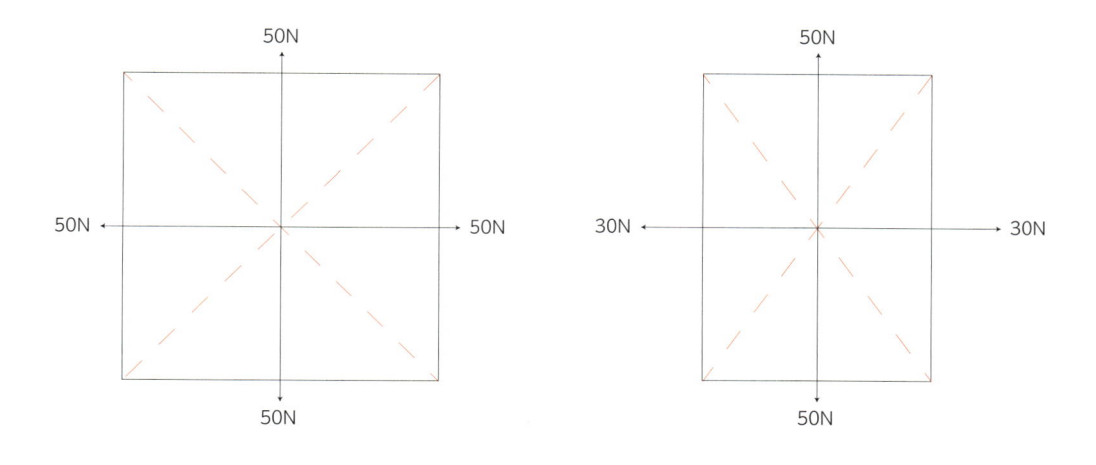

Balanced forces in equilibrium (left) and unequal forces in equilibrium (right)

The key word here is 'balanced'. All forces, whether equal or not, must be balanced. The forces cancel each other out and so add up to zero. In other words, for an object to be in equilibrium, the sum of the forces on each part of the system must be zero. This is illustrated in the drawing below.

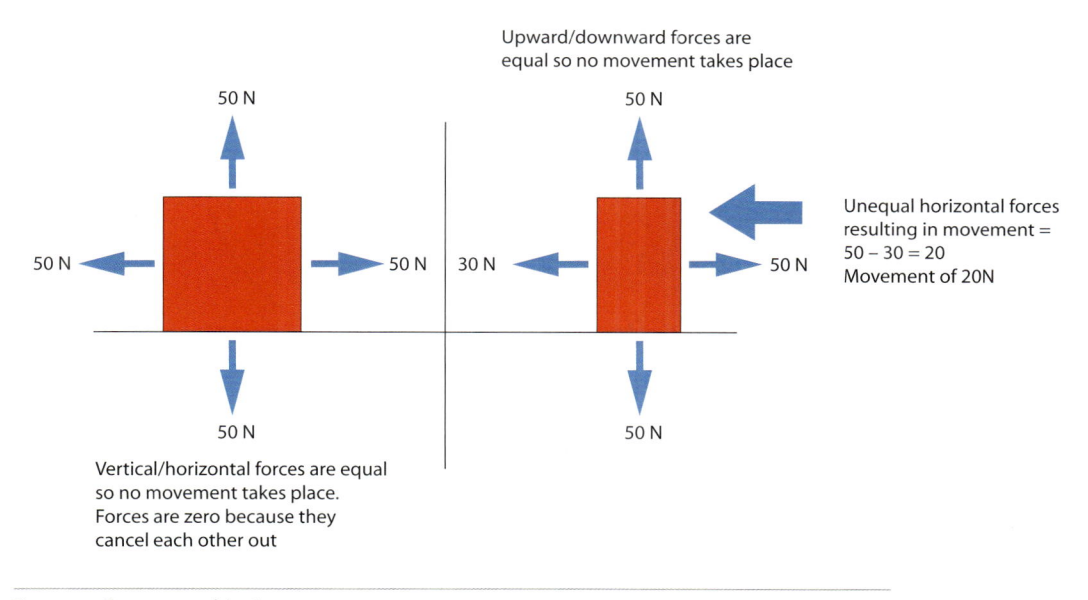

Upward/downward forces are equal so no movement takes place

Unequal horizontal forces resulting in movement = 50 – 30 = 20
Movement of 20N

Vertical/horizontal forces are equal so no movement takes place. Forces are zero because they cancel each other out

Forces acting on an object

H THE PRINCIPLES OF ELECTRICITY

Electricity is a vital part of everyday life. It powers lighting, household appliances and heating systems, but its danger cannot be overstated. We cannot see it, hear it or smell it, yet if we touch it, it can kill.

Because of the obvious dangers, it is necessary for us to have a better understanding of what electricity is and how it works.

In this section we will find out about electricity, its scientific laws and basic circuitry.

H1 THE BASIC PRINCIPLES OF ELECTRON FLOW

Everything is made up of molecules and these in turn are made up of atoms. Atoms are not solid and they consist of even smaller particles. At the centre of every atom is a nucleus, which is made up of protons and neutrons. Protons have a positive (+) charge and neutrons have a neutral (0) charge.

The neutrons hold the nucleus together. Without them, the nucleus would simply fly apart. They prevent the protons from repelling each other. Their role is simply to keep the nucleus together. Revolving in orbit around the nucleus are the electrons. They have a negative (–) charge.

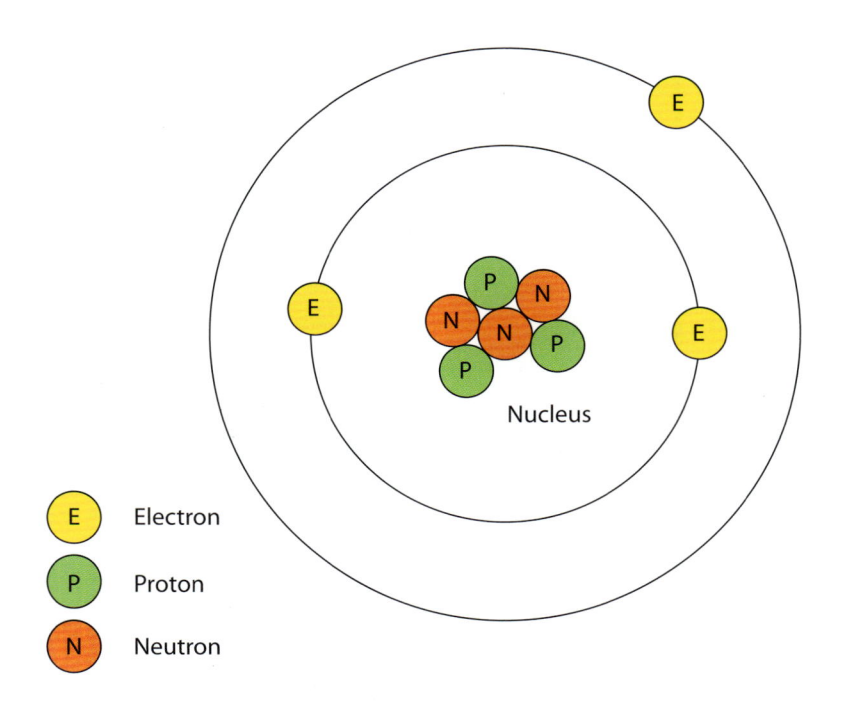

Molecules
The smallest particle of a specific element or compound that retains the chemical properties of that element or compound.

Atom
A fundamental piece of matter made up of three kinds of particles called subatomic particles: protons, neutrons and electrons.

Protons, neutrons and the nucleus

All atoms possess equal numbers of positively charged protons and electrons and these, effectively, cancel each other out, leaving the atom electrically neutral. It is possible in some cases, however, to add an electron to, or remove one from, an atom to make it either negatively or positively charged. In this case, the atom is known as an ion.

As can be seen from the diagram, the atom is like a micro-solar system whereby the electrons orbit the nucleus in the same way as the planets orbit the sun.

Electrons are arranged at varying distances from the nucleus; the further they are away, the less they are attracted to the atom and are easily deflected from their orbits to be attracted by other atoms. This constant 'to-ing and fro-ing' of electrons from one atom to another is the structure that makes electricity possible. Materials that allow the movement of free electrons are known as conductors and those that restrict their movement are known as insulators.

H2 THE MEASUREMENT OF ELECTRICAL FLOW

Electricity is measured in two ways:

1 By the amount of current. This is the number of electrons flowing per second, measured in amperes.
2 By the push or pressure that causes current to flow, measured in volts.

The push or pressure that causes current to flow is known as voltage and is caused by the fact that the electrons repel each other because of their negative (−) charges. When the electrons are concentrated in one place they will flow freely away from the area of concentration if the path is clear for them to move and it is the pressure (or voltage) that makes them move. If there are many electrons in one place, this is high voltage and many electrons will flow, provided there is a path (or a conductor). The more electrons that flow through the conductor, then the higher the current will be.

H3 THE UNITS OF ELECTRICAL MEASUREMENT

When we think about electricity we think in terms of voltage, amperage, resistance and power. But what do these terms mean and what do they do? In this section, we will investigate the various electrical units and their interaction with each other, and how we can calculate one if two of the others are known (Ohm's Law).

The most common terms associated with electricity are given in the table below:

Parameter	Unit	Symbol	Description
Voltage	volt	V or E	Unit of electrical potential $V = I \times R$
Current	ampere	I or i	Unit of electrical current $I = V \div R$
Resistance	ohm	R or Ω	Unit of resistance $R = V \div I$
Conductance	siemen	G	Reciprocal of resistance $G = I \div V$
Capacitance	farad	C	Unit of capacitance $C = Q \div V$
Charge	coulomb	Q	Unit of electrical charge $Q = C \times V$
Power	watts	W	Unit of power $P = V \times I$

Voltage

When there are more electrons in one part of the circuit than the other, the electrons will flow from where they are concentrated to the area where they are missing. The difference in electron concentration is known as 'potential difference' or voltage. The higher the voltage, the greater the imbalance in electrons; the greater the imbalance, the harder the electrons repel each other, and so the greater the current or flow (amps) in the circuit.

Voltage can be calculated by:

> Current (I) × Resistance (R)

Resistance

Resistance is the opposition to movement of electrons through a conductor. All electrical circuits will have resistance, but some will have more than others. Resistance in some circuits is necessary to ensure that not too many electrons flow, and in others as little resistance as possible is required so that high current will flow.

There is a definite interaction between current (electron flow), voltage (electrical pressure) and resistance. As the electrical pressure (voltage) increases, more electrons flow. Increasing the voltage also increases the amperes of current but if resistance is also increased, this decreases the flowing current, thus reducing the amperes.

These relationships between current, voltage and resistance are the theory behind Ohm's Law, which will be looked at in detail later in this section.

Resistance can be calculated by:

> Voltage (V) ÷ Current (I)

Amperage

In the UK, voltage is supplied at 230V, but different appliances need different amounts of electricity to work effectively. The rate at which electricity flows through an appliance is known, in electrical units, as amperage, often shortened to amps. If we consider that water at a certain pressure with a certain size pipe will deliver a set amount of water, if we increase the pipe size, then the pressure stays the same but the flow rate increases. In electrical terms, if voltage is the pressure then amps is the flow rate; the bigger the cable, the bigger the flow rate or amperage.

The ampere symbol (I) is the SI unit of electric current, and is defined in terms of the coulomb:

> 1 ampere = the amount of electric current (flow rate of electricity) carried by a charge of 1 coulomb flowing in 1 second.

Amps can be calculated by:

Voltage (V) ÷ Resistance (R)

Power

The rate at which electric energy is converted to other forms of energy, such as heat, light or mechanical, is called power (P) and is equal to the product of the current and the voltage. With an electrical shower rated at, say, 8kW, the electrical power of 8kW is converted into heat to raise the temperature of the water. Electrical power is, therefore, the rate at which electricity is consumed and can be defined as the amount of electric current flowing due to the voltage.

Electrical power is measured in watts (W). The formula is:

amps × volts = watts

H4 THE TYPES OF ELECTRICAL CURRENT

There are two types of electrical current. These are:

- direct current (DC)
- alternating current (AC).

Direct current

In a direct current (DC) circuit, the electrons always flow from the negative (−) pole towards the positive (+) pole. The polarity or direction of the electrons never reverses. Direct current can be produced from a number of sources, including electrochemical or photovoltaic cells and batteries. Direct current can be stored in batteries and cells.

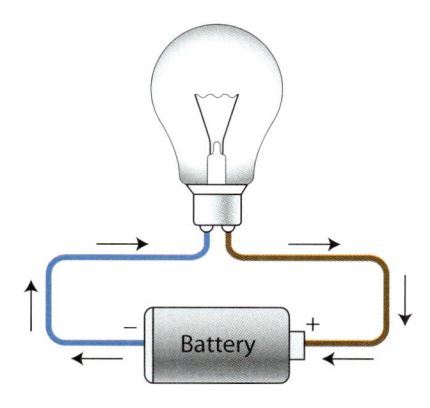

Direct current symbol

A simple direct current circuit

Hertz (Hz)
The SI unit of frequency, measuring the number of cycles per second in alternating current.

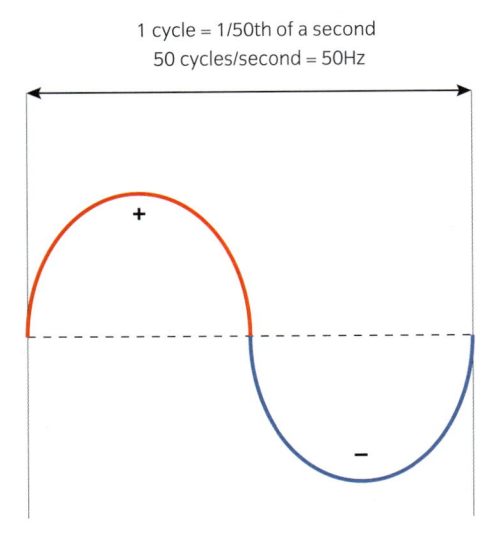

Alternating current

Alternating current (AC), unlike direct current, does not travel in a constant direction. It alternates; in other words, it reverses its direction of travel constantly and uniformly throughout the circuit at 50 times a second. This rapid movement (50 times/second) is called the frequency and is measured in hertz. In the UK, the voltage and frequency of alternating current for power and lighting in domestic properties is 230V, 50Hz. It is the current used for power and lighting in almost all domestic properties in the United Kingdom.

1 cycle = 1/50th of a second
50 cycles/second = 50Hz

AC cycle

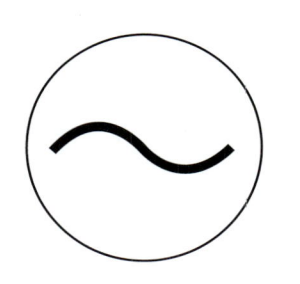

The symbol for alternating current

The advantage that alternating current has over direct current is that AC voltages can be easily transformed to higher or lower voltages. DC voltages are difficult to transform. Changing AC voltages is done by the use of a transformer, which uses the properties of AC electromagnets to change the voltages.

Another advantage is that AC can be easily transported over long distances without excessive voltage loss and is, therefore, much more efficient than DC.

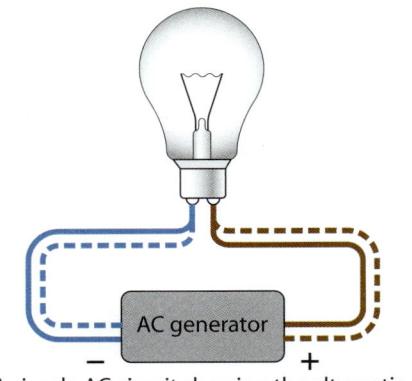

A simple AC circuit showing the alternating direction of electron flow

A simple AC circuit

004 SCIENTIFIC PRINCIPLES

Alternating current is generated at power stations and portable electricity generators. It cannot be stored.

H5 MATERIAL CONDUCTIVITY AND RESISTANCE

As we have already seen, the atom is orbited by electrons. The electrons carry a negative charge and can move from atom to atom. The direction of movement between atoms is random unless a force causes the electrons to move in one direction. This directional movement of electrons due to an electromotive force (EMF) is known as electricity. How well a material allows electron movement is called conductivity and how well it resists electron flow is known as resistivity.

Conductivity

Electrical conductivity is a measure of how well a material accommodates the movement of an electric charge. This means that any electrical conductor is one that has many free electrons. A good conductor allows the free movement of electrons whereas a poor conductor (known as an insulator) restricts this free movement. As a general conductor, copper is the most commonly used because it is cheap, reasonably flexible, reasonably light, is the second best conductor in terms of electrical resistance, and is the best conductor per unit weight.

Table of common conductors

Element	Density (kg/m^3)
Silver	10,490
Copper	8960
Gold	19,300
Aluminium	2700
Iron	7150
Chromium	7860
Lead	11,340
Titanium	4506

Resistivity

Electrical resistivity is the opposite of conductivity. It is the opposition of a material to the flow of electrical current through it, resulting in a change of electrical energy into heat, light or other forms of energy. For example, when electricity passes through the heating element of an immersion heater, the element resists the flow of electrical current therefore generating heat. The same effect occurs in a light bulb. The lighting filament offers resistance to the flow of electricity and 'glows' with the heat generated. By including an electronic variable resistor in the light switch, the brightness can also be resisted (ie a dimmer switch). The amount of resistance depends on the type of material.

H6 OHM'S LAW

So far we have looked at voltage, current, resistance and power. Here we will investigate how these are related to each other by the use of Ohm's Law.

Ohm's Law states:

> 'The current through a conductor between two points is proportional to the voltage across the two points, and inversely proportional to the resistance between them.'

It defines the relationships between (P) power, (V) voltage, (I) current, and (R) resistance. One ohm is the resistance through which one volt will maintain a current of one ampere.

Ohm's Law says that if we have any two electrical values, we can always find a third. For example, if I = current, V = voltage, R = resistance and P = power, then:

I = V ÷ R

V = I × R

R = V ÷ I

P = V × I

All of these variations of Ohm's Law are mathematically equal to each other and can be easily remembered using the Ohm's Law triangle and the power triangle.

KEY POINT

Remember:

(I) Current is what flows in a wire or conductor. Current is measured in (A) amperes or amps.

(V) Voltage is the difference in electrical potential between two points in a circuit. It is the push or pressure behind current flow and is measured in volts.

(R) Resistance determines how much current will flow through a component. Resistors are used to control voltage and current levels. Resistance is measured in ohms (Ω).

(P) Power is the amount of current multiplied by the voltage at a given point. It is measured in watts.

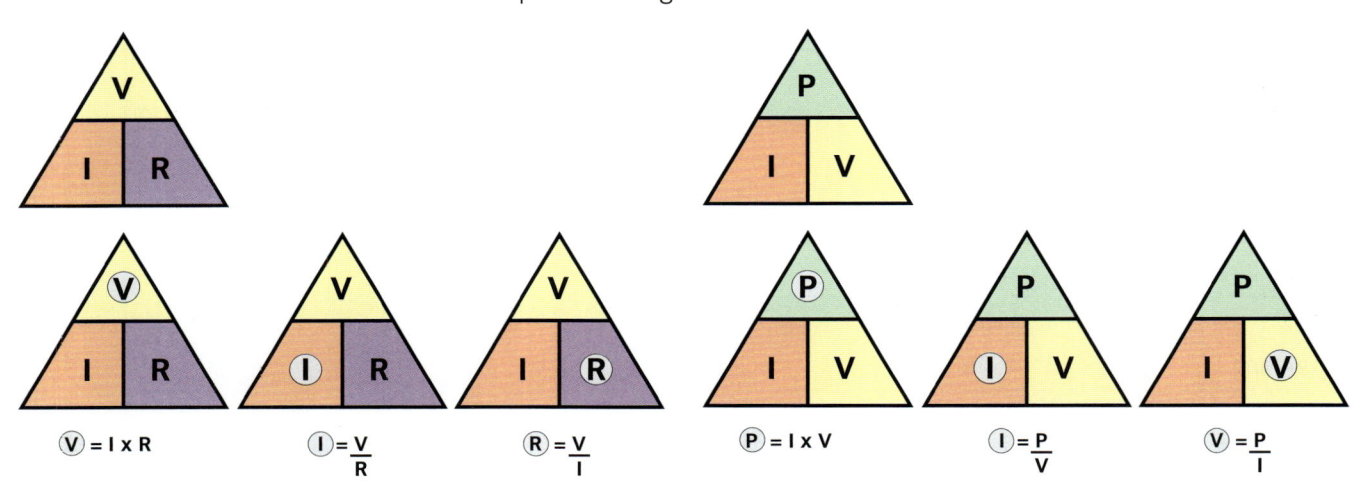

SmartScreen Unit 004 handout 5

Ⓥ = I × R   Ⓘ = V/R   Ⓡ = V/I   Ⓟ = I × V   Ⓘ = P/V   Ⓥ = P/I

Ohm's Law triangle [left] and power triangle [right]

SUGGESTED ACTIVITY...

Transposing the formula below, calculate the current flowing to a 3kW electric fire fed by a 110 volt AC supply.

Current (I) = Power (P) ÷ Voltage (V)

Answer: 27.27 amps

Ohm's Law calculations

Example 1:

What size of over current protection device will be needed to protect a circuit that has a 3kW immersion heater installed on a 230V supply?

The formula for this is shown in the power triangle as: I = P ÷ V, therefore:

First, convert the kilowatts to watts by multiplying by 1000:

3kW = 3000

We can now complete the calculation:

3000 ÷ 230 = **13 amps (I)**

Example 2:
Using the formula power (P) = voltage (V) × current (I), calculate the current flowing to a 6kW shower fed by a 230 volt AC supply.

Look at the power triangle. Since we only know the power and the volts, we will need to transpose the equation to find the amps (I). Thus the equation becomes:

I = P (in watts) ÷ V

6kW = 6000 watts, therefore:

6000 ÷ 230 = 26 amps (I)

Example 3:
Using the Ohm's Law formula (shown in the Ohm's Law triangle): voltage = current × resistance, calculate the voltage in a circuit which has a resistance of 115 ohms and a current of 2 amps.

The formula for this is:

V = I × R, therefore:

115 × 2 = 230 volts

> **SUGGESTED ACTIVITY...**
> Transposing the formula below, calculate the voltage to a 10kW shower with a fuse rating of 45 amps.
> Current (I) = power (P) × voltage (V)
>
> Answer: 27.27 amps

> **SUGGESTED ACTIVITY...**
> Transposing the Ohm's Law formula, voltage = current × resistance, calculate the resistance in a circuit that has a voltage of 230 volts and a current of 15 amps.
>
> Answer: 15.333 ohms

H7 VOLTAGE, CURRENT AND RESISTANCE IN SERIES AND PARALLEL CIRCUITS

Series circuits
A series circuit is one where there is only one path from the source through all the loads back to the source. This means that the current must flow through all the loads.

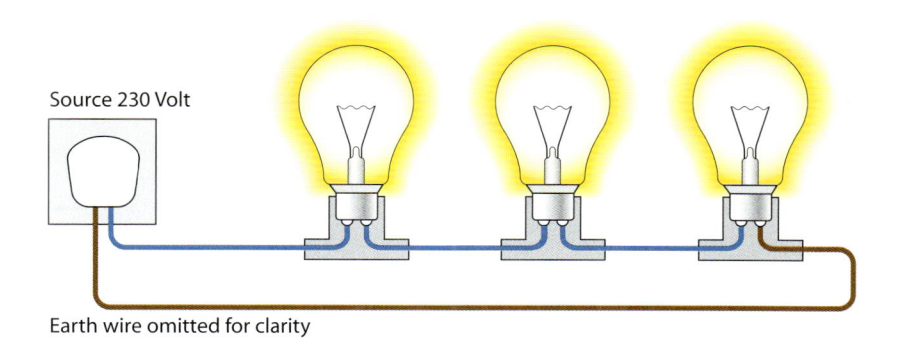

Source 230 Volt

Earth wire omitted for clarity

A simple series circuit

In the circuit of light bulbs, shown above, if one of the light bulbs blows the whole circuit will become open, the circuit will stop operating and all the light bulbs will go out. So, how does a series circuit operate?

Consider a basic series circuit with one 40 watt light bulb connected to a 230 volt electricity supply. The bulb glows at full brightness as it receives a full 230 volts. If Ohm's law is applied, the resistance in the circuit is as follows:

$$I = \frac{P}{V} \text{ (from the power triangle) to find the current (I)} = 40 \text{ watts} = 0.174A$$

$$R = \frac{V}{I} = \frac{230}{0.174} = 1322.6\Omega$$

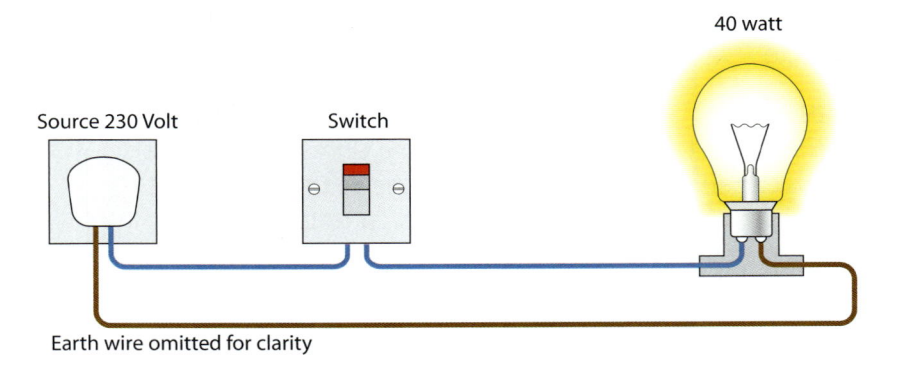

A simple series circuit with one resistor

If a second light bulb with the same wattage is added to the circuit, the resistance in the circuit doubles and the current flow is half of what it was when there was only one bulb. The voltage is now only 115 volts to each bulb because of the reduced current flow and the bulbs glow with much less brightness. Since both bulbs have the same wattage, they both have equal voltage drop.

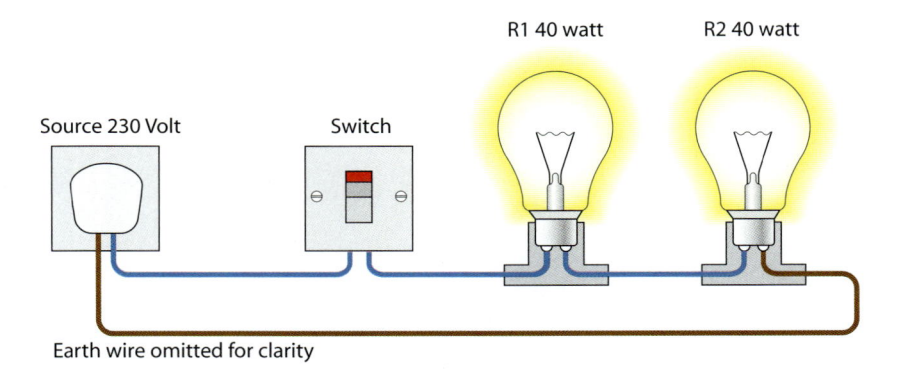

A simple series circuit with two resistors

Since each 40 watt has the resistance previously calculated (1322.6Ω) then the total resistance in the circuit is 2645.198 ohms. To find the voltage supplied to each bulb:

Total resistance for the circuit = 2645.2Ω

Therefore the current = $\dfrac{\text{volts}}{\text{ohms}}$ = $\dfrac{230}{2645.2}$ = 0.0869A

V across R1 = I × R = 0.0869 × 1322.6 = 114.93V
V across R2 = I × R = 0.0869 × 1322.6 = 114.93V

But what if a bulb is added of lower wattage?

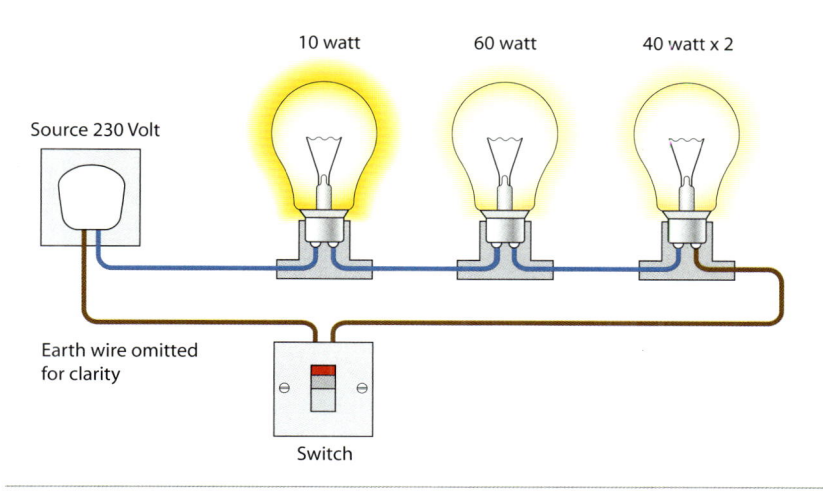

A simple series circuit with three resistors of unequal watts

The third bulb added is a 230V, 10W bulb, so we first need to calculate its resistance.

I = $\dfrac{P}{V}$ (from the power triangle) to find the current:

(I) = $\dfrac{10 \text{ watts}}{230 \text{ volts}}$ = 0.0435 amps

R = $\dfrac{V}{I}$ = $\dfrac{230}{0.0435}$ = 5290Ω

Total resistance in the circuit = 5290 + 1322.6 + 1322.6 = 7938.2Ω

Therefore: I = $\dfrac{V}{R}$ = $\dfrac{230}{7938.2}$ = 0.0290 amps

V at R1 = I × R = 0.0290 × 529.95 = 153.29V

V at R2 = I × R = 0.0290 × 1322.6 = 38.35V

V at R3 = I × R = 0.0290 × 1322.6 = 38.35V

Total Volts = 229.99V (230V)

This shows that the bulb with the highest resistance (10W) would draw more of the voltage than the other two bulbs and would glow almost at full brightness, whereas the other 40 watt bulbs would hardly glow at all.

Parallel circuits

A parallel circuit is one that has at least two independent paths in the circuit.

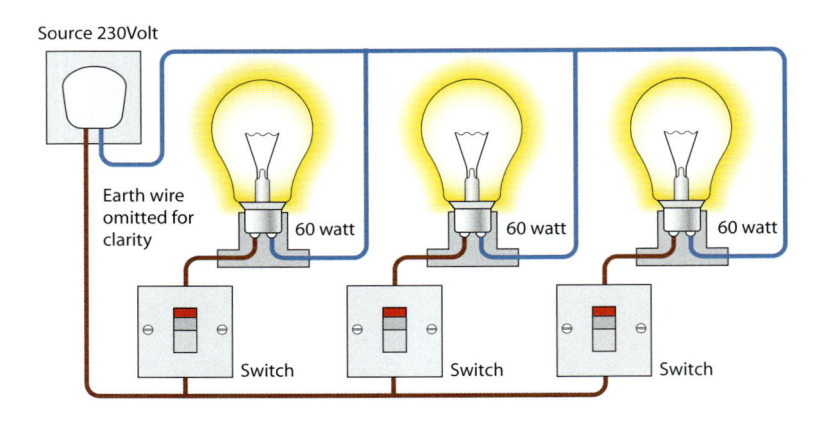

A simple parallel circuit with three resistors

Since each light bulb has its own independent closed circuit to the power source, one is unaffected by the other and voltage is equal across all components within the circuit. However, the amperage increases when more light bulbs are added and if too many are added, the circuit will overload.

To find the amps at each light bulb: $\frac{P}{V}$

R1 = $\frac{40}{230}$ = 0.173 amps

R2 = $\frac{60}{230}$ = 0.260 amps

R3 = $\frac{100}{230}$ = 0.434 amps

Total current in the circuit: IR1 + IR2 + IR3

0.173 + 0.260 + 0.434 = 0.867 amps

To find the resistance of each light bulb: $\frac{V}{I}$

R1 = $\frac{230}{0.173}$ = 1329.47 ohms

R2 = $\frac{230}{0.260}$ = 884.6 ohms

R3 = $\frac{230}{0.434}$ = 529.9 ohms

	R1	R2	R3	R4
Volts	230	230	230	230
Amps	0.173	0.260	0.434	0.867
Ohms	1329.47	884.6	529.9	265.28
Watts	40	60	100	200

The table above shows that the voltage remains constant and that the total resistance in the circuit decreases as the watts and amps increase.

H8 THE REQUIREMENTS FOR EARTHING OF ELECTRICAL CIRCUITS

The ground that we stand on is a better conductor of electricity than the copper wires in domestic electrical installations and circuitry. Electricity will always travel the least line of resistance, even if that path to earth is a human being. If you touch a live source of electricity, the current will always flow through you to the earth, causing an electric shock and possibly even death. The proper earthing of electrical circuits is, therefore, of paramount importance for the safe use of electricity.

Earthing

To prevent damage caused by an electrical fault, electrical installations (lighting and power) must be earthed. This is done by means of a dedicated earth wire, which permanently connects each socket, light fitting and switch to a metal earthing block in the consumer unit (fuse box). When an electrical fault occurs, the current is carried safely away to earth and the change in the electrical flow will cause the fuse to blow or the residual current device (RCD) to trip out, cutting off the supply of electricity. The earthing cable is always coloured green/yellow and must be installed in 10mm^2 cable.

In electrical appliances, the earth wire in the flex is covered in yellow and green plastic, and should be connected to the earth terminal of the plug. In the electrical installation (lighting and ring main), the earth wire is an unsheathed copper wire included in the cable between the neutral and live wires. A yellow/green coloured sheath has to be placed over it.

When earthing the mains cold water and gas pipes, 10mm^2 green and yellow single-core earth wire should be used to clamp the wires within 600mm of the meter or stop valve.

Plastic pipes do not require earthing.

Equipotential bonding

All metal fixtures in a domestic property, such as hot and cold water pipes, central heating pipes and gas pipes, radiators, stainless steel sinks, steel and cast iron baths, and steel basins, must be equipotentially bonded. This means that sinks, basins, radiators and pipework should be bonded using an earth bonding clamp connected by a yellow and green earth wire to ensure a path through to earth. When attaching the clamp to a pipe, always make sure that it makes a good connection by removing any paint or corrosion and then cleaning the pipe. If plastic connectors have been used on copper pipework, the bonding connection must be made across them to permanently cross bond the copper pipework.

When bonding metallic sinks, basins, baths and shower trays, 4mm² single-core earth wire should be used.

Plastic pipework does not require bonding.

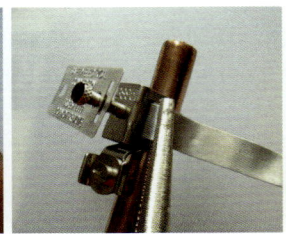

Earth bonding clamps

Equipotential bonding

CONCLUSION

In this chapter, we have seen how even simple actions, such as hammering a nail or using a screwdriver, have a scientific explanation. The actions we perform and the materials we use employ the laws of physics and chemistry to useful effect that allow us to install systems of plumbing safely and professionally. We have also investigated the limitations of some materials and how we must always be aware of what we are using and how we use it, if problems of corrosion and poor workmanship are to be avoided. These are points that will become clearer as we move forward through the following chapters of this book.

004 TEST YOUR KNOWLEDGE

SmartScreen Unit 004 revision sample questions

1. Complete the following table:

Measure of	Base SI unit	Symbol
Length		
Mass		
Time		
Electric current		
Thermodynamic temperature		

2. Name four pure metals.

3. What is an alloy?

4. Natural gas has a relative density (specific gravity) of 0.6–0.7, but what does this mean?

5. Polybutylene, polyethylene and polypropylene are which kind of plastic?

6. Give a brief definition of the term **tensile strength**.

7. Name the two types of hard water. Which one causes scaling of pipes?

8. How many kilojoules would it take to heat 140 litres of water from 4°C to 80°C?
The formula for this is: L × Δt × SHC of water

9. According to Boyle's Gas Law, what happens to a given volume of gas if the pressure is doubled?

10. When water is heated to its boiling point, the physical rise in temperature is due to sensible heat but where and how would we see the presence of latent heat.

11. What are the three methods of heat transfer?

12. What is meant by the term **static head**?

13. Name the SI unit of force.

14. What six factors determine the flow rate of cold and hot water systems?

15. Newton's third law of motion states that: **For every action there is….**

16. In physics, what is a moment of force and how do we apply it in our work?

17. Define the term **direct current**.

18. What is electrical resistivity?

19. A float-operated valve has a flow rate of 0.20 l/s. How long would it take to fill up a 6 litre cistern?

20. What is equipotential bonding?

005
UNDERSTAND AND CARRY OUT SITE PREPARATION AND PIPEWORK FABRICATION TECHNIQUES FOR DOMESTIC PLUMBING AND HEATING SYSTEMS

A plumber's job is to install the systems of hot and cold water, central heating, sanitation and gas in a professional manner, using materials safely, economically and correctly. This involves planning and setting out the work and using installation techniques that not only satisfy the requirements of the customer and protect their property but also comply with the relevant regulations, British Standards and codes of good practice. In this chapter we will explore the wide variety of tools that plumbers use, the range of materials available and the correct methods of working needed to install them.

IN THIS CHAPTER, YOU WILL COVER:

A Hand and power tools used in the plumbing industry
A1 Hand tools
A2 Plumbing-specific tools
A3 Other hand tools
A4 Hand tool safety and maintenance
A5 Power tools
A6 Power tool safety and maintenance

B Materials used in domestic plumbing and heating installations
B1 Copper tubes to BS EN 1057:2006 + A1: 2010B2
B2 Bending copper tube
B3 Jointing methods and fittings for copper tube
B4 Low carbon steel pipes to BS EN 10255:2004
B5 Bending low carbon steel pipe
B6 Jointing methods and fittings for low carbon steel pipe
B7 Plastics
B8 Polyvinyl chloride (PVC) and acrylonitrile butadiene styrene (ABS)
B9 Jointing methods and fittings for PVCu, muPVC and ABS
B10 Polypropylene (PP)
B11 Jointing methods and fixings for polypropylene
B12 Polyethylene (PE)
B13 Jointing methods and fixings for medium density polyethylene (MDPE)
B14 Polybutylene (PB-1)
B15 Bending polybutylene
B16 Jointing methods and fixings for polybutylene

C Fixings for masonry, timber and plasterboard
C1 Nails
C2 Screws
C3 Heavy-duty fixings
C4 Plasterboard and light structure fixings
C5 Plastic wall plugs

D Installation processes of domestic plumbing systems
D1 The design process of plumbing installations
D2 Planning the work and the delivery of materials
D3 Pre-installation activities on new and existing installations
D4 Preparation of the work area
D5 Installation activities on new and existing installations
D6 Protection of the building fabric and its surroundings
D7 Testing and commissioning procedures
D8 Decommissioning procedures
D9 Maintenance activities
D10 Making good the building fabric

E Drawing symbols of plumbing valves and appliances

Test your knowledge

A HAND AND POWER TOOLS USED IN THE PLUMBING INDUSTRY

In this first section of the chapter, we will look at the wide range of tools a plumber will use and how we can keep them in good working order to ensure a long and trouble-free life.

A1 HAND TOOLS

Screwdrivers

There are many different types of screw head, which all have different uses. As a result, a plumber should have a wide selection of screwdrivers available. Some have specialist applications and uses, such as insulated electrical screwdrivers and long-bladed types. The common head types are shown below:

Flat blade For use with slotted screws. Care should be taken to ensure the correct blade size for the screw slot.	⊖
Phillips head Originally designed in the 1930s to intentionally 'ride-out' of the screw head to prevent over-tightening.	✚
Pozidriv head Similar to the Phillips head but has an eight-pointed star shape for better grip. Not compatible with Phillips screws.	✚
Star head Not often used, except in specialist installations and appliances. Also known as Torx screwdrivers.	✡
Hexagon head (Allen key) Mainly used in the gas industry for appliance servicing and installation.	⬡

Screwdrivers have a particular use and when used correctly should give a long-lasting service. Problems often occur if they are mistreated or used improperly. Always remember:

- A screwdriver is not a chisel and should not be used as such.
- Use the correct screwdriver for the screw, ie a Pozidriv screw needs a Pozidriv screwdriver not a Phillips screwdriver.
- Never over-tighten the screw as this can damage the screw head, making it difficult to withdraw the screw in future.
- Choose the right-sized blade for slotted screws – using too small a blade will result in the screwdriver slipping out of the head, causing damage.
- Keep fingers behind the blade.
- Use an insulated screwdriver when working with electricity.

Hammers

There are two primary types of hammers used by plumbers:

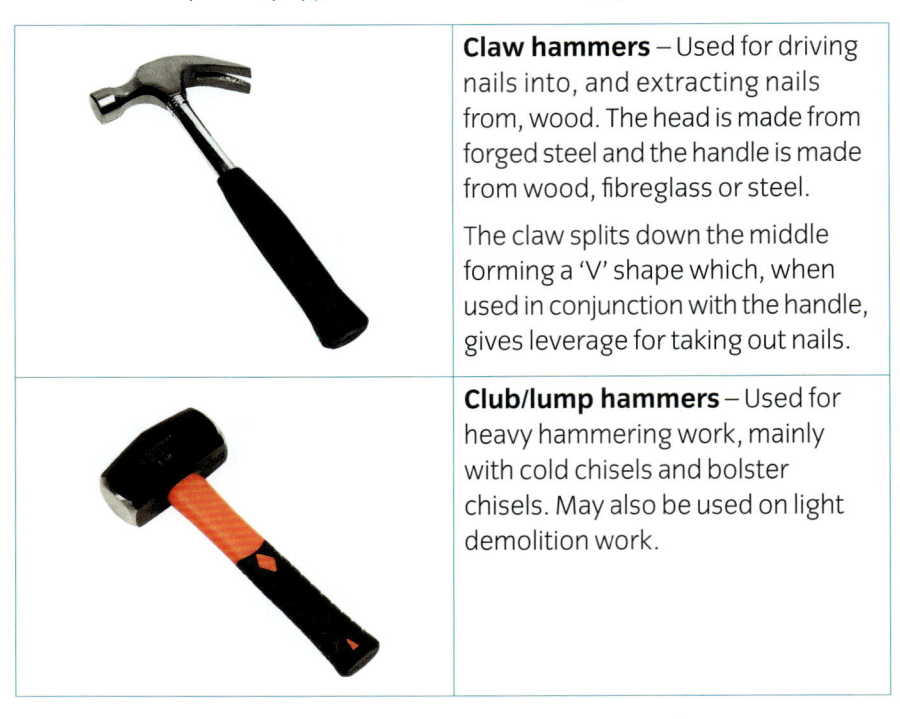

	Claw hammers – Used for driving nails into, and extracting nails from, wood. The head is made from forged steel and the handle is made from wood, fibreglass or steel. The claw splits down the middle forming a 'V' shape which, when used in conjunction with the handle, gives leverage for taking out nails.
	Club/lump hammers – Used for heavy hammering work, mainly with cold chisels and bolster chisels. May also be used on light demolition work.

There are two types of chisel that a plumber will use. Both types have very different uses:

- Cold chisels are used for breaking and cutting masonry and concrete. These include:
  - bolster chisels
  - plugging chisels
  - flat chisels.
- Wood chisels.

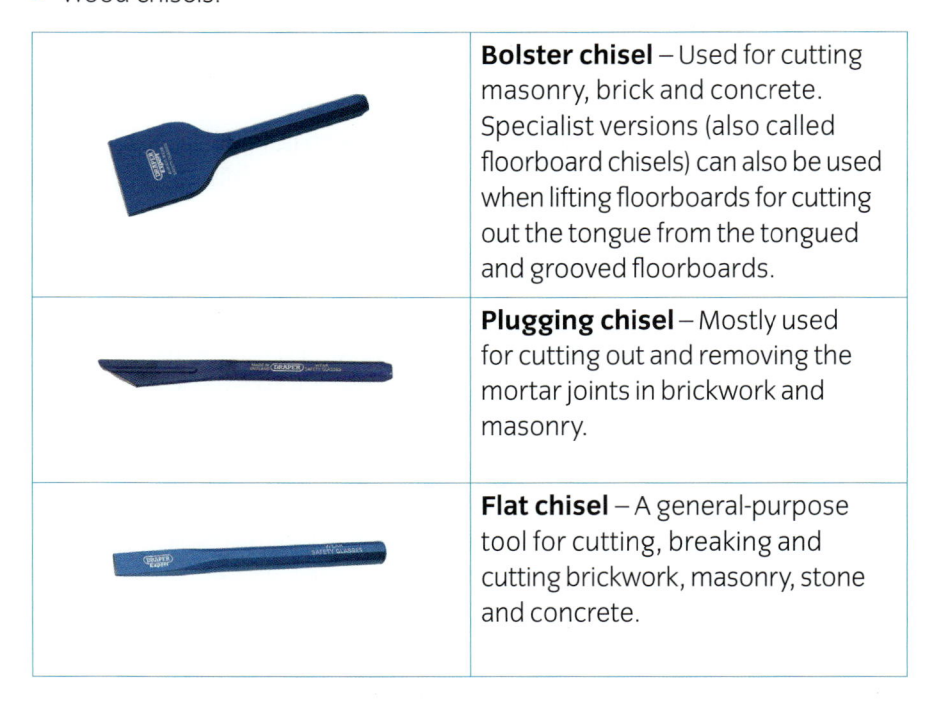

	Bolster chisel – Used for cutting masonry, brick and concrete. Specialist versions (also called floorboard chisels) can also be used when lifting floorboards for cutting out the tongue from the tongued and grooved floorboards.
	Plugging chisel – Mostly used for cutting out and removing the mortar joints in brickwork and masonry.
	Flat chisel – A general-purpose tool for cutting, breaking and cutting brickwork, masonry, stone and concrete.

Wood chisel – There are many different types of wood chisel, including flat-bladed and bevelled-edge chisels. These are mainly used by plumbers for notching joists. Care should be taken when using these, as the blades can be extremely sharp.

Below are some important points to remember about the use of chisels:

- Eye protection must be worn when using any type of chisel because of the risk of flying debris.
- Always keep the cutting blade sharp and well ground.
- Watch out for 'mushrooming' on the heads of cold chisels. This is where the metal begins to fold over and split due to being repeatedly hit with a hammer. Mushrooming should be removed by grinding on a grinding wheel.
- Always wear gloves when using cold chisels. They help to protect the hands from cuts and abrasions.
- Always keep fingers away from the cutting blade, especially when using very sharp wood chisels.

Spanners

A plumber's toolbox should contain a variety of spanners. Different types will be needed, depending on the type of work:

	Adjustable spanners – General-purpose spanners, used for tightening compression joints, radiator valve unions and nuts and bolts. Three pairs of spanners of varying sizes is optimum for a plumber's toolbox. This tool should be kept oiled and clean, and always ensure that the retention screw is tight.
	Open jawed spanners – Mostly used for boiler and appliance servicing. A small set is recommended for the toolbox.
	Ring spanners – Again, mostly used for boiler and appliance servicing.
	Box spanners – The main tool for fixing taps to sanitary ware in sizes 13mm for monobloc mixers, ½ inch for sink and wash basin taps and ¾ inch for baths.
	Immersion heater spanner – A specialist tool for installing and removing immersion heaters from hot water storage cylinders and vessels.

Handsaws

There are four main types of handsaw that plumbers would use, which should be included in their toolbox:

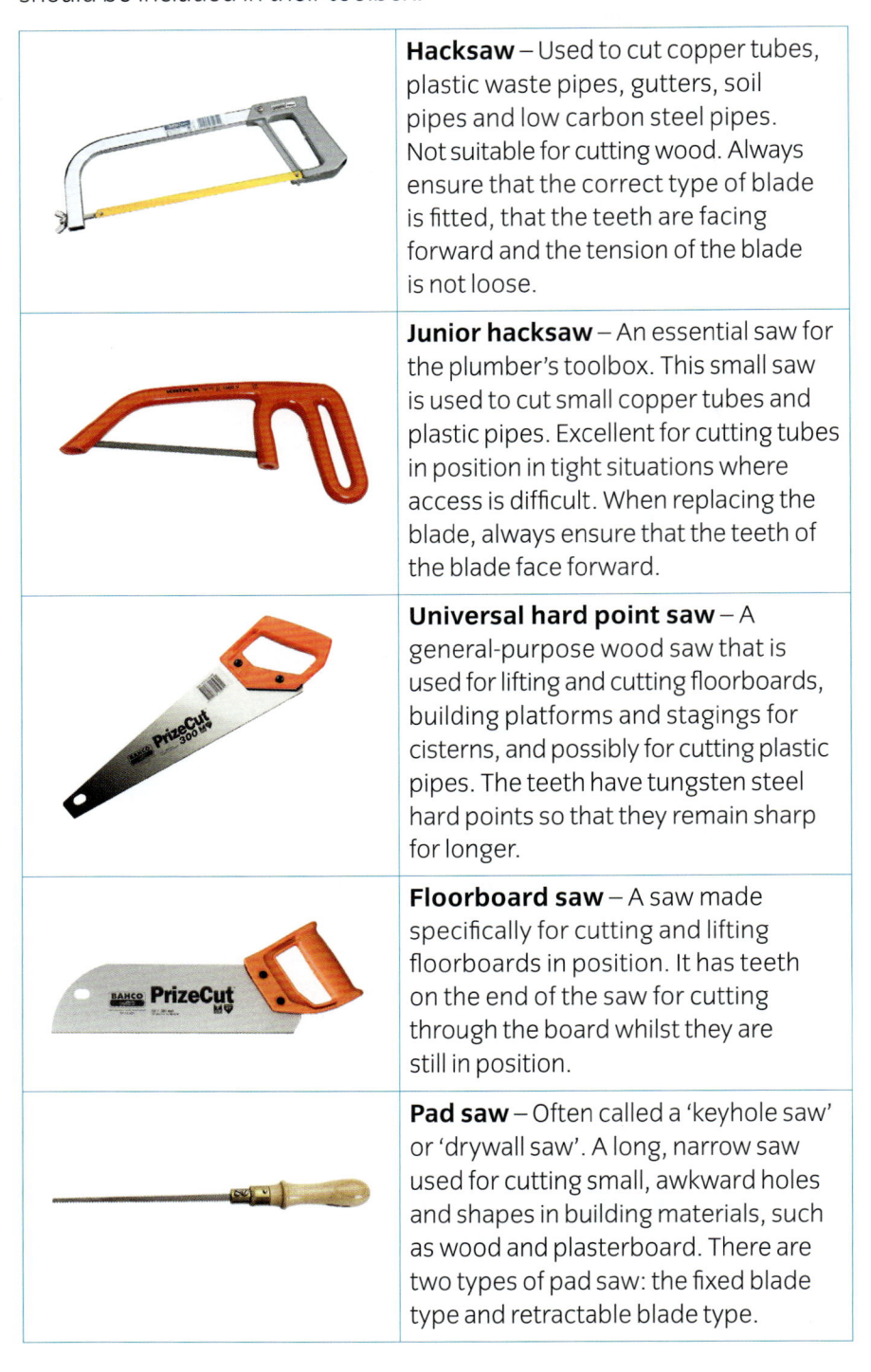

	Hacksaw – Used to cut copper tubes, plastic waste pipes, gutters, soil pipes and low carbon steel pipes. Not suitable for cutting wood. Always ensure that the correct type of blade is fitted, that the teeth are facing forward and the tension of the blade is not loose.
	Junior hacksaw – An essential saw for the plumber's toolbox. This small saw is used to cut small copper tubes and plastic pipes. Excellent for cutting tubes in position in tight situations where access is difficult. When replacing the blade, always ensure that the teeth of the blade face forward.
	Universal hard point saw – A general-purpose wood saw that is used for lifting and cutting floorboards, building platforms and stagings for cisterns, and possibly for cutting plastic pipes. The teeth have tungsten steel hard points so that they remain sharp for longer.
	Floorboard saw – A saw made specifically for cutting and lifting floorboards in position. It has teeth on the end of the saw for cutting through the board whilst they are still in position.
	Pad saw – Often called a 'keyhole saw' or 'drywall saw'. A long, narrow saw used for cutting small, awkward holes and shapes in building materials, such as wood and plasterboard. There are two types of pad saw: the fixed blade type and retractable blade type.

Grips and wrenches

Grips and wrenches are tools that are used almost every day by plumbers. They are essential tools for tightening and gripping. There are several different types:

	Water pump pliers – Although known as pliers, these are a plumber's general-purpose grips. Three pairs should be available in the toolbox – 175mm, 250mm and 300mm.
	Footprint – Another general-purpose grip (also known as an adjustable pipe wrench), used by plumbers for tightening fittings and unions. Care should be taken when using these as they can easily trap fingers if used incorrectly.
	Stillson – Used when installing low carbon steel pipe. They are available in many sizes ranging from 10in to 36in.
	Basin wrench – Used for tightening and loosening tap connections in hard-to-reach areas, such as behind wash hand basins, baths and kitchen sinks.
	Mole grips – A locking type of pliers. They give a high clamping force and can be locked to allow hands-free gripping.

Pliers

Pliers are two-handled, two-jawed hand tools used mainly for gripping, twisting and turning. The jaws meet at the tip, which means that they can grip with precision. Some types are also made for cutting cable and wire. Listed below are some of the main types of pliers used by plumbers:

	General-purpose pliers – A useful addition to the toolbox, these are used to grip and tighten small nuts and bolts. They can also be used to cut thin wire and electrical flex.
	Long-nose pliers (needle nose pliers) – Useful both for gripping small items and reaching into small, deep spaces. They are used to tighten small nuts and bend wire. They often include a wire cutter.
	Circlip pliers – These have a specific use: removing the circlips from non-rising tap spindles and shower valves. Mainly used in maintenance and repair operations.
	Wire cutters – A useful addition to a plumber's toolbox, these are used for cutting electrical cables and flex.

SmartScreen Unit 005 interactive activity 2

Level
When pipework is perfectly horizontal.

Plumb
When pipework is perfectly vertical.

Spirit levels

Spirit levels are used to ensure that appliances and pipework are installed level and plumb. They use a bubble positioned between two markers. Electronic and laser spirit levels are also available.

	Torpedo/boat Level – A 300mm level with a magnetic strip on the bottom, which makes it easier to level appliances such as boilers.
	Spirit level – Two sizes, 600mm and 120mm, are advisable for levelling large appliances such as baths and wash hand basins.

A2 PLUMBING-SPECIFIC TOOLS

So far we have seen the more common hand tools. As well as these, there are many 'plumbing-specific' hand tools that a plumber must have in their toolkit, for cutting, bending and working with pipes.

Pipe cutting tools

	Pipe slice – An essential tool for cutting copper tube. The pipe slice can be used in tight situations where junior hacksaws and adjustable pipe cutters cannot. Two sizes are available, 15mm and 22mm. Always ensure that the cutting wheel, wheel spring and rollers are oiled and free from dirt.
	Adjustable pipe cutters – Again, an essential tool that can be adjusted to cut many sizes of copper tube. Periodic maintenance of this tool is recommended, such as changing the cutting wheel and regular oiling.
	Plastic pipe cutters – Can be used to cut all forms of plastic pipe. They give a clean cut, which is essential when jointing push-fit pressure plastic pipe.

Annealing
A process that involves heating the copper to a cherry-red colour and then quenching it in water. This softens the copper tube so that the copper can be worked without fracturing, rippling or deforming.

Pipe bending tools

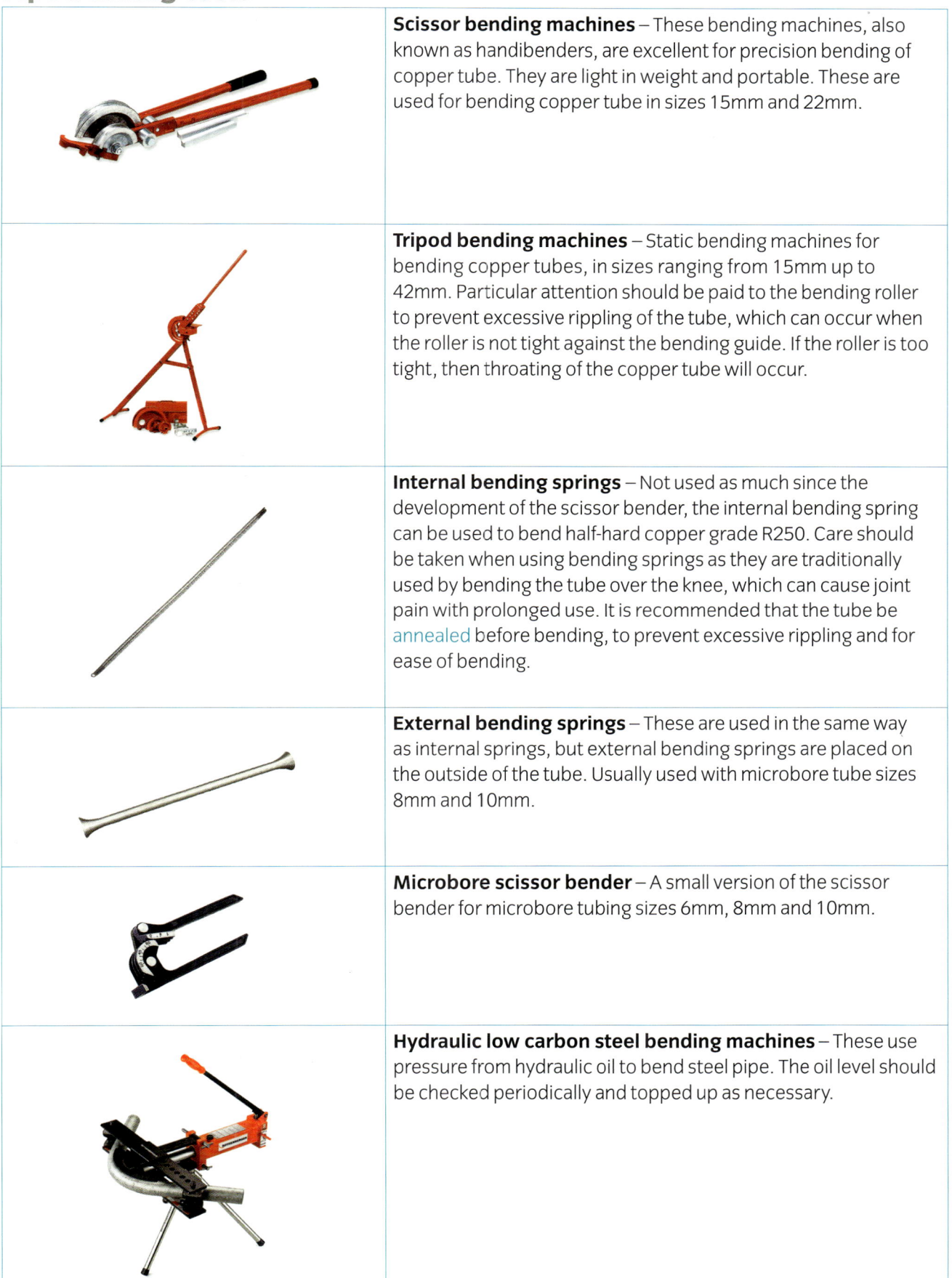

Scissor bending machines – These bending machines, also known as handibenders, are excellent for precision bending of copper tube. They are light in weight and portable. These are used for bending copper tube in sizes 15mm and 22mm.

Tripod bending machines – Static bending machines for bending copper tubes, in sizes ranging from 15mm up to 42mm. Particular attention should be paid to the bending roller to prevent excessive rippling of the tube, which can occur when the roller is not tight against the bending guide. If the roller is too tight, then throating of the copper tube will occur.

Internal bending springs – Not used as much since the development of the scissor bender, the internal bending spring can be used to bend half-hard copper grade R250. Care should be taken when using bending springs as they are traditionally used by bending the tube over the knee, which can cause joint pain with prolonged use. It is recommended that the tube be annealed before bending, to prevent excessive rippling and for ease of bending.

External bending springs – These are used in the same way as internal springs, but external bending springs are placed on the outside of the tube. Usually used with microbore tube sizes 8mm and 10mm.

Microbore scissor bender – A small version of the scissor bender for microbore tubing sizes 6mm, 8mm and 10mm.

Hydraulic low carbon steel bending machines – These use pressure from hydraulic oil to bend steel pipe. The oil level should be checked periodically and topped up as necessary.

Pipe soldering equipment

There are many types of soldering equipment available. The most popular choices are listed below:

> **KEY POINT**
> A fire extinguisher should always be available when using any form of soldering equipment.

	**Blowtorch with separate governor, hose and LPG bottle** – The traditional plumber's blowtorch. The governor can be preset or adjustable, and the nozzles on the blowtorch are interchangeable with varying sizes for different tube sizes. These are not as controllable as other torches.
	**Soldering and brazing torch** – This type of blowtorch is much more portable and gives a more concentrated flame, which is far more controllable because the set-up is smaller and easier to handle. It can be used with MAPP gas, but gas usage tends to be high.

Socket forming tools

	Socket expanding tool – These are used to form sockets in the ends of copper tube instead of using couplings. Care should be taken because the tool can split the end of the copper if used incorrectly.
	Because this type of socket can weaken the tube, it cannot be used on new water installations or on gas systems, and several authorities prohibit or discourage their use.

Socket crimping tools

	Copper pipe socket crimping tools – This is the latest addition to the plumber's list of tools. They are used for crimping special fittings onto copper tubes.

Pipe threading equipment

Although not strictly plumbing tools, pipe-threading equipment may be used occasionally when installing low carbon steel pipe. There are three main types:

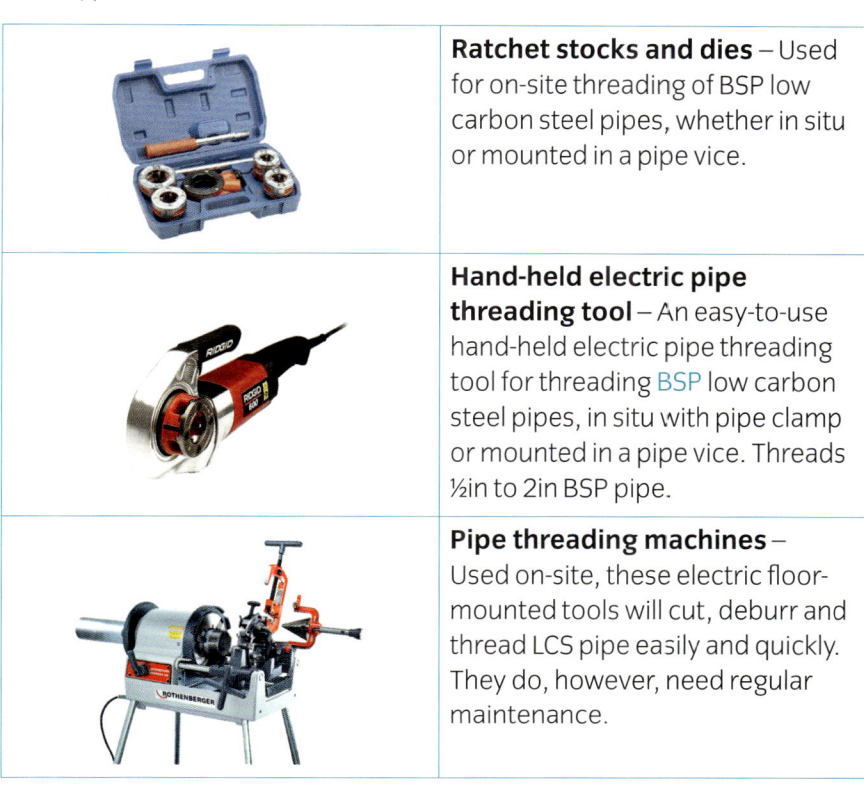

	Ratchet stocks and dies – Used for on-site threading of BSP low carbon steel pipes, whether in situ or mounted in a pipe vice.
	Hand-held electric pipe threading tool – An easy-to-use hand-held electric pipe threading tool for threading BSP low carbon steel pipes, in situ with pipe clamp or mounted in a pipe vice. Threads ½in to 2in BSP pipe.
	Pipe threading machines – Used on-site, these electric floor-mounted tools will cut, deburr and thread LCS pipe easily and quickly. They do, however, need regular maintenance.

KEY POINT

When using threading tools, plenty of oil should be used as this helps to lubricate and cool the cutting heads. Threading tools have a reversible action. This allows the cutting head to be removed from the pipe and also cleans the newly cut threads of all cut steel and excess oil (known as 'swarf').

BSP or BSPT

Stands for British Standard Pipe and British Standard Pipe Threads, and relates to the type of thread we use on screwed Low Carbon Steel pipes and fittings. Although the pipe is measured in mm, it is universally referred to in imperial measurements, eg. ½-inch BSPT (meaning ½-inch British Standard Pipe Thread).

A3 OTHER HAND TOOLS

As well as the tools we have already seen, there are other tools that a plumber may need. These are general tools that are useful additions to the toolkit, and include the following:

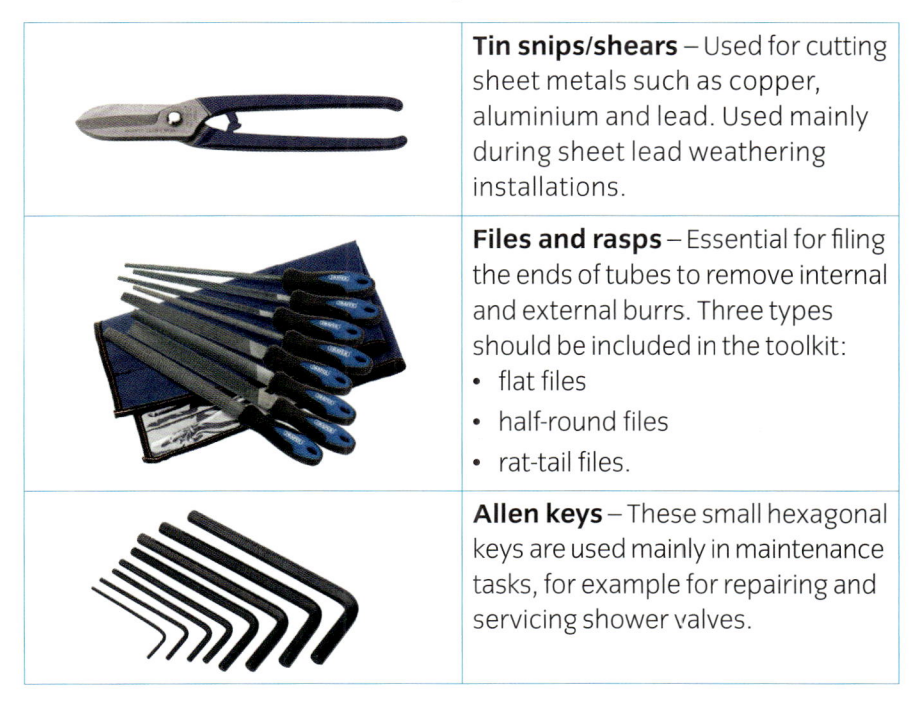

	Tin snips/shears – Used for cutting sheet metals such as copper, aluminium and lead. Used mainly during sheet lead weathering installations.
	Files and rasps – Essential for filing the ends of tubes to remove internal and external burrs. Three types should be included in the toolkit: • flat files • half-round files • rat-tail files.
	Allen keys – These small hexagonal keys are used mainly in maintenance tasks, for example for repairing and servicing shower valves.

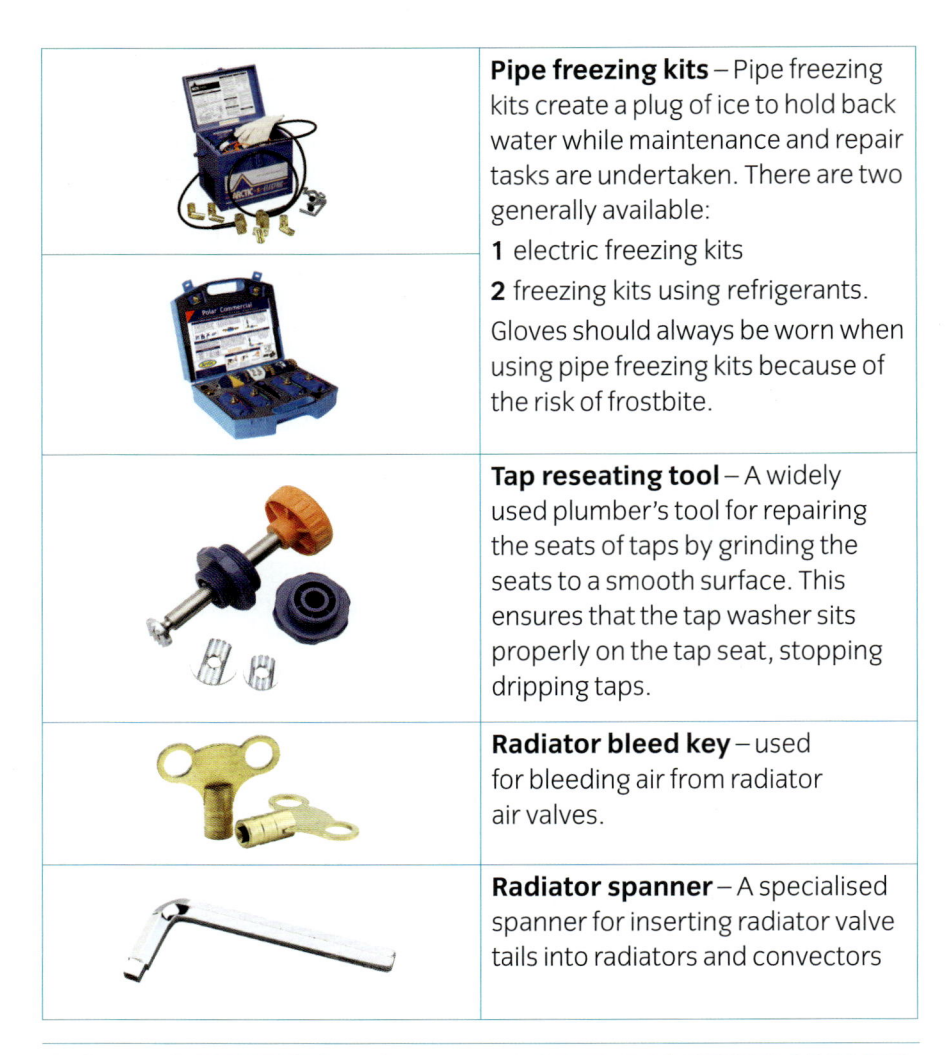

	Pipe freezing kits – Pipe freezing kits create a plug of ice to hold back water while maintenance and repair tasks are undertaken. There are two generally available:
1 electric freezing kits
2 freezing kits using refrigerants.
Gloves should always be worn when using pipe freezing kits because of the risk of frostbite.
	Tap reseating tool – A widely used plumber's tool for repairing the seats of taps by grinding the seats to a smooth surface. This ensures that the tap washer sits properly on the tap seat, stopping dripping taps.
	Radiator bleed key – used for bleeding air from radiator air valves.
	Radiator spanner – A specialised spanner for inserting radiator valve tails into radiators and convectors

A4 HAND TOOL SAFETY AND MAINTENANCE

When used properly, hand tools should present no risk to the user. It is when unsafe practices are used that they can become dangerous. A large number of accidents occur every year in the construction industry because of the incorrect use of manual and power hand tools, such as using a screwdriver as a chisel or a lever. Most accidents involving tools result from:

- using the wrong tool for the job
- using the tool incorrectly
- not wearing personal protective equipment (PPE)
- not following approved safety guidelines
- poor maintenance.

The most common tools involved in plumbing accidents are:

- chisels
- saws
- screwdrivers
- files
- snips
- hammers
- wrenches, grips and pliers.

Using hand tools is an important part of your job. They must be treated, cared for and used in a professional manner. By following these safety rules, many hand tool injuries can be avoided:

- Know the purpose of each tool in your toolbox, and use it for the task it was designed to do.
- Never use any tool unless you are trained to do so.
- Inspect tools before each use and replace or repair if they are worn or damaged.
- Always clean your tools when you have finished using them.
- Always keep cutting edges of chisels and saws sharp.
- Always keep any moving parts free from dirt and make sure they are well oiled.
- Select the right-sized tool for the job.
- When working on ladders or scaffolding, be sure that you and your tools are secure. Falling tools could injure people working or passing below.
- Do not put sharp or pointed tools in your pockets. Use a sheath or holster instead.
- Do not throw tools as they are easily damaged.
- Do not use a tool if the handle is missing or has splinters, burrs or cracks or if the head of the tool is loose.
- Do not use cold chisels that have mushroomed heads.
- When using tools such as jigsaws, chisels and drills, always wear PPE such as safety glasses/goggles, facemasks or gloves.

A5 POWER TOOLS

Apart from the hand tools we have looked at, a plumber needs certain power tools to help with the installation process.

Power drills, cordless drills and power saws

Power drills and cordless drills can be divided into three different types and these are shown in the table below. Power saws are also included as these play a very important part in the installation process of plumbing and heating systems.

Rotary hammer drills – This type of drill has a standard chuck so accessories such as metal drills and hole saws can be used. The chuck should be kept well oiled to prevent breakdown.

KEY POINT
You might like to refresh your knowledge about working at heights by checking the information in Chapter 001, page 63.

KEY POINT
Don't forget that under the Health and Safety At Work Act 1974, you have a duty of care to yourself, your employer and others who may be affected by your acts or omissions.

KEY POINT
Power tools for use on site should be 110V, which is colour-coded yellow for easy identification. You should not use 230V – 110V is a safer voltage.

	SDS hammer drills – This type of drill uses the Secure Drill System (SDS) bayonet type fixing to secure the drill bits into the chuck. This type of drill is necessary when using core bits.
	Cordless drills – Typical voltages are from 14.4V to 36V. These drills are available in many forms, from screwdriver type drills to large voltage SDS types.
	Circular saws – A very useful tool for lifting floorboards and notching joists. Care should be taken to ensure that the blade guard is in place and that the blade is securely fastened.
	Jig saws – Used for cutting out sinks and wash hand basins in worktops in kitchens and bathrooms. Always ensure that the blade guard and blade are securely in place.

Drill bits, core drills and hole saws

There are many types of drill bits that should be included in a plumber's toolkit. Each one has a specific job:

	Masonry drill bits – The tip of this drill bit is made from tungsten carbide steel to enable the bit to penetrate masonry, concrete and stonework.
	Wood drill bits – Also known as a 'spur points' or 'dowel bits', these have a central point and two raised spurs that help keep the bit drilling straight.
	Metal drill bits – Also known as 'twist bits', these can be used on timber, metal and plastics. Most twist bits are made from high speed steel (HSS), and are suitable for drilling most types of material. When drilling metal the HSS stands up to the high temperatures.

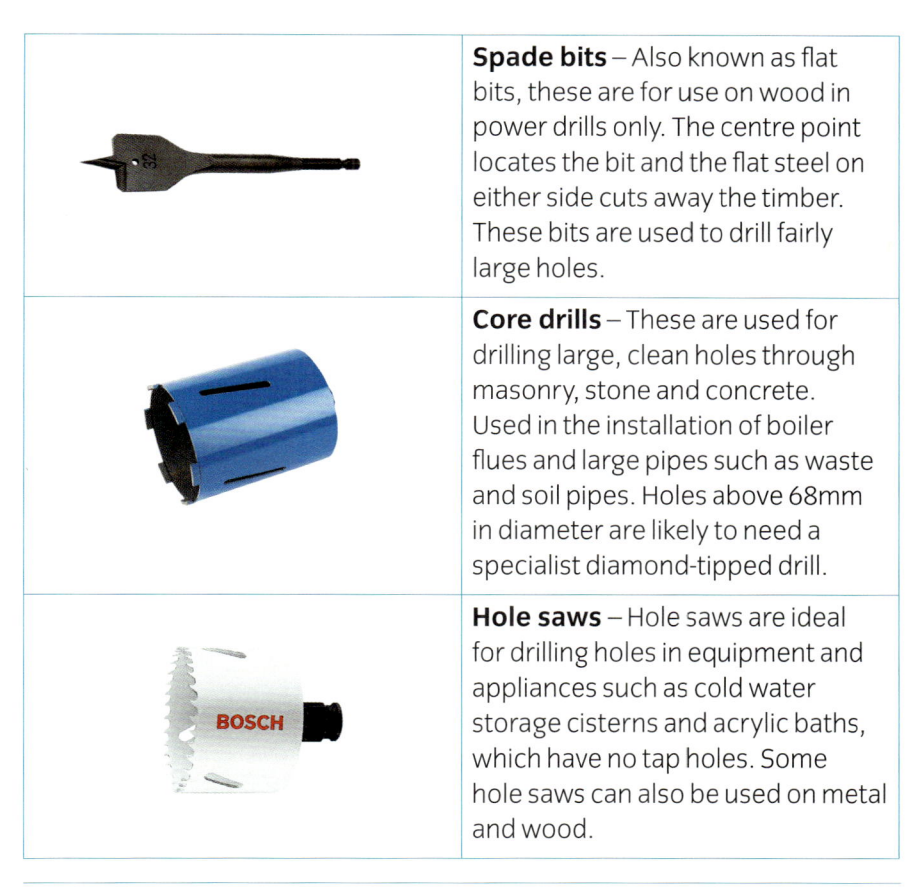

	Spade bits – Also known as flat bits, these are for use on wood in power drills only. The centre point locates the bit and the flat steel on either side cuts away the timber. These bits are used to drill fairly large holes.
	Core drills – These are used for drilling large, clean holes through masonry, stone and concrete. Used in the installation of boiler flues and large pipes such as waste and soil pipes. Holes above 68mm in diameter are likely to need a specialist diamond-tipped drill.
	Hole saws – Hole saws are ideal for drilling holes in equipment and appliances such as cold water storage cisterns and acrylic baths, which have no tap holes. Some hole saws can also be used on metal and wood.

A6 POWER TOOL SAFETY AND MAINTENANCE

As with hand tools, power tools need regular inspection and maintenance. There are certain points that should be followed:

- Power tools should be PAT tested every three months (see Chapter 001, pages 49–51).

- Always wear safety goggles or safety glasses when using power tools.

- Use a dust mask in dusty conditions and wear hearing protection if the tool is being used for an extended period of time. Prolonged use of hammer type power tools can cause vibration injury.

- Always check the tool, the cord and the plug before use for any signs of wear or damage. Damaged tools should be taken out of use, tagged and sent for repair.

- Always check to make sure that the tool is the correct voltage for the power supply.

- Make sure that all appropriate safety guards are in place and never remove a safety guard.

- Make sure that cutters or blades are clean, sharp and securely in place. Never use bent or broken blades or cutters.

- Make sure that extension cords are the correct type and don't use cords designed for inside use outside.

- Make sure that the work area is clean and free of debris that might get in the way and always make sure that the work area has plenty of light.

SmartScreen Unit 005 handout 1

SmartScreen Unit 005 worksheet 1

SmartScreen Unit 005 interactive activity 1

- Never use power tools in wet or damp conditions.

- Never drag the tool or the cord across the floor, or lift or lower it by its cord.

- When using hand-held power tools, always grip with both hands. Anti-vibration gloves may be required with some vibrating power tools.

- Always turn off and unplug the tool before any adjustments or change of blades takes place.

- Never overreach when using a power tool and always take care when using power tools at height.

- Always unplug, clean and store the tool in a safe, dry place when the job is finished.

B MATERIALS USED IN DOMESTIC PLUMBING AND HEATING INSTALLATIONS

In this part of the chapter, we will take a brief look at the pipe materials that a plumber uses in their everyday installation work: copper, low carbon steel and plastics. We will see how the different methods of jointing, bending and installation practices dictate the methods of working we need to employ.

B1 COPPER TUBES TO BS EN 1057:2006 + A1: 2010B2

Copper tube has been used in the UK since the 1940s and today still accounts for around 60 per cent of all new installations. The type of copper used in the manufacture of tubes is phosphorus de-oxidised copper, with a minimum copper content of 99.90 per cent.

De-oxidised copper tube can be safely soldered, welded or brazed. The density of copper is 8900 kg/m³, it has a melting point of 1083°C and its coefficient of linear expansion is 0.000016 per °C (between 20 and 100°C).

The standard for copper tubes for water, gas and sanitation installations is BS EN 1057, which is available in three grades or tempers:

Grade	Description
R220	This is softer copper tube, full annealed and supplied in coils. It is thicker-walled than other grades of copper tube. Used for underground water services (sizes 15, 22, 28mm) and microbore central heating systems (sizes 6, 8 and 10mm).
R250	This is the most widely used grade of copper tube for plumbing and heating applications. Supplied in straight lengths of 3m or 6m in sizes 15, 22, 28, 35, 42, 54mm, it is known as half-hard tempered.
R290	This grade is hard tempered, thin-walled and totally unsuitable for bending. Not normally used in the UK.

KEY POINT

During your time as a plumber you will come across many materials, and each one will have its own unique working properties, including different melting points and expansion rates. It is important that we recognise these properties so that we can choose the correct material for a given installation. You will come across other such properties as you work through this book.

Temper

The temper of a metal refers to how hard or soft it is.

Tubes in half-hard (R250) and hard tempered (R290) condition are supplied in straight lengths of 3 or 6 metres. Tubes in the soft, fully annealed (R220) condition, up to 28mm outside diameter (OD), are supplied in coils. The length of the coils is between 10 and 50 metres, depending on the diameter.

Copper tubes are generally used in buildings for the following services:

- domestic hot and cold water supplies under pressure, usually up to mains pressure (typically up to 4 bar but can be up to 10 bar in some parts of the UK) or head pressure from a storage cistern
- sanitary waste pipe installations
- central heating systems (with radiators/convectors)
- existing underfloor heating systems
- natural gas installations for heating and cooking
- oil installations for heating
- medical gases (when degreased).

Copper tube is available chromium-plated for situations where there are aesthetic considerations and PVC or plastic coated in various colours where protection from corrosion is necessary.

B2 BENDING COPPER TUBE

Bending copper tubes becomes easy with practice. The two methods used to correctly bend copper tubes are:

1 **Machine bending** – the preferred method of bending copper tubes.

2 **Spring bending** – using a bending spring. Not so widely used now since scissor benders have become available.

Bending copper tube using a bending machine

Bending copper tubes using a bending machine is an economical method of installation, especially where lots of bends or changes of direction are required. There are many types of bending machine available for copper tubes up to 42mm diameter, all of which are worked by hand. For larger diameters, ratchet-action machines are required. The most useful type of machine for 15 and 22mm tube is the portable type or scissor bender (see page 215), which is light in weight and requires no adjustment before use.

The advantages of machine bending over spring bending can be summarised as follows:

- bends can be formed quickly
- multiple bends can be formed easily
- bends can be formed close to the end of the tube
- bend radius, quality and accuracy are consistent.

Producing accurately positioned bends depends on determining the bending point and the position of the tube in the machine. This will become apparent in the series of illustrations that follows.

90° bends method 1

1. Measure the distance required and mark it on the tube.

2. Measure back from that mark four diameters of the tube size being used, ie if 15mm tube is being used, measure back 60mm and mark the tube.

3. Place the 4d mark on the 'start-of-the-bend' mark on the bending machine.

4. Bend the tube to 90° and check the measurement using a set-square and a rule.

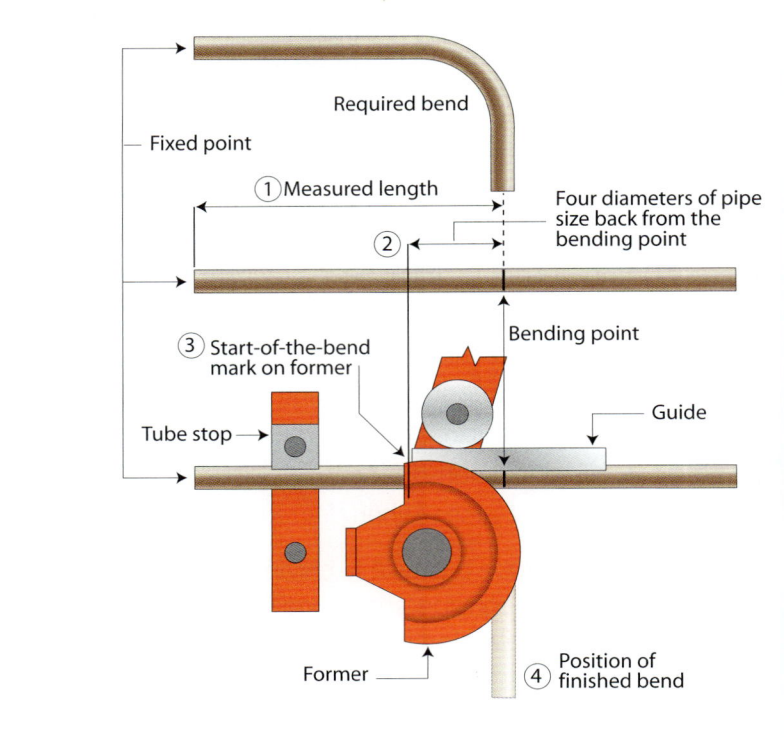

90° bends method 1

90° bends method 2

1. Measure the distance required and mark it on the tube.

2. Place a scrap piece of tube in the bending former in the finished bend position and position the mark on the tube to be bent so that it lays in the centre of the scrap tube.

3. Place a set-square against at the junction between the scrap and the tube to be bent and make sure that it is at 90° using a set-square. The mark on the tube must still be in the centre as shown in the drawing.

4. Bend the tube to 90° and check the measurement using a set-square and a rule.

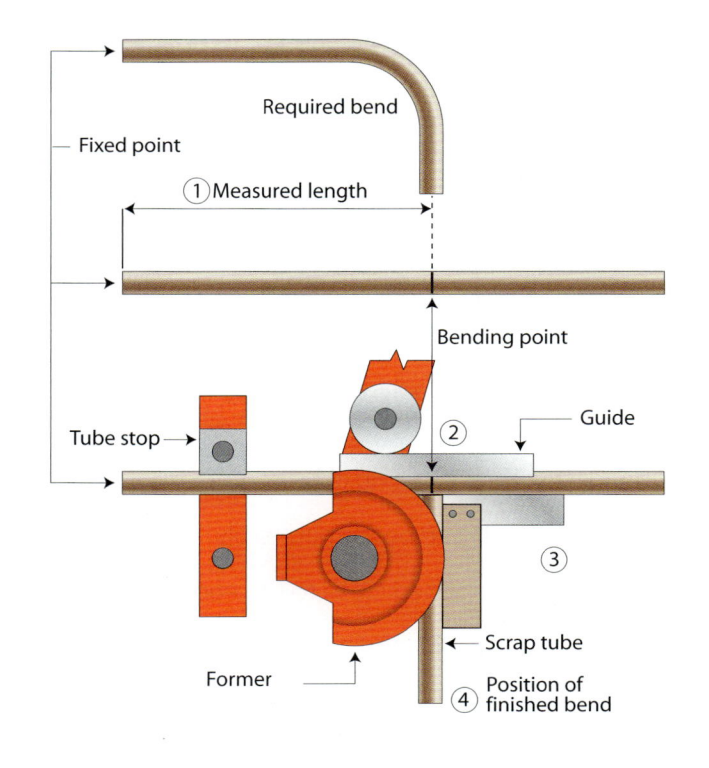

90° bends method 2

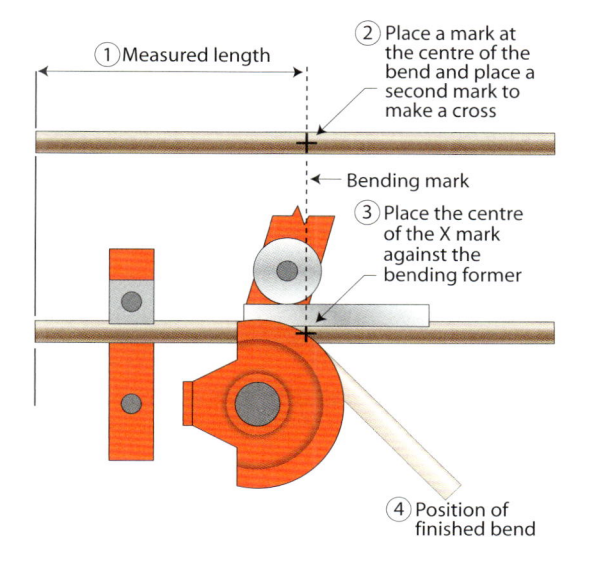

① Measured length
② Place a mark at the centre of the bend and place a second mark to make a cross
Bending mark
③ Place the centre of the X mark against the bending former
④ Position of finished bend

45° sets

45° sets

1 Measure the distance required and mark it on the tube.
2 Place a second mark on the tube to form a cross.
3 Place the centre of the cross at the edge of the bending former, ensuring that the cross is at 90° to the former.
4 Pull the bend to 45°.

Offset measurement
①

Offset measurement
50mm
Angle required
600mm rule
②

③

Bending mark
Straight edge
Offset measurement

Offset bends

1 To find the correct offset angle, the size of the offset should be deducted from the 600mm and the 600mm folding rule opened to the measurement, ie offset 50mm. 600 – 50 = 550.
2 Bend the tube to the angle set by the rule.
3 Remove the tube from the machine and mark for the second bend measuring from the inside edge of tube using a straight edge.
4 Reposition tube in the machine so that the mark forms a tangent to the former.
5 Reposition the rule to give the correct angle for the second bend.

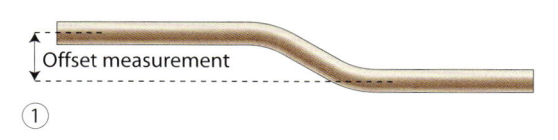

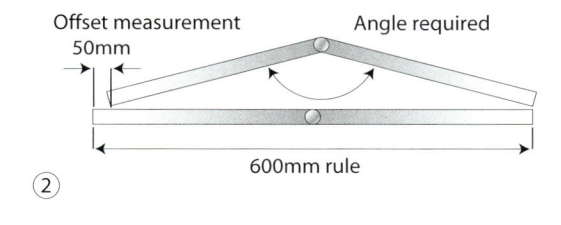

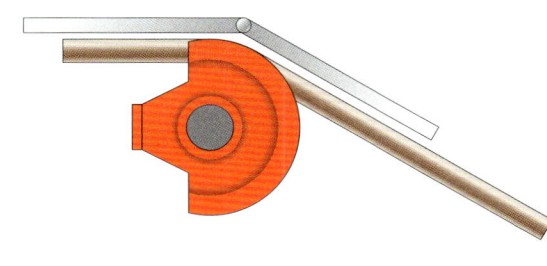

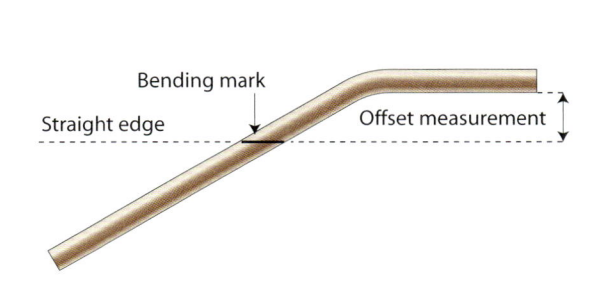

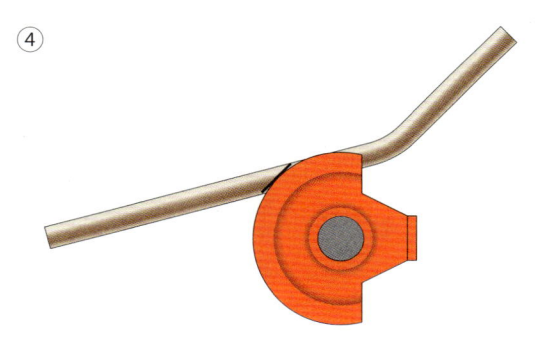

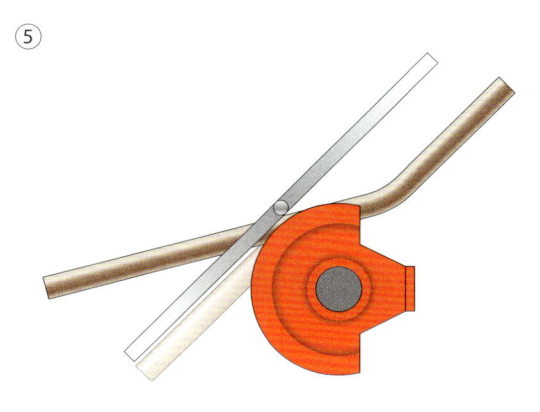

④

⑤

Offset bends

Passover bends

① **Fixed point**

Fixed point to centre of obstruction

The bench mark on the first bend is determined by adding ¼ of the diameter of the obstruction to the measurement from the fixed point to the centre of the obstruction

— Bending mark

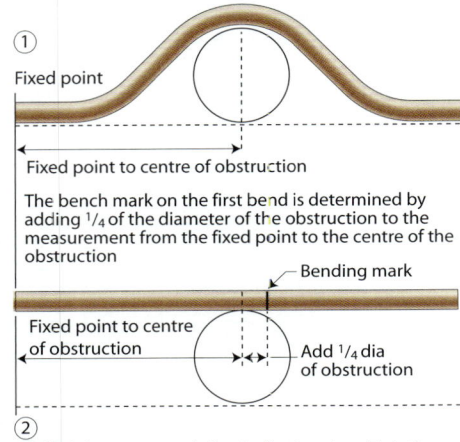

Fixed point to centre of obstruction | Add ¼ dia of obstruction

② To find the correct angle for the first bend, multiply the diameter of the obstacle by three and close the folding rule by this amount then position the tube in the machine so that the bending mark and the centre of the angle align

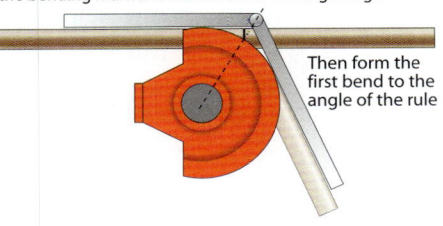

Then form the first bend to the angle of the rule

③ Making sure that the bend clears the obstruction place a straight edge over the tube and mark the bending marks on both sides

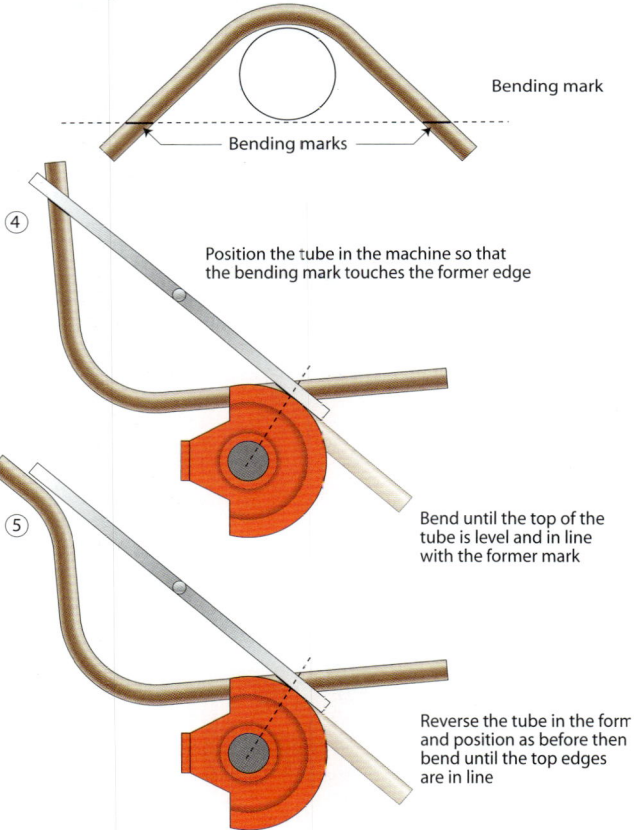

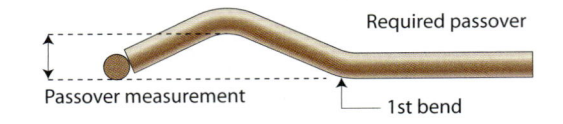

Bending marks — Bending mark

④ Position the tube in the machine so that the bending mark touches the former edge

Bend until the top of the tube is level and in line with the former mark

⑤ Reverse the tube in the former and position as before then bend until the top edges are in line

Passover bends

Partial passover bends

Required passover

Passover measurement — 1st bend

① Close a folding rule down to the passover measurement to obtain the angle for the first bend

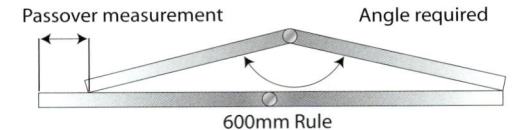

Passover measurement | Angle required

600mm Rule

② Bend tube to the angle required by the folding rule

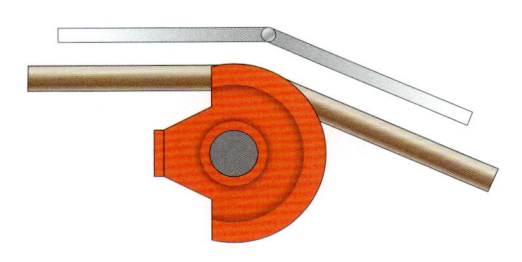

③ Close folding rule down to twice the passover measurement to obtain the angle for the second bend

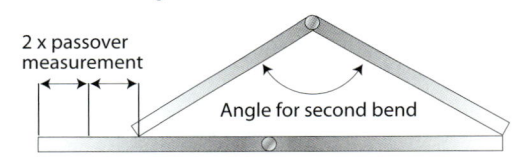

2 x passover measurement | Angle for second bend

④ Mark for the second bend by measuring from the inside edge of tube

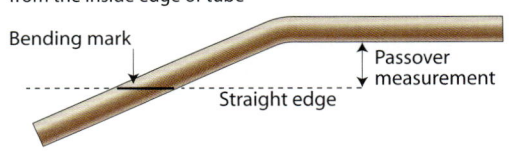

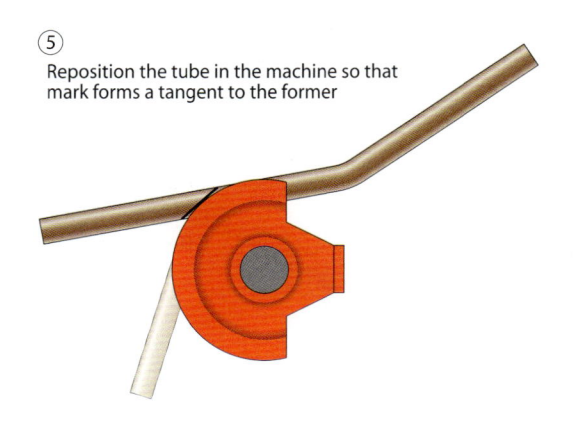

Bending mark | Passover measurement | Straight edge

⑤ Reposition the tube in the machine so that mark forms a tangent to the former

Partial passover bends

005 SITE PREPARATION AND PIPEWORK FABRICATION TECHNIQUES

226   THE CITY & GUILDS TEXTBOOK

Rippling or throating of tube in machine-made bends

Bending machines are designed to produce a smooth bend. They are designed so that the former and the bending guide supports the throat (the inside face) and sides of the tube against collapse. Ripples will occur in the throat of a bend if the pressure of the roller on the guide is insufficient or in the wrong place. The correct pressure point is slightly in front of the bending position, where the tube touches the former before the actual bending process occurs. If pressure is exerted too far forward of the bending point, then ripples will occur. If the roller is tightened too much, the pressure point will be too far back and the tube will be excessively 'throated' or made oval in section.

With scissor benders, rippling occurs with use. This is because the bending former, being made of aluminium, stretches over time and, because the pressure roller is fixed, it cannot be tightened or repositioned to give the correct bending pressure. If ripples appear when using fixed-position scissor benders, the pressure point can be readjusted by inserting a thin piece of strip steel (the thickness of a hack-saw blade) between the guide and the roller to cure the problem.

Rippled copper tube

Bending copper tube using a bending spring

Bending springs are used to support the walls of the tubes against collapse while the bend is being formed. The British Standard for bending springs is BS 5431, and they are available for copper grades R220 and R250. It is important that the correct-sized spring is used or wrinkling and even snapping of the tube may occur.

As a rule, the bend radius should be four to five diameters of the tube. This is slightly more than for a bending machine. However, choosing a four-diameter bend helps in simplifying the marking out process.

One advantage that a spring bend has over a machine-made bend is that the bend radius can be varied because it is not fixed by a bending former. This allows the tube centres to be carried around bends. In other words, the radii of the bends can be enlarged so that the aesthetic appearance of the bends is enhanced and the gap between the tubes remains even.

This will, again, become apparent in the following illustrations.

SUGGESTED ACTIVITY...

To find out how much tube is used in a machine bend, we have to know the radius of the bend. Bending machines usually bend at a radius of four times the diameter of the tube. So, for a 90° bend on a bending machine using 15mm pipe, the radius of the bend will be 60mm.

If we use the following formula, we can find the length of the bend:

Radius × 1.57 = length of pipe

Therefore:

60 × 1.57 = 94.2mm

So, a bend with a radius of 60mm using 15mm pipe uses 94.2mm of pipe. This we can round up to 95mm.

Now attempt these examples:

a What is the length of the bend of 15mm tube with a radius of six diameters of tube?

b What is the length of the bend of 22mm tube with a radius of four diameters of tube?

c What is the length of the bend of 22mm tube with a radius of five diameters of tube?

Answers: a.141.3mm, b. 138.16mm, c. 172.7mm

KEY POINT

When bending copper tube using a bending machine, the tube appears to gain length. This is called pipe gain and we have to take it into account when precision bending. The pipe gain on a 90° bend is 1.5 times the diameter of the pipe. For instance, if using 15mm tube, the pipe gain will be 22.5mm. Let's say we have to put a 90° bend on a piece of 22mm copper tube so that the finished measurements are 150mm end to centre and 250mm end to centre. The length of pipe needed appears to be 400mm but because the bend cuts the corner, we can deduct a certain amount of pipe. If pipe gain is 1.5 times the diameter and the diameter is 22mm, we can deduct 33mm, so the actual pipe length needed for the bend is 367mm.

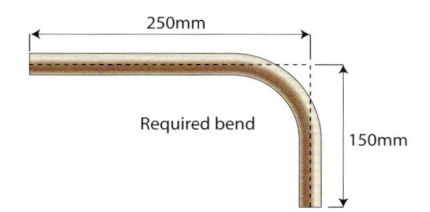

Pipe gain only occurs with 90° bends.

Bending a 90° bend by spring

1. Measure the distance required and mark it on the tube.
2. Measure back from the first mark 4d of the tube size and mark it on the tube. This is the start of the bend.
3. Measure forward 2d of the tube size and mark it on the tube. This is the end of the bend.
4. Anneal the tube between marks 2 and 3. This will soften the tube and prevent creasing.
5. Insert the spring and, starting at mark 2, carefully pull the tube over the knee, moving the bend through to mark 3.
6. Check that the angle is 90° with a set square and check the measurements with a rule.

KEY POINT

Remember to anneal the copper before attempting to bend the copper tube as this will prevent the tube from rippling, creasing or snapping.

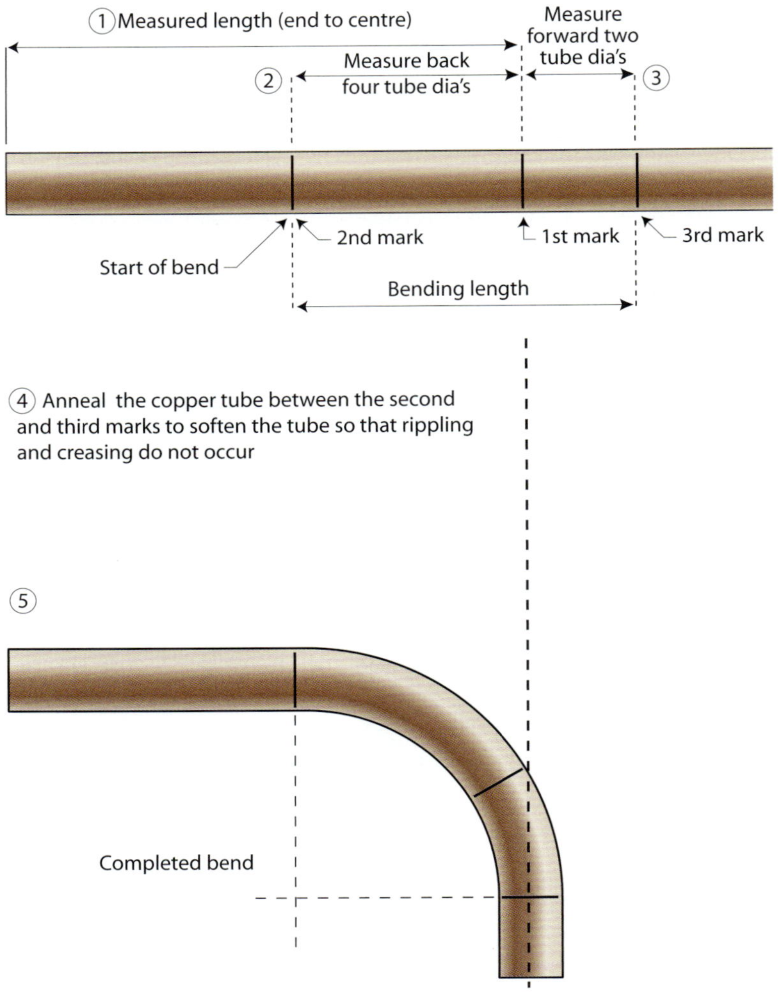

① Measured length (end to centre)
② Measure back four tube dia's
③ Measure forward two tube dia's

2nd mark — 1st mark — 3rd mark
Start of bend
Bending length

④ Anneal the copper tube between the second and third marks to soften the tube so that rippling and creasing do not occur

⑤
Completed bend

Bending a 90° bend by spring

Bending an offset bend by spring

1. Measure the distance required and mark it on the tube.
2. Measure back from the first mark 2d of the tube size and mark it on the tube. This is the start of the bend.
3. Measure forward 1d of the tube size and mark it on the tube. This is the end of the bend.
4. Anneal the tube between marks 2 and 3. This will soften the tube and prevent creasing.
5. Insert the spring and, starting at mark 2, carefully pull the tube over the knee, moving the bend through to mark 3. Remember to check the angle.
6. Using a straight edge measure the distance of the required offset. Measure back from the mark ½ d of tube size. Anneal the pipe and repeat the bending process.

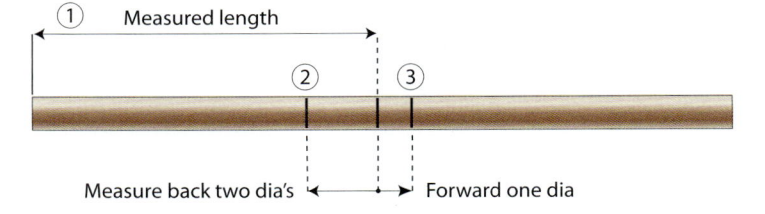

① Measured length
② ③
Measure back two dia's — Forward one dia

④ Anneal the copper tube between the second and third marks to soften the tube so that rippling and creasing does not occur

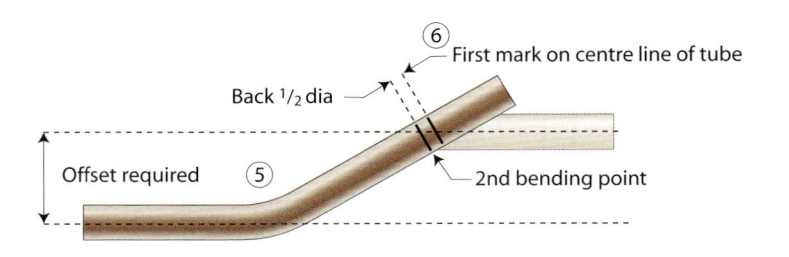

⑥ First mark on centre line of tube
Back ½ dia
Offset required ⑤
2nd bending point

Bending an offset by spring

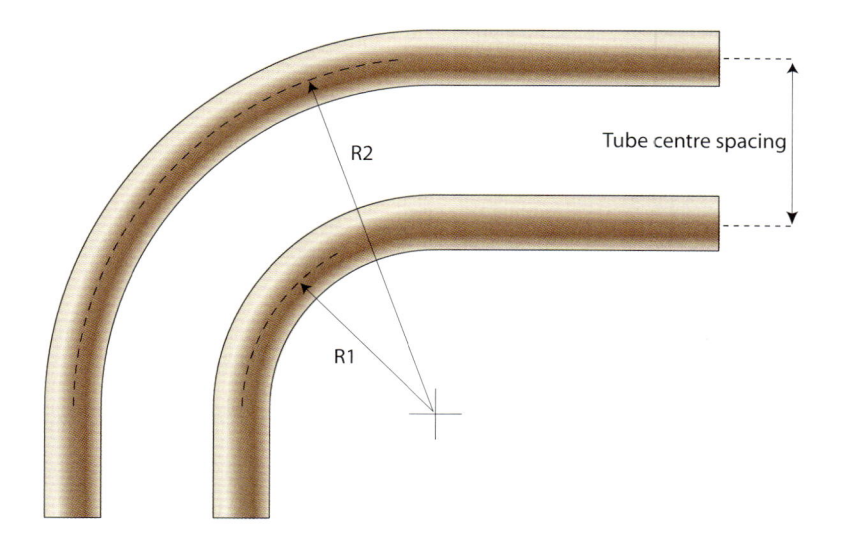

R1 = four times dia of pipe
R2 = R1 + tube centre spacing
For two 22mm diameter tubes at 80mm centres:
R1 = 4 × 22 = 88mm
R2 = 88 + 80 = 168mm

So, set out inner bend as before, then for outer bend:
Measure back distance for outer bend = 168mm
Measure forward distance = 84mm (both from first mark)

Concentric spring bends

Setting out spring bends

Spring bends should be limited to copper tube R250 up to 22mm diameter as bending tubes by hand over this diameter, although possible, is very difficult because of the thickness of the tube. The first point to remember when setting out for a spring 90° bend is that there is gain of the tube in the same way as when the bend is formed with a machine (see Key Point on page 224).

Therefore, when setting out:

- allowances have to be made for the 'gain in material'
- the bend must first be pulled in the correct position in relation to the fixed point.

Setting out step by step

1 Decide on bend radius, which is usually taken as four times the diameter of the pipe (4d) but Yorkshire Copper Tube recommend 5d. It could be any radius determined by a drawing.

The length of the pipe taken up by a 90° bend can be calculated using the formula below:

$$\text{Radius (R)} \times 1.57$$

2 Assuming that a 15mm pipe is to be bent to a radius of 5d and we need to find out how much pipe will be taken up by the bend:

> Radius of bend is 5d = 5 × 15 = 75mm
>
> Using the formula:
>
> 75 × 1.57 = 117.75mm
>
> Length of bend = 117.75mm (rounded up to 118mm)

3 Mark off the required distance from the end of the tube to the centre-line of the bend (the end-to-centre measurement).

4 Divide the calculated length of pipe by three (for 15mm tube this will be three equal measurements of about 39mm).

5 From the original measurement, mark 39mm forward and 78mm back.

6 The bend can then pulled, ensuring that it is kept within the three 39mm measurements; this will keep the centre of the bend the correct distance from the fixed point.

B3 JOINTING METHODS AND FITTINGS FOR COPPER TUBE

There are generally four methods of jointing for copper tubes for domestic purposes. These are:

1 capillary fittings
2 compression fittings
3 push-fit fittings
4 press-fit fittings.

Capillary fittings to BS EN 1254:1998

Capillary fittings use the principle of capillary action (see Chapter 004, page 166) to draw solder into the fitting when they are heated by a blowtorch. There are two different types:

1 **Integral solder ring** – This type of fitting has a band of lead-free solder housed inside a raised ring on the fitting socket and so extra solder is not needed.

2 **End feed** – This type of fitting needs solder to be fed at the end or the mouth of the fitting. It does not have solder in the fitting. It is generally used in preference to the integral solder ring type due to cost and aesthetics.

Integral solder ring

End feed

How to complete a soldered fitting

Follow these instructions when completing a soldered fitting.

STEP 1 – Cut and deburr the tube.

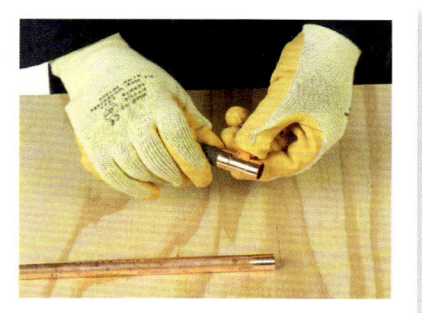

STEP 2 – Clean the end of the tube and the inside of the fitting with steel wool or cleaning pad.

STEP 3 – Apply flux to end of tube only. Do not apply the flux to the inside of the fitting. Insert the tube into the fitting. Twist the tube slightly when inserting the tube. This ensures an even spread of flux on the tube and fitting. Clean any excess flux from the outside of the tube immediately with a dry cloth.

STEP 4 – Apply heat evenly to the fitting and wait for 10 seconds. If the fitting is an integral soldered ring type, then solder will appear at the mouth of the fitting.

STEP 5 – If the fitting is an end feed type, then apply solder wire equivalent to half the diameter of the pipe (for 22mm, apply 11mm of solder wire) to the mouth of the fitting, ensuring that the solder flows all around the socket. Do not use too much heat or the fitting and flux will turn black and the fitting will not solder.

STEP 6 – When the fitting has cooled down a little, clean off any excess flux with a damp cloth.

Fluxes and solders used with capillary fittings

End feed fittings require that solder be added during the soldering process to the mouth of the fitting. For hot and cold water pipework installations, this solder must be lead free to comply with the Water Supply (Water Fittings) Regulations 1999. There are several types of lead-free solder available, the most popular being a mixture of tin and copper to EN 29453 (known as number 23 tin-based solder), which has a melting point of 230–240°C and is suitable for making end feed capillary joints on all domestic plumbing, heating and gas systems.

> **KEY POINT**
>
> For more information on fluxes and their applications, go to:
>
> http://www.wras.co.uk/PDF_Files/SoldersFluxes2.pdf

The use of leaded solder is permitted on gas and central heating installations but there is always a risk that this solder will be used on the wrong system and, if this occurs, the plumber risks a hefty fine and a criminal record if prosecuted.

Fluxes are used to clean oxides from the surface of the copper and to help with the flow of solder into the fitting. There are two basic forms of flux available. These are:

1 **Self-cleaning fluxes** – Otherwise known as 'active' fluxes because they clean the copper tube and the fitting during the soldering process. Cleaning of the tube and fittings beforehand is not necessary. Some types of active flux contain hydrochloric acid, which can be harmful if not used correctly and can promote corrosion in copper tubes if excess flux is not removed after soldering has been completed. They are, however, potable in water, which means they dissolve in contact with water and are flushed out when initial flushing of the system takes place.

2 **Traditional flux paste** – Usually made from zinc chloride and/or zinc ammonium chloride. Some fluxes contain other active ingredients such as amines. Cleaning of the tube and fitting is required with this type of flux and it is not potable. It will remain in the pipe after the soldering process has been completed and will not flush out during commissioning, so it should be used sparingly.

Compression fittings to BS EN 1254:1998

Compression type fittings are mechanical fittings that require tightening with a spanner to make a watertight joint. There are two different types:

1 **Type A – non-manipulative compression fittings.** This type of fitting consists of three main parts. The fitting body, a metal 'O' ring called an olive, and the back nut. It is called 'non-manipulative' simply because neither the tube nor the fitting need working or 'manipulating' to make the joint. When the nut is tightened, the olive is slightly compressed onto the copper tube.

To make a type A fitting, follow these steps:

STEP 1 – Cut and deburr the tube.

STEP 2 – Take apart the fitting and slip the nut and olive over the tube.

STEP 3 – Assemble the fitting and tighten by hand, then using adjustable spanners. Turn the nut clockwise one and a half turns to fully tighten the joint.

Do not over-tighten the joint as this will crush the olive onto the tube too much and may cause the fitting to leak

This joint does not require any jointing paste or PTFE tape to make the joint – these should only be used if the joint shows signs of leakage.

2 **Type B – manipulative compression fittings.** Unlike type A fittings, type B fittings require that the end of the tube is worked, or more specifically flared, with a special tool called a flaring tool, before a successful joint can be made. This type of fitting is made for jointing soft copper tube (type R220) for below-ground water services. The parts of the fitting are the fitting body, the compression nut, the compensating ring and the adapter piece.

To complete a type B compression joint, follow these steps:

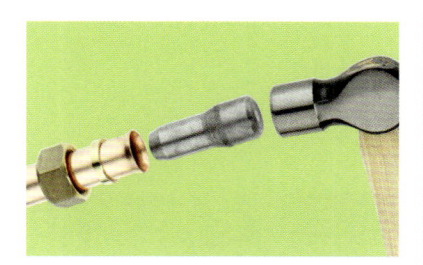

STEP 1 – Cut and deburr the tube. Slip the compression nut and the compensating ring over the tube and flare open the end.

STEP 2 – Insert the parallel end of the adaptor piece into the fitting socket and position the flared end of the copper tube over the tapered face of the adapter piece.

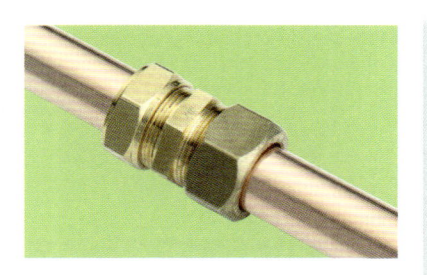

STEP 3 – Screw the compression nut and ring onto the fitting body by hand and then tighten with a spanner. One full turn should be sufficient.

Push-fit fittings for copper tube

Push-fit fittings for copper tube are made from either copper or DZR (dezincification resistant) brass, and are available in sizes 10mm to 54mm. They can be used on hot and cold water services above ground and central heating systems.

Push-fit joints rely on a stainless steel grab ring and a sealing ring to make a watertight joint. There are a number of different brands available but all use a similar method of jointing. When a piece of copper tube is pushed into the joint it passes through a release collar and then through a stainless steel grip ring. This has a number of teeth that grip onto the tube, securing it in place. It can only be released using some de-mounting tools. When the tube is pushed further into the joint it passes through a support sleeve, which helps to align the tube and compresses a pre-lubricated EPDM 'O' ring between the wall of the fitting and the tube. When the tube has passed through the 'O' ring and has reached the tube stop, a secure joint is made.

The following pressure and temperatures apply to push-fit fittings:

Temperature not exceeding	Maximum working pressure
30°C	16 bar
65°C	10 bar
90°C	6 bar

To complete a push-fit joint, follow these steps:

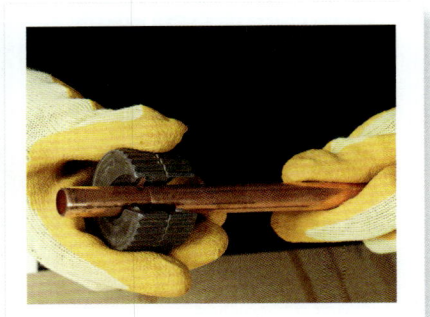

STEP 1 – Cut the tube using a pipe slice. The tube needs to be round, square cut and free from damage.

STEP 2 – Deburr the end of the tube.

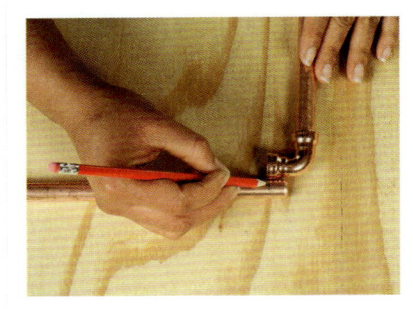

STEP 3 – Mark the socket insertion depth to provide a visual marker that the tube has been pushed fully into the socket.

STEP 4 – Keep the fitting and tube in line. Push the tube through the release collar to rest against the grab ring.

STEP 5 – Push the tube firmly with a slight twisting action until it reaches the tube stop with a 'click'.

Press-fit fittings for copper tube

Press-fit fittings require a special electrical press tool (see page 216), which crimps the fitting onto the tube to make a secure joint.

Press-fit fittings are available to suit tube sizes from 12mm to 108mm, and can be used for systems operating up to 16 bar pressure at 20°C and 6 bar pressure at 110°C. They are ideal for use where using a blowtorch is not possible. There are several different fitting types available which allow press-fit fittings to be used on hot and cold water installations, central heating systems, chilled water installations, solar hot water systems and gas installations (using a special yellow rubber 'O' ring).

A press-fit fitting consists of the fitting body, a rubber seal and a stainless steel grab ring. Press-fit fittings are packaged in separate, sealed plastic bags, and should be kept in them to prevent the lubricant from drying out.

To complete a press-fit joint for sizes up to 35mm, follow these steps:

Electrical press tool

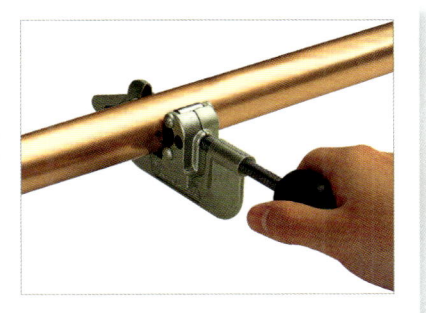

STEP 1 – Cut the tube, preferably with a tube cutter, and deburr the pipe. Care should be taken to ensure that the tube is cut square.

STEP 2 – The tube must be fully inserted into the socket. To ensure this, use a socket depth gauge to mark the depth of the socket onto the tube or, alternatively, measure and mark using a rule.

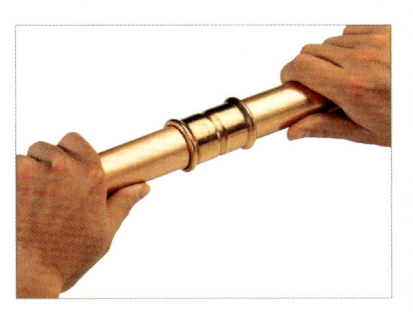

STEP 3 – Insert the tube into the fitting all the way to the tube stop. The fitting depth mark previously made on the tube will help as a guide.

STEP 4 – Place the jaws of the press-fit tool over the bead of the fitting, making sure that the jaws of the tool are well lubricated.

STEP 5 – A 90° angle between the tool and the socket must be maintained when making the joints.

STEP 6 – Press the trigger on the press-fit tool to start the jointing process, making sure that fingers are kept away from the jaws.

KEY POINT
Push-fit fittings for plastic pipe will be covered on pages 252–53 of this chapter.

Fittings recognition

Fittings recognition is a part of a plumber's job. Choosing the right fitting for the right application is a key element of a successful installation. There are four fittings that are used more than all others: couplings, equal tees, elbows and reducers. Below is a table showing these four fittings in each of the jointing types:

	Couplings	Equal tees (all three connections equal size)	Elbows	Reducers
End feed				
Integral solder ring				
Compression				
Push-fit				
Press-fit				

KEY POINT
When ordering tees with a mixture of end and branch sizes, care should be taken to ensure that the correct configuration is quoted. The method to use when ordering tees is to quote the largest end, then the smallest end, and the branch last.

In the photograph above, the tee would be ordered as:

22mm × 22mm × 15mm

As well as the fittings mentioned above, there are other common fittings, which may be used on a regular basis. These are:

- **Reducing tees** – these come in four different forms:

Reduced end	Reduced branch	Reduced end and branch	Two reduced ends (pendant tees)
One end is reduced	Branch is reduced	One end and branch are reduced	Both ends are reduced

- **Tap connectors** – used for connecting to taps and float-operated valves.

Straight tap connector	Bent tap connector

> **KEY POINT**
> Most general fittings are available in all jointing formats. Some of the more specialised fittings are only available in capillary, Type A compression and push-fit formats.

- **Cap ends** – used for blanking off the ends of the tube. Also known as stop ends.

Compression cap end	Push-fit cap end	End-feed capillary cap end

- **Tank connectors** – used for making connections to tanks and cisterns.
- **Flexible connectors** – often used instead of tap connectors on sanitary ware.
- **Central heating manifolds** – a specialist fitting used in microbore central heating systems.

Tank connectors	Flexible connections	Manifolds

Proprietary fittings

Proprietary fittings connect tubes and pipes of different materials such as copper and lead or lead and medium density polyethylene (MDPE). There are several different types of proprietary fittings, including leadlocks and Philmac.

Leadlocks	Leadlocks are specially made to connect lead pipe to copper tubes. These, however, promote galvanic corrosion (see Chapter 004, page 162) between the copper and the lead and so should only be used as a temporary connection.
Philmac	Philmac fittings are truly universal because they will connect almost all known pressure pipes and tubes together by the use of special inserts that fit into a generic fitting body.

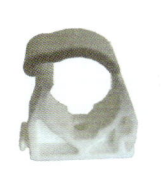

The types and uses of pipe clips for copper tubes

Copper is relatively easy to joint and bend and can produce an installation that not only looks good, but also is economical in terms of tube usage and installation costs. By adopting a systematic approach to copper tube installation, fabrication and planning, savings can be made on labour costs and material usage. A big part of installing copper tubes is the planning of pipework routes, ensuring that surface-mounted pipework, once installed, looks neat, is well clipped, is unobtrusive and performs to the design criteria.

The correct clipping of copper tube is essential. It prevents excessive noise and fittings failure from vibration, movement and water hammer and can assist in preventing accidental or intentional damage of the pipework. There are many different types of pipe clips available for copper tubes and each one has a specific use.

For most domestic installations, plastic stand-off pipe clips are preferred, the most common type being the interlocking clip lock type where several banks of pipes of different sizes can be simply clipped together. This ensures a uniformity that is often hard to accomplish with single, individually fixed pipe clips because, once assembled, all of the pipe clips have exactly the same tube centres and, provided the first clips are installed correctly, the others will be perfectly aligned. Single plastic pipe clips are also available for single runs of tube, and double pipe clips can be used when installing hot and cold pipework or flow and returns for radiator installations and central heating systems as these also ensure uniform tube centres.

Nail-on clips are also available but should be used with caution with copper tubes as the expansion of copper can loosen the clips making the copper tube vulnerable, especially in places where the pipework is hidden, such as under a suspended timber floor.

When fixing copper tubes to a skirting board, the use of copper saddle clips is recommended as the copper is fixed close to the skirting, which makes the tube a little less noticeable. Caution should be exercised with saddle clips, as they are not suitable for fixing to masonry or plastered walls. This can create corrosion of the copper due to the reaction between the copper tube and the wall surface and can also encourage condensation on the tube surface.

For installations that require a more rigid fixing such as light commercial/industrial installations, strip brass school board clips or cast brass school board clips should be used. This type of tube bracket gives more resistance to tube movement and subsequent damage.

Industrial installations require a very secure type of fixing. Brass munsen rings fastened with 10mm tapped rod and back plates are the strongest types of brackets available for copper tube installations. As well as being screwed to the building fabric, munsen rings can also be hung from the ceiling in banks of pipes using a special metal slotted channel.

Below is a table showing the clipping distances of the common sizes of copper tubes:

Tube size	Horizontal distance between the clips	Vertical distance between the clips
10mm	0.8m	1.2m
15mm	1.2m	1.8m
22mm	1.8m	2.4m
28mm	1.8m	2.4m
35mm	2.4m	3.0m
42mm	2.4m	3.0m
54mm	2.7m	3.0m

B4 LOW CARBON STEEL PIPES TO BS EN 10255:2004 (FORMERLY BS 1387:1985)

Low carbon steel pipe is used occasionally in domestic installations but its use should be restricted to wet central heating systems, gas installations and oil lines. It must not be used to supply hot or cold water supplies for domestic purposes because of the risk of rusty water being drawn from the taps. Often referred to as mild steel pipe, low carbon steel pipe is usually supplied painted red or black and can also be galvanised. Its carbon content is low.

Low carbon steel pipe is available in three grades, each grade being identified by a colour code:

Grade	Colour code
Light	Brown
Medium	Blue
Heavy	Red

The grades of low carbon steel have identical external diameters but the pipe's wall thickness will vary according to the grade, heavy grade having the thickest pipe wall and light grade the thinnest. Medium grade pipe is the most common grade used in plumbing installations but heavy grade may be used where a long system life is expected. Heavy grade pipe can also be used below ground. Light grade pipe is seldom used except in some dry sprinkler installations for fire prevention.

Low carbon steel pipe is available in 6m lengths and may be supplied with threaded ends or plain ends. It is referred to by imperial pipe sizes, which are specified as nominal bore. The common pipe sizes for domestic purposes are:

Thread size / fitting size (in)	⅛	¼	⅜	½	¾	1	1¼	1½	2
Nominal diameter (mm)	6	8	10	15	20	25	32	40	50

B5   BENDING LOW CARBON STEEL PIPE

There are two methods of bending low carbon steel pipe:

1 **By hydraulic bending machine** – This method uses a hydraulic bending machine (see page 215) to bend the pipe. It uses an oil to exert hydraulic pressure. The oil, being incompressible, exerts great force on the pipe through the bending former to bend the pipe when the handle of the machine is pumped. Steel is very tough to bend and tends to 'spring' back once the bend is formed. Because of this, bends should be over bent about 5° to allow for the bend springing back slightly. This is the method used in domestic installations.

2 **By heat** – Mainly used on industrial installations. It involves the use of oxyacetylene torches to heat the steel almost to white heat to soften the pipe. This allows the steel pipe to be bent easily by hand.

Below, we will look at how to bend a 90° bend and an offset bend using a hydraulic bending machine.

Bending a 90° bend

1. Mark a line on the pipe at the required distance from the fixed point to the centre line of the required bend.

2. From this measurement, measure back towards the fixed point one nominal bore (the internal diameter) of the pipe to point A.

3. Place point A at the centre of the correct size bending former on the bending machine.

4. Pump the handle of the bending machine until an angle of 90° + 5° (allowance for springing back) has been achieved. Make sure you are standing to the side of the machine. Never stand in front of it whilst bending is taking place.

KEY POINT

You may find it easier to judge the angle of the bend by making a template from a welding rod bent to 90° or by the use of a steel set-square. The template can be placed on the bending machine so that you can see where to stop the bend.

Bending a 90° bend

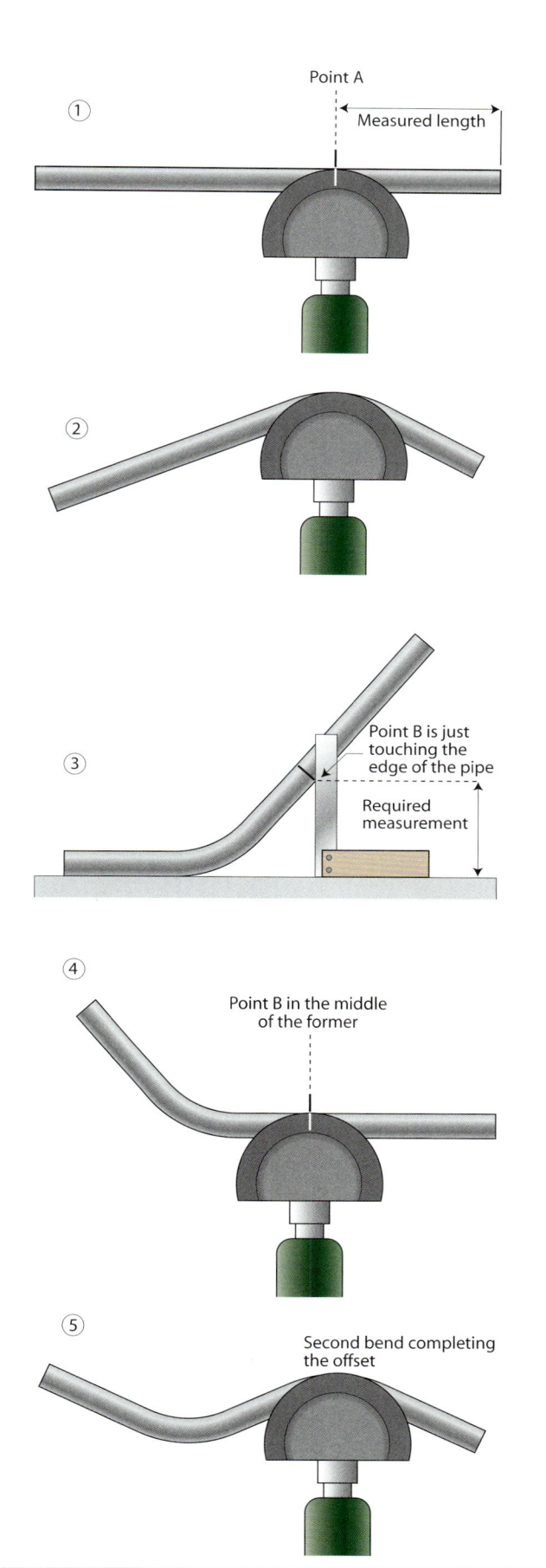

Bending an offset bend

1 Mark a line at the required measurement for the first bend onto the pipe.

2 Place the pipe in the machine but this time do not make any deduction. The mark goes directly on the centre of the former. The measurement 'A' is from the fixed end of the pipe to the centre of the set.

3 Make the first bend to the required angle and check the angle using the template.

4 Take the pipe from the machine and place a straight edge against the back of the pipe. Mark the measurement of the second bend at point 'B'.

5 Put the pipe back into the machine and line up the mark with the centre of the former. Bend the second bend and check with the template.

Bending an offset bend

KEY POINT

With the offset bend, you may find it easier to make a welding rod template bent to the required angle (say 45°). This can be used for both the first and second bends. Remember to over bend both bends by 5° to allow for spring back.

B6 JOINTING METHODS AND FITTINGS FOR LOW CARBON STEEL PIPE

There are three ways to joint low carbon steel pipe. These are:

- threaded joints
- compression joints
- welded joints.

We will look only at the first two, as welded joints are generally used on larger pipes in industrial applications and installations.

Threaded joints

Low carbon steel pipes can be jointed using threads to BS 21:1985, and BS EN 10226-1:2004, which are cut into the end of the pipes using either manual stocks and dies or electric threading machines (see page 217). There are two kinds of thread:

- **Tapered threads** – A standard thread cut onto the ends of pipes and blackheart malleable male fittings to ensure a watertight, gas-tight or steam-tight joint. The tube tightens the further it is screwed into the fitting.

- **Parallel threads** – A screw thread of uniform diameter used on fittings such as sockets.

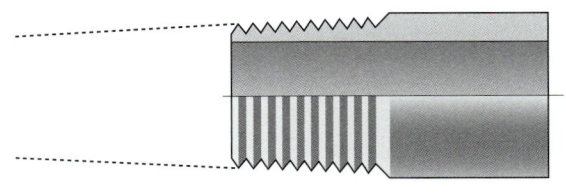

KEY POINT
Male threads are external threads; female threads are internal threads. All threads are BSPT, which stands for British Standard Pipe Threads.

Threads taper towards the end of the tube

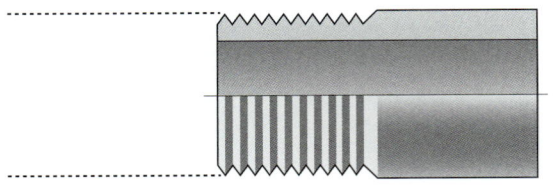

Threads remain parallel throughout the length of the tube

Tapered thread (left) and parallel thread (right)

When cutting a thread onto a length of pipe, the length of the thread should be such that, once the joint is made, one and a half to two threads should be visible when the joint is completed.

Fittings for low carbon steel pipe are made from steel and malleable iron to BS EN 10242 (formerly BS 143) and BS 1256. Steel fittings, although stronger than malleable iron, tend to be more expensive.

Malleable iron fittings for low carbon steel fall into two groups known as:

- Blackheart fittings with tapered female threads, which are identified by a square-edged bead around the mouth of the fitting. These fittings are made from cast malleable iron and are quite brittle and susceptible to splitting if over-tightened.

- Whiteheart fittings with parallel female threads, which are identified by a rounded bead around the mouth of the fittings. These fittings are made from a slightly softer type of malleable iron and therefore are more flexible and tend to stretch if over-tightened.

Blackheart (top) and whiteheart (bottom) fittings

Malleable iron fittings are suitable for smaller, domestic installations with a wide range being available.

Couplings	Equal tees	Elbows	M/F elbows
Pitcher tees	Unions	Nipples	Bushes

A variety of jointing compounds can be used with threaded joints to make leak-free joints. Each one has a specific use, although some are universal and can be used on a number of different installations. Jointing compounds include:

Linseed-oil-based compounds (boss white, hawk white and templars paste)	Can be used in conjunction with hemp on wet central heating systems and compressed air lines. Must not be used on natural gas installations.
Unsintered polytetrafluroethylene (PTFE tape)	A thin, white (or thicker, yellow-sleeved if used on gas) tape that can be used on most installations, including hot and cold water, central heating and gas installations.
PTFE-based jointing compounds (boss green)	A compound specially made for use on hot and cold water supplies. Not suitable for natural gas installations.
Heldite paste	A universal compound that can be used on many installation types such as oil, gas, hot and cold water, central heating, compressed air lines and vacuum lines. It is not suitable for potable water supplies.
Manganese paste	These are specialist compounds for use with high temperature hot water and steam installations.
Graphite paste	
Gas seal paste	A specialist compound for use with natural and liquid petroleum gas installations.

Compression joints

Compression joints for low carbon steel pipes are available from a number of manufacturers. They incorporate a rubber compression ring to ensure a leak-free joint. They tend to be rather expensive but can save time on installation costs. They are often referred to as transition fittings. They have several advantages to screwed fittings:

- very versatile connection suitable for connecting LCS pipe to different pipe materials, such as copper and lead
- quick and easy-to-make joints
- no special tools necessary
- no threads on steel pipe required.

Low carbon steel compression fittings can be used on new installations, pipe repair and pipework extensions on the following installations:

LCS compression fitting

- water (hot and cold water, central heating systems)
- gas (natural gas, LPG)
- oil
- compressed air.

The types and use of pipe clips for low carbon steel pipes

Low carbon steel pipe is a very rigid material and is heavier than most types of pipes and tubes. The clips and fastenings need to be capable of carrying the weight of the material. Because of this, the clips available tend to be very robust. The types of clips and fastenings for low carbon steel pipe are limited. For fixing to walls, cast steel school board clips are recommended.

Since most low carbon steel is used in industrial installations, the use of munsen rings and tapped rod is recommended, which can be used with backing plates or, if being hung from a ceiling, by the use of anchor bolts.

The clipping spacings for low carbon steel are listed below:

Pipe size	Horizontal	Vertical
½ in	1.8m	2.4m
¾ in	2.4m	3m
1 in	2.4m	3m
1¼ in	2.7m	3m
1½ in	3m	3.6m

Tapped rod

Anchor bolt

SmartScreen Unit 005 handout 4

B7 PLASTICS

There are two main types of plastics:

- those that can be solvent welded, such as PVCu and ABS
- those that cannot be solvent welded and use other methods of jointing (polyethylene, polypropylene and polybutylene).

B8   POLYVINYL CHLORIDE (PVC) AND ACRYLONITRILE BUTADIENE STYRENE (ABS)

Polyvinyl chloride is available in three different types:

- **Unplasticised polyvinyl chloride (PVCu) to BS 4514** – Used mainly for push-fit and solvent weld soil and vent pipes, below ground drainage, solvent weld waste and overflow pipes, gutters and rainwater pipes. It has good resistance to ultraviolet (UV) light but can suffer from photodegradation especially in light colours such as white and grey. It has a high coefficient of linear expansion. Sizes available are 110mm, 50mm, 40mm, 32mm and 21.5mm for soil/vent pipes and waste and overflow pipes. Gutters and rainwater pipes are available in a variety of sizes and styles, and these are discussed further in Chapter 009.

- **Modified unplasticised polyvinyl chloride (muPVC) to BS 5255** – Used for solvent weld waste and overflow pipes. It is more durable than

PVCu and performs better than other plastics, especially at higher temperatures. Sizes available are 50mm, 40mm, 32mm and 21.5mm.

- **Chlorinated unplasticised polyvinyl chloride (cuPVC)** – Used for solvent weld cold water service pipes in the late 1970s. Fittings are still available for repairs but pipe is increasingly difficult to find. It has a tendency to snap, especially at low temperatures and if mishandled.

- **Acrylonitrile butadiene styrene (ABS)** – Usually used for soil and waste pipes and fittings and, because of its toughness, can also be used for mains cold water pipes. It degrades quickly when exposed to UV light. ABS possesses extremely good impact strength and high mechanical strength, which makes it suitable for plumbing pipework and installations. The jointing methods used, pipe sizes and clipping distances are the same as for PVCu.

B9 JOINTING METHODS AND FITTINGS FOR PVCu, muPVC AND ABS

PVCu can be jointed in two ways:

1 **solvent weld** – used on soil/vent pipes, waste pipes and overflow pipes
2 **push fit** – used on soil and vent pipes.

To make a solvent weld joint on PVCu, muPVC and ABS soil/vent pipes:

Spigot
Another name for the plain end of a pipe. If the fitting we buy has a plain pipe end, we call this a spigot end.

STEP 1 – Cut the pipe square using a hacksaw, and deburr with a rasp.

STEP 2 – Wipe the pipe with a clean, dry cloth to remove dirt and swarf.

STEP 3 – Clean inside the socket and the pipe spigot with solvent cleaner.

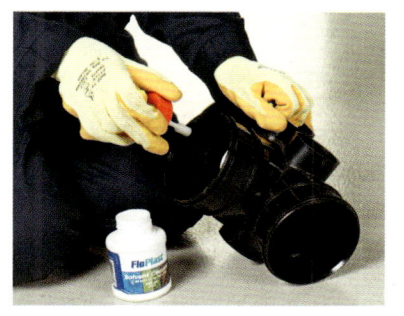

STEP 4 – Apply solvent weld cement inside the socket first and then to the spigot. This will allow a little more time to make the joint before the cement begins to dry out.

STEP 5 – Insert the pipe into the socket and twist fully into the socket.

STEP 6 – Wipe off excess cement using a dry cloth.

To make a push-fit joint on PVCu, muPVC and ABS soil/vent pipes:

STEP 1 – Cut the pipe square using a hacksaw.

STEP 2 – Chamfer the pipe using a file or rasp.

STEP 3 – Wipe the pipe with a clean, dry cloth to remove dirt and swarf.

Chamfer
To take off sharp edge at an angle. If we chamfer a pipe end, we are taking the sharp, square edge off the pipe.

STEP 4 – Lubricate the end of the pipe using silicone spray lubricant or silicone grease. Do not use liquid soap as this can adversely affect the rubber seal.

STEP 5 – Check that the seal is in the correct place in the fitting.

STEP 6 – Push the pipe all the way into the fitting and mark the pipe at the end of the fitting using a pencil or pen.

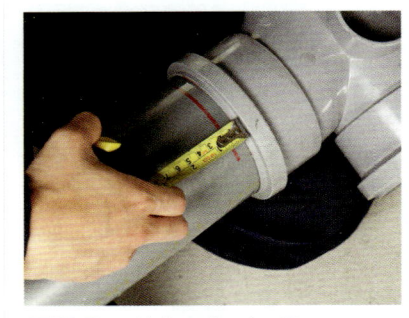

STEP 7 – Withdraw the pipe 10mm from the fitting. This is to allow for expansion of the pipe. Fittings must be supported by a pipe bracket to prevent the fitting from slipping.

Fittings for PVCu, muPVC and ABS soil/vent and waste pipe installations

Fittings for PVCu and muPVC soil/vent and waste pipes are the same size. This means that the two systems are interchangeable. Opposite is a table of the more common types of soil pipe fittings:

90° bends	45° bends	Junctions	Sockets
Boss pipes	**Strap boss**	**Access pipes**	**Boss pipe adapters**
Cages	**Pile clips**	**Drain adapters**	**Waste pipe manifolds**

PVCu 82mm, 110mm and 160mm soil pipe is available in 2.5m, 3m and 4m lengths, in a variety of colours. The pipe and fittings are manufactured to BS EN 1329-1:2000. The pipe has a socket on one end and a chamfered spigot on the other.

Below is a table of the more common types of solvent weld waste pipe fittings.

90° knuckle bends	90° bends	45° bends	Tees
Sockets	**Reducers**	**Pipe clips**	

PVCu waste pipe is manufactured to BS EN 1455-1:2000 and BS 5255 and is available in 3m lengths in sizes 21.5mm (overflow pipe), 32mm, 40mm and 50mm.

Clipping distances for PVCu soil and waste pipes are listed in the table overleaf:

	Maximum support distance		Maximum distance between expansion joints
	Vertical	Horizontal	
Pipe size – soil			
82mm	2m	0.9m	4m
110mm	2m	1m	4m
160mm	2m	1m	4m
Pipe size – waste			
32mm	1.2m	0.5m	2m
40mm	1.2m	0.5m	2m
50mm	1.2m	0.9m	2m

B10 POLYPROPYLENE (PP)

Polypropylene is a common plastic in plumbing systems. It is used to manufacture cold water cisterns, WC siphons and push-fit waste and overflow pipe. It is the waste pipe we will look at here.

Polypropylene waste pipe manufactured to BS EN 1451-1:2000, BS 5254 and BS 5255 is flexible, tough and resistant to most acids and alkalis. It melts at a relatively low temperature of 160°C and starts to soften at 100°C. For this reason, its use as a waste pipe is limited to waste water below 100°C. It is also adversely affected by direct sunlight and cannot be solvent welded. It is jointed by the use of push-fit fittings, which have a rubber sealing ring inside the fitting.

B11 JOINTING METHODS AND FITTINGS FOR POLYPROPYLENE

Polypropylene pipe is supplied in 3m lengths and in various colours including white, black, grey and brown. The most common fittings are shown below:

90° bends	90° swivel bends	45° bends	Swept tees
Sockets	Reducers	Pipe clips	

Clip distances for polypropylene push-fit waste pipes

Clipping distances for polypropylene push-fit waste pipes are listed in the table opposite:

	Maximum support distance		Maximum distance between expansion joints
	Vertical	Horizontal	
Pipe size – polypropylene waste pipe			
32mm	1.2m	0.5m	2m
40mm	1.2m	0.5m	2m
50mm	1.2m	0.6m	2m

B12 POLYETHYLENE (PE)

Polyethylene is used extensively in the plumbing industry for mains cold water pipes. Two types are used below ground on cold water services:

Medium density polyethylene is:

- manufactured in accordance with the requirements of BS EN 12056-2:2000

- a hardwearing plastic for water pipes, gas pipes and fittings

- available in a variety of colours depending on its use – blue in colour when used for cold water supplies

- used for all new domestic mains cold water supplies below ground – must not be used above ground except for temporary installations

- resistant to shock (and subsequent fractures) and has good performance in freezing weather conditions

- susceptible to UV and direct sunlight and it is recommended that a maximum of 150mm of pipe is showing when it enters the building.

For new installations, MDPE piping and pipe fittings are available in sizes of 20mm to 63mm, supplied in coils of 25m to 150m. The most common pipe size used for cold water services for domestic properties is 25mm.

High density polyethylene (HDPE) was used until the mid-1980s for mains cold water pipe until superseded by MDPE. It is still manufactured but is not used as extensively as MDPE. Coloured black, this is available in grades A, B, C and D.

B13 JOINTING METHODS AND FITTINGS FOR MEDIUM DENSITY POLYETHYLENE (MDPE)

Medium density polyethylene (MDPE) pipe can be jointed in a variety of ways. The most common types are as follows.

Compression fittings made from brass

These require a pipe insert, which can either be made of copper or nylon. The insert is placed inside the pipe to strengthen the wall of the pipe so that the fitting does not blow off under mains pressure.

Pipe insert for compression fittings (pipe support liner)

The process of making this type of compression joint fitting is as follows:

1. Measure and cut the pipe to the required length, ensuring that the cut is square. A plastic pipe cutter should be used to do this.
2. Deburr the pipe inside and out.
3. Slip the compression nut and the olive over the pipe.
4. Put the pipe insert inside the pipe. Make sure that the nut and olive are in place before you do this as placing the insert inside the pipe makes slipping the olive over the pipe difficult.
5. Put the pipe inside the fitting body and hand tighten the nut.
6. Now, using suitable spanner, fully tighten the fitting, one to one-and-a-half turns.

Compression fittings made from plastic
Known as Philmac fittings, these give the ability to connect MDPE to MDPE, or MDPE to most forms of pressure pipe, including copper tube and lead pipe.

Push-fit fittings made from plastic
These fittings are the simplest form of jointing for MDPE. They simply push onto the pipe to make a secure, watertight joint. No tightening is needed. The fitting contains a stainless steel grab ring to grab and hold the pipe and a neoprene rubber seal. A pipe insert made of nylon is required inside the pipe.

Fusion welded fittings
Large, underground water mains use fusion-welded fittings where the fitting and the pipe are welded together by heat created by electricity. A special fitting is used that has an electrical element inside the fitting body, which when subjected to electricity, generates heat that melts the fitting and the pipe together.

> **KEY POINT**
>
> For more information on Philmac fittings, visit:
>
> http://www.philmac.co.uk

Plastic push fitting

B14 POLYBUTYLENE (PB-1)
Polybutylene is the latest plastic to be manufactured into pipe for pressurised plumbing systems. It is very flexible, allowing it to be cabled easily and quickly through timber joists during the installation process. It has a high temperature and pressure resistance, low noise transmission, low thermal expansion and low thermal transmission. Its internal bore is very smooth, giving it good flow rate characteristics and it does not suffer from corrosion or scaling. It is, however, micro-porous, allowing air to leech through the walls of the pipe. When PB-1 pipe was first introduced in the late 1980s, central heating systems suffered failure due to increased black oxide sludge created by excess air in the system. This has since been cured with the introduction of barrier pipe, which has an impermeable barrier placed in the walls of the pipe. Barrier pipe is not needed for hot and cold water installations.

Barrier pipe

Polybutylene (PB) pipe is usually coloured white or grey but older PB pipe (known as 'acorn') is usually coloured brown. It can be used on hot water and cold water installations, wet central heating systems and underfloor heating. It is available in sizes 10mm, 15mm, 22mm and 28mm in straight lengths of 3m and coils of 25m, 50m and 100m. The pipe sizes are compatible with copper tubes to BS EN 1057.

The benefits of using polybutylene pipe

In recent years, polybutylene pipe has become very popular with both installers and architects for new-build installations. There are many reasons for this:

- does not affect the taste or colour of the water
- has a 50-year guarantee
- minimal internal resistance, thereby increasing flow rates
- very flexible even at very low temperatures
- highly resistant to stress
- non-corrosive
- safe installation processes (no flame needed or chemicals such as flux required during installation)
- high resistance to frost damage
- not affected by water hardness or softness
- not affected by chemical central heating inhibitors or antifreezes
- not affected by microbiological growth
- has high impact strength
- suitable for heating and cooling applications
- multitude of coil lengths for economical installation with minimal waste
- low environmental impact in terms of soil, water and air pollution.

Installing polybutylene pipe

Building Regulations Document A allows for joists to be notched or drilled for the installation of pipes and cables. On new buildings, one of the major benefits to plumbers offered by polybutylene pipe is that, during the installation process, its flexibility allows the pipe to be installed through holes drilled (with an angled drill) in the centre of the joists rather than placed in notches. This is known as cabling and it has several advantages for the building structure:

- the integrity of the joist is maintained with little or no loss of strength
- as the pipe is supplied in coils, longer runs of pipe without joints are possible, which means less likelihood of damaging leaks
- it allows the floorboards to be fitted before installation takes place, giving the building added strength and rigidity
- pipes are less likely to be damaged by nails when the floor is laid.

The benefits to the installer are:

- quicker installation leads to savings on installation costs
- push-fit joints ensure that there is no fire risk
- the use of a bending machine is not required as the pipe is flexible enough to be bent without pipe wall collapse
- testing can begin immediately after the installation is completed.

B15 BENDING POLYBUTYLENE

Polybutylene pipe can be bent by hand without the use of a bending machine. However, the use of cold forming bend fixtures is recommended. These are preformed metal braces that hold the pipe in a 90° position.

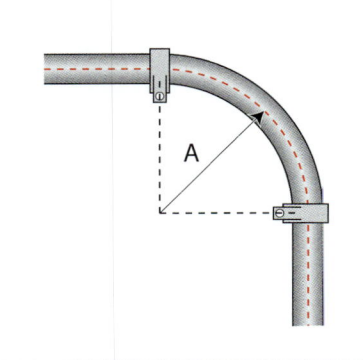

Alternatively, it is possible to brace the bend using pipe clips, ensuring that the radius of the bend is not less than the following:

Diameter of pipe (mm)	10	15	22	28
Radius dimension A (mm)	80*	120*	160*	224*
*depending on the pipe manufacturer				

Calculating dimension A

B16 JOINTING METHODS AND FITTINGS FOR POLYBUTYLENE

Polybutylene pipe can be joined in two ways, as detailed below.

Push-fit fittings

These have a stainless steel grab wedge to hold the pipe firm and a neoprene rubber 'O' ring to make a watertight joint.

A pipe insert usually made from either plastic or stainless steel (depending on the pipe manufacturer) must be placed inside the pipe before the joint is made. The procedure for making a push-fit joint on polybutylene pipe is:

1. Cut the pipe using a scissor-type plastic pipe cutter. This ensures a clean cut to the pipe end. Do not use a hacksaw.
2. Push the pipe insert into the pipe.
3. Most pipe manufacturers put marks on the pipe at fitting depth distance. This helps to visually ensure that the pipe is pushed fully into the joint.
4. Some manufacturers recommend that you lubricate the end of the pipe with silicone spray lubricant – check their literature.
5. Push the pipe fully into the fitting until the fitting stop is felt.

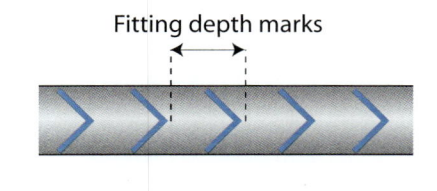

Depth marks on pipe

There are many different manufacturers of polybutylene pipe and each one has their own type of push-fit fitting. The features are almost always the same:

- a fitting body
- a rubber 'O' ring to make the watertight seal
- a stainless steel grab ring or grab wedge to hold and lock the pipe into the fitting body
- a spacer washer between the 'O' ring and the grab ring
- most of the fittings are demountable, in other words, they can be taken off the pipe and reused.

Below is a table of the most common styles of push-fit fitting:

Hep20	Speedfit	Polyplumb

Standard Type A non-manipulative type compression fittings to BS EN 1254:1998

Because polybutylene is manufactured to the same pipe sizes as copper tubes, type A compression fittings can be used. Again, if using a compression fitting, a pipe insert must be pushed inside the tube. This is because the polybutylene pipe is too soft to support the olive being crushed onto it. The pipe insert (or liner) supports the pipe wall.

Clipping and supports for polybutylene pipe

Unlike copper tubes and low carbon steel pipe, polybutylene is very flexible and can sag if not clipped correctly. Because of its flexible qualities, polybutylene pipe should be clipped at the distances shown in the table below:

Pipe diameter (mm)	Horizontal spacing (m)	Vertical spacing (m)
10	0.3	0.5
15	0.3	0.5
22	0.5	0.8
28	0.8	1.0

If the pipework is adequately supported or is run in concealed spaces, such as through joists on a suspended timber floor, pipe clips need not be fitted, provided that:

- the pipe is not part of an open vent connected to a heat source or an appliance, such as a boiler or hot water storage cylinder where the pipework is liable to become hot
- the pipe is not part of a distribution pipe or circuit where poor pipe alignment may affect the venting of air
- any hot water or heating pipe will not come into contact with a cold water supply pipe
- there is no risk of the pipe coming into contact with sharp or abrasive edges.

Pipe insert

SmartScreen Unit 005 worksheets 5 and 6

SmartScreen Unit 005 interactive activity 5

C FIXINGS FOR MASONRY, TIMBER AND PLASTERBOARD

In this section of the chapter, we will take a brief look at the various fixings for brickwork, concrete, stone, timber and plasterboard that plumbers use in their work.

C1 NAILS

There are many different types of nail that are used for a variety of jobs. It is not important that we know every type of nail but it would be beneficial to become familiar with some types, such as floor brads and oval nails.

Nails are usually described by their purpose and their dimensions in mm, eg '150 × 4' is 150mm long and 4mm in diameter. Below are listed some of the different nail types you may use:

	Masonry nails – Used for making fixings to masonry. Normally made of hardened zinc.
	Copper nails – Used by plumbers to fix sheet lead. They are made of copper to prevent corrosion between the lead and the nail. As they do not rust, they have a long life.
	Floor brads – Used to fasten floorboards. Generally, these are 50mm long. In cases where a floorboard gives access to pipework underneath, it is best practice to use screws and not floor brads to make access easier.
	Galvanised clout nails – Used for fixing slates and roof tiles.
	Round bright wire nails – Used generally for rough joinery work where strength is more important than appearance.
	Oval bright wire nails – Suitable for joinery work where appearance is important. The head is lost when driven into the timber.

C2 SCREWS

There are many types of screws available for different applications:

- countersunk
- crosshead/Pozidriv
- raised countersunk
- round head
- mirror
- coach
- chipboard.

Screws can be made from steel, stainless steel or brass and come with a range of screw head types (see section on screwdrivers, page 209). They can be coated with corrosion protection such as bright zinc and black japanned coatings. Screws are specified by their length in millimetres or inches, and their gauge. The most common lengths used in plumbing range from 15mm for fixing copper saddle clips to skirting boards to 50mm × 10 for fixing radiator brackets.

Screw length and gauge

See SmartScreen Unit 005 worksheet 4

	Countersunk screw – Used for general work. The head sinks flush or a little below the wood surface. Needs a flat blade-type screwdriver.	
	Crosshead/Pozidriv screw (countersunk type) – Used for general work and very similar in appearance to the countersunk screw, but needs a crosshead screwdriver. This ensures that the screwdriver does not slip out of the screwhead. Ideal for pipe clips.	
	Raised countersunk screw – Used for fixing decorative fittings with countersunk holes. The head is designed to be visible.	
	Round head screw – Used for fixing copper saddle clips.	
	Mirror screw – Used for fixing mirrors and bathroom fittings such as bath panels. The chromed cap threads into the screw head to hide the screw.	
	Coach screws – These usually come with purpose-made wall plugs. They are used for fixing heavy constructions such as boilers. Can be tightened with a spanner but some have Pozidriv screw heads.	
	Chipboard screw – Used for securing chipboard and medium density fibreboard (MDF). Various types of heads are available.	

005 SITE PREPARATION AND PIPEWORK FABRICATION TECHNIQUES

255

C3 HEAVY-DUTY FIXINGS

There are two heavy-duty fixings that plumbers use occasionally:

	Coach bolts – Not usually used for plumbing but can be useful for carrying heavy loads such as cold water cisterns and hot water cylinders on building structures and platforms. They are usually made from galvanised steel.
	Rawl bolts – Also known as a heavy-duty expansion anchors. These are easy to use with good load-carrying capacity and can be used in concrete, brickwork and stone for fixing heavy appliances and large diameter pipework.

C4 PLASTERBOARD AND LIGHT STRUCTURE FIXINGS

Light structure fixings are used when working on a lightweight wall, such as a plasterboard stud wall. Plasterboard is extremely difficult to fix to. Generally, if a fixing is required, it is better to ask the joiner to put a wood nogging in the wall before it is plasterboarded and skimmed with the plaster topcoat. When working in existing properties, this is not always possible without damaging the wall's surface and decoration. In this situation, plasterboard fixings are the only option. There are several different types:

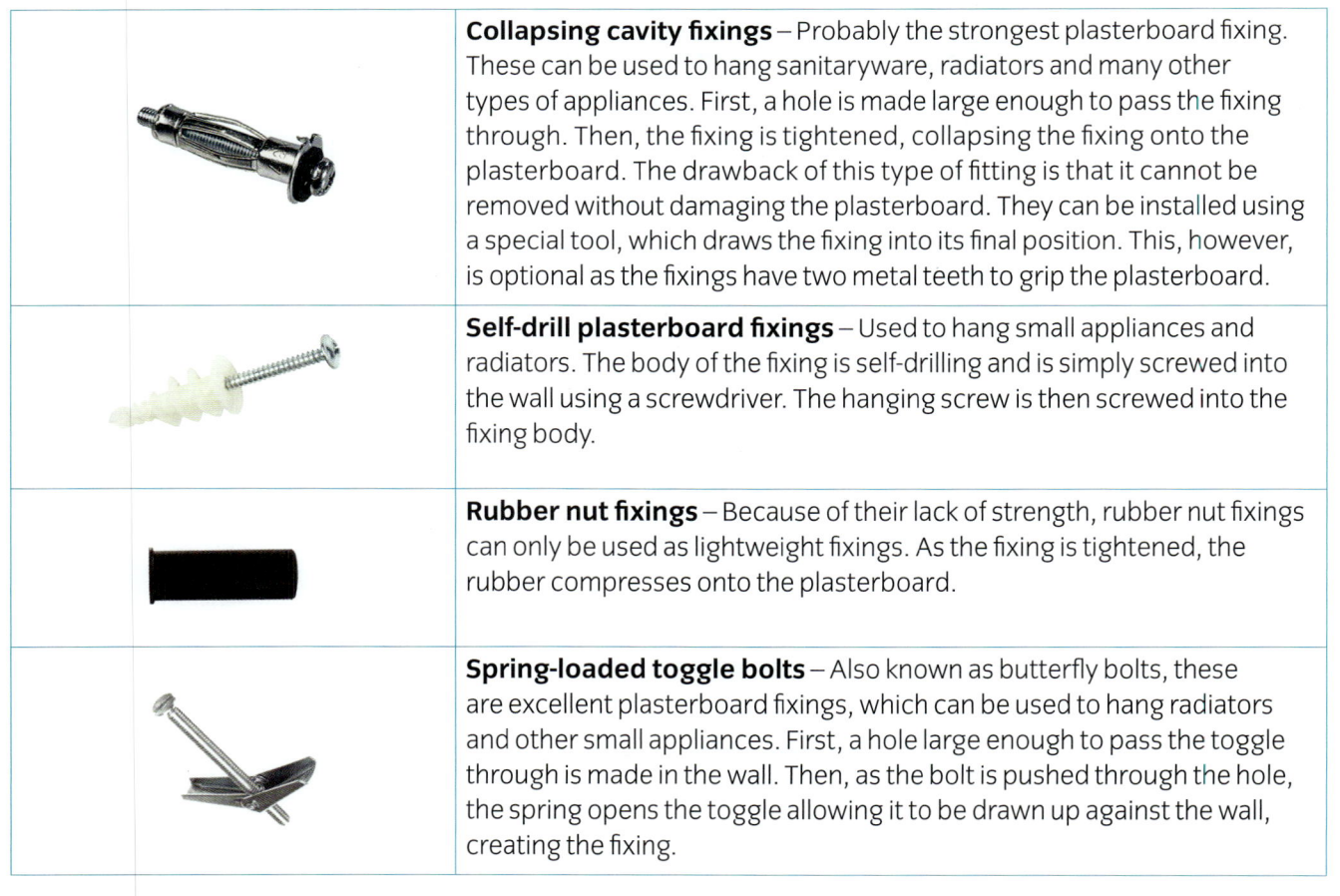

Nogging
Term often used on site to describe a piece of wood that supports or braces timber joists or timber-studded walls. They are particularly common in timber floors as a way of keeping the joists rigid and at specific centres but they can also be used as supports for appliances such as wash hand basins and radiators that are being fixed to plasterboard.

	Collapsing cavity fixings – Probably the strongest plasterboard fixing. These can be used to hang sanitaryware, radiators and many other types of appliances. First, a hole is made large enough to pass the fixing through. Then, the fixing is tightened, collapsing the fixing onto the plasterboard. The drawback of this type of fitting is that it cannot be removed without damaging the plasterboard. They can be installed using a special tool, which draws the fixing into its final position. This, however, is optional as the fixings have two metal teeth to grip the plasterboard.
	Self-drill plasterboard fixings – Used to hang small appliances and radiators. The body of the fixing is self-drilling and is simply screwed into the wall using a screwdriver. The hanging screw is then screwed into the fixing body.
	Rubber nut fixings – Because of their lack of strength, rubber nut fixings can only be used as lightweight fixings. As the fixing is tightened, the rubber compresses onto the plasterboard.
	Spring-loaded toggle bolts – Also known as butterfly bolts, these are excellent plasterboard fixings, which can be used to hang radiators and other small appliances. First, a hole large enough to pass the toggle through is made in the wall. Then, as the bolt is pushed through the hole, the spring opens the toggle allowing it to be drawn up against the wall, creating the fixing.

C5 PLASTIC WALL PLUGS

Plastic wall plugs (Rawlplugs) are used in conjunction with screws to fasten appliances, sanitaryware and many other pieces of equipment to masonry, concrete and stone walls. They are available in different sizes to match screw gauges and are colour coded for easy identification. The wall must be drilled with a masonry drill bit (see page 220) of a specific size for the colour of the plug.

Plastic wall plugs are installed in the following way to hang a radiator bracket:

1 Measure, mark and drill the hole, using the correct sized drill to the correct depth for the chosen screw.

2 Insert the plug into the hole and push it slightly below the wall's surface.

3 Using the previously selected correct screw for this job, fasten the radiator bracket to the wall using a screwdriver.

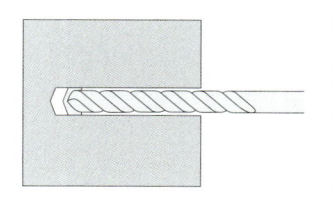

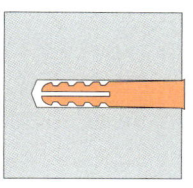

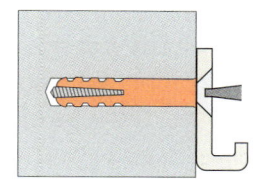

Wall plugs – hanging a radiator

D INSTALLATION PROCESSES OF DOMESTIC PLUMBING SYSTEMS

The successful installation of a domestic hot and cold water system or a domestic central heating system is the result of a series of processes. In this section of the chapter, we will look at the processes that are involved when working on new and existing installations, from the initial design of new systems to permanent decommissioning of existing installations. We will also investigate the associated skills we need for some of those jobs that are outside a plumber's skill base and look at how we can care for and protect the customer's valuables and possessions.

D1 THE DESIGN PROCESS OF PLUMBING INSTALLATIONS

The design of plumbing and heating installations involves the calculation of flow rates, heat losses from the building fabric, capacities of stored water required and pipe sizes. On large contracts it may also involve the planning of pipe routes based upon the Architect's drawings. Design processes can be divided into three distinct groups:

- **Large industrial/commercial contracts** – On large contracts, a Building Services Engineer (see Chapter 002, page 84) will design the hot and cold water installations, heating systems, above- and below-ground drainage and air conditioning systems.

- **New-build domestic installations** – With this type of installation, the company that has won the contract to supply the heating boilers and equipment may undertake the design of the central heating system. The hot and cold water and sanitation design will be completed by the plumbing company, who may employ their own designer/estimator.

- **Existing dwellings** – The design of systems for existing dwellings is usually completed by an experienced operative who has knowledge of the procedures for completing basic heat loss, flow rate and pipe sizing calculations.

The design of a system is the first important step towards a successful installation. The calculations completed (using information specified in the manufacturer's literature) must allow the finished installation to deliver the desired flow rates, heat outputs and component performances. It is at this stage that the type of materials to be used for the installation will be chosen. This will be based upon the type of building, its uses, the type of system being installed and cost.

D2 PLANNING THE WORK AND THE DELIVERY OF MATERIALS

Once the terms of the contract along with the specification have been agreed and signed by the customer, the next step in the process is planning. Planning involves timescales for completion, and logistics for delivery and storage of tools and materials. Here are some points to remember regarding delivery and storage of materials:

- When working on any large site, make sure that all tools and materials are locked away in a secure lock-up when not in use. Materials that are left uninstalled in unoccupied work areas may not be covered by theft insurance. Materials should not be left in the open and all unused materials should be returned to the store.

- Ensure that materials such as sanitaryware, boilers and radiators are stacked to a safe height and are covered to prevent damage.

- Keep large pieces of equipment and tools in a separate part of the store. This can help to prevent accidental damage to fragile materials such as wash basins and WC pans.

- Have a materials requisition system in place so that materials can be booked out of the stores for use and any unused materials can be booked back in. This ensures that a close check can be kept on the stock of pipes, tubes and fittings, which can help to prevent theft and over-ordering.

- Keep on file all delivery and advice notes so that a check can be made against the stock of materials delivered and the materials used.

- When undertaking work in a private dwelling, the delivery and storage of materials should be agreed with the customer so that they can be delivered at a convenient time and stored in a place that will cause as little disruption to the day-to-day activities of the household as possible.

- Partially installed items, such as baths, wash basins and WCs should be protected from damage. Any protective tape or plastic coverings on sanitaryware should be removed before installation so that they can be visually checked for damage that occurred in manufacture and transit.

D3 PRE-INSTALLATION ACTIVITIES ON NEW AND EXISTING INSTALLATIONS

Working on site in a new-build house, compared with working in an occupied dwelling, requires a completely different style of working. While many of the working practices we use on site can be used in an occupied dwelling, the overriding emphasis is on care and attention to detail. There are three additional concerns when working in an occupied dwelling:

- protecting the customer's property
- protecting the building fabric
- installing in accordance with the customer's wishes while maintaining the quality of the installation against the regulations in place.

Many instances have occurred in the past where a good installation has been spoiled by the carelessness of the plumber and failing to liaise with the customer. This may result in disputes, withholding of money owed and, occasionally, court action.

Working in private houses

Good service means not only doing the job well but also keeping the customer fully informed throughout. Before an installation takes place, ensure that:

- The customer knows what day and time you will be arriving. Agree a start time with the customer and stick to it: early morning arrivals are not always welcome, and lateness causes inconvenience and gives a bad impression.

- Walk around the house with the customer. Point out any existing damage to furniture, fixtures, carpets and wall coverings and if necessary take photographs for reference. This will prevent any misunderstandings regarding damage and marks already in place.

- Point out which carpets and pieces of furniture will need to be removed before you begin work and ask the customer to remove them. If you are going to be working outside, ask the customer politely to move any of their vehicles or equipment before you begin work so that they do not get damaged.

- Cover with clean dust sheets all furniture, carpets and fixtures that cannot be removed in the area where you are going to work.

SmartScreen Unit 005 worksheet 2

SmartScreen Unit 005 interactive activity 3

- Agree with the customer the position of radiators, boilers and all visible pipework at planning stage, well before work begins. When fitting sanitaryware, make sure that you are fitting the appliances in the position that the customer wants.
- Keep customers informed about any problems that arise that may need their decision.
- Let the customer know when any services, eg water, gas, electricity, or appliances such as the WC, are going to be turned off or taken out of service. Ensure that the customer has collected enough water for the period of temporary decommission. If working on a central heating system, ensure that they have access to other forms of heat, especially during cold weather.

D4 PREPARATION OF THE WORK AREA

Much of the work in occupied and existing dwellings involves installing pipework under floors, and in and through walls. We will look at the procedure for lifting floorboards, notching and drilling joists and chasing walls to allow the installation of pipework.

Lifting floorboards

To lift floorboards using power tools, carry out the following steps:

1 Decide on the boards to be lifted and mark them with a pencil.

2 Locate the position of the joists. This can be done by searching for the row of nails holding the board to the joist.

3 Mark the centre of the joist where the board is to be cut. If this is not possible, a cut can be made inside the joists and supporting cleats fitted before the board is replaced. Number the boards as this makes replacement easier.

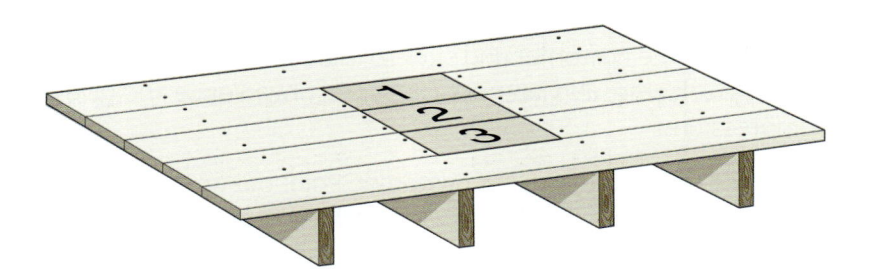

4 Using a nail punch, punch the floorboard nails below the surface of the board.

5 Ensuring that the correct type of blade is used, set the depth on the circular saw to just less than the depth of the board. This is to ensure that any cables or services already installed are not damaged.

The blade of the circular saw just less than the thickness of the floorboard

6. Run down the length of each board to be lifted with the circular saw to cut the tongue of the board.

7. Using the marks on the joist previously made, carefully cut across the board at the joist using the circular saw.

8. The board can now be lifted using a bolster chisel to prise up the board.

To lift a floorboard using hand tools, carry out the following steps:

1. Follow points 1–4 of the method for lifting floorboards using power tools, as shown above.

2. Break the tongue of the board. This can be done by either carefully driving the bolster chisel through the tongue with a claw hammer or cutting down it with a hand floorboard saw.

3. Now, using the marks on the joist previously made, carefully cut across the board at the joists using a hand floorboard saw.

4. The board can now be lifted using a bolster chisel to prise up the board.

Chipboard flooring is quite different from timber floorboards. It is laid in large sheets measuring 2m × 0.6m and glued at the tongued and grooved joint. The sheets break very easily if mishandled. When part of a board is lifted, unlike timber boards, the rest of the board needs support at every edge, including the edges where there are no joists. Lifting chipboard flooring is best done with a circular saw.

To lift chipboard flooring, carry out the following steps:

1. Determine which boards are to be lifted and mark them with a pencil.

2. There is no need to mark the joists with chipboard as the long joints indicate where the joists are. All that is needed is to mark the area of the board that needs lifting.

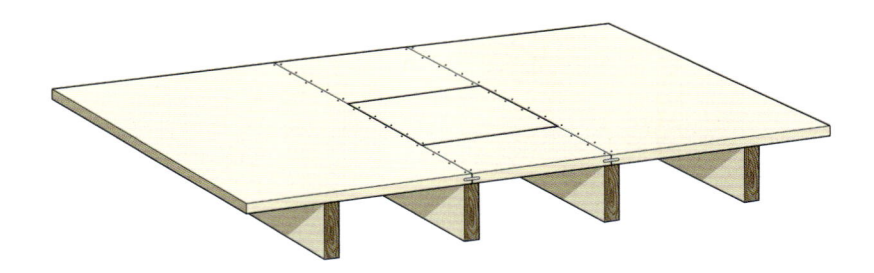

3 Using a nail punch, punch the nails below the surface of the board.

4 Set the depth on the circular saw just less than the depth of the board. This is to ensure that any cables or services already installed are not damaged.

5 Run down the length of the boards to be lifted with the circular saw to cut the tongue of the board.

6 Now, using the marks previously made, carefully cut across the board using the circular saw.

7 The board can now be lifted using a bolster chisel to prise up the board.

8 When replacing the board, the edges need to be supported by wooden cleats. This can be done as shown in the drawing below:

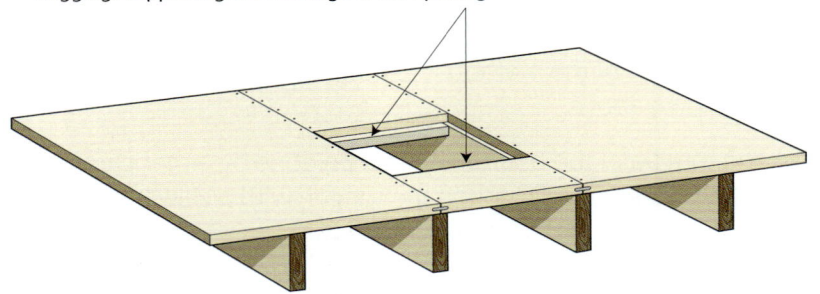

Noggings supporting the free edge of the opening

Notching and drilling joists

Many installations require the notching and drilling of timber joists to accommodate pipes and fittings under the floor. Holes or notches that are made too close together, or holes drilled too near the end of a joist and incorrectly positioned too near to the centre of the joist span, can weaken joists to the point where they become useless as structural supports. In essence, the strength and the stiffness of the joist must not be compromised.

Notches must be made as shown in the diagram below:

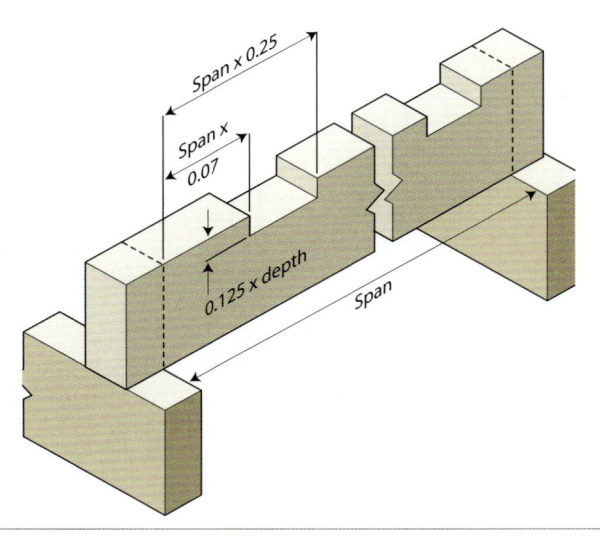

Notching measurements

To find out where notches can be made in a joist, use the following procedure:

1 Measure the span of the joist from wall to wall.

2 Multiply the span measurement by 0.07. This will give a measurement equal to 7 per cent of the span.

3 Measure from the wall the 7 per cent measurement and mark it on the joist. No notches must be made within this mark.

4 Multiply the span measurement again by 0.25. This measurement is equal to 25 per cent of the joist's span.

5 Measure from the end of the joist again and find the 25 per cent distance and mark it on the joist.

6 All notches must be within the 7 per cent and 25 per cent marks.

SmartScreen Unit 005 handout 3

To put this into practice we must look more closely at the calculation:

Length of span of the joist	= 4m
7 per cent of the span	= 4 × 0.07 = 0.28 = 280mm
25 per cent of the span	= 4 × 0.25 = 1 = 1m

Holes drilled or cut into joists follow a similar procedure. A hole must not begin within 25 per cent of the span, measured from the end of the joist and must stop at a point equal to 40 per cent of the span, again measured from the end. The size of the hole must not exceed a measurement equal to 25 per cent of the depth of the joist when measured from the centre line. This is illustrated in the diagram below:

Holes must be at least three diameters (centre to centre) apart and no holes must be within 100mm of a notch

SUGGESTED ACTIVITY...

1 The span of a joist measures 4.5m long and 200mm in depth. Using the calculations shown as a guide, calculate:
   a the area where notches can be made
   b the maximum depth of those notches.

2 The span of a joist measures 3m long and 250mm in depth. Using the calculations shown as a guide, calculate:
   a the area where notches can be made
   b the maximum depth of those notches.

3 The span of a joist measures 3.6m long and 300mm in depth. Using the calculations shown as a guide, calculate:
   a the area where notches can be made
   b the maximum depth of those notches.

Answers: 1a. 315mm to 1125mm, 1b: 25mm, 2a: 210mm to 750mm, 2b: 31.25mm, 3a: 252mm to 900mm, 3b: 37.5mm

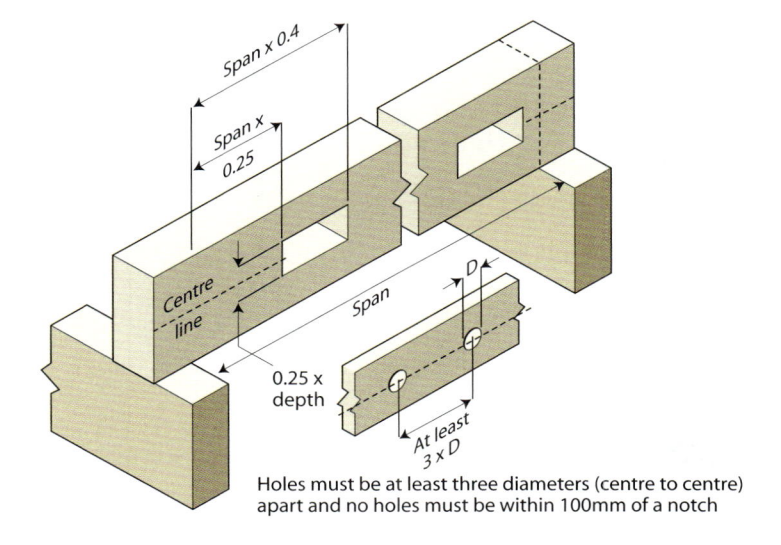

Again, to understand this fully we must look at the calculation:

Length of span of the joist	= 4m
25 per cent of the span	= 4 × 0.25 = 1m
40 per cent of the span	= 4 × 0.4 = 1.6m

> **SUGGESTED ACTIVITY...**
> The span of a joist measures 4.5 m long and 200mm in depth. Using the calculations shown as a guide, calculate:
>
> a the area where holes can be made
>
> b the maximum size of those holes.
>
> Answers: a. 1125mm to 1800mm, b. 50mm

Centre to centre
Measuring from the centre line of one pipe to the centre line of another so that all the tube centres are uniform. This ensures that the pipework will look perfectly parallel because all of the tubes will be at equal distance from each other.

Therefore, holes drilled or cut in the joist must start 1 metre from the end of the joist and must finish 1.6m from the end of the joist. All holes required must be made within a distance of 600mm. Again, this can be done from both ends of the joist so that two sets of holes can be made.

To calculate the size of the holes:	
Depth of the joist	= 250mm
25 per cent of the depth	= 250 × 0.25 = 62.5mm

This measurement must be taken equally either side of the centre line of the joist. No holes can be drilled in a joist within 100mm of a notch and circular holes must be at least three diameters of the hole size apart, measured centre to centre.

Cutting chases in walls

Occasionally, it may be necessary to cut a chase in a wall to conceal pipework, for example burying pipes for a downstairs radiator. This will involve the use of an angle grinder or wall-chasing machine to first cut the outline of the chase onto the wall and then carefully removing the unwanted masonry from between the cuts with a bolster chisel and a lump hammer. Caution should be exercised, as detailed below.

- It is advisable not to carry out this operation in a room containing carpets and furniture. If this is unavoidable, ensure that all furniture and carpets are placed to the far side of the room and covered with dust sheets, and that all doors are closed and taped up with masking tape. An angle grinder produces excessive amounts of dust and this must, wherever possible, be prevented from escaping from the room you are working in to the rest of the house. If possible, open a window, to allow some of the dust to leave.

- Always wear the correct PPE. This task will require the use of safety goggles (not glasses), gloves, overalls and a very good dust mask of the correct type to avoid breathing in the dust.

- Always check the angle grinder beforehand to ensure that:
  - it is in good condition and carries an in-date PAT test certificate
  - the correct masonry cutting wheel is installed
  - the wheel is secure and the wheel guard is in place.

Be wary of installing hot and cold water pipes in a wall where they are going to be concealed or tiled over. It is best practice to surface mount these pipes to allow access.

Chases cut in walls must be cut to no more than the following depths:

- Horizontal chases must not be deeper than ⅙ of the wall thickness.

- Vertical chases must not be deeper than ⅓ of the wall thickness.

D5 INSTALLATION ACTIVITIES ON NEW AND EXISTING INSTALLATIONS

Much thought should be given with regard to the positioning of pipework because not all of the pipes we install can be hidden. The golden rule is that visible pipework needs to be as neat as possible. A pipe that is not plumb or level looks unsightly and the eye is drawn to it immediately. Many people believe that surface-mounted pipework is an eyesore and customers will often ask the plumber to hide pipes wherever possible (which can be done with chasing or boxing). There are occasions, however, where because of the constraints of regulations and approved good practice this cannot be done and surface-mounted pipes are the only solution. Correct positioning, marking and installation of pipework is therefore essential.

Positioning of surface-mounted pipework

The routes taken by surface-mounted pipework should be well planned to take the shortest practicable route but to not be intrusive. There should also be as little marking out as possible so as not to deface the customer's decorations. The area must be well protected by dust sheets and coverings.

Select an appropriate pipe clip. Large, sturdy pipe clips in a domestic dwelling would look obtrusive while plastic pipe clips would not stand up the everyday knocks if used on large commercial/industrial installations. If a number of pipes are to be installed in one place, say in an airing cupboard, they can be arranged in banks, so that all the pipe clips are in a neat line.

The use of machine-made bends over elbows should be considered wherever possible as these not only provide a visually attractive installation but also aid better flow rates. The finished pipework should be as aesthetically pleasing as possible with even spaces between the pipe clips and supports and even gaps between different lines. The pipe tube should installed plumb and level or installed with the correct fall where this is needed.

Finally, make sure that when the pipework is in position, it is wiped down with a damp cloth. This will ensure that any flux that has run down the pipe during soldering operations is removed. Where possible, also remove any setting out marks and fingerprints from the wall with a damp, soapy cloth.

Prefabrication of pipework

Prefabrication of pipework often takes place on large housing contracts where many houses of the same type and style will be built. Prefabrication of pipework can often save time in this situation as the pipes can be bent beforehand to fit a particular part of the job. This will save time and installation costs and can be of benefit where hot working (ie the use of blowtorches) is forbidden. It can also be used where making joints in the fitted position may be difficult .

Prefabrication of pipework involves precise marking, cutting and forming with measurements taken either on site or from a drawing and then fabricated in a workshop and delivered to site ready for use. In this way, many units can be made at the same time and stored on site ready for installation.

On-the-job working will also involve some prefabrication of pipework. Precise measurements, cutting and bending are essential if the pipework is to look good. Consider the drawing below:

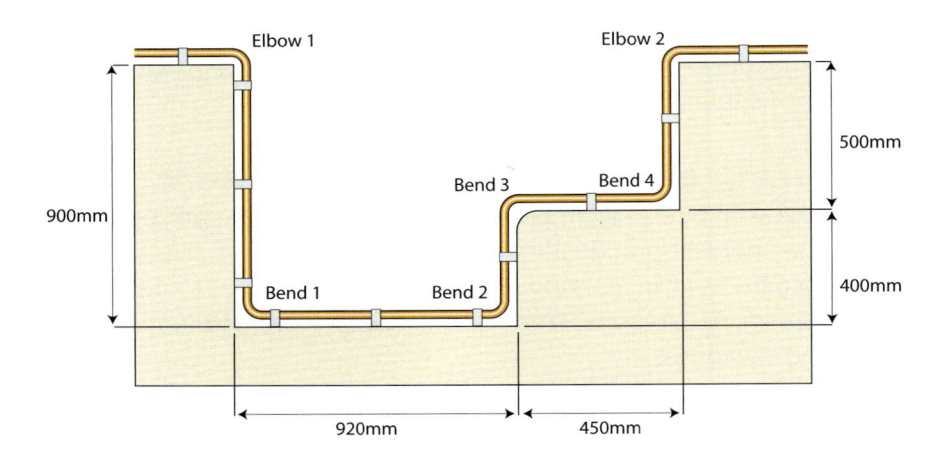

Pipework layout drawing

The pipework is to be fabricated on site from one piece of tube from elbow 1 to elbow 2 using measurements taken on site.

The 'X' dimension is measured from the tube stop to the centre of the socket at 90°

Method

When calculating and marking out tube for one-piece bending, there are several pieces of information we require:

- the 'X' dimension of any fittings
- the distance to the centre of the clip
- the measurements of the space where the tube is going to be installed
- the pipe gain of any machine bends.

Look at the drawing above. We can see that the tube has to fit in an alcove. Elbows will be required at elbow 1 and elbow 2, simply because the wall has sharp corners at those points. All other changes of direction can be achieved using machine-made bends. For this example, we will assume that:

The 'X' dimension of a 15mm elbow = 12mm

Distance to the centre of the clip = 15mm

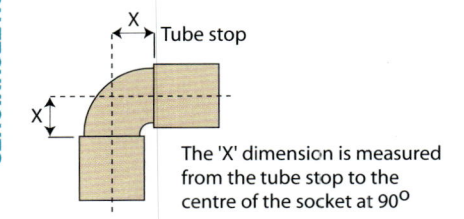

Total up the amount of tube required for the one-piece bend:

From elbow 1 to bend 1. The distance is 900mm and because pipe clips are present at elbow 1 and bend 1, the distance is the same. However, because we need to make an end-feed elbow joint, we have to deduct the 'X' dimension of the elbow:

900 − 12 = 888mm

So, the measurement 1 = 888mm. Therefore, bend 1 can be marked and bent at this distance.

From bend 1 to bend 2. The distance here is 920mm but the bends are fixed between clips either side, so deduct the distance to the centre of the clip each side.

920 − (15 + 15) = 890mm. Therefore, bend 2 can be marked and bent at this distance.

From bend 2 to bend 3. The distance here is 400mm and because there are clips at both bends, the distance between the bends does not change.

Bend 3 can be marked and bent at the distance of 400mm.

From bend 3 to bend 4. The distance here is 450mm and because there are clips at both bends, the distance between the bends does not change.

Bend 4 can be marked and bent at the distance of 450mm.

From bend 4 to elbow 2. The distance is 500mm and because there is an elbow at the end, 12mm should be deducted for the 'X' dimension.

500 − 12 = 488mm

Therefore, length of pipe:

888 + 890 + 400 + 450 + 500 = 3128mm or 3.128m

There are four machine bends on the 15mm pipe and, as we have seen, these have a pipe gain of 22mm each. Therefore:

22 × 4 = 88mm. This can be deducted from the total length:

3128 − 88 = actual tube length = 3040mm or 3.040m

Many of the appliances we fit arrive on site, also prefabricated. Boilers, hot water storage cylinders and some sanitaryware can be manufactured 'pre-plumbed' so that only the final connections have to be made when the unit is put into position.

Prefabrication techniques can be carried out on most fixed pipework types, including copper, low carbon steel and plastic soil and waste pipes. The techniques will differ depending on the material used.

Sleeving of pipework through walls

Pipes passing through masonry, stone and concrete should be sleeved by a piece of tube one size larger than the pipe being installed to allow for expansion and pipe movement and to prevent damage to the pipe by building movement. The sleeve should then be sealed with an approved sealant to prevent the rain, insects and vermin from getting through.

Special consideration needs to be given to gas pipework as all gas work is covered under the Gas Safety (Installation and Use) Regulations 1998. All pipework passing through a wall must be sleeved and take the most direct route. Internal pipework must be sealed at both ends with a fire resistant, non-setting compound. Pipework that passes through an external wall must be sealed internally only. It must be remembered that to work on gas installations you must be properly trained and Gas Safe registered.

Working on existing installations: in-situ working

In-situ installation operations include:

- cutting fittings, such as isolation valves and tees, into an existing hot or cold water pipe
- capping off existing pipework
- removing existing bath, wash basin and sink taps
- changing WC pans and cisterns and other bathroom equipment
- boiler swaps on existing central heating installations.

Working on existing installations is challenging. There is always a risk of disturbing joints and causing further problems. Situations often occur where it is necessary to cut into existing pipework, and this should be treated with care. Problems can occur when connecting to old imperial sized pipework when the pipe sizes differ from new metric fittings and tubes, which were introduced in 1973. Because of this, particular care should be taken in planning and working. Conversion from imperial to metric pipework can be done using special metric to imperial adapters.

D6 PROTECTION OF THE BUILDING FABRIC AND ITS SURROUNDINGS

We now understand how a customer's personal belongings should be protected from dust and damage from the installation process. But there are other ways that we can protect the building and its surroundings.

When soldering is taking place in the building, the risk of fire is ever present. To protect the building fabric, a heatproof soldering mat should be used. It should be remembered, however, that these would not protect if the flame were directly on the mat. A certain amount of angle should be applied to the blowtorch, if possible, to deflect the heat away from the wall/floor/ceiling/skirting board. There are three different types of mat available, which will resist temperatures of 600°C, 1000°C and 1300°C.

SmartScreen Unit 005 worksheet 3

SmartScreen Unit 005 interactive activity 4

Solder mat

One other way we can protect against heat is to use heat-dissipating spray gel. It offers protection against the scorching of wallpaper and paint and loosening of existing joints. It also reduces the risk of fire by protecting surfaces and dissipating heat.

When working outside the building, protect the customer's garden by the use of walk boards across flowerbeds and protective sheeting across grass lawns. Do not dig ladders into lawns.

Before drilling a wall, check it first with a cable/pipe detector to ensure that there are no services already in the wall. Also, to prevent blowing the surface of the backside of the wall you are drilling, first drill a small pilot hole and drill from both sides. This will ensure that the wall surface around the hole is not damaged.

SmartScreen Unit 005 handout 2

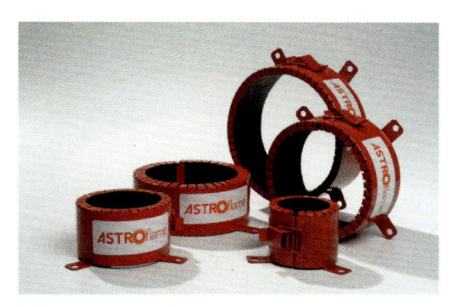

When removing an old radiator, there is a risk of spilling dirty water. To prevent this, turn the radiator upside down so that the valve tails are at the top. Always remember to protect those carpets that cannot be removed during simple maintenance operations such as these.

Fire stopping

Where pipes (including soil and vent pipe) pass between floors, the holes around the pipe must be fire stopped to prevent the spread of fire, smoke and hot gases. This can be done in two ways:

1. **By the use of an intumescent collar** – This is a collar that is placed around the pipe that expands in the presence of heat to stop the spread of fire.

2. **By the use of intumescent sealant** – This is sealant that acts in the same way as intumescent collars.

Correctly used, these techniques will help to contain fire in the room where it started, reducing the damage to the rest of the property and keeping escape routes open.

Intumescent collar

D7 TESTING AND COMMISSIONING PROCEDURES

Testing of installations is the first time we see whether the installation is watertight. For pressure systems and sanitary systems, testing procedures are set out in the relevant British Standards and Regulations.

Pre-testing checks

Before commissioning takes place:

- Walk around the installation, checking that you are happy that the installation is correct and meets installations standards.

- Check that all open ends are capped off and all valves are isolated.

- Check that all capillary joints are soldered and that all compression joints are fully tightened.

- Check that enough pipe clips, supports and brackets have been installed and that all pipework is secure.

> **KEY POINT**
>
> Hot and cold water systems testing is detailed in BS 6700:2006 + A1:2009, central heating systems testing is detailed in BS 5449:1990, and above-ground sanitation systems should be tested in accordance with Document H of the Building Regulations.

Testing procedures

Testing procedures differ depending on the type of pipework installed. The process involves filling the system with water to a specific pressure, letting it stand for a period of time to temperature stabilise and then checking it for pressure loss.

- Copper tubes and low carbon steel pipes – Systems installed in copper tube and low carbon steel pipes should be tested to 1.5 times normal operating pressure. They should then be left for a period of 30 minutes to allow for temperature stabilisation and then left for a period of 1 hour with no visible pressure loss.

- Plastic (polybutylene) pressure pipe systems – These are tested rather differently from rigid pipes. There are two tests that can be carried out. These are known as test type A and test type B and are detailed in BS 6700:2006 + A1:2009.

    - **Test type A** – Slowly fill the system with water and raise the pressure to 1 bar (100kp). Check and re-pump the pressure to 1 bar if the pressure drops during this period provided that there are no leaks. Check for leaks. After 45 minutes, increase the pressure to 1.5 times normal operating pressure and let the system stand for 15 minutes. Now release the pressure in the system to one third of the previous pressure and let it stand for a further 45 minutes. The test is successful if there are no leaks.

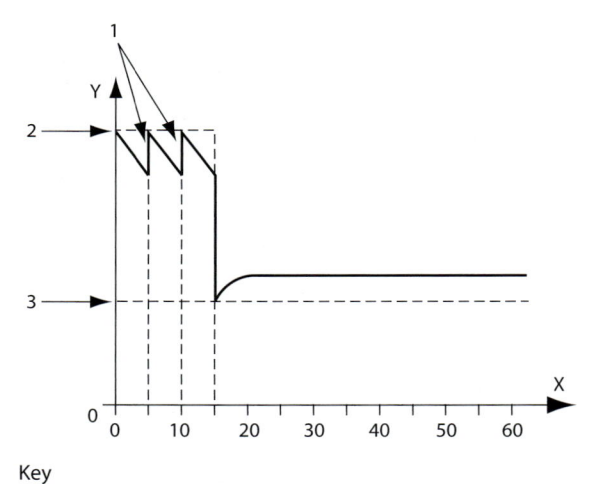

Key
1 Pumping
2 Test pressure 1.5 times maximum working pressure
3 0.5 times maximum working pressure
X Time (minutes)
Y Pressure

Test type A pressure test graph

    - **Test type B** – Slowly fill the system with water and pump the system up to the required pressure and maintain the pressure for a period of 30 minutes and note the pressure after this time. The test must continue with no further testing.

    Check the pressure after a further 30 minutes. If the pressure loss is less than 0.6 bar (or 60kPa), the system has no visible leakage.

Visually check for leakage for a further 120 minutes. The test is successful if the pressure loss is less than 20kPa (0.2 bar).

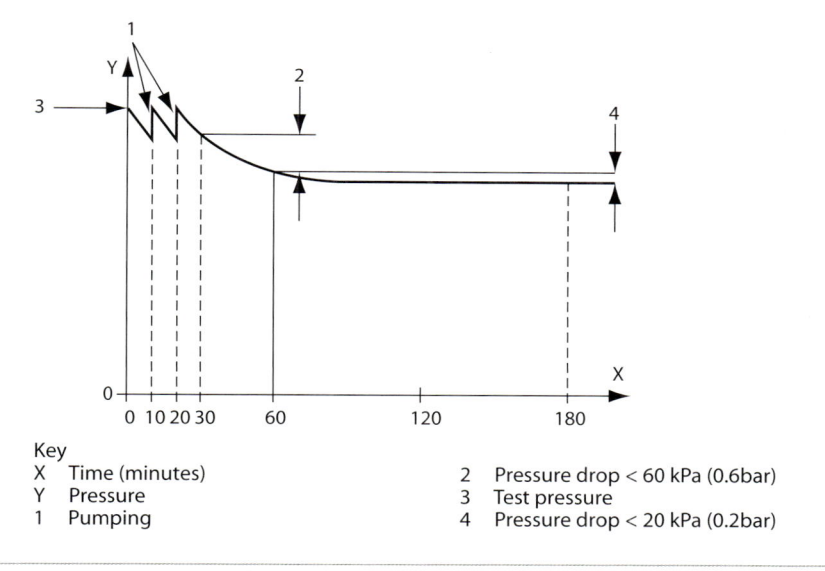

Key
X Time (minutes)
Y Pressure
1 Pumping
2 Pressure drop < 60 kPa (0.6bar)
3 Test pressure
4 Pressure drop < 20 kPa (0.2bar)

Test Type B pressure test graph

- **Above-ground sanitation systems** – These should be tested to a pressure of 38mm water gauge (w/g) for a period of three minutes with no pressure loss.

Commissioning

Commissioning is the part of the installation where the system is filled and run for the first time. This is where you establish whether or not there is a leak in the pipework. The first task is to fill the system and check for leaks at the appliances. This is best carried out in stages so that sections of the installation – ie cold water, hot water, central heating – can be filled and tested separately. At each stage of the filling process, the system should be checked for leaks before moving on to the next section.

Once the systems have been filled they should be drained down and flushed through with clean water and refilled. The water levels in WC cisterns, cold water storage cisterns and feed and expansion cisterns (if fitted) should be checked for compliance with the relevant regulations.

Check the flow rates at all taps to see if they deliver the flow rates demanded by the manufacturer's literature and check the operation of all controls, including thermostats and motorised valves. Set the temperature of any cylinder thermostats and let the water reach full temperature.

Using a thermometer, check the temperature of all radiators and check the temperature of the hot water.

Gas installations should be checked for tightness and central heating systems should be run up to full operating temperature before being drained down while they are still warm. Refill the system and add inhibitor before running the system again.

Benchmarking

At this stage of the installation, it is time to benchmark the system. Here the boiler and any hot water cylinder installed are checked for compliance with the manufacturer's instructions, including:

- hot water flow rates
- flow and return temperatures
- hot water temperature
- operation and types of controls
- gas rates.

The benchmark certificate should then be signed by the commissioning engineer.

Building Regulations Compliance certificates

From 1 April 2005 the Building Regulations demanded that all installations must be issued with a Building Regulations Compliance certificate. This is to ensure that all Building Regulations relevant to the installation have been followed and complied with. This includes:

- the heating installation
- the sanitation system
- the hot and cold water systems
- the gas installation.
- any electrical controls.

Certificates are issued by the competent registered installer.

Handover to the customer

When the system has been tested, commissioned and benchmarked, it can then be handed over to the customer. The customer will require all documentation regarding the installation:

- all manufacturers' installation and servicing instructions for boilers, electrical controls, taps, sanitaryware and any other equipment fitted to the installation

- the benchmarking certificate

- the Building Regulations Compliance certificate, which is done online by the installer with the certificate sent directly to the customer.

The customer must be shown around the system and shown how to use any controls, thermostats and time clocks or integrated programmable thermostats. Isolation points on the system for gas, water and electricity should be pointed out and a demonstration should be given of the correct isolation procedure in the event of an emergency. Explain to the customer how the systems work and ask if they have any questions. Finally, highlight the need for regular servicing of the appliances and leave emergency contact numbers. Also, remind them to complete and return their boiler warranty registration.

D8 DECOMMISSIONING PROCEDURES

Decommissioning a system or an appliance simply means taking it out of service. It falls into one of two categories:

1 **Temporary decommissioning** – Where a system or an appliance is taken out of service for a period of time for repairs, replacement or maintenance. The customer must be kept informed of when the system is being shut down, the expected length of time of the decommission and the expected reinstatement time. If the period of time is considerable, ensure that the customer has access to vital services, ie gas, water end electricity. The key here is to keep the customer informed.

2 **Permanent decommission** – Where a system or appliance is permanently disconnected and/or removed. This will involve disconnection and making safe of any services. Pipes should be cut back and capped and, if necessary, tested for soundness. All electrical disconnections should be made by a qualified operative or an electrician.

D9 MAINTENANCE ACTIVITIES

Maintenance falls into two categories:

1 **Planned preventative maintenance** – On larger installations, it may be necessary to have a planned maintenance schedule so that systems and equipment can be serviced and checked at regular intervals to ensure optimum performance. Maintenance activities should be recorded in a logbook, together with the results of any tests performed. Planned preventative maintenance operations include:

- checking and repairing float-operated valves and setting water levels in cisterns
- cleaning out cold water cisterns of all sediment as required
- routine boiler maintenance
- checking and re-washering taps as required
- routine testing of above-ground drainage systems
- checking the operation of any safety valves
- checking the operation of all external controls and isolation valves, including:
  - stop taps
  - gate valves
  - isolation valves
  - motorised valves
  - thermostats.

2 **Breakdowns, repairs and emergencies** – These are unplanned maintenance activities that can occur at any time and include:

- burst pipes
- boiler breakdowns

- running overflows
- blockages
- dripping taps
- WC cistern problems.

D10 MAKING GOOD THE BUILDING FABRIC

During the installation process, there will be many occasions where the building fabric will need to be worked on. Holes will need to be drilled or broken through with a hammer and chisel, chases will need to be made to accommodate pipework, and floorboards will need to be lifted and replaced. Unless it is specified in the contract that these will be repaired by other tradespersons on site, they will have to be repaired by the plumber.

Making good involves having a few basic skills of another, associated trade such as a bricklayer, plasterer and carpenter/joiner. We have already seen the methods of lifting and replacing floorboards (see pages 260–62) and here, we will look at making good the holes we have made in walls.

By far the easiest holes to repair are those made by drills and masonry bits. These will require pointing with a 4:1 mortar (four amounts of sand to one amount of cement) mixed to a fairly stiff consistency. A pointing trowel should be used for this. Larger holes may need the replacement of broken or half bricks. Any new bricks used should match the existing wall bricks. The finished wall should be pointed with a pointing trowel and cleaned with a soft brush.

Patching plaster can be a tedious task. The type of plaster used will depend on the wall surface. Sand and cement rendering will need a smooth finish plaster, and plasterboard will need a plasterboard finish plaster. Board finish dries much quicker and so is harder to 'skim' to a smooth finish.

At the end of the making-good procedures, make sure that the area is cleared of all waste materials and cleaned.

E DRAWING SYMBOLS OF PLUMBING VALVES AND APPLIANCES

Working drawings for plumbing and heating installations often contain symbols that represent pipes, valves and appliances. It is important that these symbols are recognised for systems to be installed properly. All symbols shown will be in accordance with BS 1192:2007.

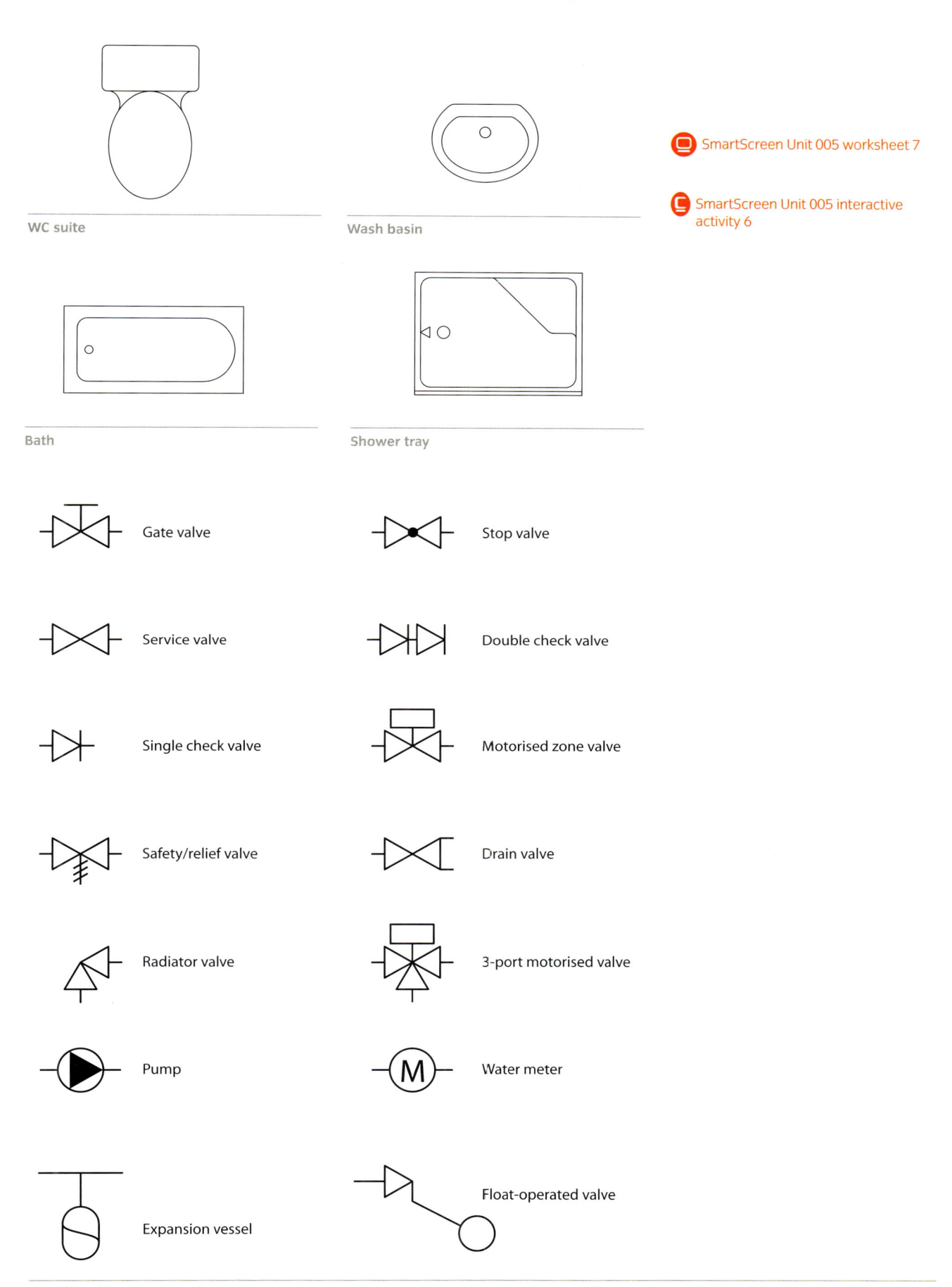

WC suite

Wash basin

🔲 SmartScreen Unit 005 worksheet 7

🔲 SmartScreen Unit 005 interactive activity 6

Bath

Shower tray

Symbol	Name	Symbol	Name
	Gate valve		Stop valve
	Service valve		Double check valve
	Single check valve		Motorised zone valve
	Safety/relief valve		Drain valve
	Radiator valve		3-port motorised valve
	Pump		Water meter
	Expansion vessel		Float-operated valve

Plumbing drawing symbols

005 SITE PREPARATION AND PIPEWORK FABRICATION TECHNIQUES

CONCLUSION

During this chapter we have explored the tools we need, the materials we use and the installation practices we need to master to enable us to install good, working systems that meet, not only the requirements of the regulations, but also the customers' needs and expectations. Good working practices at the start of your plumbing career will serve you well as you broaden your experience, gain knowledge and improve your skills.

005 TEST YOUR KNOWLEDGE

SmartScreen Unit 005 revision sample questions

1. Identify which tools we would use for the following tasks:

   a bending a copper tube offset onsite
   b tightening a compression fitting
   c cutting polybutylene pipe
   d cutting a floorboard
   e drilling a hole for a boiler flue
   f cutting low carbon steel pipe.

2. State the checks to be made on a power tool before use.

3. Name the three grades of copper tube. Which one is used for most plumbing installations?

4. There are two types of capillary fittings. What are they and how do they differ?

5. What is the purpose of flux?

6. There are two types of compression fittings. What is the difference between the two?

7. What do the initials LCS stand for?

8. Describe a tapered thread?

9. Which two types of plastic pipe can be solvent welded?

10. Which solvent weld plastic pipe suffers from UV degradation?

11. What is the bend radius of a 15mm polybutylene pipe?

12. When working in a private dwelling, what should be your first task before beginning work?

13. If a joist is 200mm in depth, what is the maximum depth allowed of any notches made?

14. If a joist has a span of 5m, at which point should notches start and finish?

15. What is an intumescent collar and what is it used for?

16. Which two certificates should the customer keep on completion of a central heating system?

17. What is permanent decommissioning?

18. Describe planned preventative maintenance.

19. Draw the symbol for a water meter.

20. Draw the symbol for a double check valve.

006 UNDERSTAND AND APPLY DOMESTIC COLD WATER SYSTEM INSTALLATION AND MAINTENANCE TECHNIQUES

The supply of fresh, wholesome cold water to people's homes is a basic human requirement. As a domestic plumber, it is your job to get the water from the main external stop valve to the taps in a clean and fit state for human consumption. To do this, you will need to understand the processes that occur before the water reaches this point, the regulations that govern the plumbing industry and the practices that enable you to work safely and correctly on domestic cold water systems at every stage.

IN THIS CHAPTER, YOU WILL COVER:

A Cold water supply and treatment
- A1 The origin, collection and storage of water
- A2 The treatment of water
- A3 The distribution of wholesome water
- A4 The Water Supply (Water Fittings) Regulations 1999

B Cold water systems – inside the building
- B1 Entering the property
- B2 Domestic systems of cold water supply
- B3 Cold water storage cisterns
- B4 Frost protection

C Prevention of backflow and back siphonage
- C1 What is backflow and back siphonage?
- C2 The fluid categories
- C3 Air gaps used as a method of backflow prevention
- C4 Mechanical backflow prevention devices

D Taps and valves
- D1 Isolation valves
- D2 Float-operated valves
- D3 Terminal fittings
- D4 Shower mixing valves
- D5 Scale reduction and water treatment in domestic properties

E Installation of pipework
- E1 Choosing the right materials
- E2 Preparation, planning and positioning of pipes
- E3 Installation, testing and commissioning of cold water systems
- E4 Existing installations

F Maintenance of cold water systems
- F1 Planned preventative maintenance
- F2 Unplanned/emergency maintenance
- F3 Simple maintenance tasks
- F4 Decommissioning of systems

G System fault-finding and correction
- G1 Noise
- G2 Airlocks on low-pressure systems
- G3 Corrosion
- G4 Leakage

Test your knowledge

A COLD WATER SUPPLY AND TREATMENT

The first part of this chapter investigates how water is recycled in nature, time and again, and how it is collected and cleaned ready for use. As a plumber, you need to understand where water comes from and who is responsible for its cleanliness.

A1 THE ORIGIN, COLLECTION AND STORAGE OF WATER

Water is a simple compound made up of two hydrogen atoms attached to a single atom of oxygen, with the chemical symbol H_2O. Water is tasteless and odourless, and in small quantities it is colourless – in large quantities it possesses a light-blue hue.

Water molecule

The rainwater cycle

There is no 'new' water on Earth. All water is about 4.2 billion years old, whether it is seawater (saline), river or stream water, ground water, fossilised water or water from the polar ice caps.

Water moves constantly, in what is scientifically known as the hydrological cycle. We know this by its more common name – the rainwater cycle. This is a natural process where water is continually exchanged between the atmosphere, the Earth's surface, the ground and plants.

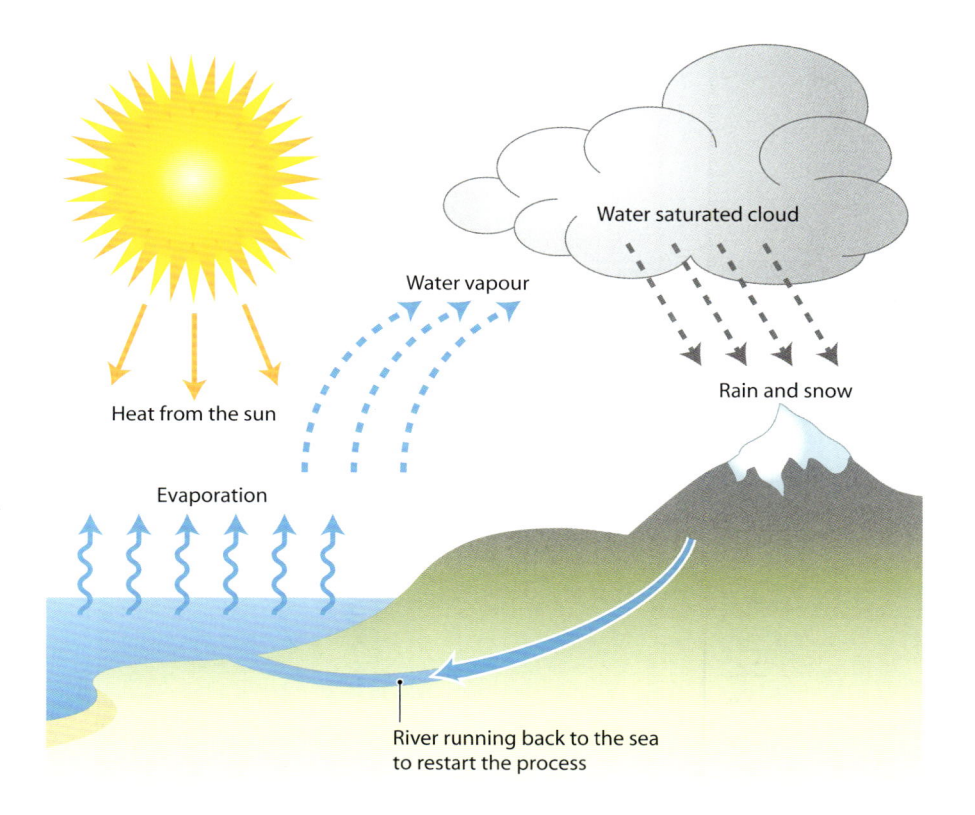

The rainwater cycle

As the sun warms the Earth, surface water evaporates and plants lose water through transpiration. This vapour rises through the air and is carried by the prevailing winds. If the vapour passes over land, some of it condenses to form clouds. As more water vapour is attracted to the cloud or the ground beneath it rises (in hills or mountains), it becomes saturated to the point where it can no longer hold all the moisture. The vapour is then released from the cloud in the form of rain, sleet, snow or hail.

On reaching the ground, the water may re-evaporate back into the atmosphere; it may be absorbed by the ground, where it will travel towards the water table or aquifer; or it may remain on the surface where it will eventually find its way into rivers, streams, lakes or the oceans. Here, the process begins again.

Sources of water

If we look at all the water on Earth, 97 per cent is saline (seawater) and only 3 per cent is fresh water. Of fresh water, nearly 69 per cent (or 2.07 per cent of the Earth's total water resources) is trapped in the polar ice caps and glaciers and 30.1 per cent (0.9 per cent of the total water resources) is ground water. It is this ground water that the Earth's population relies on for its drinking water supply.

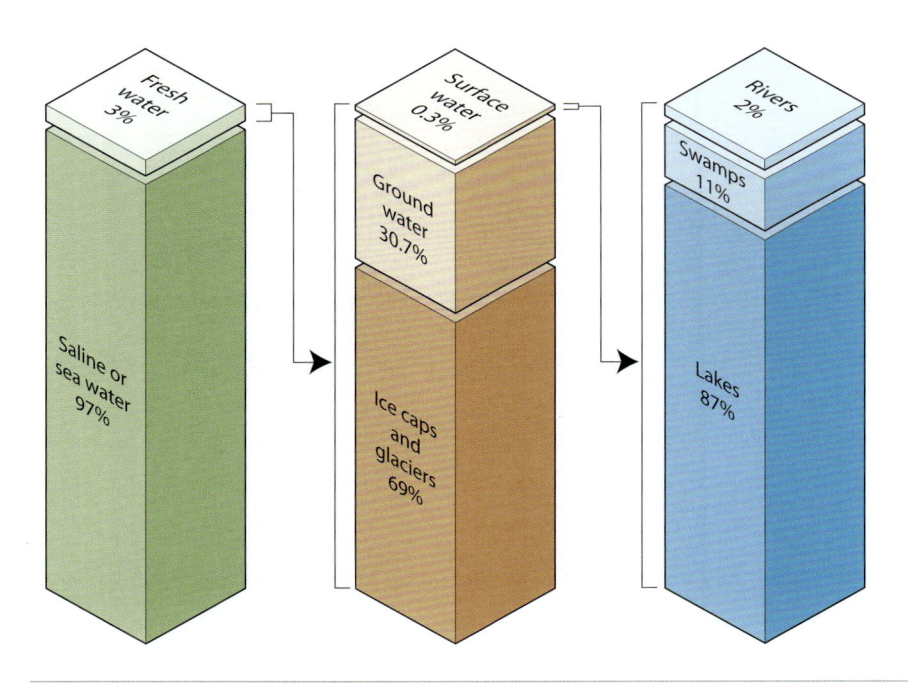

Distribution of earth's water

The total supply of fresh water for the world is in the region of 1350 trillion litres, the majority of which is stored on the ground, where it is available in reservoirs, streams, rivers, lakes, etc. A further 13,650 trillion litres is also available in the form of water vapour, which will eventually fall as rain. Conversely, about 1100 trillion litres of water evaporates into the atmosphere worldwide every day.

Sources of water in the United Kingdom

Of the rain that falls on the United Kingdom annually, only 5 per cent is collected and stored in reservoirs for drinking water supply. The rest flows in rivers to the sea or is filtered down to the natural water table or aquifers that exist below the ground surface. The main sources of water in the UK are shown in the table below:

Source of water	Description	Properties of water from this source
Deep well	A machine-dug well that draws its water from below the shallow, impervious strata.	Usually good quality, as it is extracted from below the Earth's surface.
Shallow well	A well dug by hand or excavator that only penetrates the first water-bearing strata, or aquifer, in the Earth's surface	Must be considered dangerous because it may be contaminated with water from cesspits or broken drains, etc.
Upland surface	Water that has collected in upland lakes and rivers without passing through the Earth's strata.	Naturally soft and acidic, this water is not contaminated with salts or minerals. It is the main water source for the north-west of England.
Spring	A naturally occurring flow of water from the Earth's surface.	The purity of spring water is highly dependent on the distance it has travelled from the source.
River	A large, natural flow of water usually starting as a small stream on high ground, which enlarges with distance travelled. Rivers usually terminate at the sea and may be tidal, for example the river Avon.	Usually poor quality due to industrial pollution. The cost of treatment is high.
Canal	Most canals are a product of the Industrial Revolution and for many years fell into dereliction and disrepair. Many have now been cleaned and reopened and are now sites of natural beauty.	Very poor quality, generally only used for industrial purposes and irrigation.
Aquifer	Naturally occurring water-bearing strata, often deep beneath the Earth's surface. Mostly consists of permeable rock, such as sandstone, gravel silt or clay, which soaks up water like a sponge.	Very good quality, but prone to contamination by nitrates from farming.
Artesian well and spring	Water that rises from underground water-bearing rock layers under its own pressure, but only if the well head is below the level of the water table.	Usually very good quality, as the water is filtered naturally through layers of rock.
Borehole	Man-made well drilled directly to a below-ground water source; the water is extracted for use when connection to a water main is extremely difficult.	Very high-quality water that, in most cases, is cleaner than in the water undertaker's water main. Filtering and chlorination are not necessary although the quality should be monitored.

KEY POINT
Visit www.groundwateruk.org for more information on the use and distribution of aquifers in the UK.

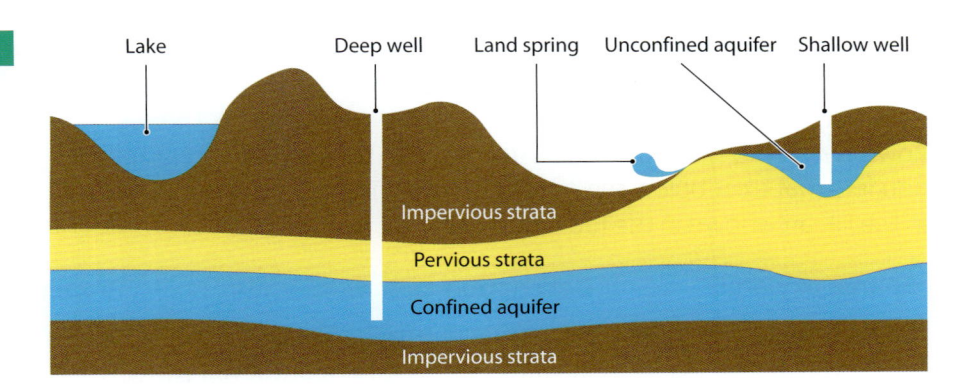

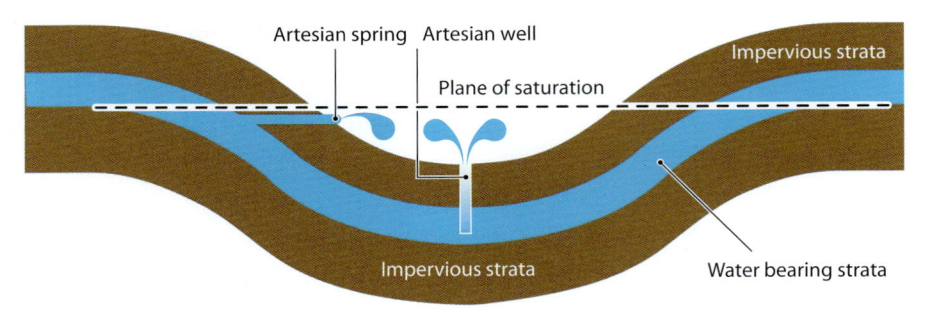

Sources of water

KEY POINT
The Environment Agency is the overseeing authority of all water courses in the UK. They sample about 7000 river and canal sites 12 times per year to test their chemistry and nutrients, and to see whether there are any pollutants present and if they need to target areas for improvement.

Types of water

When water leaves the cloud as rain, hail or snow, it is almost pure, and contains very few impurities. As it falls to the ground it absorbs some of the carbon dioxide (CO_2) present in the atmosphere to become a weak solution of carbonic acid (H_2CO_3), and if there is sulphur trioxide (SO_3) also present, it can become very weak sulphuric acid (H_2SO_4). When this acidic solution reaches the ground, it dissolves some of the salts it comes into contact with, which will affect its pH value. A pH value below 7 leads to soft, plumbo-solvent water that attacks lead in plumbing systems (see Chapter 004, page 166). A pH value above 7 may indicate temporarily hard water, which can be softened by boiling and leaves

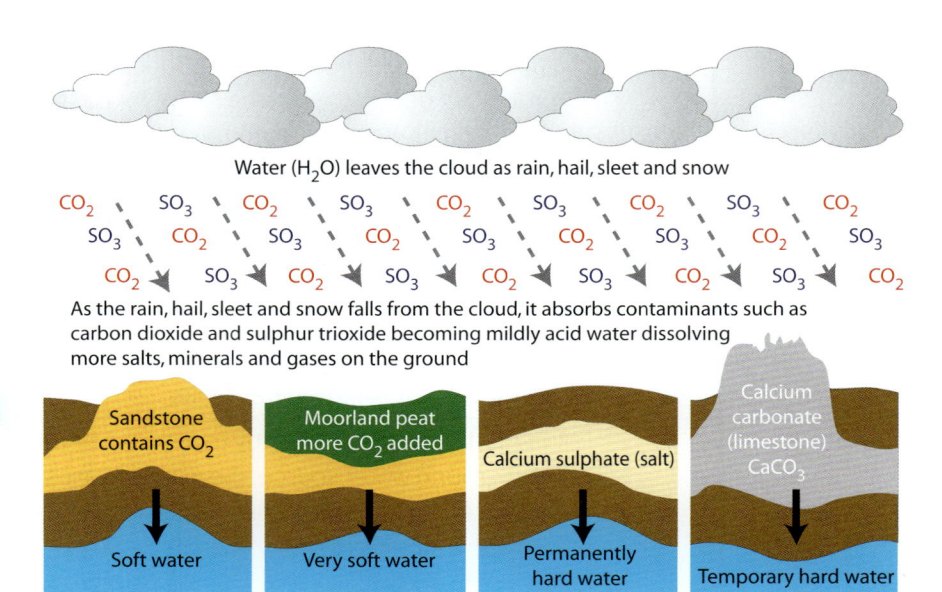

KEY POINT
Rainwater that contains sulphur trioxide has the chemical symbol H_2SO_4. It is better known as very weak sulphuric acid, and is responsible for the 'acid rain' that destroyed the Scandinavian pine forests.

Types of water

006 COLD WATER SYSTEMS

282 THE CITY & GUILDS TEXTBOOK

limescale in kettles and pipework, or permanently hard water, which cannot be softened by boiling.

Removing water hardness

Temporarily hard water can cause many problems within plumbing systems. If hard water is heated above 65°C, it loses its calcium. If you look at a kettle in a hard-water area, you'll see that a layer of limescale has built up on the inside of the kettle. When limescale builds up in plumbing and heating systems, the following problems can occur:

- The limescale insulates the water from the heat source, resulting in bad heat transfer, which wastes fuel and means a longer heating-up time.

- The limescale may block heat exchangers and pumps, causing localised boiling as the water cannot move away from the heat exchanger fast enough and the water boils momentarily. This is known as 'kettling'.

For this reason, hard water is often treated using a water softener, which reduces hardness by the use of the 'base exchange' process. This technique removes the ions that cause the water to be hard – in most cases, calcium and magnesium ions. A water softener collects hardness minerals within its conditioning tank and from time to time flushes them away to the drain, replacing them with sodium and potassium salts through the exchanger reservoir.

A2 THE TREATMENT OF WATER

This part of the chapter looks at the way the water we use every day for drinking, washing and cooking is filtered, cleaned and sterilised to ensure that it is fit for human consumption. Fresh, clean water is described as 'wholesome' or potable.

Potable
(pronounced poe-table)
Water that is fit to drink.

The UK water undertakers

The supply of wholesome water to consumers is the responsibility of the UK water authorities, often referred to as the 'water undertakers'. They are responsible for the collection of water, carrying out any treatments necessary to make the water fit for human consumption, and maintaining the supply of water to dwellings and other consumers.

All water undertakers are public limited companies (PLCs) owned by shareholders. There are 21 companies supplying water in England and Wales. Water rates differ across the UK, and you cannot switch your water supplier. There are water-only companies and combined water and sewerage companies. You may pay the company a standing charge for water or have a meter fitted. Scotland has only one water undertaker; Northern Ireland also has only one and domestic customers there do not have to pay water rates.

SUGGESTED ACTIVITY...
Find out which water undertaker is responsible for the delivery of water in your area. On what occasions might you need to contact them during your work as a plumber?

KEY POINT
The water industry in the UK is governed by OFWAT. Its job is to ensure that the water companies provide a high-quality service and maintain a good standard of drinking water throughout the UK. Find out more through their website: www.ofwat.gov.uk.

Sedimentation, filtration and sterilisation of water

Before it is considered wholesome, the water undergoes several stages of treatment to ensure its cleanliness and quality: sedimentation, filtration and sterilisation. These stages are the responsibility of the water undertaker.

Sedimentation

The rainfall is initially collected and stored in lakes and reservoirs, where it is allowed to remain undisturbed. Here, the solid impurities such as grit, mud and decaying vegetation sink to the bottom. This is known as 'primary sedimentation'. Storage also has the effect of reducing the bacteriological content of the water. From here the water is pumped from the storage reservoir through coarse strainers and is held in sedimentation tanks where further sedimentation takes place before being filtered

Filtration

Slow sand filters provide the most common method of filtration. The water flows over a graded sand bed. The top of the bed is provided with large colonies of minute vegetable algae growth, which form naturally. The algae assist the purification of the water by living off any contaminants. As the algae builds up, so the filtering process slows down and the water becomes clean. Rapid sand filters speed up the process by operating under pressure; they require frequent maintenance and backwashing, although this is automated by mechanical plant.

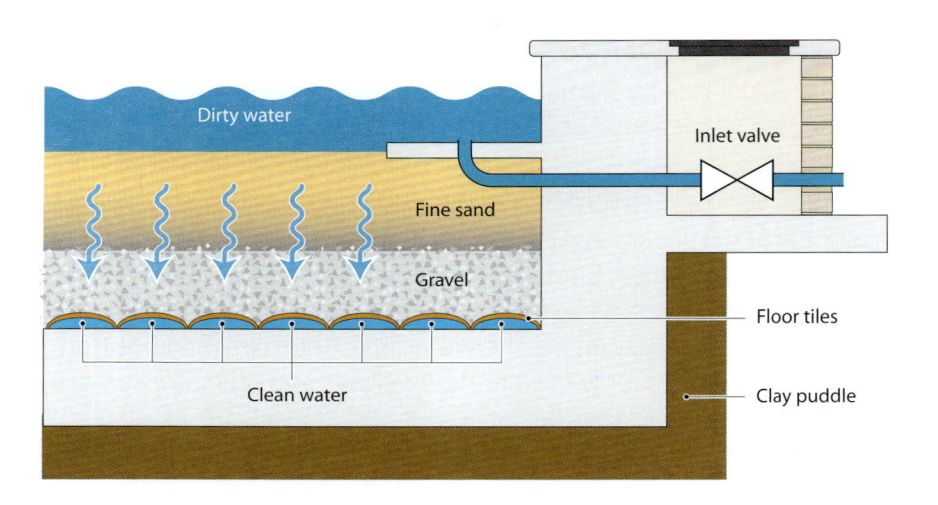

Slow sand filtration

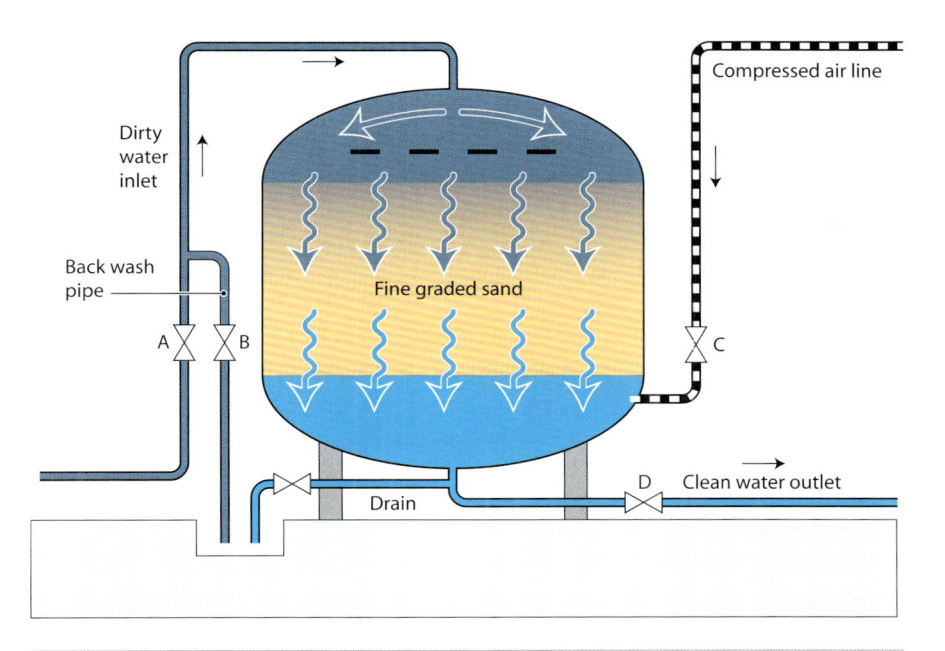

Rapid sand pressure filtration

Sterilisation

Finally, water is treated with chlorine and ammonia before it enters the water supply. This will kill off any bacteria missed by the water filters. Fluoride is still added in some parts of the UK but only in minute quantities.

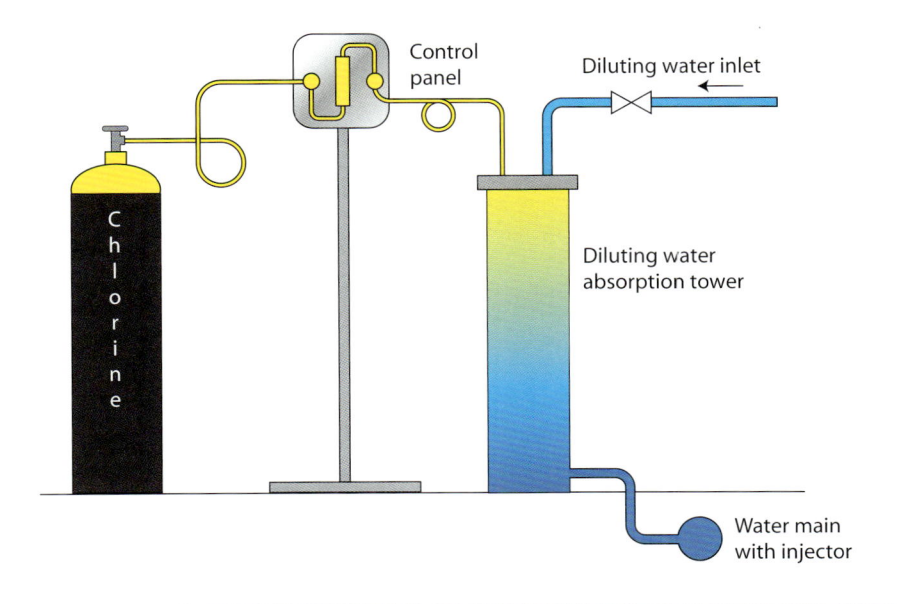

Sterilisation by injection of chlorine

Alternative private sources of wholesome water supply

A private water supply is a wholesome water source that is not provided by a licensed water undertaker. This is typically from a local well, borehole, spring, lake, river or stream. The water quality is the responsibility of the owners of the property from where the water source is extracted and/or used, and should be closely monitored.

The majority of the UK population is supplied with water from a water undertaker, but about 1 per cent of the population is supplied from a private supply: there are around 140,000 private suppliers. Water from a single private water supply may be used for one or more premises, including dwellings, businesses, holiday homes, caravan parks or hotels.

The revised Private Water Supply Regulations 2009 came into force in January 2010 – do not confuse these with the Water Supply (Water Fittings) Regulations 1999, which are covered in detail later in this chapter. The 2009 Regulations identify the powers and responsibilities of the Local Authority to enforce compliance of the regulatory requirements by the owner or person responsible for the water supply.

Boreholes

Water boreholes are the modern equivalent of the well, but are smaller, less intrusive and easier to maintain. As water passes through the ground and into the water table it flows through layers of rock and chalk, which act as natural filters. This produces water that is usually far cleaner and purer than the water provided by most water companies.

> **KEY POINT**
> Because boreholes are classed as private water sources, they are not subject to the Water Supply (Water Fittings) Regulations 1999. It is an offence to cross-connect water from a borehole with water from a water undertaker's main.

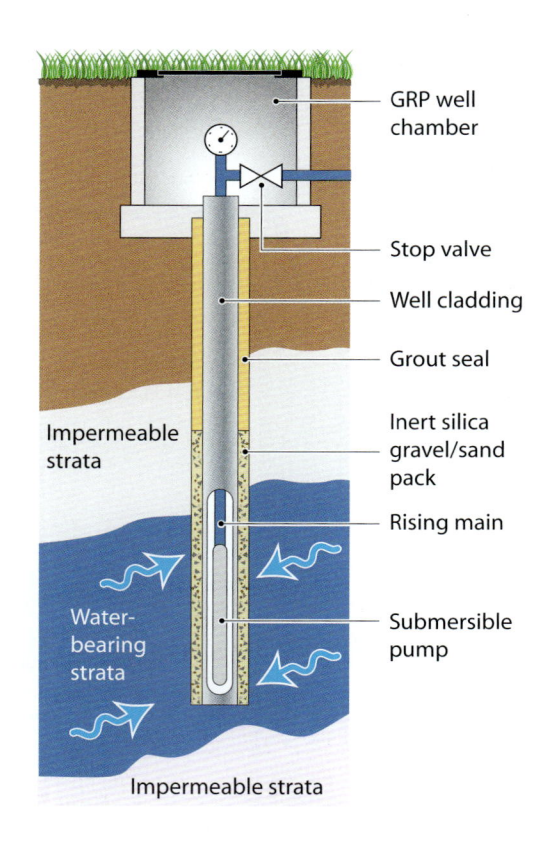

A typical modern borehole

A single borehole has over 40 times the water capacity that an average home would need. Twenty thousand litres of water per day can be extracted from a borehole without the need for any permissions or licences. An average four-bedroom house uses approximately 50,000 litres of water over a three-month period; this amount can legally be extracted from a borehole in two and a half days.

Sources of recycled, unwholesome water supply in domestic dwellings

Over the past 20 years, demand for water has increased dramatically in the UK. The average person in the UK uses 150 litres of water a day for washing, flushing the WC, drinking, cooking, gardening and other household tasks. Government targets set out in the Code for Sustainable Homes aim to reduce this to 80 litres per day. There are many ways in which water usage can be reduced in a dwelling, from simple rainwater collection in water butts for garden use, to more complex systems such as rainwater harvesting and grey water recycling for clothes washing and flushing the WC. Remember that this type of water is not fit for human consumption; the pipework must be marked in a way that makes identification easy. For example, BS 1710 suggests colour coding the pipework, using green adhesive tape with a black stripe to denote reclaimed water. Any installation of this type in a dwelling must not cross-connect with the mains cold water supply.

There are two types of water recycling:

1 **Grey water** – this is water from wash basins, showers, washing machines and baths that is reused, after being filtered and cleaned, to flush WC cisterns only. It is stored in separate storage cisterns away from wholesome water supply cisterns.

2 **Rainwater harvesting** – rainwater is collected, filtered and cleaned before being used to flush WC cisterns and in some clothes-cleaning operations.

These systems are described in detail in Chapter 003, pages 148–50.

KEY POINT

For more information about private water sources, see the United Kingdom Water Treatment Association website at www.ukwta.org/privatewatersupplies.php#what.

Black water

This is water and effluent from WCs and kitchen sinks that can only be treated by a water undertaker at a sewage works. In some domestic systems that are a distance away from the main sewer, black water will flow into a septic tank or cess pit where it is collected, usually at three-month intervals, by a specialist waste water collection company.

KEY POINT

There are over 2000 reservoirs used for drinking water in the UK. These are the responsibility of the Environment Agency.

A3 THE DISTRIBUTION OF WHOLESOME WATER

There are two methods of water supply distribution used in the UK. These are:

1. **Gravity distribution** – collected upland surface water is impounded in reservoirs on high ground. Here the water is filtered by slow sand filters and chlorinated before being fed to homes and factories by gravity. No pumping is required.

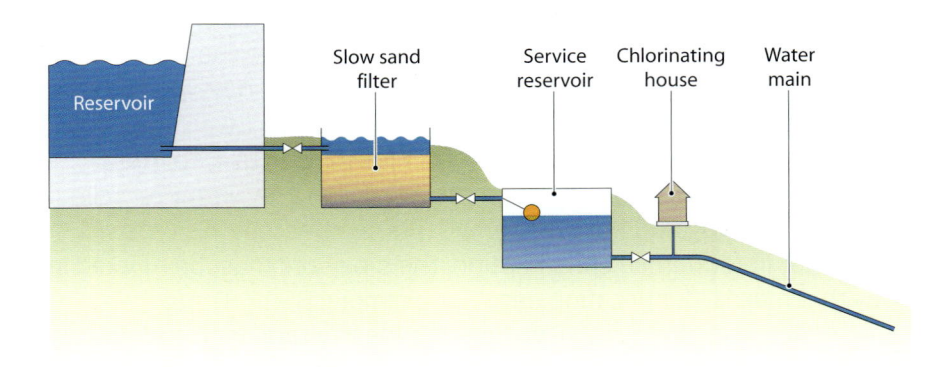

Gravity water distribution

2. **Pumped distribution** – water taken from a river is pumped directly to a settlement tank, where all of the heavier impurities sink to the bottom. It is then passed through a slow sand filter to remove any organic matter and chlorinated to wholesome water standard before being pumped to a water tower. From the tower it flows via gravity to the water main.

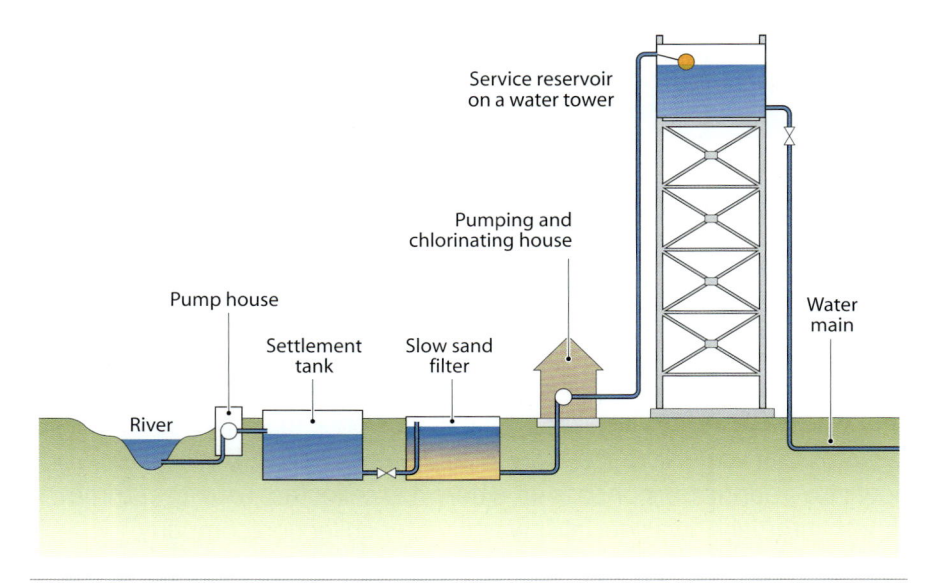

Pumped water distribution

006 COLD WATER SYSTEMS

288 THE CITY & GUILDS TEXTBOOK

The distribution of water in cities, towns and villages

Water is supplied to homes via a grid-system network of pipes known as 'trunk mains', a phrase dating back to a time when the mains were constructed from hollowed-out tree trunks. Trunk mains vary in diameter depending on the purpose of the main and the likely demand of the supply. Pipes that transfer water to the various points in the distribution system can vary in diameter from 75mm to 2.3m. The size of the water main depends on the size of the community that it serves.

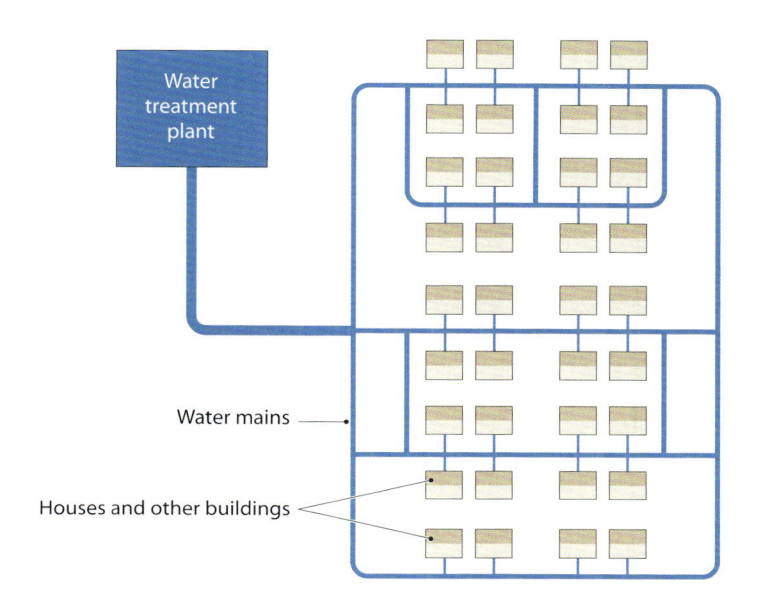

Water mains supply

Town population	Size of main (metres)
500,000	1.05 to 1.20
200,000	0.75
5000–20,000	0.2 to 0.3

When a new house has to be connected to the water supply, the supply pipes are usually 25mm in diameter. At the boundary to the dwelling, a screw-down stop valve is installed, so that the supply to the house can be isolated, if necessary, while any repairs are carried out.

Methods of connection to the water main

The water main runs underneath the road. The connection between the water supply and the dwelling is made via a brass ferrule, which is the responsibility of the water undertaker. The ferrule is a type of shut-off valve that allows the connection to be isolated for maintenance and repair. The connection can be made in a number of ways, depending on the material from which the water main is made. For instance, if the water main is made from cast iron then a self-drilling and tapping machine is used. With this tool, the water main can be drilled and threaded and a ferrule inserted while the main is still under pressure, so that the supply to other properties is not disrupted.

If the main is made from PVCu or cementitious-lined asbestos then a brass strap-type ferrule is used.

Mains water tapping machine

A strap-type ferrule water main connection (above) and standard gunmetal ferrule (below)

From the water main to the building

The water supply from the water main into the building is comprised of two separate pipes:

- the communication pipe, owned and maintained by the water undertaker
- the service pipe, owned and maintained by the owner of the building.

The communication pipe is installed by the water undertaker, and runs from the ferrule on the water main to the main external stop valve (also known as the boundary stop valve because it is usually located at the boundary of the property). It incorporates a gooseneck bend to allow for any settlement of the roadway or pavement. It is the sole responsibility of the water undertaker to install, repair and maintain the communication pipe and main external stop valve.

The service pipe runs from the main external stop valve to the dwelling and is the responsibility of the house owner. It must be installed at a minimum depth of 750mm and a maximum depth of 1350mm. It must terminate within the building with an isolation valve so that the system may be isolated for repair and maintenance.

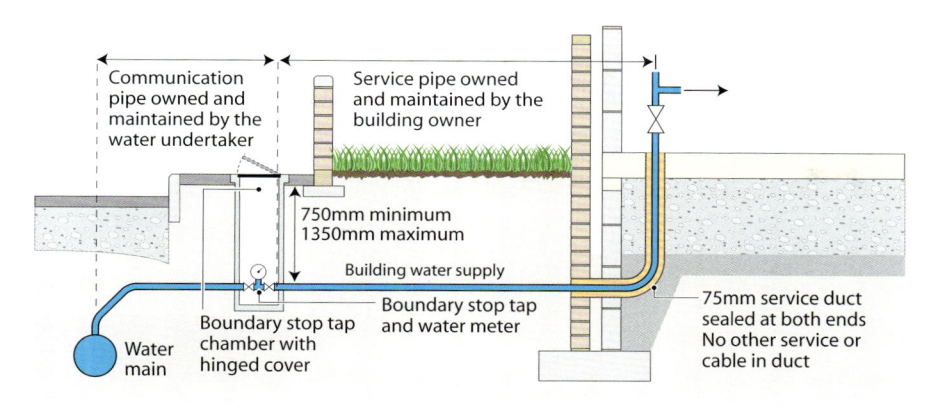

Communication and service pipes

The water supply to buildings can be arranged in numerous ways. In each case, separate dwellings supplied must have a controlling stop valve in a position that will allow the water supply to be turned off in an emergency without affecting any other property.

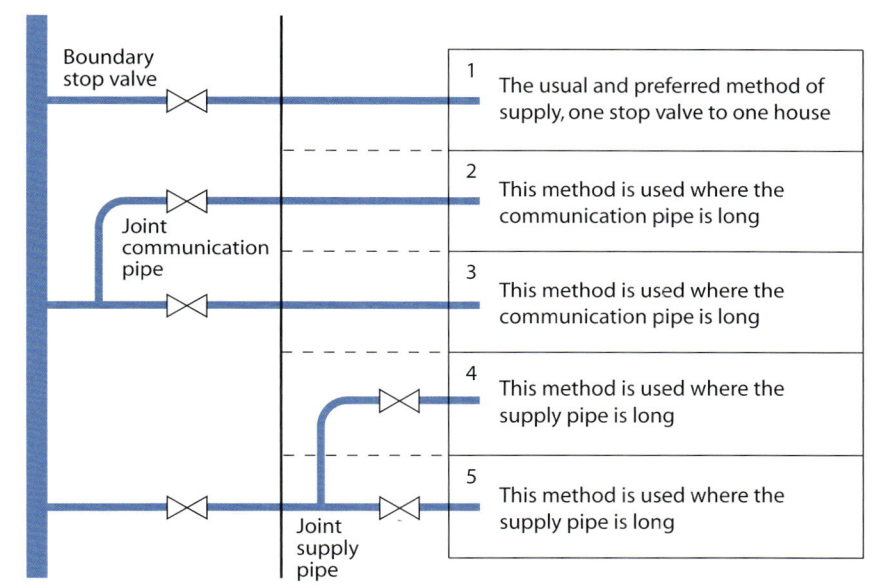

1. The usual and preferred method of supply, one stop valve to one house
2. This method is used where the communication pipe is long
3. This method is used where the communication pipe is long
4. This method is used where the supply pipe is long
5. This method is used where the supply pipe is long

Water suppliers will normally insist on individual supplies to properties and do not favour joint supplies (commonly called communal supplies).

Alternative methods of supplying more than one dwelling

Most water supplies in modern dwellings and industrial premises are piped in medium density polyethylene (MDPE) pipe, which is coloured blue to show mains cold (wholesome) water, and is commonly known as 'blue poly'. The minimum pipe size for modern dwellings is 25mm. Soft copper pipe (conforming to BS EN 1057 R220) may also be used.

Boundary water meter boxes

With new installations, a water meter is either fitted at the boundary to the property in the external stop-valve chamber or in an external groundbreaker type meter box on the side of the property. This is so that the customer does not have to be present when the meter is being read and to prevent illegal tampering with the water meter.

Water meters inside the dwelling are usually fitted to existing water supplies. They must be fitted between two stop valves with a drain-off valve fitted after the meter but before the upper stop valve.

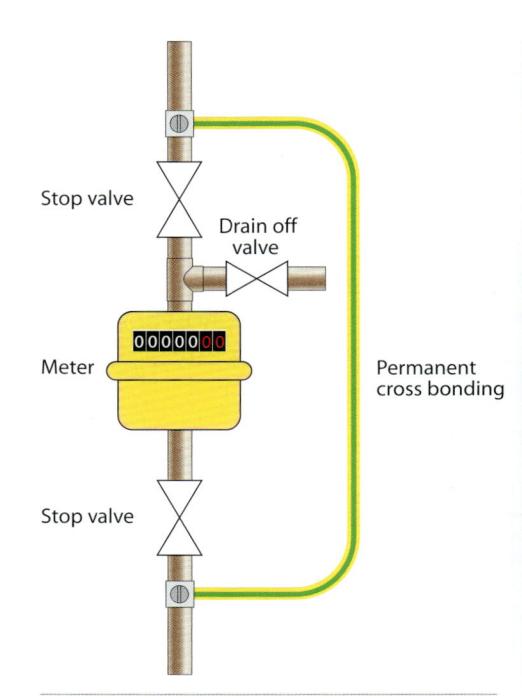

Installation of an internal water meter

Groundbreaker type meter box

From the external main stop valve is where the plumber's involvement with the Water Supply (Water Fittings) Regulations 1999 begins. From here on, everything that you do to the cold and the hot water systems in the building is regulated to ensure that the water supplied is fit for its intended purpose.

A4 THE WATER SUPPLY (WATER FITTINGS) REGULATIONS 1999

Before 1999, each water undertaker had its own set of water byelaws that were based on the 101 Model Water Bye-laws issued by the UK Government in 1986. The Water Supply (Water Fittings) Regulations 1999 were introduced by the government to provide a common standard throughout the UK. They are linked to British Standard BS 6700:2006+A1:2009, which is discussed below.

Simply put, the Water Supply (Water Fittings) Regulations 1999 were put in place to ensure that the systems that plumbers install and maintain prevent the following:

- contamination of water
- wastage of water
- misuse of water

- undue consumption of water
- erroneous metering of water.

A free copy of the Water Supply (Water Fittings) Regulations 1999 can be downloaded from www.opsi.gov.uk/si/si1999/19991148.htm

Complying with the Regulations

The plumbing systems you install and maintain must comply with the Regulations. You must prevent contamination of wholesome water, and you must give advanced notification of installation work. You may benefit from becoming an Approved Plumbing Contractor once you have completed a Water Regulations Advisory Service (WRAS)-sponsored water regulations training course.

The WRAS has produced the Water Regulations Guide, which has become the essential guide for plumbers to the legislation affecting water supply installations in the UK. It incorporates government and water industry recommendations and explains how plumbers can avoid costly mistakes.

Water Regulations Guide

> **KEY POINT**
>
> The aim of the Water Regulations Advisory Service (WRAS) is to promote and provide knowledge of the Water Regulations to the plumbing industry. It is funded by all of the UK water undertakers. It provides advice on how to understand and comply with the Regulations, much of which can be found on the WRAS website: www.wras.co.uk.

BS 6700:2006+A1:2009 – design, installation, testing and maintenance of services supplying water for domestic use within buildings and their curtilages

British Standard 6700 covers the system of pipes, fittings and connected appliances installed to supply any building, whether domestic or not, with water for drinking, culinary, domestic laundry, ablutionary, cleaning and sanitary purposes. By following the British Standard, you will satisfy and comply with the Water Regulations. Like all British Standards, it gives recommendations only and is not a legal requirement. This document is referred throughout the rest of this chapter.

B   COLD WATER SYSTEMS – INSIDE THE BUILDING

Each dwelling should have a wholesome water supply, the most important location for this being at the kitchen sink. In most premises it is likely that people will drink water from all taps. This means that water to all taps should be connected to the mains supply or from a protected storage cistern. Wholesome water should also be provided in convenient locations in offices and other buildings, especially where food is being eaten or prepared.

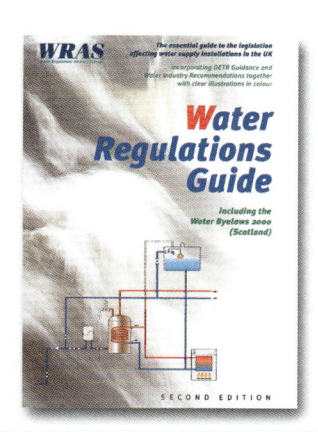

B1 ENTERING THE PROPERTY

When the water supply enters the property, it should terminate with a screw-down stop valve to BS 1010 or BS 5163, a lever action, spherical plug valve to BS 6675 or, on large installations, a flanged gate valve to BS 5163. The Water Regulations are very specific:

> 'As far as is reasonably practicable:
> A stop valve should be located inside the building; and,
> Be located above floor level; and,
> As near as possible to the point where the supply enters the building; and,
> Be so installed that its closure will prevent the supply of water to any point in the premises.'

The water undertakers recommend that no more than 150mm of blue MDPE pipe is exposed above the floor level of the building. This is to minimise the amount of MDPE pipework visible; it decomposes under persistent exposure to the ultraviolet light present in daylight.

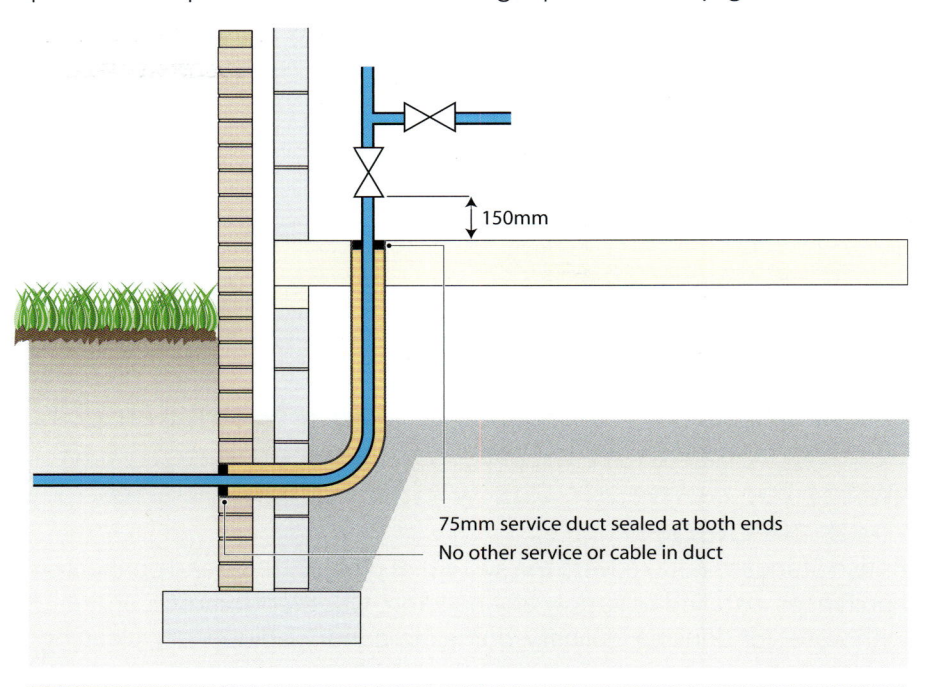

The entry of water supply to a property

There are many different stop valve/tap styles available, all manufactured to BS 1010-2. Stop taps that include an integral drain-off valve are also available but these would not be acceptable for use as the lower stop tap of a water meter installation, because this would allow water to be drawn off before the meter, leading to erroneous metering.

Any stop valve used above or below ground must be made from either gunmetal or corrosion-resistant brass to prevent dezincification of the stop valve (which could result in water contamination or waste). All fittings that are made of a copper alloy, such as brass or gunmetal, should carry either CR or GM markings on the fitting body to show that they are corrosion resistant.

A drain-off valve conforming to BS EN 1254 should be installed immediately above any stop valve/valve to allow draining of the system.

B2 DOMESTIC SYSTEMS OF COLD WATER SUPPLY

There are two basic systems of cold water:

- direct system
- indirect system.

The direct system of cold water

With this system, all cold water taps are fed directly from the mains supply, and therefore all taps supply wholesome water. Storage is only required for supplying cold water to the hot water cylinder, typically via a 150-litre cistern. A feed cistern will not be necessary if the hot water is supplied via an instantaneous hot water heater or 'combi' boiler, unless incoming pressure is unusually low.

The direct system is the most common type of cold water system used in domestic properties because there is usually a relatively high pressure of supply available from the mains and it is a cost-effective installation.

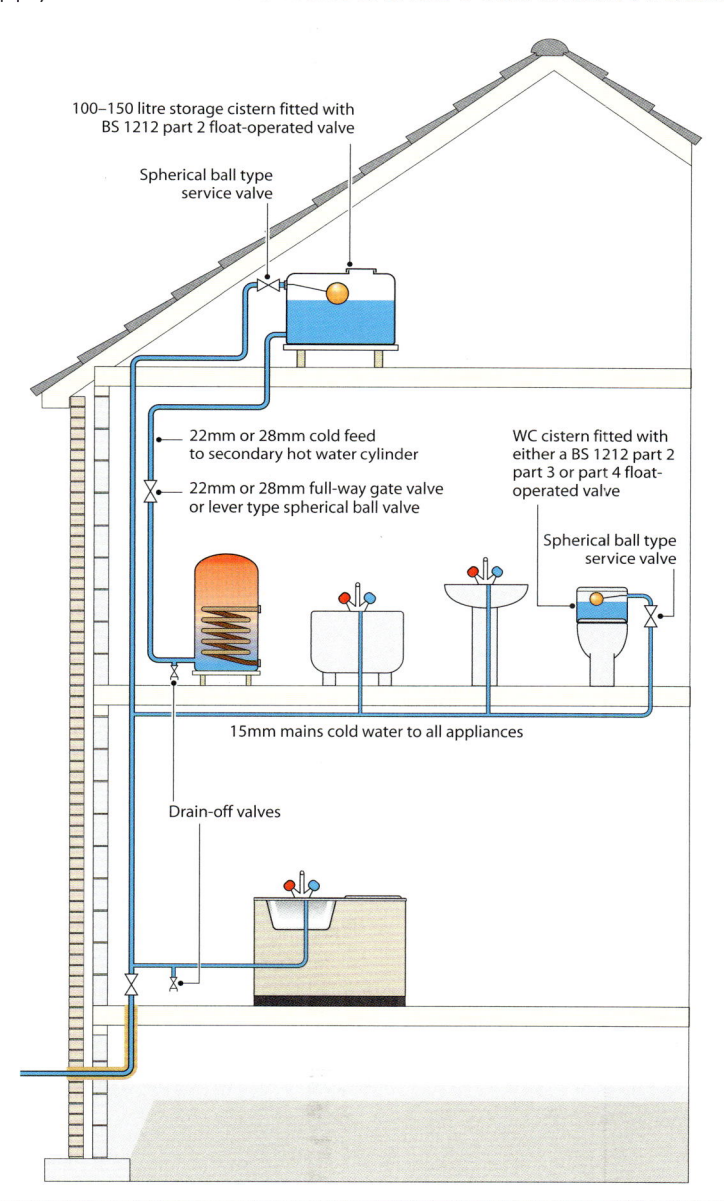

SmartScreen Unit 006 worksheet 1

The direct system of cold water supply

SmartScreen Unit 006 animation 2

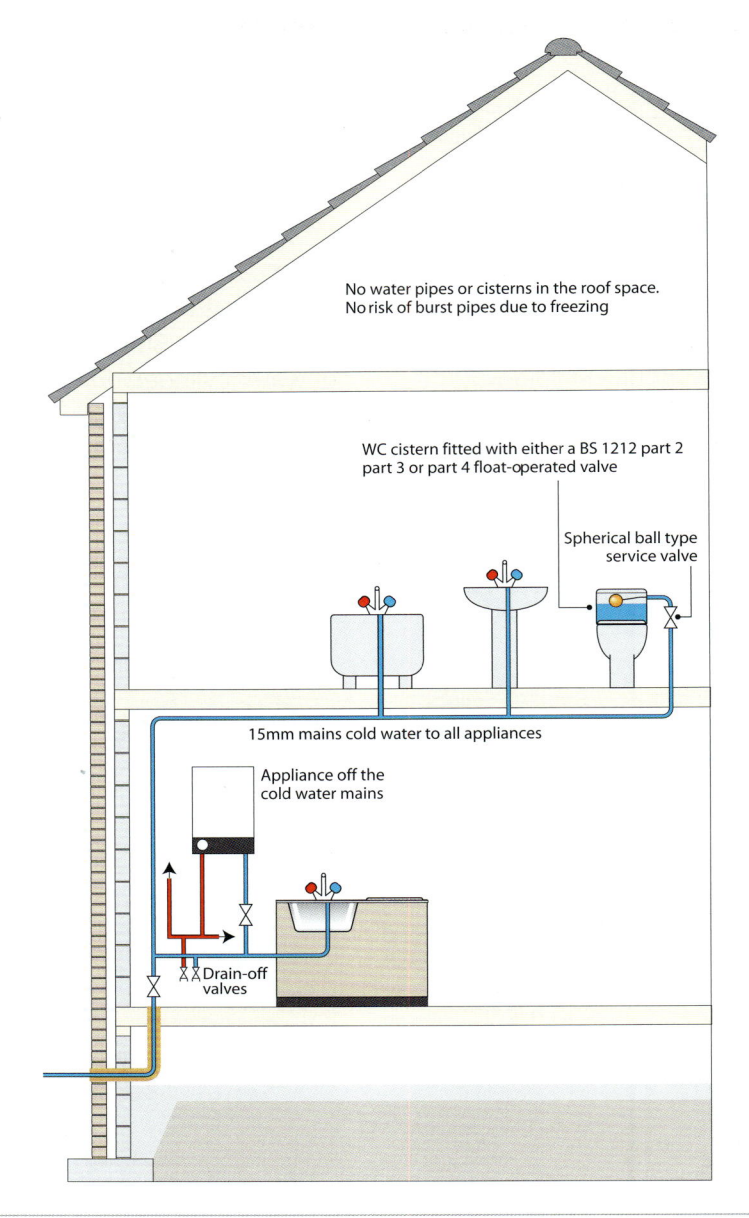

Diagram labels:
- No water pipes or cisterns in the roof space. No risk of burst pipes due to freezing
- WC cistern fitted with either a BS 1212 part 2 part 3 or part 4 float-operated valve
- Spherical ball type service valve
- 15mm mains cold water to all appliances
- Appliance off the cold water mains
- Drain-off valves

The direct cold water system with combi boiler or instantaneous hot water heater

Pipe sizes for the direct system

Pipe size depends on the system design but, generally speaking, a 15mm rising main will be large enough to supply most cold water demands for a three- or four-bedroom house, with all cold water outlets being supplied in 15mm, including the bath. If a hot water storage vessel is to be installed, then a 22mm cold feed pipe is needed to the hot water storage vessel supplied from a 150 litre cold water feed cistern in the roof space. On larger installations, a 22mm rising main may be required but this will depend on the water needs of the household.

If a combi boiler or instantaneous water is installed, then a 15mm mains cold water supply should be sufficient, depending upon the supply pressure and flow rate.

The advantages and disadvantages of the direct system

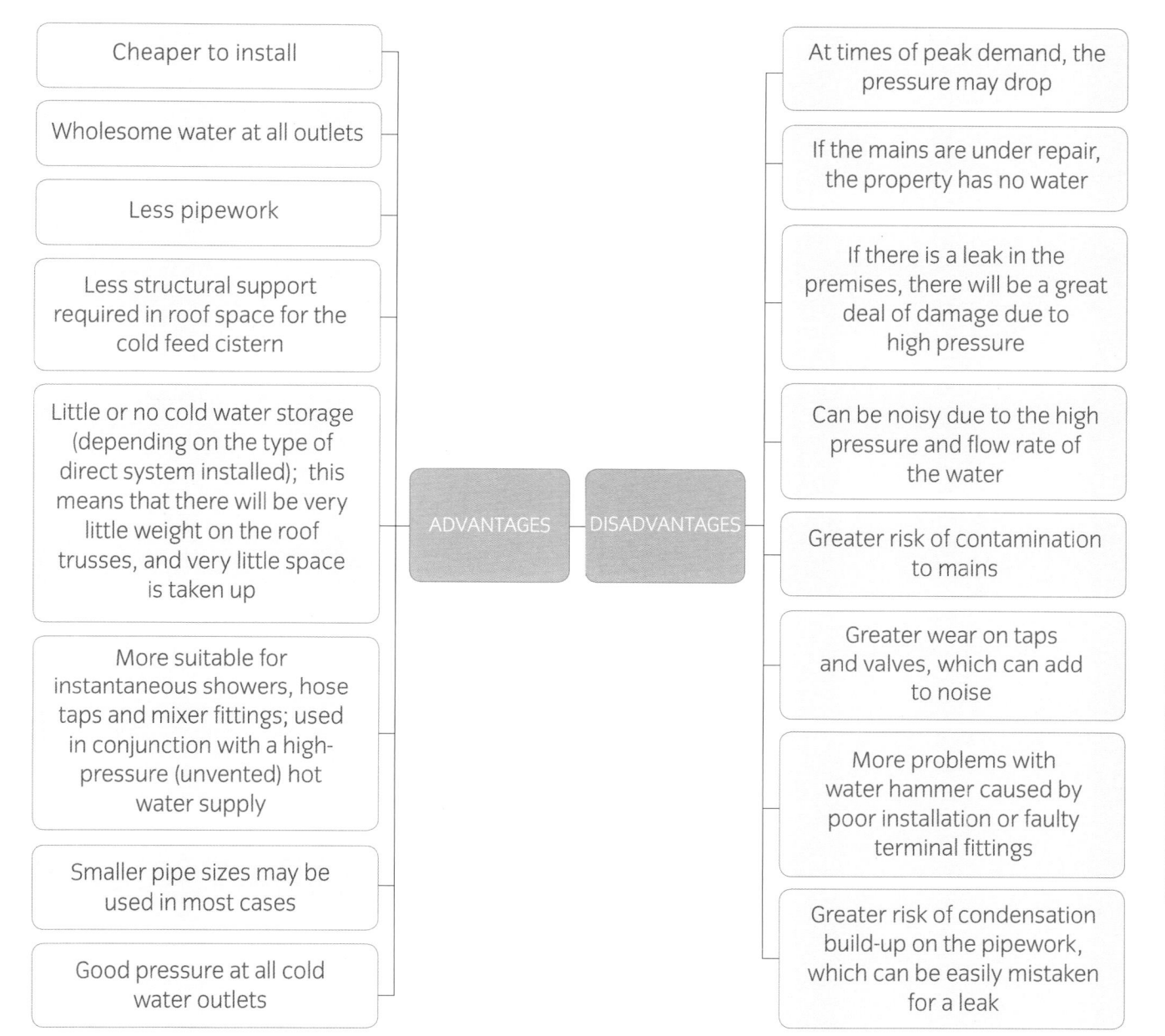

ADVANTAGES
- Cheaper to install
- Wholesome water at all outlets
- Less pipework
- Less structural support required in roof space for the cold feed cistern
- Little or no cold water storage (depending on the type of direct system installed); this means that there will be very little weight on the roof trusses, and very little space is taken up
- More suitable for instantaneous showers, hose taps and mixer fittings; used in conjunction with a high-pressure (unvented) hot water supply
- Smaller pipe sizes may be used in most cases
- Good pressure at all cold water outlets

DISADVANTAGES
- At times of peak demand, the pressure may drop
- If the mains are under repair, the property has no water
- If there is a leak in the premises, there will be a great deal of damage due to high pressure
- Can be noisy due to the high pressure and flow rate of the water
- Greater risk of contamination to mains
- Greater wear on taps and valves, which can add to noise
- More problems with water hammer caused by poor installation or faulty terminal fittings
- Greater risk of condensation build-up on the pipework, which can be easily mistaken for a leak

The indirect system of cold water supply

With this system of cold water supply, only the kitchen sink and the cold water storage cistern are fed directly from the mains cold water supply. This is a requirement of the Water Supply (Water Fittings) Regulations 1999 for food preparation areas. The other appliances are fed indirectly with wholesome water via the cold water storage cistern in the roof space. A large amount of water will, therefore, need to be stored to supply both cold water and hot water to appliances and fittings from a minimum of 230 litres of water stored in the cistern. The system is designed for use in low-pressure water areas where the mains supply pipework is not capable of supplying the full requirement of the system. This type of system also has a reserve of stored water in the event of mains failure. The cistern should be installed as high as possible to increase the system pressure (while complying with Building Regulations).

SmartScreen Unit 006 animation 3

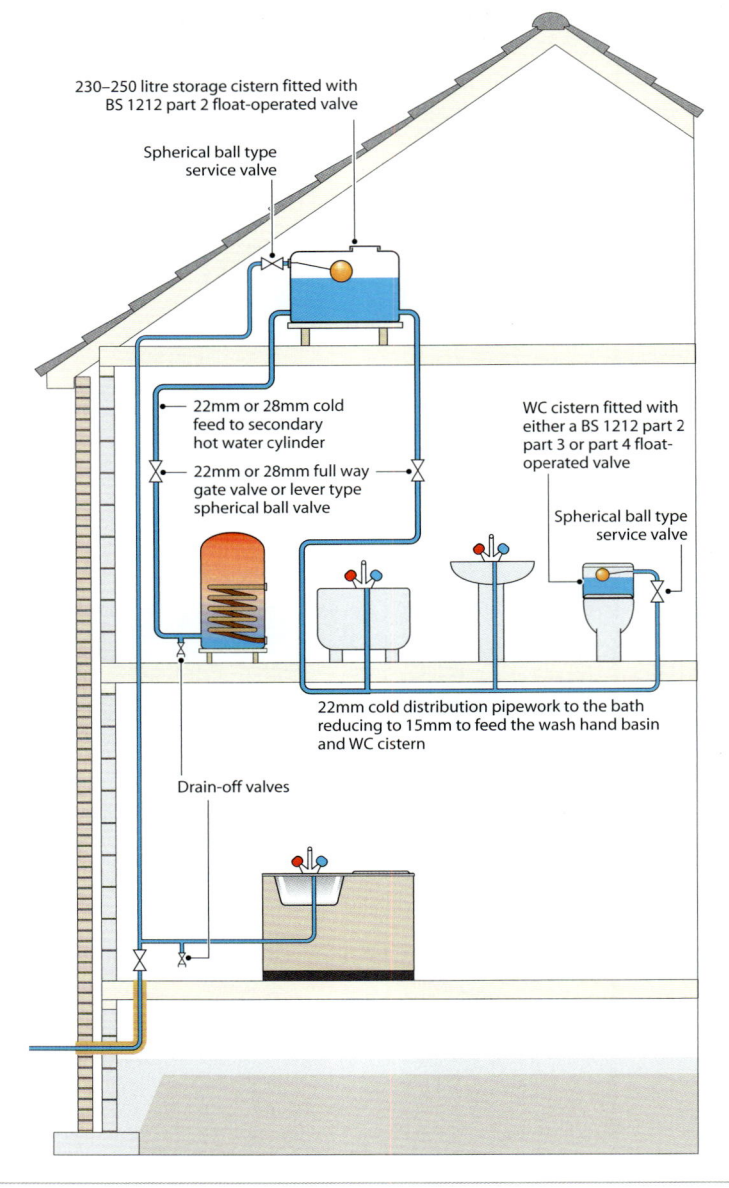

The indirect system of cold water supply

Pipe sizes for the indirect system

As before, in most cases a 15mm rising cold water main will be large enough to supply most cold water demands for a three- to four-bedroom house. The kitchen sink should be supplied with water direct from the cold water main, and 15mm pipework is adequate for this. The cold water storage cistern can also be supplied via 15mm pipework. A cold water distribution pipe (22mm minimum) distributes cold water from the cistern to the wash basin, WC and bath. The bath should be supplied from 22mm pipework because of the lack of pressure, but all other appliances can effectively be supplied from 15mm pipework. A minimum size of 22mm cold feed pipe is needed to supply the hot water storage vessel. This system is ideal when mixing valves and taps require equal pressure and flow rate as both hot and cold supplies are fed from the same source: the cold water storage cistern.

The advantages and disadvantages of the indirect system

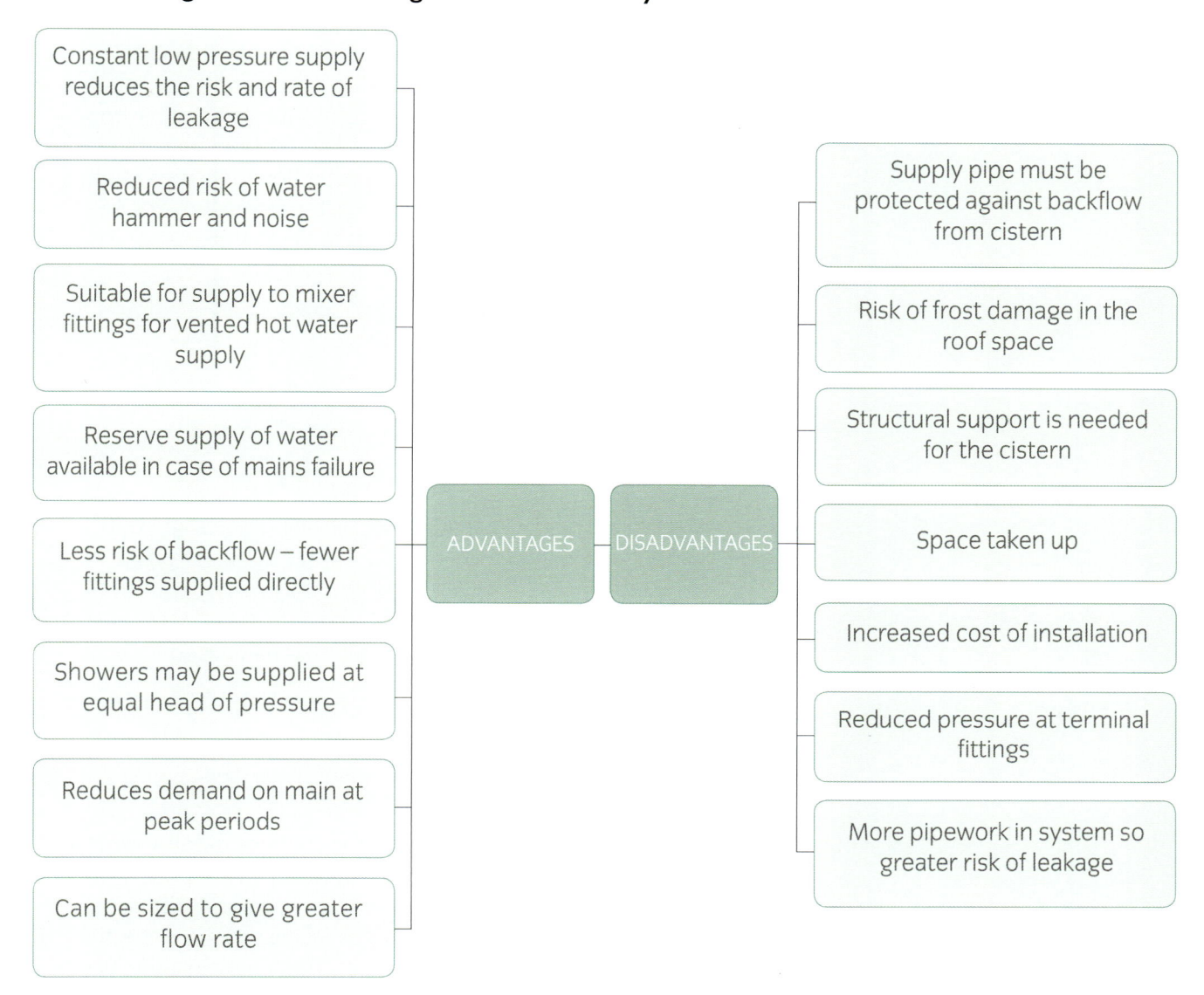

Advantages:
- Constant low pressure supply reduces the risk and rate of leakage
- Reduced risk of water hammer and noise
- Suitable for supply to mixer fittings for vented hot water supply
- Reserve supply of water available in case of mains failure
- Less risk of backflow – fewer fittings supplied directly
- Showers may be supplied at equal head of pressure
- Reduces demand on main at peak periods
- Can be sized to give greater flow rate

Disadvantages:
- Supply pipe must be protected against backflow from cistern
- Risk of frost damage in the roof space
- Structural support is needed for the cistern
- Space taken up
- Increased cost of installation
- Reduced pressure at terminal fittings
- More pipework in system so greater risk of leakage

For larger buildings, eg office blocks, factories, hotels, etc, it is preferable for all water, except wholesome water, to be supplied indirectly via a protected storage cistern or cisterns. This is necessary to relieve the water consumption on the water main.

Summary of cold water systems

In some cases a combination of both methods of supply may the best arrangement. In a house, for example, the ground floor outlets and any outside tap could be supplied under mains pressure while all other cold water outlet fittings could be fed from a storage cistern.

The performance of any cold water system is dependent upon the pressure of the incoming supply and its flow rate. Direct systems require a good pressure and flow rate because all of the appliances use mains cold water supply and, in some cases, mains-fed instantaneous hot water supply as well. Indirect cold water systems, where low-pressure supply is used, must have correctly sized pipes to ensure that the system meets the design specification; lack of pressure is compensated for by using a larger pipe size to increase flow rate.

SmartScreen Unit 006 worksheet 2

B3 COLD WATER STORAGE CISTERNS

Storage cisterns and the Water Regulations

Schedule 2, paragraph 16 of the Water Regulations tells us that a storage cistern supplying low-pressure cold water to sanitary appliances, or feeding a hot water storage system, should be capable of supplying wholesome water. Various protection measures must, therefore, be included in the design of the cistern to ensure that the water supply does not become contaminated or unwholesome. To comply with the Water Regulations, cisterns must:

- be fitted with an effective inlet control device to maintain the correct water level, ie a float-operated valve (FOV)
- be fitted with service valves on inlet and outlet pipework connections to allow for maintenance and repair activities
- be fitted with screened warning/overflow pipes
- be covered with a rigid, close-fitting lid, which is not airtight but excludes light and insects
- be insulated against freezing or undue warming
- be installed so that the risk of contamination is minimised
- be arranged so that water can circulate, preventing stagnation
- be supported to avoid distortion or damage, which could lead to leaks
- be accessible for maintenance and cleaning.

Cisterns must be manufactured and installed to these requirements if the problems of contamination are to be avoided. In the past, when cisterns were fitted with unscreened overflows and poorly fitting lids, insects and small mammals could easily gain access to the water that the cisterns contained. Some insects, like mosquito larvae, need water to complete their life cycle and these must be excluded at all costs.

Correct installation will eliminate the problems mentioned, especially when avoiding the problem of stagnation of water.

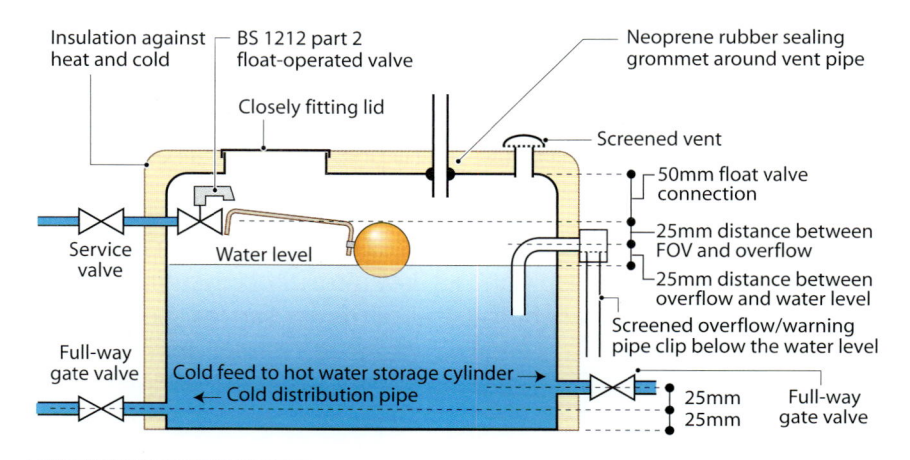

SmartScreen Unit 006 animation 4

Cistern complying with Schedule 2, para 16 of the Water Regulations

Types of cistern

Storage cistern – this is designed to hold a supply of cold water to feed appliances fitted to the system. Storage cisterns are used on indirect cold water systems, and supply cold water only.

Feed cistern – this can be identical to the storage cistern. However, it only holds the water required to supply the hot water storage vessel. In other words, it supplies cold water via the cold feed pipe to a hot water storage system.

Combined storage and feed cistern – this is a combination of the previous two. It is used on an indirect system of cold water where only the drinking supply is taken directly from the main and the rest of the water is supplied from a cistern. It stores water for the domestic hot water system and the indirect system of cold water to the appliances, wash hand basin, bath, WC, washing machine etc.

Feed and expansion cistern – used to feed a vented central heating system and also allows expansion of water into the cistern when the system is hot.

WC and automatic urinal flushing cisterns – used to clear the contents of a WC or urinal. The water they contain is not considered wholesome. They will be covered in more detail in Chapter 010.

Capacities of cisterns

Clause 5.3.9.4 of BS 6700 recommends the following capacities for houses:

Dwelling type	Cistern capacity
Small house	**Cold water only** – 100/150 litres **Hot and cold water** – 230 litres minimum **Hot water only** – at least the capacity of the hot water storage vessel
Large house	100 litres per bedroom

The size of cold water cistern installed will depend on what it is feeding. For cisterns that only supply cold water to baths, wash basins and WCs, then 100–150 litres of stored water is sufficient. If the cistern is supplying cold water to a hot water storage vessel, then the cistern should have at least the same amount of storage as the hot water storage vessel. If the cistern is supplying cold water for both hot and cold supplies then the minimum recommended storage capacity of the cistern should be at least 230 litres.

Kitchen sinks cannot be supplied from a protected cistern. Their supply must come direct from the mains cold water supply.

General installation requirements for cisterns

Water is heavy. At 4°C it weighs 1kg per litre, so 230 litres will weigh 230kg – almost a quarter of a tonne! A cistern full of water will therefore need adequate support, especially if placed in a roof space. Normal practice would be to try to place the cistern over a load-bearing supporting wall, but if this is not possible then the platform must be big enough to support the weight of the cistern and the water it contains by spreading the load across the roof joists.

The platform that the cistern sits on should be covered with 21mm tongued and grooved boarding or moisture-resistant plywood. The platform should be at least as big as the base area of the cistern and, if possible, 150mm larger all the way around.

Access to the cistern once it has been installed is vital for cleaning, inspection and maintenance. The minimum access allowance will depend on the size of the cistern. For cisterns of less than 1000 litres, 350mm must be allowed to permit access to the FOV for removal and replacement.

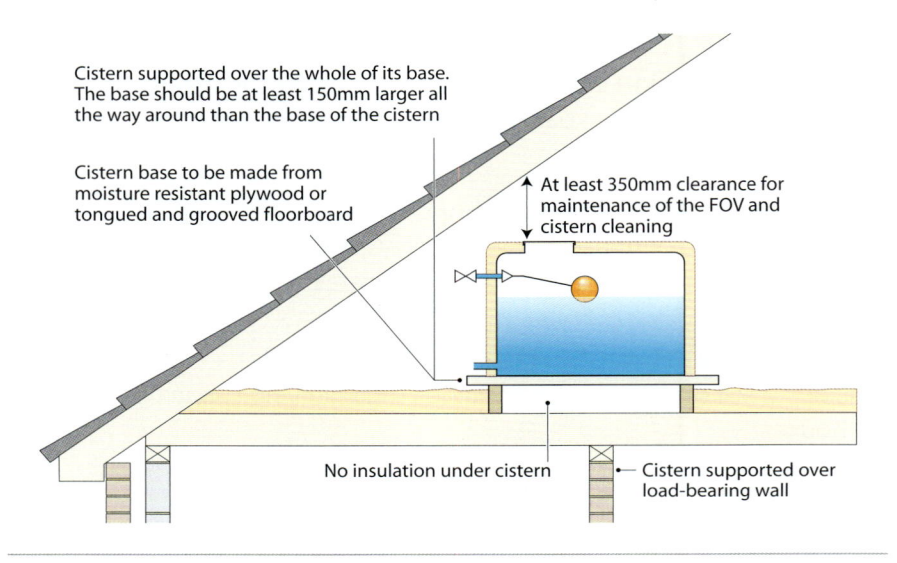

Cistern location and access arrangements

Inlet requirements for storage cisterns

The inlet requirements state that all cisterns will be fitted with an adjustable water inlet control device. These devices are usually float-operated valves (FOV) that must conform to British Standard 1212:1990. These will be covered in detail on pages 317–19.

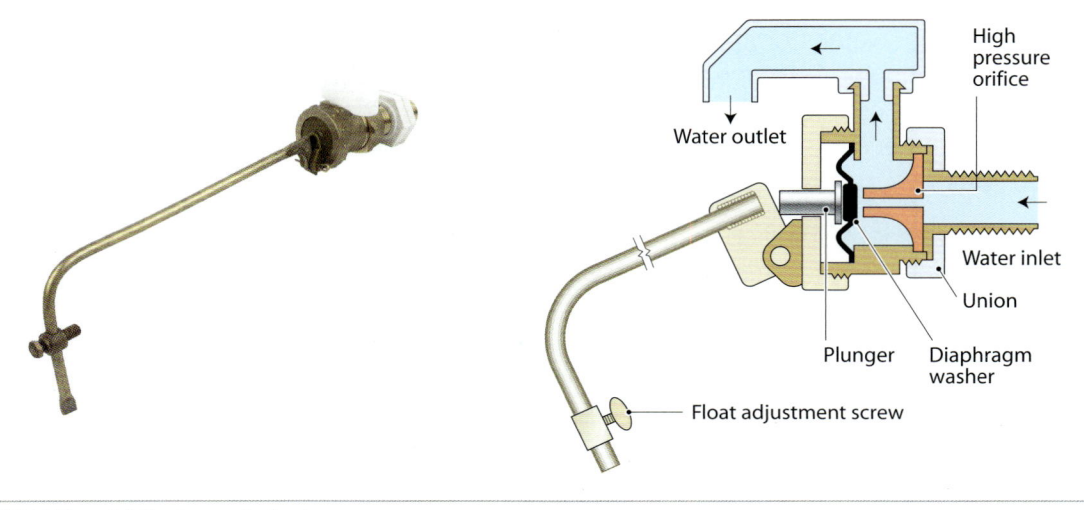

A BS 1212 part 2 float-operated valve

Outlet requirements for storage cisterns

Outlets from a cistern include indirect cold water distribution pipes and cold feed pipes to hot water storage systems.

Connection of the cold feed pipe and cold water distribution pipe to the cistern should preferably be positioned at the base of the cistern, although positioning these connections on the side of the cistern is also acceptable. The distance between the cold feed and the cold distribution pipework should not be less than 25mm, with the cold feed pipe higher than the cold distribution pipe.

The cold feed for the hot water system should always be higher than the cold distribution pipe connection. This is so that, in the event of mains cold water failure, the hot water will run out first, which will prevent any potential scalding situation if any mixing valves (showers, bath mixers or monobloc wash basin mixers) are installed on the system.

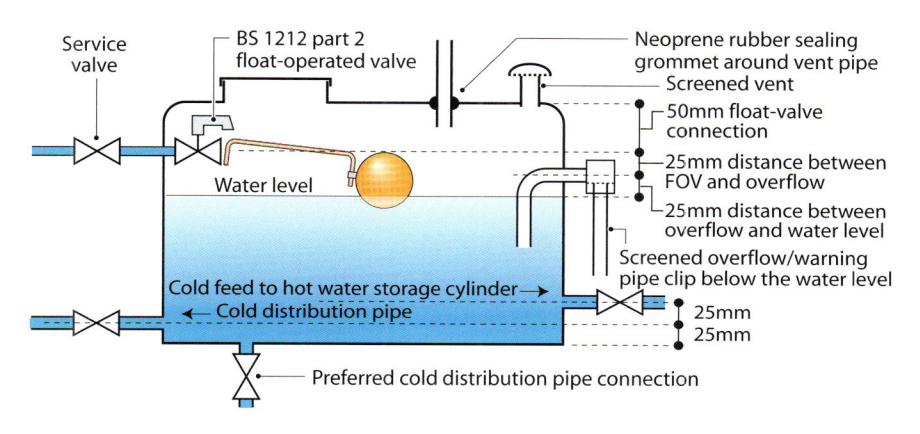

Cistern connections

Prevention of stagnation

Pipework connections to cisterns should be positioned so as to avoid stagnation by ensuring an even flow of water through the cistern. A single outlet should be positioned on the opposite side to the FOV. Where two connections are required, they should be arranged so that the connection with the greatest water usage is positioned furthest away from the FOV. This ensures that stagnation of the water does not occur.

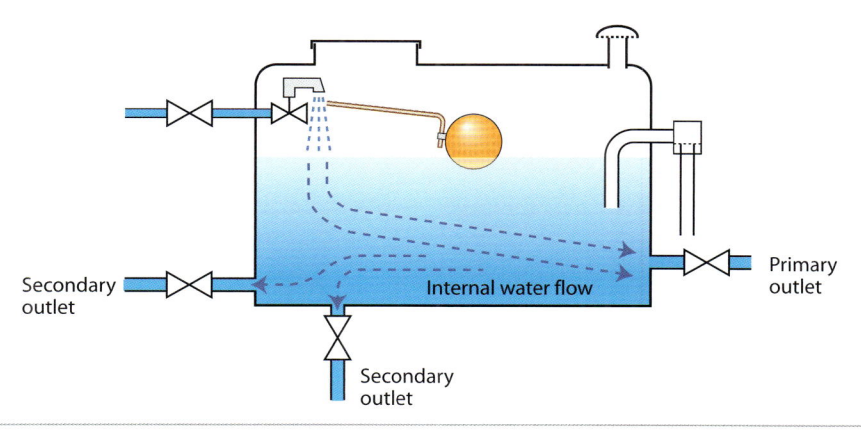

Flow of water to prevent stagnation

> **KEY POINT**
> For more information on warning and overflow pipes see section G16.8 of the Water Regulations Guide.

Warning and overflow pipes

A warning pipe is smaller in diameter than an overflow pipe. It acts as a warning that the FOV has developed a problem and the cistern is about to overflow. It is installed in a lower position on the cistern than the overflow pipe.

An overflow pipe is larger in diameter than the incoming water supply and must be able to cope with the flow of water in the event of complete FOV failure. This is to ensure that the FOV does not become completely submerged in water at any time during critical velocity, when the water is running at full bore.

Warning and overflow pipes should be installed so that the water is taken to the outside of the property, and positioned in a conspicuous location.

The requirements for warning and overflow pipes differ depending the volume of the cistern. Cisterns below 1000 litres capacity do not require separate warning and overflow pipes. A single combined warning/overflow pipe only is required. The bottom of the combined warning and overflow pipe should be a minimum of 25mm above the water level of the cistern.

Interconnection of cisterns

On occasion, to provide large quantities of stored water, or due to space restrictions, two or more cold water storage cisterns may need to be linked. If the cisterns are not interconnected correctly then stagnation of the water may occur and this could lead to the growth of legionella bacteria (legionnaire's disease).

To avoid legionella bacteria, cisterns should:

- be small enough to ensure rapid turnover of water to prevent stagnation
- have inlet and outlet connections at opposite ends
- be regularly inspected and maintained in a clean condition
- conform to the requirements of the Water Regulations.

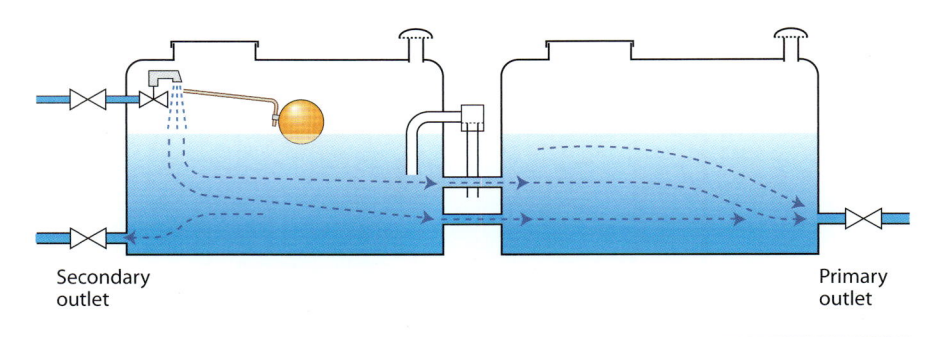

The interconnection of cisterns

In practice, cisterns should be interconnected to allow free movement of water from one cistern to the other. They should be connected at the bottom and the middle so that water passes evenly through them. The primary outlet connection should be made on the opposite cistern to the FOV to encourage water movement, with the secondary connection made on the cistern with the FOV installed. The overflow/warning pipe should be fitted on to the same cistern as the FOV. Both cisterns must be of the same size and capacity.

When connecting two or more cisterns together, care should be taken to ensure that the water movement is regular and even across all cisterns. In this situation, it is a good idea to install FOVs on all cisterns with appropriate service valves as detailed in the Water Regulations:

> 'Every float-operated valve must have a service valve fitted as close as is reasonably practicable.'

Wherever an FOV is fitted, an overflow/warning pipe must accompany it. These should terminate in a conspicuous, visible position outside the building. On no account should they be coupled together. There should be service/gate valves positioned to allow for isolation and maintenance of the cisterns without interrupting the supply. The diagram shows how any two of the four cisterns can be decommissioned, leaving two in operation. This ensures continuation of supply.

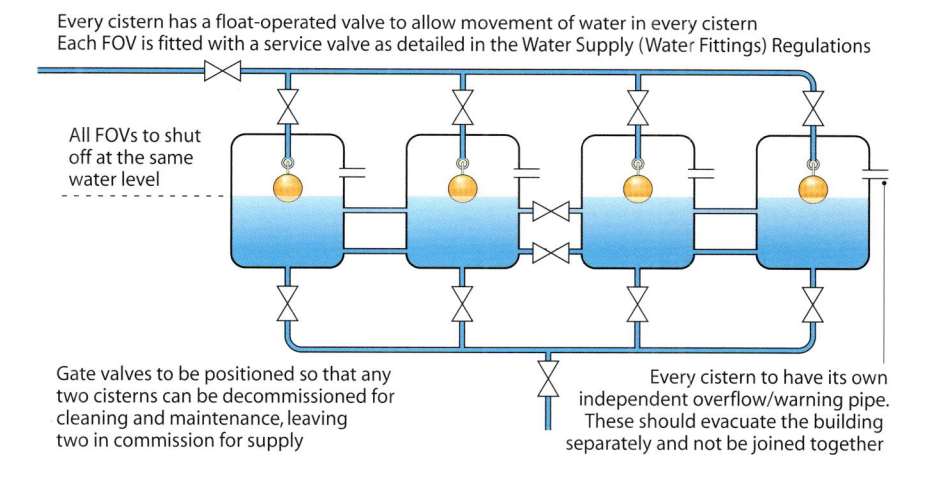

Every cistern has a float-operated valve to allow movement of water in every cistern
Each FOV is fitted with a service valve as detailed in the Water Supply (Water Fittings) Regulations

All FOVs to shut off at the same water level

Gate valves to be positioned so that any two cisterns can be decommissioned for cleaning and maintenance, leaving two in commission for supply

Every cistern to have its own independent overflow/warning pipe. These should evacuate the building separately and not be joined together

Installing three or more cisterns

Materials for cisterns

Almost all new installations use cisterns made from plastics such as polyethylene, polypropylene and glass reinforced plastic (GRP). Most cisterns manufactured today are made from polypropylene, because this allows:

- lightweight construction
- strength
- hygiene
- resistance to corrosion
- flexibility; cisterns pass easily through roof space openings.

- SmartScreen Unit 006 handout 3
- SmartScreen Unit 006 worksheet 4

Holes for pipe connections should be cut out using a hole saw. A 17mm hole saw should be used for ½-in thread connections such as FOVs, and a 25mm hole saw for ¾-in thread connections such as tank connectors.

The joint between the cistern wall and the fitting should be made using plastic washers (known as polywashers). On no account must any oil-based jointing compounds be used, as they can break down the plastic and provide a culture where microbiological growth such as legionella pneumophila (legionnaire's disease) can occur.

Galvanised steel cisterns were used for many years but these were notorious for corrosion. Galvanised steel cisterns are still manufactured and can still be used provided the inside of the cistern is protected by the use of a special paint which is registered in the Water Fittings and Materials Directory as safe to use with wholesome water.

B4 FROST PROTECTION

You can never fully protect against freezing temperatures. No matter how much insulation is wrapped around pipes and fittings, if the weather gets cold enough, the pipes will freeze. Therefore, you can merely delay the freezing process for as long as possible by applying insulation. When insulating pipes, you are not attempting to 'keep the cold out'. The idea of insulation is to keep in the heat that is already there. In other words, to retain the 'heat energy' already present in the water for as long as possible. The greater the thickness of insulation, the longer the heat energy will be retained.

The Water Regulations Guide G4.2 states:

> 'All cold water fittings located within a building but outside the *thermal envelope*, or those outside the building must be protected against damage by freezing.'

In practice, this means that all pipes and fittings that are installed in vulnerable locations inside and outside a building, such as unheated cellars, roof spaces, under ventilated suspended floors, garages and outbuildings, must be insulated. Where pipework is installed in a roof space, the pipes should still be insulated, even if they are placed below the roof insulation. This is to avoid unnecessary warming by heat from the rooms below. The thickness of the insulation will depend on the size of the pipe: the smaller the pipe, the greater the thickness of insulation needed. Where pipes are located outside the dwelling, the insulation should be to external standards and waterproof.

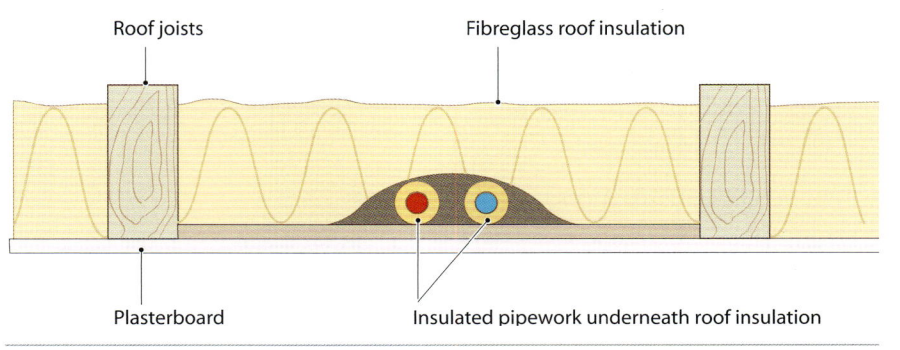

Insulation around pipes

Thermal envelope
That part of a building that is enclosed within walls, floor and roof, and that is thermally insulated in accordance with the requirements of the Building Regulations.

Pipes in roof spaces

Cisterns should be insulated using the jacket supplied with the cistern. This is a PVC jacket with fibreglass insulation inside, which is simply wrapped around the cistern and tied in place. Any roof insulation below the cistern should be removed to allow some warmth from the room below to filter upwards to help keep the cistern free of freezing temperatures.

Insulation materials and their effectiveness

Pipework insulation should be of the closed-cell type complying with BS 5422 and installed in accordance with BS 5970. It must be resistant to rain, moisture, water, mechanical damage and vermin. The recommended materials for pipe insulation are:

- rigid phenolic foam (less than 0.020 W/mK)
- polyisocyanurate foam (0.020–0.025 W/mK)
- PVC foam (0.025–0.030 W/mK)
- expanded polystyrene, extruded polystyrene, cross-linked polyethylene foam and expanded nitrile rubber (0.030–0.035 W/m²K)
- expanded synthetic rubber, cellular glass and standard polyethylene foam (0.035–0.040 W/m²K)

BS 6700 gives the following table of minimum thicknesses for copper pipe insulation:

Calculated minimum thickness of insulation to protect copper pipes fixed inside premises for domestic cold water systems

		Thermal conductivity of insulation material at 0°C in W/mK			
		0.025	0.035	0.045	0.055
Outside diameter of pipe	Inside diameter (bore) of pipe	Thickness of insulation			
15mm	13.6mm	30mm	62mm	124mm	241mm
22mm	20.2mm	12mm	20mm	30mm	43mm
28mm	26.2mm	8mm	12mm	17mm	23mm
35mm	32.6mm	6mm	9mm	12mm	15mm
42mm	39.6mm	5mm	7mm	9mm	11mm

> **KEY POINT**
>
> All materials lose heat (measured in watts). W/m²K is the amount of watts transmitted through a substance that is measured in m² for every degree temperature difference measured in Kelvin (or Celsius). This is also known as the u-value.
>
> So, if a material has a u-value of 0.020, it means that it will lose 0.020 watts per square metre for every degree of temperature difference. You are likely to come across these terms when choosing insulation materials.

The larger the pipe size, the less insulation is required. This is because larger pipes contain a greater amount of water and, therefore, a greater amount heat energy and so they require less insulation to keep the heat in. The smaller the pipe size, the greater the amount of insulation needed, because the small amount of water it contains loses heat more quickly.

C PREVENTION OF BACKFLOW AND BACK SIPHONAGE

Two of the major causes of contamination in plumbing systems are backflow and back siphonage. The Water Regulations specifically target the avoidance of contamination of wholesome water, so it is crucial to understand its causes and, more importantly, the methods by which you can prevent it from occurring.

C1 WHAT IS BACKFLOW AND BACK SIPHONAGE?

Backflow is simply the flowing of water in the wrong direction due to loss of system pressure. If, for example, the mains cold water were shut off in the street for repairs, then the cold water in your plumbing system would backflow (or flow backwards) into the water main. This is sometimes called 'back pressure'.

Back siphonage occurs when the pressure at the supply end of the water main (ie in the street) drops suddenly. This causes a vacuum that can literally suck water backwards (see Chapter 004, page 184). The dangers of back siphonage can be understood from the following scenario: a watering can of weedkiller is being filled with water from a hosepipe when the pressure suddenly drops due to a burst water main or the fire service using water from the main to fight a fire. The contents of the watering can (including the weedkiller!) will be sucked up the hose, through the tap and down into the water main, contaminating the wholesome water with weedkiller solution.

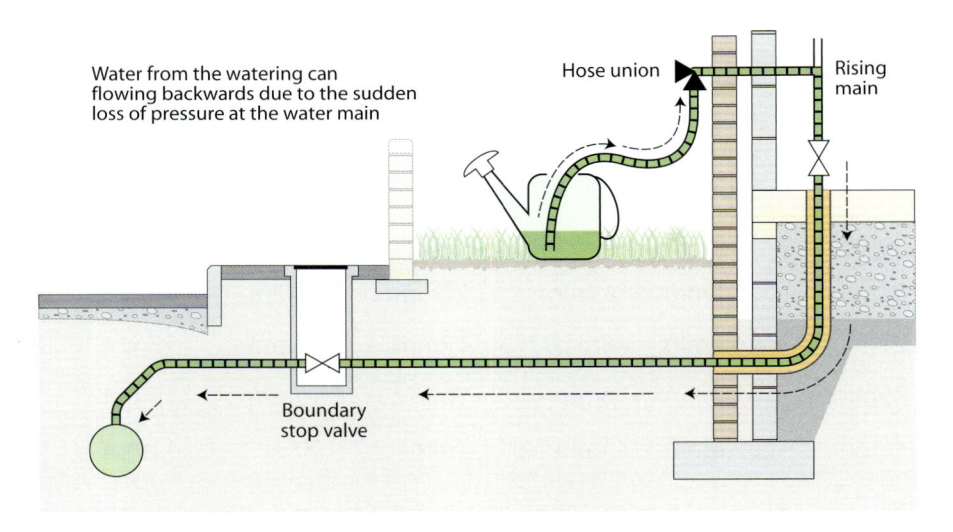

The effects of back siphonage on plumbing installations

To avoid these potentially dangerous situations in plumbing work, there are several methods that you can employ. You can prevent backflow and back siphonage by:

- the use of a physical air gap
- the use of a mechanical backflow prevention device

- eliminating or protecting against potentially dangerous cross-connections.

The choice of method will depend upon the severity of the situation and the fluid category of the potential hazard.

C2   THE FLUID CATEGORIES

Any water that is not cold wholesome water supplied by a water undertaker can be classed as a potential hazard. The Water Supply (Water Fittings) Regulations 1999 list five fluid categories:

Fluid category	Description
1	Wholesome water supplied by a water undertaker and complying with the requirements of regulations made under section 67 of the Water Industry Act 1991
2	Water in fluid category 1 whose aesthetic quality is impaired owing to a change in its taste, colour, odour or temperature (this includes water in a domestic hot water system)
3	Fluid that represents a slight health hazard because of the concentration of substances of low toxicity
4	Fluid that represents a significant health hazard due to the concentration of toxic substances
5	Fluid that represents a serious health hazard because of the concentration of pathogenic organisms, animal remains, human waste, radioactive or very toxic substances
For more information on fluid categories see http://www.sms-environmental.co.uk/fluid_categories.html	

As you can see from the table, fluid category 1 is clean, cold, wholesome water direct from the water undertaker's main and no other fluid category must come into contact with it or contamination may occur.

Backflow and back siphonage risks in the home

There are many instances in the home where backflow and back siphonage could present contamination risks. These will need to be considered during any planning, design and installation of hot and cold water supplies and central heating systems. First consider the risks of plumbing appliances and systems:

> **SUGGESTED ACTIVITY...**
> Try this simple experiment to see the siphonic process in action:
>
> Get two empty soft drinks bottles and fill one with water. Now insert a short piece of hose and, while placing the full bottle higher, suck on the hose and quickly put the hose end in the bottom bottle.

Appliance or system	Potential contamination	Risk
Kitchen sink	May contain animal remains from food preparation	Fluid category 5
WC	Contains human waste	
Bidet (over rim type)	May contain human waste	
Grey water and rainwater harvesting systems	May contain bacteria and disinfectants	
Washing machines and dishwashers	Contain soap and other detergents and chemicals from dishwashing and clothes cleaning	Fluid category 3
Bath	May contain soap and other detergents from personal hygiene	
Wash hand basin		
Shower valves and instantaneous showers	At risk from soap and other detergents from personal hygiene	
Hose union bib taps (outside tap)	At risk from gardening and other activities such as watering, weed killing, car washing, irrigation, etc	
Combi boilers	The water in the heating system is often contaminated with dissolved metals, flux and some form of chemical inhibitor	Fluid category 3 or 4 (depending on boiler size)
Hot water system	Contains hot water	Fluid category 2

SmartScreen Unit 006 worksheet 5

There are many potential contamination risks in every dwelling, and the bigger the building the more risks there are.

Eliminating the risk of contamination of wholesome water

The Water Regulations and, more specifically, the Water Regulations Guide, can help you choose the right course of action based upon the risk. The manufacturers too, help in this regard by designing and manufacturing their appliances, taps and valves to conform to the Water Regulations. For example, most kitchen and bidet taps are designed and made with a fluid category 5-level risk in mind, and most bath and wash-basin taps are designed and made with fluid category 3 in mind.

In most cases, where baths, wash basins, bidets and kitchen sinks are concerned, a simple air gap will protect the mains cold water supply. The size of the air gap, however, is dependent on the size of the tap, appliance type and its likely contents.

C3 AIR GAPS USED AS A METHOD OF BACKFLOW PREVENTION

An air gap is simply a physical unrestricted open space between the wholesome water and the possible contamination, the simplest form of which is the gap between a tap outlet and the spill-over level of, say, a kitchen sink.

With this type of air gap, contamination cannot occur because the outlet is a specific minimum distance away from the contents of the sink, and so neither backflow nor siphonage can occur. The minimum size and type of air gap would depend on the size of the pipe feeding the tap and the risk from the water in the appliance, for example:

Air gap type	Appliance	Size of tap	Gap distance	Figure
AUK 1	WC	n/a	300mm (min) from the overflow on the cistern to the spill-over level of the WC pan	
AUK 2	Bath	¾-in BSP	25mm	
	Wash basin	½-in BSP	20mm	
AUK 3	Kitchen sink Bidet	½-in BSP	At least twice the bore of the pipe	

Most domestic clothes- and dishwashing machines are designed and manufactured with a built-in backflow prevention device or arrangement, usually a built-in air gap, suitable for fluid category 3 and so no additional backflow measures are required. However, some domestic installations and multiple installations (more than one machine) will require a double check valve to be fitted on the cold supply.

SmartScreen Unit 006 animation 5

Where an air gap is not possible, then the use of a mechanical backflow prevention device (a specific fitting or valve type) will have to be used, such as a single check valve or double check valve, which only allows water to flow in one direction.

C4 MECHANICAL BACKFLOW PREVENTION DEVICES

There are many different backflow prevention devices on the market (the Water Regulations Guide should be consulted with regard to any specific contamination risk), but two of the most common are:

- single check valves
- double check valves.

Check valves, whether double or single, only allow water to flow in one direction. If backflow or back siphonage occurs, the springs in the valves shut the valve off, preventing contamination.

Check valves may be:

- **verifiable** – they can be checked to see if they are working correctly because they have a test point built into the valve
- **non-verifiable** – they cannot be checked because they do not have a test point.

Both verifiable and non-verifiable check valves do the same job, but in some situations verifiable may be specified over non-verifiable.

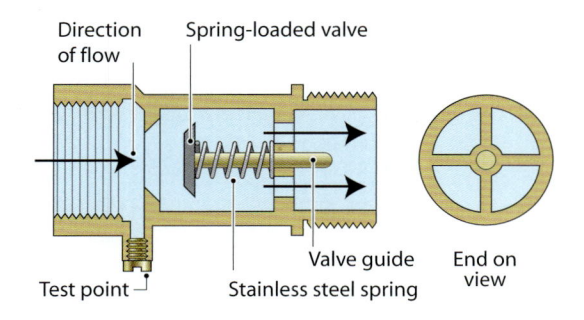

Single check valve

Verifiable single check valve (type EA air gap)

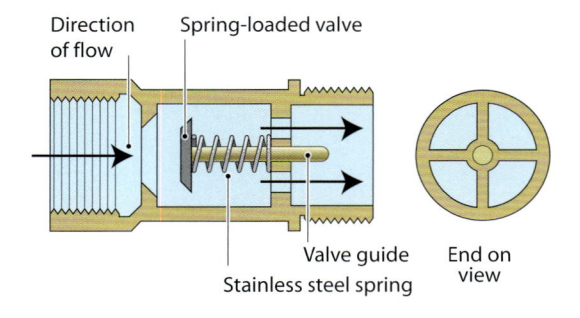

Non-verifiable single check valve (type EB air gap)

Cross-connection
When one fluid category connects with another, for example, within a mixer tap.

KEY POINT
Double and single check valves are only necessary in this situation when the tap is a true mixing valve – where the hot and cold water mix in the valve body before flowing from the outlet and both hot and cold have unbalanced supplies, ie cold from the water main and hot from a vented hot water storage vessel. If both supplies are balanced, ie from a cistern or mains, or the tap is a bi-flow mixer, then check valves are not required.

Cross connection and the use of single check valves

The single check valve is the most basic of all the backflow prevention devices. It is simply a spring-loaded one-way valve that can only be used where the wholesome water supply is at risk from a cross-connection with fluid category 2, such as hot water above 20°C.

If any other fluid category cross-connects with fluid category 1, the mains cold wholesome water supply must be protected or the cross-connection eliminated completely. For example, when a wash hand basin monobloc mixer tap is connected to both hot and cold supplies, then both the hot (fluid category 2) and cold (fluid category 1) pipes must include a single check valve arrangement to stop hot water flowing down the cold water pipe. This is because the mixer tap acts as a cross-connection between the hot and cold supplies. The single check valve only allows the water to flow in one direction and stops the hot water backflowing down the mains cold water supply in the event of mains failure.

The same method can be applied to shower mixer valves when the hot water supply is under mains water pressure from an unvented hot water storage unit or a combi boiler/instantaneous hot water heater.

Monobloc mixer acting as a cross-connection

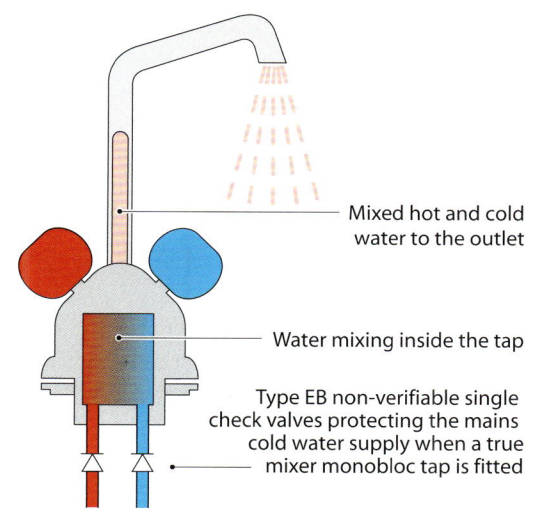

Shower mixer valve acting as a cross-connection

Double check valves

Double check valves offer greater backflow protection than single check valves and for this reason can be used when the fluid category risk is higher. Double check valves can be used up to fluid category 3.

Appliance	Backflow prevention device
Shower valve and instantaneous shower	Double check valves on the hot and cold supplies, or a wall-mounted shower hose retention ring
Hose union bib tap	Double check valve
Combi boiler	Double check valve and an AUK3 air gap

Non-verifiable double check valve

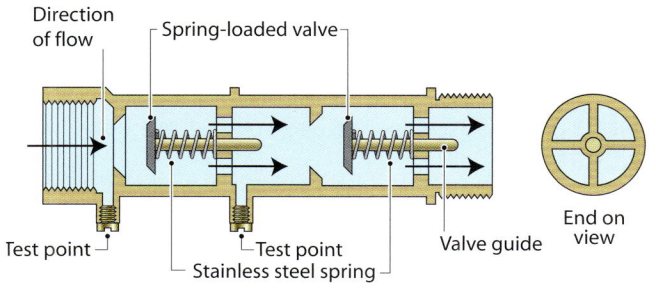

Verifiable double check valve (type EC air gap)

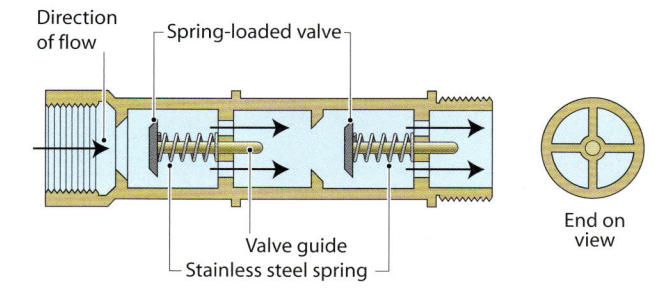

Non-verifiable double check valve (type ED air gap)

Shower valves and instantaneous showers have the option of either a double check valve or a wall-mounted shower hose retention ring. The retention ring prevents the shower spray head from entering the bath water or shower tray and therefore creates an air gap, so a check valve is not required. Shower manufacturers now provide retention rings as part of their installation packages.

SmartScreen Unit 006 handout 4

SmartScreen Unit 006 worksheet 6

Hose union bib taps fitted outside or in a garage for garden use must have a double check valve installed inside the building, together with an isolation valve. This is to prevent freezing of the valve during cold weather. Hose union bib taps with integral double check valves (called an HUK1 air gap) are not permitted except in garages.

Combi boilers require two forms of protection: a double check valve and an AUK3 air gap. Water in a heating system is classified as fluid category 3, but because the heating water is under pressure, an AUK3 air gap must also be provided. This is conveniently supplied in the form of a removable filling loop, and once the heating system is filled with fluid category 1 water from the water main, the filling loop must be removed, creating an AUK3 air gap. The double check valve protects the cold water main during the filling process. This is covered in more detail in Chapter 008.

D TAPS AND VALVES

Taps and valves can be divided into four separate categories:

- stop valves, servicing valves and drain valves
- float-operated valves (FOVs)
- terminal fittings (taps)
- shower mixer valves.

This section of the chapter looks at each one in turn including the types available, their uses and their working principles.

D1 ISOLATION VALVES

Isolation valves turn off (isolate) either complete systems, parts of systems or appliances. They can be divided into four distinct types:

- those that isolate high-pressure systems, such as stop valves
- those that isolate low-pressure systems, such as full-way gate valves
- those that isolate appliances and terminal fittings on either high- or low-pressure systems
- those that are used for draining down systems.

Stop valves (high-pressure isolation) to BS 5433 or BS 1010

Stop valves are designed for isolation of high-pressure cold water systems, and because of their restrictive internal design should not be used on low-pressure supplies. They are manufactured to either BS 5433 or BS 1010 for domestic use.

They consist of a brass valve body, a head gear with a rising spindle, a packing gland and a re-washerable loose jumper plate. Stop valves have an arrow on the valve body that shows the direction of flow of the water.

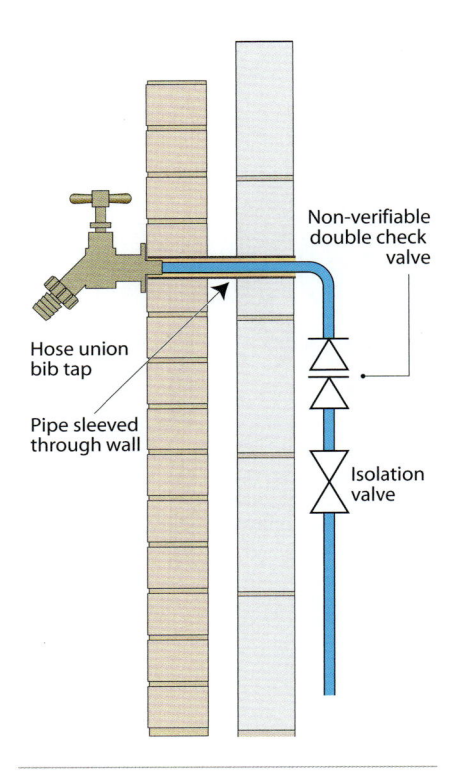

Installation of a hose union tap

Some stop valves have a drain valve built into the stop valve body, but care should be taken with this type when installing internal water meters as the drain valve position may allow water to be drawn from the main without being metered.

Stop valves are available with either capillary or compression connections to suit copper tubes to BS EN 1057, compression connections for MDPE and push-fit connections for polybutylene pipe.

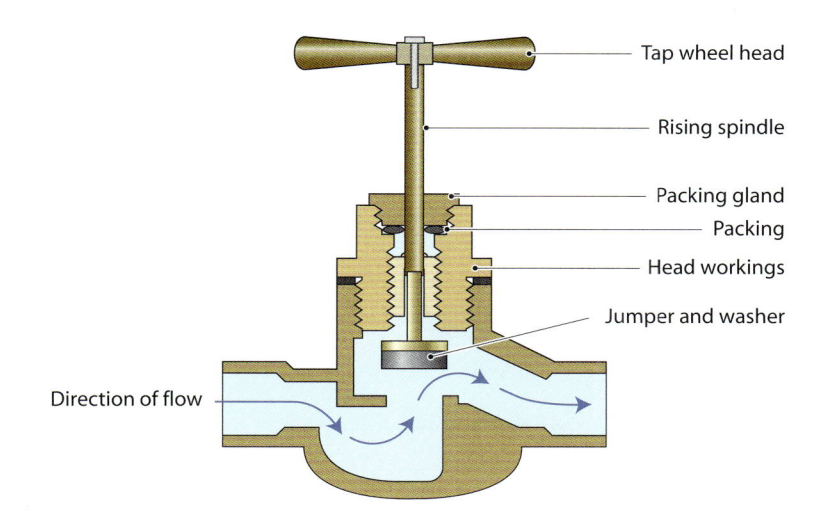

SmartScreen Unit 006 animation 6

Internal workings of a stop valve

Full-way gate valves (low pressure isolation) to BS 5154

Gate valves are used on low-pressure installations such as the cold feed to vented hot water storage cylinders and the cold distribution pipework for indirect cold water systems. They do not have a washer; instead they use a brass, wedge-shaped gate that rises inside the valve.

They are known as 'full-way' gate valves because the design allows water to flow at full bore without much restriction to the flow rate. However, they should not be used on high-pressure supplies as they tend to allow water to pass by the gate when the valve is under pressure. They consist of a brass valve body and a headgear with a non-rising spindle.

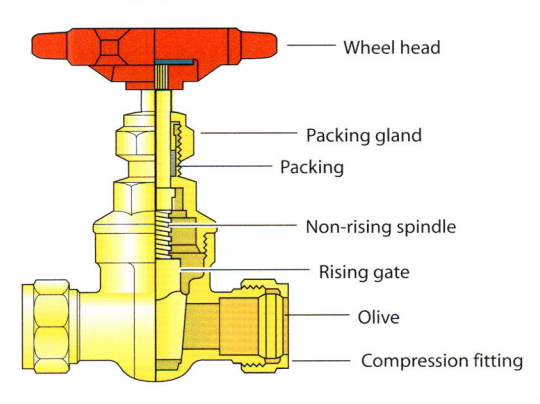

SmartScreen Unit 006 animation 8

Internal workings of a gate valve

A lock-shield gate valve

Gate valves are also available with a lock-shield head to prevent the valve being tampered with.

Spherical plug valves

Spherical plug valves (also known as service valves) are used for isolation of appliances and terminal fittings such as taps and FOVs. There is a variety of styles available including with or without a handle (the latter use a screwdriver slot to isolate the water) or for use with an appliance such as a washing machine.

The internal design of the valve allows water to be isolated by turning a ball through 90 degrees. The ball has a hole through it, which when it is in line with the direction of water flow, allows water to pass through it. It is isolated when the hole is at 90 degrees to the flow of water.

SmartScreen Unit 006 animation 7

A spherical plug valve

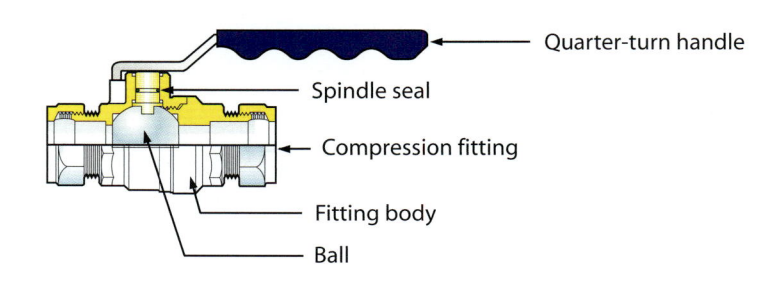

Internal workings of a spherical plug valve

Drain-off valves

These are small valves that are strategically placed at low points in the installation to allow draining down of the system. Several types are available:

- with a male BSP thread to allow connection to low carbon steel pipes and fittings
- with a spigot end to facilitate connection to either copper capillary fittings or compression fittings
- with or without a packing gland.

Drain-off valves should be positioned in accordance with the Water Supply (Water Fittings) Regulations 1999: above ground and nowhere where they could be submerged in water.

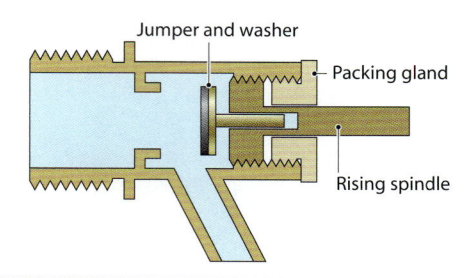

SmartScreen Unit 006 animation 9

Internal workings of a drain-off valve

The types of isolation valves and their uses

As a general rule of thumb, remember that:

- stop valves are high-pressure only valves and should not be used on low-pressure supplies
- gate valves are for low-pressure installations
- servicing valves are primarily for terminal fixture isolation.

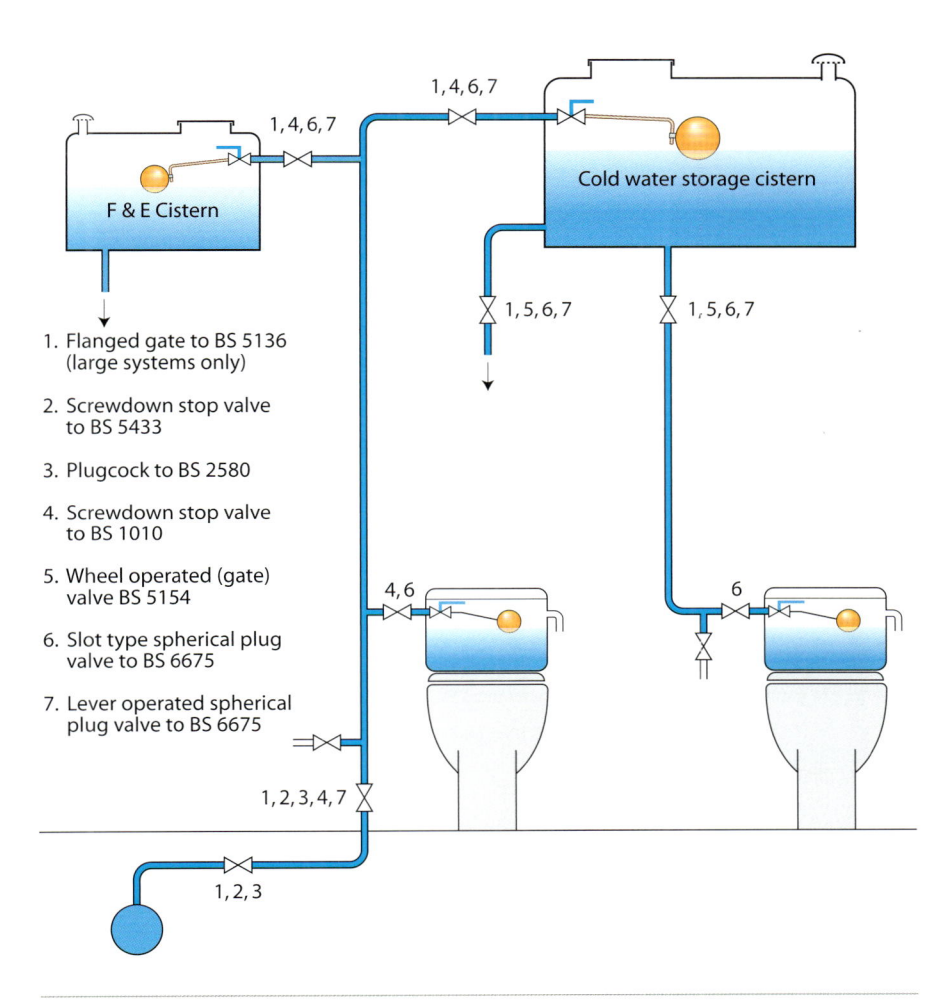

1. Flanged gate to BS 5136 (large systems only)
2. Screwdown stop valve to BS 5433
3. Plugcock to BS 2580
4. Screwdown stop valve to BS 1010
5. Wheel operated (gate) valve BS 5154
6. Slot type spherical plug valve to BS 6675
7. Lever operated spherical plug valve to BS 6675

The type and position of valves

D2 FLOAT-OPERATED VALVES

FOVs are used to control the flow of water into cold water storage and feed cisterns, feed and expansion cisterns and WC cisterns. They are designed to close when the water reaches a pre-set level. They are made to BS 1212 and it is important that plumbers recognise the different types. There are four basic FOV types:

SmartScreen Unit 006 interactive activity 1

SmartScreen Unit 006 interactive activity 2

- **BS 1212 part 1** – Portsmouth pattern and Croydon pattern (can only be used with some form of backflow prevention such as a double check valve, fitted before the FOV)
- **BS 1212 part 2** – Diaphragm type
- **BS 1212 part 3** – Diaphragm type (plastic)
- **BS 1212 part 4** – Torbeck equilibrium type (WC cisterns only).

FOVs can either be high-pressure or low-pressure depending on the type of orifice fitted. The orifice is the part of the valve where the water passes through. A high-pressure orifice is white in colour and has a small hole for the water to flow through, whereas the low-pressure orifice is coloured red with a larger hole. The orifice is universal for part 1, 2 and 3 FOVs.

A low pressure orifice

SmartScreen Unit 006 animation 10

BS 1212 part 1: Portsmouth-pattern FOV

The Portsmouth-type FOV discharges water from the bottom of the valve, which makes it susceptible to back siphonage should the valve become submerged in water. It should not be fitted on new installations without some form of backflow protection device, although existing Portsmouth-type valves can be repaired and maintained.

Portsmouth FOVs have moving parts, which will come into contact with water, and this makes them vulnerable to failure and noise.

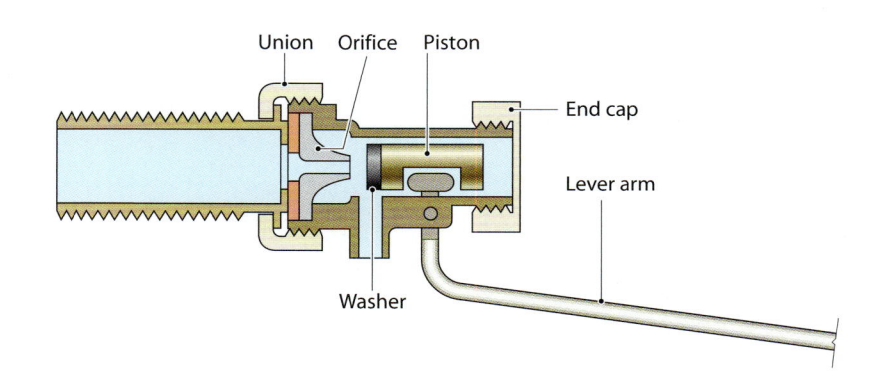

A BS 1212 part 1 Portsmouth-type FOV (diagram)

BS 1212 part 1: Croydon-pattern FOV

The Croydon-type FOV is less common than the Portsmouth type. Like the Portsmouth, it discharges water from the bottom of the valve but it is easily recognisable by its vertical piston and by the fact that it delivers water into the cistern in two streams of water. This type of FOV is very noisy and no longer manufactured but it may still be used in some older WC cisterns.

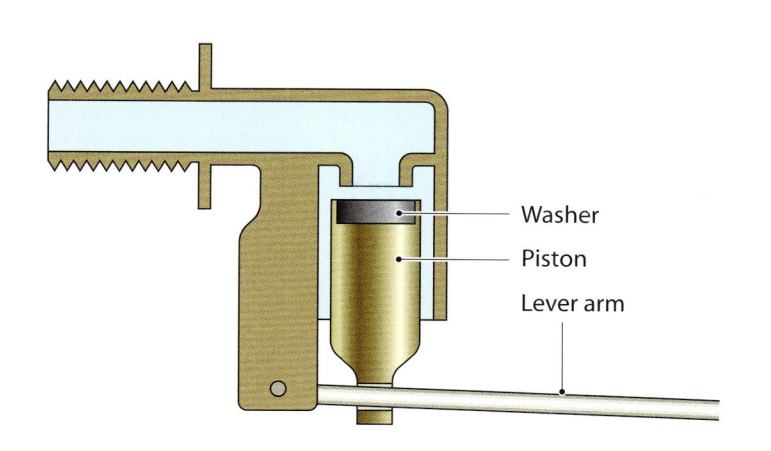

A BS 1212 part 1 Croydon-type FOV (diagram)

KEY POINT

Water hammer is caused by a rapid opening and closing of the float-operated valve. As the water nears the water level in the cistern, the ball valve can begin to bounce quickly up and down and from side to side. This causes the noise to travel down the pipework, resulting in reverberation or a whining noise. This can also be caused by a faulty washer or diaphragm.

BS 1212 part 2 and 3 (plastic) diaphragm FOVs

These FOVs use a diaphragm rather than a washer to control the flow of water (see page 302) and, unlike part 1 FOVs, they discharge water over the top of the valve. This makes them less susceptible to being submerged in water when the overflow runs and so less likely to cause a contamination issue. They also have fewer moving parts and this makes the valve quieter in operation and less likely to cause water hammer and reverberation of the pipework.

The main difference between a part 2 FOV and a part 3 FOV is that the part 2 is made of brass and the part 3 is made of plastic. They are almost identical in all other respects. It should be noted that plastic FOVs are not recommended for cisterns other than WC cisterns because of the risk of freezing and subsequent splitting of the plastic.

BS 1212 part 4 equilibrium diaphragm FOV

The equilibrium FOV is a diaphragm valve that works on the principle of equal pressure in front and behind the diaphragm when the valve is open. No moving parts come into contact with the diaphragm. It closes the valve when a build-up of pressure occurs in front of the diaphragm due to the float arm closing the pressure relief orifice on the front of the valve. Although quieter in operation than other FOVs, the positive 'snap' closing action can lead to problems of banging and reverberation in some systems.

The valve can be used on either high or low pressures by the insertion of either a low-pressure or high-pressure flow restrictor in the valve stem. Some valves also have a filter to remove any minute solid impurities in the water, which could cause malfunction.

The most common type of equilibrium valve, the Torbeck-type valve, can only be used on WC cisterns.

A BS 1212 part 2 diaphragm FOV

SmartScreen Unit 006 animation 11

SmartScreen Unit 006 animation 12

BS 1212 part 4 Torbeck equilibrium FOV

D3 TERMINAL FITTINGS

Terminal fittings, better known as taps, are fitted to sanitary appliances, such as baths and wash basins. There are several different types:

Pillar taps for baths, wash basins and bidets – these are available for baths (¾-in tails), wash basins and baths (½-in tails)	
High-necked pillar taps for kitchen sinks – similar internal design to pillar taps but designed with a high stem to provide an AUK3 air gap at kitchen sinks	
Bi-flow mixer taps including monobloc mixers – two taps in a single body. A bi-flow mixer has a single spout that is divided down the middle so that the water does not mix until it has exited the tap. It is not a true mixer tap	
True mixer taps – allow the hot and cold water supplies to be mixed inside the body of the tap. Caution should be exercised as these taps can provide a cross-connection between low-pressure hot (fluid category 2) and high-pressure cold (fluid category 1)	
Bib taps and hose union bib taps – bib taps are mostly fitted to the wall above cleaners' sinks and Belfast sinks. Hose union bib taps are specifically designed for garden use so that a hose may be connected	

SmartScreen Unit 006 animations 16 and 17

Terminal fittings fall into three categories:

- taps with a rising spindle to BS 1010
- taps with a non-rising spindle to BS 5412
- ceramic disc taps.

Taps with a rising spindle to BS 1010

BS 1010-type taps have a rising spindle attached to a jumper plate and a washer. When the tap is turned on, the spindle rises, allowing the pressure of the water to push the jumper plate and washer upwards to start the flow of water. Older BS 1010 taps may appear slightly different from modern taps because some taps were made with a loose jumper plate and others with a fixed jumper plate.

- Taps with loose jumper plates are for high-pressure supplies such as mains cold water. The loose jumper acted as a very crude backflow prevention device to stop contamination of the mains cold water supply by dropping to close the tap seat when water began to flow backwards.

- Taps with fixed jumper plates are for low-pressure supplies such as indirect cold water installations and vented hot water supplies. Most new BS 1010 taps are this type.

Both types have a packing gland designed to stop water leaking through the spindle. The design of BS 1010 taps is generic across most manufacturers. This means that the head workings of one tap will almost certainly fit the tap body of another manufacturer, including stop-valve heads.

SmartScreen Unit 006 animation 13

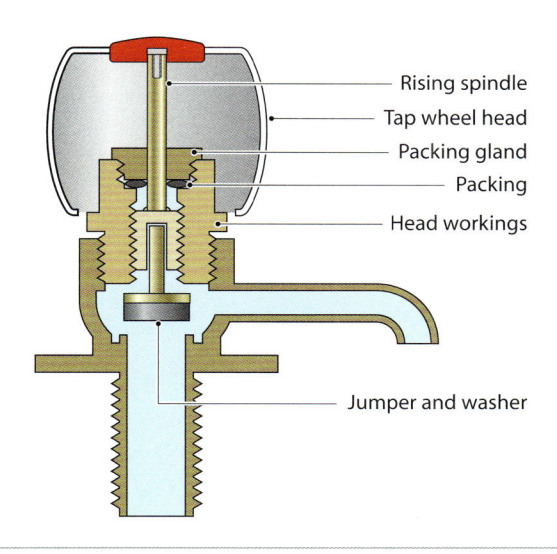

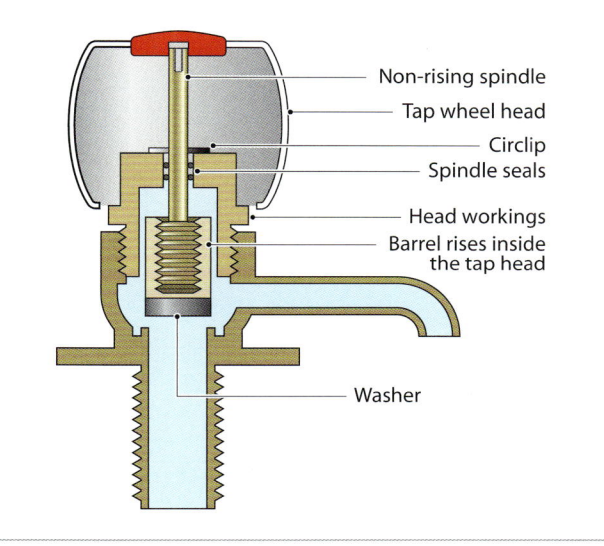

BS 1010 pillar tap

BS 5412 pillar tap

Taps with a non-rising spindle to BS 5412

Unlike BS 1010 taps, these taps do not have a rising spindle. Instead, the spindle has a thread at the end, which lifts a hexagonal barrel with a rubber washer attached, inside the valve-head workings. The spindle is fixed in the head workings by a circlip.

There are many different styles and types of BS 5412 taps and each manufacturer has their own style of conforming to the British Standard. The result of this is that very few of the head workings are interchangeable between manufacturers.

BS 5412 taps are available as pillar taps for wash basins and bidets, high-necked pillar taps for kitchen sinks, mixer taps for baths (¾-in thread) and kitchen sinks (½-in thread), monobloc mixer taps and bib taps.

SmartScreen Unit 006 animation 14

BS 5412 tap head workings showing the circlip

Ceramic disc taps

Unlike washer-type taps, ceramic disc taps use two thin, close-fitting, slotted ceramic discs in place of rubber washers. One of the discs is fixed while the other is turned by the handle of the tap a quarter of a turn through 90 degrees.

Ceramic disc tap heads are 'handed'. This means there are specific hot tap head workings that turn to the left and specific cold tap head workings that turn to the right, and they are usually colour coded for easy identification (red for hot, blue for cold).

Ceramic disc taps are not universal. If replacement head workings are required during maintenance operations, the correct type for the make of tap will be needed.

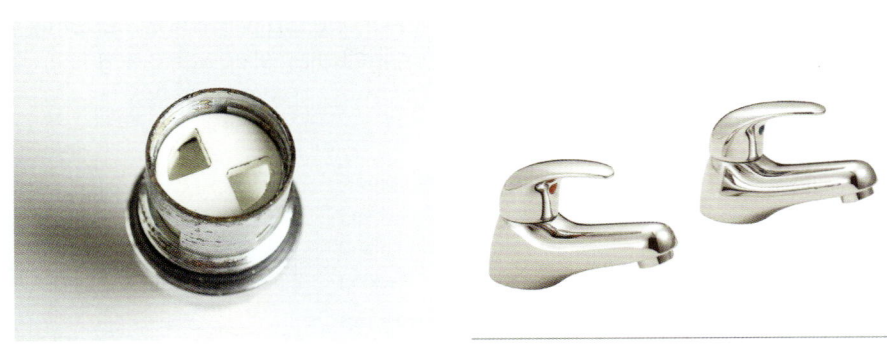

SmartScreen Unit 006 animation 15

Ceramic discs

Ceramic disc taps

D4 SHOWER MIXING VALVES

These mix water from cold water and hot water installations and discharge the mixed water from a shower head. The requirement of all shower valves is that they blend hot and cold water to the required temperature. They come in four types:

- manual mixing valves
- venturi boost mixing valves
- pressure compensating mixing valves
- thermostatic mixing valves (wax capsule type or bi-metal coil type).

Shower valves are available in three styles:

- **exposed, surface-mounted valves** – mounted on the surface, generally with concealed pipework

- **concealed valves** – the valve and all the pipework are concealed with only the controls on show

- **bar valves** – a recent addition, an exposed-type shower valve designed to be thin and modern-looking.

Externally, all mixing valves appear very similar in style and most have common distances of 150mm between the hot and cold connections. The difference is in the internal workings of the shower.

Manual mixing valves

Manual mixing valves do not have thermostatic control. They rely wholly on the hot and cold supplies having balanced pressure and flow rate. Once the temperature of the blended water has been adjusted, the setting remains fixed and does not adjust to fluctuations in flow rate, pressure or temperature. For this reason, the temperature of the hot water needs to be stable.

Although manual mixing valves can be used on high-pressure supplies, they are best suited to low-pressure installations to avoid pressure fluctuations. They are not recommended for fitting to systems that contain instantaneous water heaters or combi boilers.

Venturi boost mixing valves

The venturi mixing shower valve is specifically designed for installations that do not have balanced hot and cold supplies, such as mains fed cold water and low-pressure hot water. For the valve to work correctly, the mains cold water must have a pressure of at least 1 bar and a maximum pressure of 3 bar. Pressures in excess of this will require a pressure-reducing valve to be fitted.

The venturi mixing valve uses the extra pressure of the cold water supply to increase the pressure of the hot water supply by using the Bernoulli effect (see Chapter 004, page 185). The operating principle is as follows:

As the cold water passes though the venturi tube within the valve, its velocity increases and its pressure is slightly reduced. At this point, the hot water is drawn into the cold supply and mixed. As the mixed water leaves the venturi, the pressure reverts to almost as high as the initial cold supply, giving a fairly powerful shower.

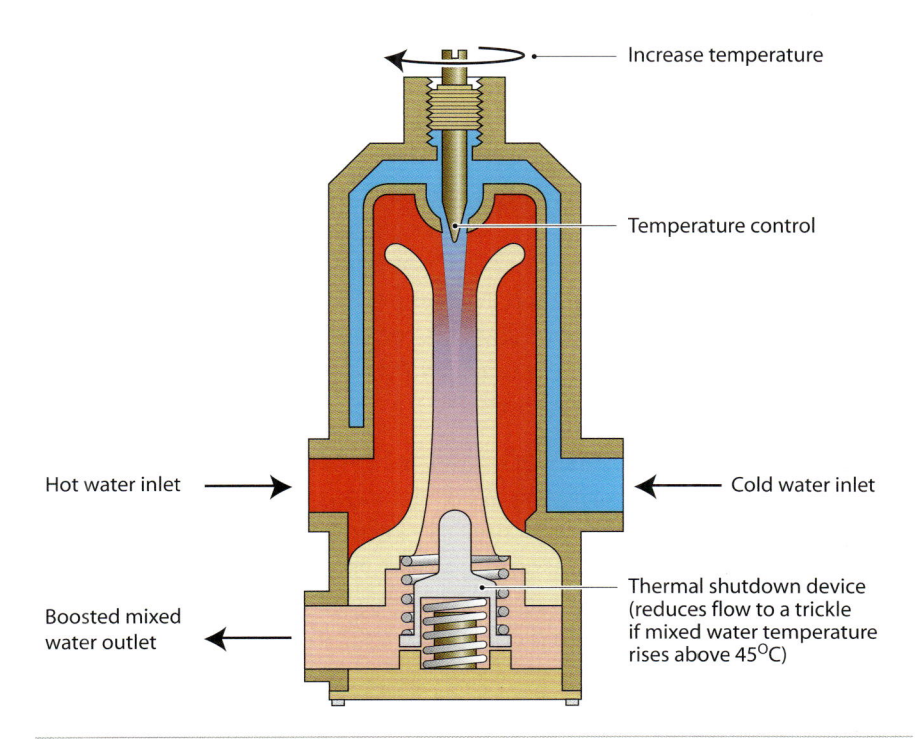

The working principles of a venturi boost shower mixing valve

Pressure-compensating mixing valves

This type of mixing valve gives a greater temperature stability compared to manual mixing valves. Some valves can be used on both high- and low-pressure systems, while others are specifically designed for high-pressure system use.

These valves are manufactured with either:

- **Sequential control** – starting the shower at a low temperature and progressively turning the control towards the hot gradually increases the temperature and maintains a steady flow rate. When the temperature is set, a balancing diaphragm reacts to subtle changes in water pressure and maintains the correct hot/cold mix.

- **Dual control** – these have a separate flow control and temperature control mechanism. The temperature control mechanism consists of a metallic shuttle that moves backwards and forwards inside a plastic mixing tube. The hot and cold water is regulated as the water flows through the tube. If there is a drop in pressure on either supply, the shuttle is moved inside the mixing tube, increasing the flow on the reduced pressure side and decreasing the flow on the opposite supply. This maintains an even showering temperature when pressure fluctuations occur. They do not, however, react to changes in temperature.

Pressure-compensating shower mixing valve

Thermostatic mixing valves

Thermostatic mixing valves give the best overall temperature control of all the shower valves currently available. In most cases, the maximum temperature is preset by the manufacturer, with a manual override for the end user. Incoming hot water temperature averages about 55°C and the cold supply at 15°C, giving a showering temperature of between 38°C and 42°C. There are two different types:

- **Wax capsule type** – a copper capsule containing a mixture of fine metal particles and a heat-sensitive wax is positioned in the mixing chamber of the valve. The wax expands with heat. As the wax expands, it forces a metal piston to activate a shuttle, which effectively controls the flow of hot and cold water into the valve by restricting the flow rate of the hot and cold water. If the temperature of the hot water is very high then the hot flow is restricted allowing more cold water into the valve to compensate for the high temperature. When a cooler shower is required, then the reverse happens and the shuttle moves backwards as the wax contracts, aided by a spring pushing against it.

- **Bi-metal coil type** – this works on the bi-metallic coil principle, where two metals with differing expansion rates are bonded together. When heat is applied, the two metals expand, but one faster than the other, causing the metal coil to distort. In the case of a shower valve, the bi-metallic coil is fastened at one end to a shuttle, which controls the inflow of hot and cold water to the mixing chamber of the valve.

Wax capsule

Bi-metal coil

Shower pumps

Low-pressure shower valves can have boosted hot and cold supplies by the use of shower boosting pumps. There are two types available:

- **Single impeller outlet pump** – this type of pump is installed after the mixing valve to boost the mixed water to the shower head.
- **Twin impeller inlet pump** – this is fitted before the mixing valve and boosts the individual hot and cold supplies to the valve where the water is mixed. It has a single electric motor that drives two impellers (hot and cold).

The installation of showers and shower pumps is covered in more detail in Chapter 007.

It should be remembered that shower pumps may not be fitted on mains cold water installations. The Water Supply (Water Fittings) Regulations 1999 prohibit the use of pumps on mains cold water, except when special permission has been granted by the water undertaker.

Single impeller shower pump

Twin impeller shower pump

D5 SCALE REDUCTION AND WATER TREATMENT IN DOMESTIC PROPERTIES

As well as the treatment given to the water by the water undertaker, many domestic properties, especially in hard water areas, employ alternative methods to condition the water so that scaling problems do not occur.

Scaling occurs in hot and cold water systems and central heating installations when the water contains salts and minerals such as calcium carbonate that re-form in the water as a hard limescale that sticks to the inside of pipes and appliances. This process is known as precipitation. The resulting limescale reduces the appliance's efficiency and can, in some cases, make it unusable.

There are several methods used to prevent precipitation from occurring, including the use of:

- water conditioners
- water softeners
- water filters.

Water conditioners

These alter the chemistry of the precipitation process by suppressing limescale formation, thereby reducing the rate of scaling. Their benefits include:

- reduction in scale formation in pipes
- reduction of limescale on taps
- easier cleaning of shower heads and places where limescale may form.

There are many different types of water conditioner, using a wide variety of conditioning methods, including:

- **Magnetic** – these prevent scale build-up by influencing the type of calcium crystals precipitated, which ensures that only needle-like

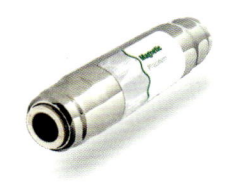

Magnetic water conditioner

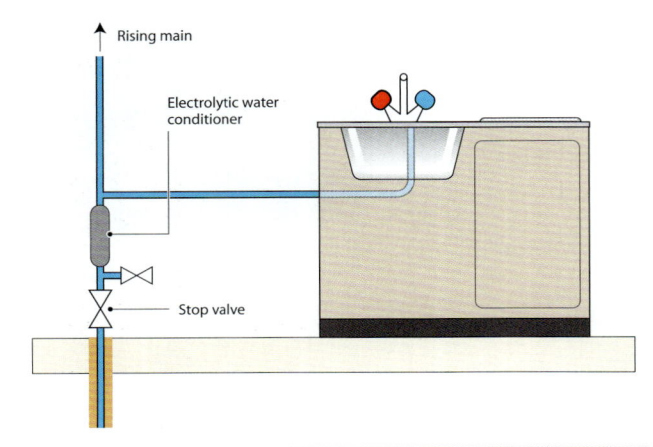

Electrolytic conditioner

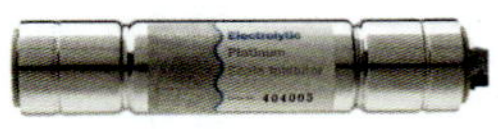

Electrolytic water conditioner

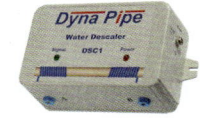

Electromagnetic water conditioner

aragonite crystals are formed. These are less able to stick to smooth surfaces than the normal calcium crystals. These are for individual appliance protection only, such as a combi boiler. They are installed on the cold main to the appliance.

- **Electrolytic** – these add a minute amount of zinc to the water, which suppresses the formation of calcium crystals. Any crystals that are formed are washed away by the flow of water. They can be used for whole-house protection.

- **Electronic** (electromagnetic) – these cause dissolved hardness salts and minerals to cluster together rather than forming on surfaces.

- **Electrochemical** – these conditioners contain a cartridge filled with ceramic beads, which cause the magnesium and calcium crystals to precipitate. The conditioner units are usually quite large and require an electrical supply.

Water softeners

A water softener is an appliance that is fitted directly to the water supply to a domestic dwelling or a commercial building and is specifically designed to remove water hardness. Water softeners are usually installed as close to water main entry into the building as possible. Most modern softeners are very compact and can easily be fitted under a kitchen sink.

Water softeners use a process called ion exchange. The softener contains a column, which is filled with special resin beads. The beads remove the dissolved calcium and magnesium salts by replacing them with sodium as the water passes through them. Once a day, the unit automatically washes the beads with brine (salt water) to remove calcium and magnesium ions, taking the solution to drain. Every month, the unit has to be refilled with salt in the form of granules, tablets or blocks. Use of a water softener generally reduces the hardness of the water from 350mg/l (milligrams per litre) to less than 10mg/l.

When installing a water softener, there must remain at least one unsoftened cold water outlet in the dwelling (the kitchen draw-off).

Drinking water filters

Drinking water filters alter the water composition to improve its taste, odour and appearance for drinking and cooking purposes. There are two common types:

- **jug filters** – filled from a tap and stored in a fridge
- **plumbed-in filters** – usually sited underneath the kitchen sink with a separate drinking water tap installed at the kitchen sink.

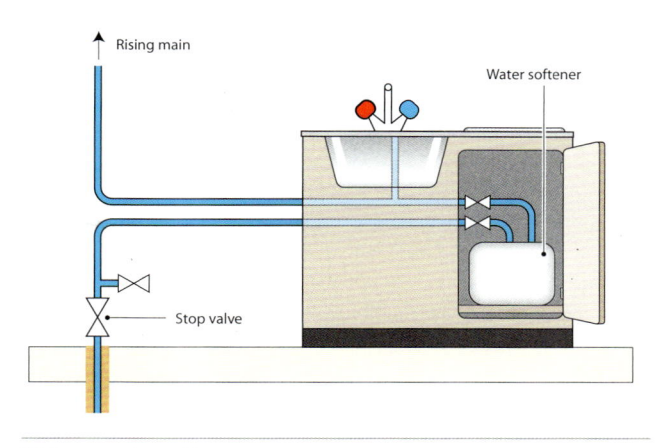

Drinking water filters can usually be supplied in six different forms:

- **Activated carbon filter** – used to reduce taste and odour such as chlorine. The carbon filter has a large surface area that attracts and absorbs organic substances from the water. The carbon is usually in powder, granular or block form.
- **Ion exchange** – used to reduce limescale formation and other metal-ion contaminants such as lead. It takes the form of tiny granules that work by replacing the mineral or contaminant ions with hydrogen ions.
- **Sediment filter** – designed to remove fine particles from the water. These filters comprise a mesh through which the water passes, trapping the sediment. The smaller the holes in the filter, the smaller the particles that can be removed.
- **Reverse osmosis** – these type of filters work under pressure to remove most of the dissolved mineral content by passing the water through a very fine membrane.
- **Distillation** – this process removes the mineral content of the water by boiling it and condensing the steam back to water vapour.
- **Disinfection** – this process is used to reduce the bacteria content and other microorganisms either by ultraviolet light or a using very fine sediment filter (usually ceramic or membrane).

Fitting a water softener

Installation of a water filter

E INSTALLATION OF PIPEWORK

Many of the requirements for pipework installation were covered in Chapter 005. This section looks at techniques specific to cold water installations.

E1 CHOOSING THE RIGHT MATERIALS

Cold water supply in domestic dwellings is strictly regulated by the Water Supply (Water Fittings) Regulations 1999. This means that the choice of materials for cold water installations is limited to:

- **Copper tubes and fittings** – (see Chapter 005, pages 222–39). Copper has a proven record for cold water installations. It is light, rigid, has many jointing techniques available and requires only minimal clipping. It is highly resistant to corrosion and has a minimum life, in ideal conditions, of 50 years. It does, however, take great skill to fabricate and install it properly. There is a fire risk when using soldering equipment and it requires many specialist tools to successfully complete an installation.

- **Polybutylene pipe and fittings** – (see Chapter 005, pages 250–53). Manufacturers state that polybutylene has a life expectancy of 50 years. It is light and extremely flexible and requires regular clipping when fixed on the surface. It is easy to install and can be cabled through joists easily and quickly. Push-fit joints make installation quick; total time can be reduced by 40 per cent with no fire risk. Soundness testing techniques are more complicated and time-consuming than for copper tube.

E2 PREPARATION, PLANNING AND POSITIONING OF PIPES

The installation of cold water systems needs to comply with the Water Supply (Water Fittings) Regulations 1999 and you should also always consider the recommendations of BS 6700. Manufacturer's instructions must be followed with regard to the appliances installed and materials used. Design flow rates and operating pressures will need to be considered at the system design stage for any installation to operate effectively.

The installation procedures will vary depending on the property. For instance, the methods used on new buildings will differ from an occupied dwelling where the customer's possessions will need to be taken in to account.

Irrespective of the property type, pipework runs need to be planned carefully. It is advisable to avoid positions where frost and heat could cause a problem, such as on outside walls, and in cellars and unheated roof spaces. Wherever possible, pipework should be positioned out of sight and boxed in where appropriate.

Pipes in suspended timber floors

Notching and drilling of joists should be done carefully, following the advice in Chapter 005 (see pages 262–64). Notching or drilling of joists should not be carried out on joists or rafters 100mm deep or less. Notches should not be too tight for the pipes, or creaking and 'ticking' noises may become a problem as the pipes expand and contract. Pipes in notches should be covered with joist clips to prevent excessive movement and floorboards should be screwed (not nailed) when they are repositioned.

There are many different styles of suspended floors, including engineered timber joists, laminated strand beam, parallel strand beam and concrete block-and-beam systems. The diagram opposite shows the installation requirements for some of these systems.

Cold water pipes should not be installed in the same notch as hot water and central heating pipes. There must, where possible, be a minimum horizontal distance of 300mm between cold water pipes and any hot water/central heating pipes to prevent radiated heat from warming the cold, wholesome water. Where there is a significant risk of cold water pipes being warmed by other pipework, the cold water installation should be insulated. To eliminate the risk of contamination from undue warming, the cold water pipework must never be allowed to exceed a temperature of 25°C.

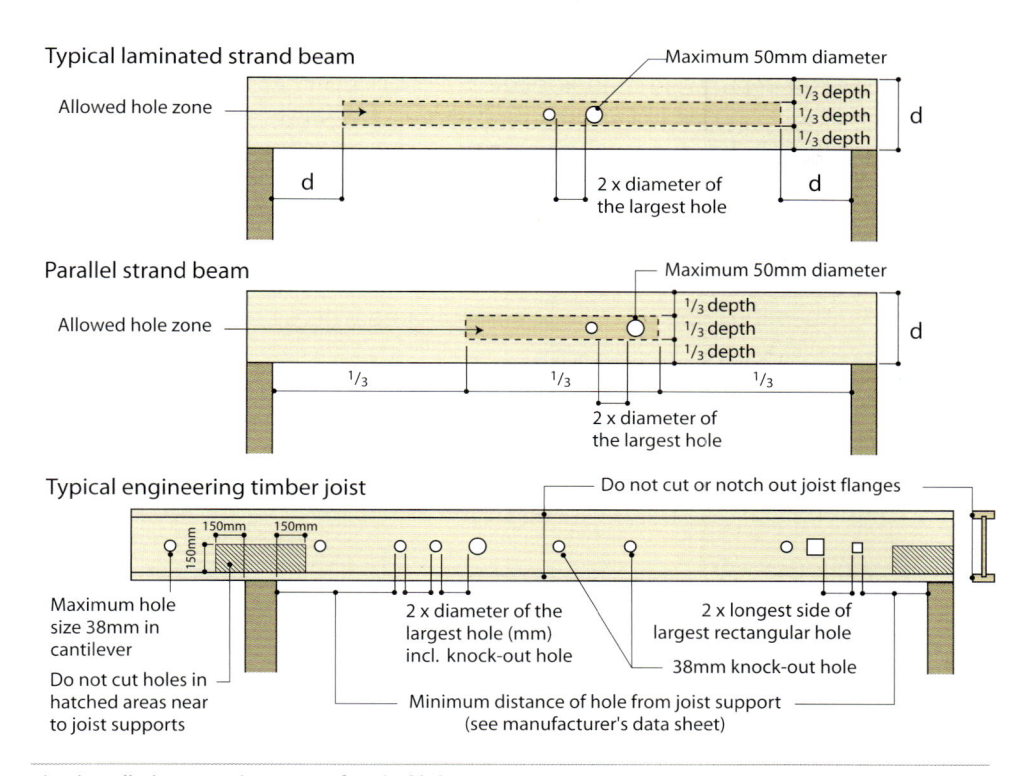

Typical laminated strand beam
- Allowed hole zone
- Maximum 50mm diameter
- 1/3 depth, 1/3 depth, 1/3 depth
- d
- 2 x diameter of the largest hole

Parallel strand beam
- Allowed hole zone
- Maximum 50mm diameter
- 1/3 depth, 1/3 depth, 1/3 depth
- d
- 1/3, 1/3, 1/3
- 2 x diameter of the largest hole

Typical engineering timber joist
- 150mm, 150mm, 150mm
- Do not cut or notch out joist flanges
- Maximum hole size 38mm in cantilever
- Do not cut holes in hatched areas near to joist supports
- 2 x diameter of the largest hole (mm) incl. knock-out hole
- Minimum distance of hole from joist support (see manufacturer's data sheet)
- 2 x longest side of largest rectangular hole
- 38mm knock-out hole

Pipe installations requirements of typical joist systems

Pipes in walls and sleeved through walls

Pipes laid in walls should be accessible. The Water Regulations Guide states that:

> 'Unless they are located in an internal wall which is not a solid wall, a chase or duct which may readily be removed or exposed, or under a suspended floor which may be readily removed and replaced, or to which there is access, water fittings shall not be:
>
> - located in the cavity of a cavity wall
> - embedded in any wall or solid floor
> - installed below a suspended or solid floor at ground level.'

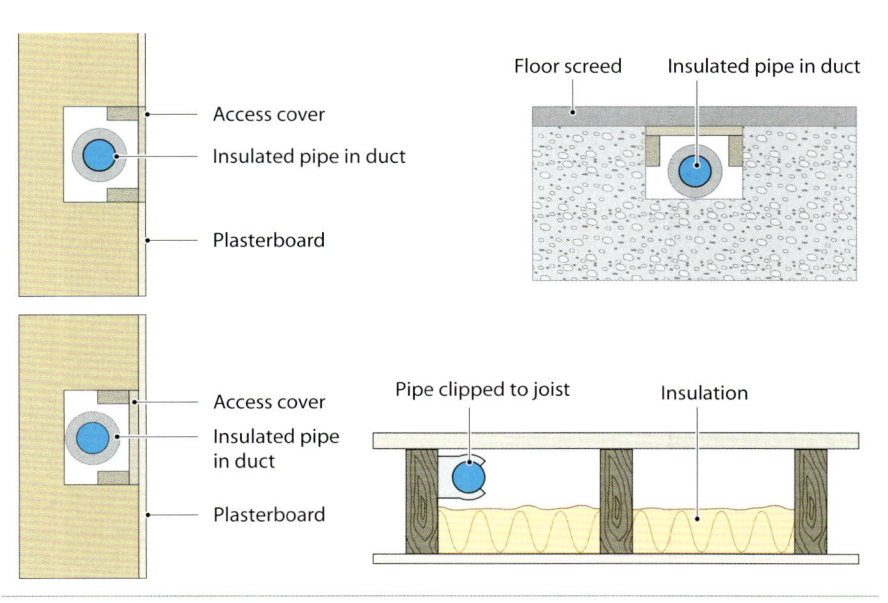

- Access cover
- Insulated pipe in duct
- Plasterboard

- Floor screed
- Insulated pipe in duct

- Access cover
- Insulated pipe in duct
- Plasterboard

- Pipe clipped to joist
- Insulation

Pipes in walls and floors

Where the laying of pipes in walls and floors is unavoidable, they should be placed in purpose-made ducts that have an accessible, removable cover. Pipes laid in chases must have adequate room for expansion and contraction and should be sufficiently insulated or protected.

Pipes passing through walls should be sleeved to allow for expansion and to protect the pipe from building settlement and the corrosive effects of the masonry on the pipe. The sleeve should be sealed at both ends. The pipe should be thermally insulated where necessary.

Preparing to install

Chapter 005 discussed taking care of the customer's property and possessions during the installation process, and the use of various methods to protect the customer's environment. There are also other steps you can take before you start the installation, to save time:

- **Walk the job** – take the time to walk around the job and plan the routes that you intend your pipework to take. Do a visual risk assessment for the different activities involved and make sure that you have the correct PPE and equipment available.

- **Prepare the job** – use this time to lift floorboards and cut notches in preparation for the pipework installation. The floorboards can be replaced temporarily so that the customer is not inconvenienced by holes in the floors. Remember to clear any mess as you go along and always ensure that the customer's property is protected with dust sheets. Don't leave it all until the last minute and never leave cleaning to the customer.

- **Mark out** – if you have decided on the routes that you intend to take, mark out the clip positions of any surface-mounted pipework and drill any holes that you need to drill. Chases in walls and floors can also be marked at this point. Don't forget to use the correct PPE, such as protective goggles, when carrying out drilling and chasing procedures.

- **Keep the customer informed** – let the customer know where you are going to be working and how long you plan to be in that area.

Keep entrances and exits clear at all times and don't leave trip hazards such as cables and tools lying around the work area.

E3 INSTALLATION, TESTING AND COMMISSIONING OF COLD WATER SYSTEMS

The fabrication of pipework, installation techniques, commissioning and testing were discussed in Chapter 005 (see pages 265–72). However, these important subjects require reinforcement. The key aspects to remember for cold water installations are as follows:

- Keep all exposed pipework as neat as possible. Use the recommended clipping distances and protect the building fabric when making soldered joints.

- Prefabrication techniques for copper tube can save time and money on installations. Try to use machine bends wherever possible as these help with the flow rates in the finished installation.

- If the installation is an existing system, leave the final connections to the system until last. This will help to keep the decommissioning and turnover time of the system as short as possible.

- Pipework installed in floors and walls should be placed in properly prepared and accessible chases and ducts. Protect the customer's property at all times with dust sheets when cutting in chases.

- Hot and cold pipework should not be installed together unless the cold water can be protected from undue warming from the other surrounding services. If possible, when pipework is to be fixed on wall surfaces, the hot water pipework should be installed above the cold water pipework, and when installed in a floor cavity, a gap of at least 300mm horizontally should be maintained.

- Cisterns should be marked and drilled for pipe connections in accordance with BS 6700, and all holes drilled with a holesaw. Installation requirements should be in accordance with the Water Supply (Water Fittings) Regulations 1999.

- Holes and notches in joists must be carried out in line with the Building Regulations.

Testing

Before testing takes place, walk around the job and check that all joints have been made correctly, that there are a sufficient number of pipe supports and clips and that you are happy that the installation conforms to the Regulations. Any problems, such as insufficient clipping of pipes and missing or incorrectly installed service valves, should be rectified before testing begins. Close any open ends of pipes with cap ends.

Pressure testing of the completed installation will depend on the materials used:

- **copper tubes** – testing as detailed in BS 6700 and on page 270.
- **plastic (polybutylene PB-1)** – this will depend on which test is being performed. The requirement for both test A and test B are detailed in BS 6700 and on pages 270–71.

Testing should be performed using a hydraulic pressure testing pump.

Hydraulic pressure testing pump

SUGGESTED ACTIVITY...
Refresh your memory of testing procedures by rereading Chapter 005, pages 270–72.

Commissioning cold water systems

The commissioning and flushing procedure should be undertaken with fresh wholesome water direct from the water undertaker's main.

- Check that all pipework is secure, all taps and tank connectors are fully tightened and all drain-off valves are turned off.

- Check the inside of any cisterns installed to ensure that they are free of debris and that all connections are tight. Ensure that all isolation valves and terminal fittings are off.

- Open the kitchen cold tap and slowly open the mains cold water stop valve. Allow the water to flow into the kitchen sink to clear any debris that may have collected in the pipework.

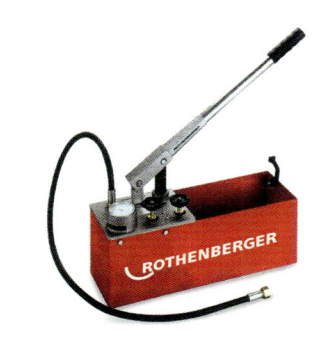

- Close the cold tap on the kitchen sink and allow the system to fill to full standing pressure.

- Turn on the cold taps one at a time until the water runs clear, and check for leaks.

- Turn on the isolation valves to the FOV in the WC cisterns and allow the cistern to fill to the water line. Adjust the water level as necessary. Flush the WC and check for leaks.

- Fill any cisterns in the roof space and adjust the water level at the FOV as necessary.

- Open any taps and terminal fittings fed from the cistern and clear any air in the system. Allow the water to run to clear any debris.

- Allow the system to stand and check for any leaks throughout the system.

- Isolate at the mains cold water stop valve and completely drain the system to flush it through. This should clear any flux residue and swarf from the system.

- Refill the system and test for standing and running pressure at all mains outlets using a pressure gauge.

- Check that all flow rates meet the specification and any manufacturer's instructions using a flow meter or a weir gauge.

- Recheck the system for leaks.

Commissioning cold water storage cisterns

When commissioning cisterns you should go through a visual inspection to ensure that:

- there is adequate support to the cistern base
- the lid is correctly fitted
- the cistern is sufficiently insulated
- the FOV and all pipework is correctly positioned
- a service valve is fitted and is turned off ready for commissioning
- all filters are fitted to the overflow and air vent
- the vent pipe is sealed with a grommet.

Conduct soundness testing to check that:

- all connections to the cistern have been correctly made and are watertight
- no leakage is visible.

Next, flush and disinfect to BS 6700:

- Ensure that any debris in the cistern is removed before turning the water on, and remove any debris that may get washed into the system.

Pressure gauge

SUGGESTED ACTIVITY...
Check your knowledge of commissioning systems by referring to Chapter 005, page 271.

Weir cup and thermometer

SUGGESTED ACTIVITY...
Remember! Water pressure is measured in bar pressure. 1 bar is the equivalent of 10 m head of water or 100kPa.

Now, using the above figures, calculate:
a 38 m in bar pressure
b 4.5 bar in kilopascals (kPa)
c 150kPa in bar pressure
  (Hint: check back to chapter 004, page 183)

Answers: a. 3.8 bar, b. 450kPa, c. 1.5 bar

SmartScreen Unit 006 worksheet 3

- Fill the cistern with wholesome water from the water main and flush through the installation to remove any flux residue.
- For larger installations the system should be disinfected as described in BS 6700 (not applicable for domestic systems).
- Adjust the FOVs to the correct shut-off level.

Conduct performance testing:

- Check that all taps and valves are operating correctly.
- Ensure that all outlets are served with water.
- Ensure that pressures and flow rates are checked with a flow meter/pressure meter and a weir cup to confirm that all appliances fitted comply with the manufacturer's flow rate and pressure specifications.

E4 EXISTING INSTALLATIONS

Existing systems can be notoriously difficult to work on and the older the system, the more difficult it can be. Over the years, there have been a variety of materials used for the installation of cold (and hot) water systems and each of them brings its own unique set of problems when jointing new pipework into it:

- **Lead pipes** – there are still hundreds of installations that contain lead pipe and there are situations where making a joint on lead pipe is unavoidable. Joints using leaded grade D solder were banned in 1986. This means that only proprietary joints, such as leadlocks and Philmac fittings (see page 237) can be used to joint the old lead pipe to workable copper tubes or polybutylene pipes. Even so, you must still exercise caution, as brass fittings such as leadlocks can cause galvanic corrosion to occur (see page 162), which could lead to water contamination downstream of the fitting. Wherever possible, lead pipe should be removed and replaced.

- **BS 659 copper tube** – this type of copper tube was introduced in the 1950s and has a much thicker pipe wall compared to modern copper tube. Jointing techniques were very similar to those of today with both compression and capillary fittings being used. However, the tube sizes are imperial and so converters are required for some sizes. ½-in tube will fit modern 15mm tube, although it is a tight fit. ¾-in tube is much smaller than modern 22mm tube and so must be converted, and 1-in tube is extremely tight when used with 28mm fittings, so a converter fitting is also recommended. Both capillary and compression converter fittings are available.

- **Red band thin wall copper tube** – this kind of copper tube is identifiable by a red line running down the length of the tube and is mostly of German origin. It was used in the early 1970s when copper tube was scarce due to a copper shortage. It is very susceptible to pinhole corrosion. Only capillary joints should be made on this type of tube. Tube sizes are imperial.

- **Stainless steel** – stainless steel tube was used extensively in the early 1970s due to a copper shortage. Unfortunately, the tube was

---

SUGGESTED ACTIVITY...

p.57: Water flow rate is measured in litres per second or litres per minute. To convert from litres per second (l/s) to litres per minute (l/m) simply multiply the l/s by 60, eg

0.3l/s = 0.3 × 60 = 18l/m

To convert from litres minute per (l/m) to litres per second (l/s) simply divide the l/m by 60, eg

25 l/m = 25 ÷ 60 = 0.41l/s

Now try the following calculations:

a 30 l/m into l/s
b 0.25 l/s into l/m
c 12 l/m into l/s
d 0.12 l/s into l/m

Answers: a. 0.5 l/s, b. 15 l/m, c. 0.2 l/s, d. 7.2 l/m

SmartScreen Unit 006 interactive activity 3

manufactured from low-grade stainless steel and this has led to many problems of corrosion. Compression joints can be made onto this type of tube but care should be taken; it requires harder tightening because stainless steel is a much harder metal than copper. Tube sizes are imperial.

- **High density polyethylene (HDPE)** – HDPE was used for underground service pipes from the external stop valve (boundary stop valve) to the dwelling. It is black in colour and comes in four grades (A, B, C and D). Compression fittings are still available for this type of pipe but it should be noted that the grades have different wall thicknesses and so it is important that the correct type of pipe insert is used when making joints. Conversion to blue MDPE is a fairly simple task when the correct fitting is used.

- **Chlorinated unplasticised polyvinyl chloride (cuPVC)** – better known as 'PolyYork', this is a plastic pipe that is suitable for cold water supplies only. It was used extensively in some parts of the UK during the early 1970s for cold water systems inside domestic dwellings. Fittings used a solvent cement system that, once a joint was made, had to be left for 24 hours before testing could take place. It is very susceptible to fracture and fitting blow-off. Care should be taken when this pipe is encountered as it is extremely easy to fracture a fitting just by turning the water supply off!

- **Acorn (polybutylene)** – this is an early version of polybutylene pipe that first appeared in the mid 1980s. It is compatible with all new polybutylene pipes and fittings and copper tubes and compression fittings. However, a special pipe insert is required.

F MAINTENANCE OF COLD WATER SYSTEMS

Maintenance tasks on cold water services, appliances and valves are essential to ensure the continuing correct operation of the system. The term used when isolating a water supply during maintenance operations is 'temporary decommissioning'. As a plumber, you will deal with emergency maintenance calls, as well as scheduled maintenance designed to prevent emergencies from arising in the first place.

F1 PLANNED PREVENTATIVE MAINTENANCE

Planned preventative maintenance is usually performed on larger systems and commercial/industrial installations. It is performed to a prearranged maintenance schedule, which may mean out-of-hours working if the supply of water cannot be disrupted during normal working hours. It is designed to stop problems from occurring by catching faults in their early stages. Planned preventative maintenance could include:

- periodic system inspection – checking for leaks
- re-washering of FOVs
- re-washering and reseating of terminal fittings and taps

- inspection and cleaning of cisterns
- readjustment of water levels in cisterns
- re-washering of drain valves
- cleaning of filters and strainers
- maintenance of water softeners
- checking correct operation of stop valves
- checking flow rates at all outlets.

When a maintenance task involves isolating the cold water supply, a notice will need to placed at the point of isolation stating 'System off – do not turn on' to prevent accidental turn-on of the system. In most systems, it will be possible to isolate specific parts of the installation without the need to have the whole supply turned off. Where no such isolation exists, it may be of benefit to use a pipe freezing kit so that total system isolation is not undertaken.

A record of all repairs and maintenance tasks completed will need to be made on the maintenance schedule at the time of completion, including their location, the date when they were carried out and the type of tests performed. This will ensure that a record of past problems is kept for future reference.

Where appliance servicing is carried out, the manufacturer's installation and servicing instructions should be consulted. Any replacement parts may be obtained from the manufacturer.

Do not forget to keep the householder/responsible person informed about the areas that are going to be isolated during maintenance tasks and operations, and always ask customers if they need to draw off a temporary supply of water (kettle, saucepans, bucket, etc) for use during the short period of system isolation.

F2 UNPLANNED/EMERGENCY MAINTENANCE

Unplanned maintenance and emergency maintenance occur when a fault suddenly develops, such as a burst pipe, or when a small problem suddenly becomes a larger issue, such as a dripping tap or sudden loss of water. Unplanned and emergency maintenance can include repairs to:

- burst pipes and leaks
- running overflows
- dripping taps
- loss of low-pressure, cistern fed cold water supply due to faulty FOVs
- poor past installation practices, such as incorrectly positioned overflow pipes.

Many maintenance practices involve the decommissioning of systems so that parts and pipes can be replaced.

F3 SIMPLE MAINTENANCE TASKS

Here, some of the basic maintenance tasks you may have to perform on cold water system components, such as taps and FOVs, are covered.

Re-washering and reseating a BS 1010 rising spindle tap

BS 1010 taps are probably the easiest of all taps to maintain. During the maintenance operation, taps should be reseated as well as re-washered. This involves using a special tool (a tap reseating tool), which grinds the seat of the tap to remove any pits that have occurred due to water passing between the seat and the tap, ensuring that the washer sits evenly on the tap seat.

Start by ensuring that the water supply is isolated and then opening the tap to relieve the pressure. Put the plug into the sink. This will ensure that any dropped small screws and nuts do not disappear down the sink waste and into the waste pipe trap.

A tap reseating tool

STEP 1 – Locate the screw that holds the tap head onto the spindle and carefully remove with a small screwdriver.

STEP 2 – Carefully remove the tap head. Many BS 1010 taps are cross-top heads, which can prove difficult to remove. Take care to prevent damage to the appliance that the tap is fixed to.

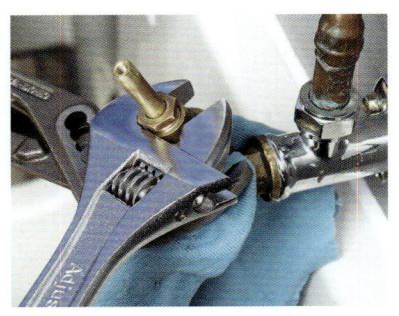

STEP 3 – With the head removed, break the joint between the tap head workings and the tap body using an adjustable spanner. This may involve using a pair of water pump pliers to counteract the force of the adjustable spanner on the head workings. Ensure that a cloth is used to protect the tap body from the effects of the jaws of the water pump pliers.

STEP 4 – Remove the jumper plate and washer from the spindle. A little force may be needed from the flat blade of a screwdriver if the jumper plate is fixed. Some rubber tap washers are held onto the jumper plate by a small brass nut.

STEP 5 – Carefully remove the nut and replace the existing rubber washer with a new rubber washer of the correct size, then replace the washer nut. Do not over-tighten the washer nut as it may break.

STEP 6 – Remove the packing gland nut and the spindle by fully winding in a clockwise direction and pushing the spindle through the packing gland.

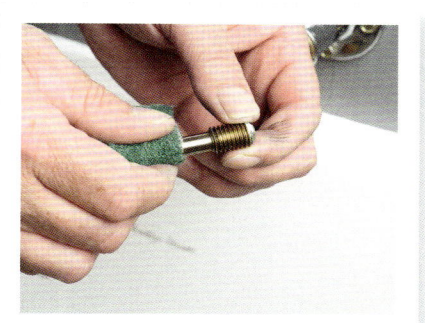

STEP 7 – Check the spindle for any signs of wear and remove any scale that may have gathered on the spindle shaft. A non-metallic fittings cleaning pad is ideal for this. Re-grease the spindle using silicone grease.

STEP 8 – Push the spindle back through the packing gland and wind until the tap spindle is in the fully open position.

STEP 9 – Check the packing in the packing gland and replace with a PTFE grommet where necessary. Squeeze a small amount of silicone grease into the packing gland before replacing the packing gland nut. Do not tighten the nut at this stage.

STEP 10 – Reinsert the jumper plate into the spindle.

STEP 11 – Check the seat of the tap by shining a torch into the tap body. If the tap requires reseating, use the tap reseating tool with the correct size of grinding head and reseat as necessary.

STEP 12 – Check the fibre sealing washer on the head workings. These tend to break when the tap head is removed. If the fibre sealing washer needs replacing, this can be done using PTFE tape.

STEP 13 – Replace the head workings into the tap body (ensuring the head workings are fully open) and retighten into the tap.

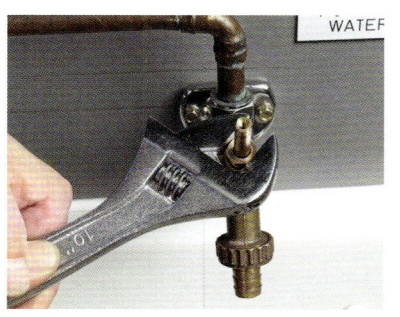

STEP 14 – Tighten the packing gland nut, taking care not to over-tighten or the tap will be difficult to open.

STEP 15 – Replace the tap head but do not secure with the screw at this point. Turn on the water with the tap open. This will ensure that any debris from reseating will be washed out of the tap. Turn off the tap and check for any drips. Finally, replace the tap head securing screw.

Re-washering and reseating a BS 5412 non-rising spindle tap

BS 5412 taps have a non-rising spindle (see page 322). Some of the problems that can occur with these taps include the following:

- The barrel, which rises inside the tap head workings, can become dislodged causing the tap to seize in the closed position, preventing the tap being opened. Often, this is a result of it being over-tightened, compressing the washer.

- The circlip, which holds the non-rising spindle in position, very often breaks.

There is no packing gland with BS 5412 taps and so maintenance is a little easier. Start by ensuring that the water supply is isolated and opening the tap to relieve the pressure. Put the plug into the sink. This will ensure that any dropped small screws and nuts do not disappear down the sink waste and into the waste pipe trap.

STEP 1 – Carefully remove the cap on the tap head to gain access to the screw.

STEP 2 – Locate the screw that holds the tap head onto the spindle and carefully remove with a small screwdriver. Some tap heads simply pull off the spindle.

STEP 3 – With the head removed, break the joint between the tap head workings and the tap body using an adjustable spanner. This may involve using a pair of water pump pliers to counteract the force of the adjustable spanner on the head workings. Use a cloth to protect the tap body from the effects of the jaws of the water pump pliers.

STEP 4 – Fully unwind the spindle until the hexagonal barrel can be removed from the head workings.

STEP 5 – Carefully remove the rubber washer and replace with the correct size of washer. A tap washer kit may be of benefit here, as there are many different sizes and styles of washer for a BS 5412 tap.

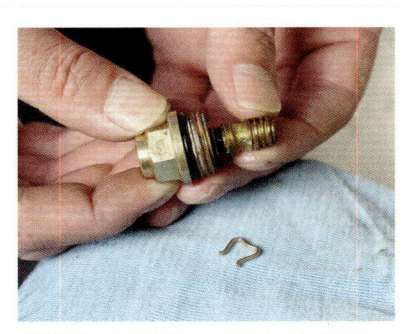

STEP 6 – Carefully remove the circlip with circlip pliers and push the spindle downwards and out of the head workings.

STEP 7 – Check and replace the spindle 'O' ring seals as necessary.

STEP 8 – Re-grease the spindle with silicone grease.

STEP 9 – Reinsert the spindle into the head workings and replace the circlip.

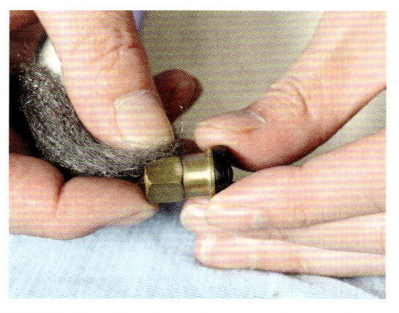

STEP 10 – Check the hexagonal barrel for any signs of scale, and clean with a cleaning pad as necessary.

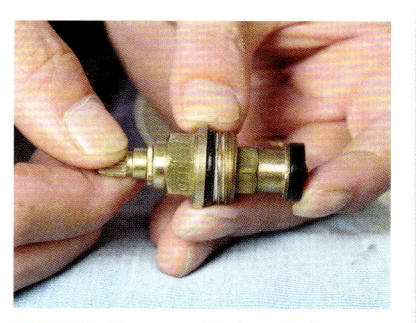

STEP 11 – Re-grease the barrel using silicone grease and very carefully rewind back into the head workings. Ensure that the tap head is in the fully open position.

STEP 12 – Check the tap seating and reseat using the tap reseating tool with the correct-size grinding head, as required.

STEP 13 – Check the rubber 'O' ring on the tap head workings. This washer seals the head workings to the tap body. Replace as required.

STEP 14 – Replace the head workings into the tap body (ensuring that the head workings are fully open) and retighten into the tap.

STEP 15 – Replace the tap head but do not secure with the screw at this point. Turn on the water with the tap open. This will ensure that any debris from reseating will be washed out of the tap. Turn off the tap and check for any drips. Finally, replace the tap head securing screw and cap.

006 COLD WATER SYSTEMS

Replacing a ceramic disc tap

A ceramic disc tap does not have a washer to replace. Instead, these taps use two very thin plates, or discs, of a ceramic material to allow water to flow through the tap. Most ceramic disc taps are not repairable. The tap head workings will need to be replaced with a like-for-like unit, which can be obtained from the manufacturer or from the local merchant or stockist. There are a wide variety of ceramic disc sets available and the correct one for the tap must be obtained. When ordering the part, the type of head workings, ie hot or cold, will need to be stated as they open and close in opposite directions.

Repairing a BS 1212 part 1 FOV (Portsmouth type)

Portsmouth-type FOVs (see page 318) may only be fitted on new installations if some form of backflow prevention device is installed before the FOV; usually this would be a double check valve (see page 313). However, if a Portsmouth valve is part of an existing installation, then repair is permissible. To repair a Portsmouth valve, follow the steps listed below:

1 Turn off the water supply to the FOV at the isolation valve.
2 Remove the FOV from the cistern by unscrewing the union nut.
3 Remove the end cap on the valve body.
4 Remove the cotter (split) pin holding the float arm to the valve body and remove the float arm.
5 Remove the piston from the valve body.
6 The piston is generally made from one of two materials – brass or nylon. For brass pistons, the FOV washer is held in the end of the piston by a retaining cap, which will need to be unscrewed to allow the washer to be removed. To remove the retaining cap:

   a Place a flat-blade screwdriver in the slot for the float arm and unscrew the retaining cap using a pair of pliers.
   b Remove the washer and replace with a like-for-like washer.
   c Replace the retaining cap and tighten.
   d Check the piston for any signs of scale and remove with a cleaning pad.

7 For nylon pistons, simply push the washer out of the gap in the side of the washer housing and replace the washer.
8 Remove the orifice from the FOV body and check to ensure that there are no cracks or splits visible. Replace as necessary.
9 Reassemble the valve making sure that the washer is towards the spindle.
10 Replace the cotter pin and open to ensure that it does not fall out.

11 Reinstall the valve into the cistern, making sure the fibre sealing washer is in place.

12 Retighten the union and turn on the water

13 Check the operation of the valve, adjusting the water level as necessary.

Repairing a BS 1212 part 2/3 FOV (diaphragm type)

Diaphragm-type FOVs discharge water over the top of the valve (see page 319). They have a large diaphragm-type washer that is easily accessible for repair and replacement. To replace the diaphragm washer, follow the steps listed below:

1 Turn off the water supply to the FOV at the isolation valve.

2 Remove the FOV from the cistern by unscrewing the union nut.

3 Unscrew the large washer-retaining union and float arm arrangement at the front of the valve and withdraw the washer.

4 Replace the washer, ensuring that it is fitted the correct way. These washers must be inserted correctly for the FOV to operate properly.

5 Replace the large washer-retaining union and float arm arrangement, ensuring that it is engaged into the retaining notch at the top of the front plate, and hand tighten the union.

6 Check that the orifice is in good order with no cracks or splits. Replace as necessary.

7 Reinstall the valve into the cistern, making sure that the fibre sealing washer is in place.

8 Retighten the union and turn on the water.

9 Check the operation of the valve, adjusting the water level as necessary with the float arm adjustment screw.

F4 DECOMMISSIONING OF SYSTEMS

Occasionally, systems will require isolation for repairs, renewal of appliances and extensions to systems, or when systems or appliances are being permanently removed. This is known as decommissioning. Decommissioning takes two forms:

- **Temporary decommissioning** – this is where systems are isolated for a period of time so that work can be performed. Eventually the system will be recommissioned and put back into normal operation.

- **Permanent decommissioning** – when a system or an appliance is taken out of use, it has to be permanently decommissioned. This will require the system to be isolated at the mains cold water stop valve (or the nearest isolation point) and then drained. The appliance(s) can then be removed and the pipework cut back, removed and capped with a stop-end to the nearest connecting branch point on the branch supply line to prevent stagnation of water in a live cold water supply.

G SYSTEM FAULT-FINDING AND CORRECTION

Cold water systems can suffer from many different faults. Some occur over a period of time, such as wear and tear on taps, valves and appliances, and others are a result of poor installation techniques or defective materials. Here, some of the common faults that occur with cold water installations are discussed, as well as ways of putting them right.

G1 NOISE

System noise can take many forms, from a squealing tap washer to violent pipework reverberation and water hammer. Most noise within direct systems of cold water is a direct result of the high pressure and flow rate that can occur within this system. It should be remembered that whenever a system has a mixture of high pressure and high flow rate, there will always be a certain amount of noise within that system. Sometimes, however, the noise can be excessive and this may be attributable to:

- **Faulty tap washers** – these tend to make a humming or squealing noise when the tap is opened. It is usually because the tap washer is either worn or split, and re-washering the tap cures the problem in most cases.

- **Faulty FOV washers** – this can cause a very loud hum throughout the pipework. Unfortunately, the noise is amplified if the cistern is in the roof space. Re-washering the FOV generally cures the fault. One way of testing to see if it is the FOV washer is to turn on a cold tap when the noise begins. If the noise stops or goes quieter, it is probably the FOV washer.

- **Loose or incorrectly supported pipework** – this can be the cause of very violent banging within the system. Every time the pipework reverberates, it is equal to twice the incoming mains pressure. If the supply is at 3 bar, then each bang is the equivalent of 6 bar. This can eventually lead to fittings failure and leakage. The best course of action is to try to find where the pipework is loose, and refix it. If this is not possible, the installation of a water hammer arrester fitted near to the main stop valve inside the property may cure the problem.

Water hammer arresters

G2 AIRLOCKS ON LOW-PRESSURE SYSTEMS

Airlocks on low-pressure systems can be a constant nuisance, especially during the commissioning stage. Airlocks stop the flow of water due to air trapped in the pipework, and there is insufficient water pressure from the cistern to push the air out. They usually occur because the cold distribution pipe rises as it leaves the cistern rather than falling towards the appliances, and this causes a high spot where air collects. It is often a result of poor installation or design.

Curing an airlock is not easy. Usually, the best course of action is to leave the system to settle. Most airlocks eventually move, allowing water to flow. This can be problematic if the system is new and at the commissioning stage, because the system cannot be tested properly until the airlock clears. To be sure that airlocks do not occur, ensure that distribution pipes from cisterns have a slight but constant fall towards the appliances.

G3 CORROSION

Corrosion, as we have seen in previous chapters, can take place in almost all systems that contain some form of metal piping. Common corrosion faults include:

- **Copper tube** – although not a ferrous metal, copper is relatively soft and can suffer from pitting corrosion if flux residue is allowed to remain on the tube after soldering activities. Erosion corrosion can also occur, especially in high-pressure, high flow rate systems, on fittings where there is a change of direction such as an elbow and the tube has not been properly deburred; the burr in the tube causes a rapid change in the direction of the water.

- **Brass fittings** – these can suffer from electrolytic corrosion in some instances, especially when the wrong type of brass fitting has been used on the installation. This leads to dezincification and eventual failure of the fitting. When a brass fitting has dezincified, it becomes brittle and susceptible to sudden fracture.

- **Galvanised steel cisterns** – these were used extensively on installations in the past. They are notorious for corroding from the inside of the cistern, leaving a thick layer of rust on the cistern bottom. The metal will eventually corrode until it is paper-thin and will suddenly give way, leading to leaks. Do not attempt to clean the rust from the bottom of galvanised cisterns. When the cistern has reached this stage of corrosion, it must be replaced.

G4 LEAKAGE

Leakage is a common problem in cold water systems. It can take three main forms:

- **Leakage from the cold water service pipe below ground before it enters the property** – this is quite difficult to detect. The main signs of leakage are loss of water pressure and flow rate and a constant distant sound of running water. To find out whether the leak is before or after the external (boundary) stop valve, the external stop valve must be turned off; if the water supply has stopped but the sound of running water remains, the communication pipe is leaking and this must be repaired by the water undertaker. If the sound of running water stops when the external stop valve is turned off, the leak is on the service pipe to the property and this is the responsibility of the property owner.

- **Leakage from the internal cold water system pipes and fittings** – this can cause a lot of damage to the property. It is fairly easy to detect the source by isolating the mains cold water stop valve. If the water stops, it is on the mains cold water supply. If the water continues to run, it is on the distribution pipework. By isolating the mains internal stop valve and opening the hot and cold water taps in the property, the system will drain quickly, allowing repairs to be carried out.

- **Leakage from taps and FOVs** – dripping taps are an annoyance but they can also waste quite a lot of water if they are dripping for a long time. If the property is on a water meter, they can make a significant impact on the water bill. Dripping FOVs are detected when the overflow to the cold water storage cistern or the WC cistern begins to run. This can first show itself by the overflow running only at night when the pressure of the water main rises. Gradually, it will start to run all the time and will need to be repaired (see pages 340–41).

CONCLUSION

During this chapter, we have investigated water supply from the cloud to the tap and have set out best practice of system installation, including the choice of materials and components. As you gain experience as a plumber, it will become apparent that there are a multitude of different systems, materials and fittings; variations on those looked at in this chapter. By seeing these different systems in operation, you will soon become proficient at identifying the correct methods of working and continue to enhance and develop your knowledge.

006 TEST YOUR KNOWLEDGE

SmartScreen Unit 006 revision sample questions

1. Briefly describe the key stages of the rainwater cycle.

2. Name three sources of water.

3. What is meant by the term **wholesome water**?

4. Which chemical is used to sterilise the water before it enters the mains cold water supply?

5. What are trunk mains and who is responsible for them?

6. Label the following drawing:

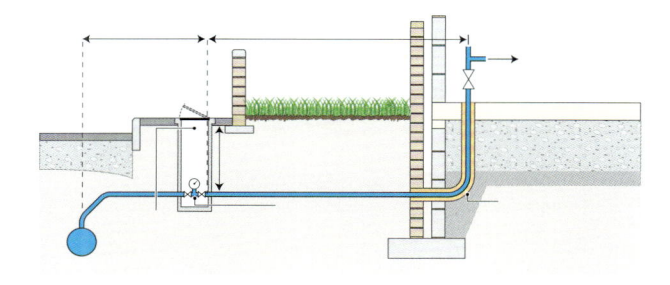

7. What should be placed immediately above the internal stop valve as the water supply enters the dwelling?

8. Briefly describe a direct system of cold water supply.

9. Name the system that is being described: **'A system of cold water supply where only the kitchen sink is fed directly from the water main. All other appliances are fed from a protected storage cistern in the roof space'**

10. In an indirect system of cold water supply, what is the distribution pipe and where would it be fitted?

11. Water for cooking and food preparation should be fed from where?

12. What is the minimum recommended capacity for a storage cistern serving an indirect system of cold water supply?

13. Describe the function of an overflow/warning pipe.

14. What is backflow?

15. Which backflow prevention device is shown below?

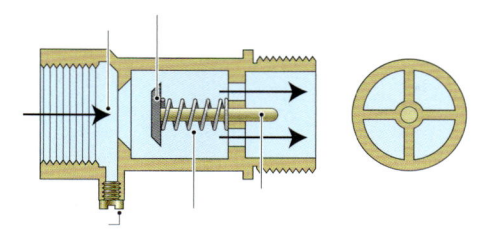

16. A BS 1212 part 1 float-operated valve can only be used on new installations if it is accompanied by a backflow prevention device. Which type of device is recommended?

17. What identifies the difference between BS 1212 part 2 float-operated valves and part 3 float-operated valves?

18. Which plumbing component would contain a ceramic disc?

19. A maintenance task in a domestic dwelling requires that the cold water supply to the property be isolated for most of the day. What should we advise the customer to do?

20. Name two common faults that can occur on domestic cold water supplies.

007

UNDERSTAND AND APPLY DOMESTIC HOT WATER SYSTEM INSTALLATION AND MAINTENANCE TECHNIQUES

A supply of hot water is essential. We use it every day for personal hygiene, cooking and washing clothes. It is a vital resource for combating germs and bacteria, but it can also cause harm if the temperature of the water is not controlled. This chapter investigates the many methods of supplying hot water to the home. It looks at the systems of hot water supply, the installation methods that should be employed, the appliances used to generate hot water and the ways in which its temperature can be controlled within safe, usable limits. It also explores some of the common hot water-related faults that occur and looks at ways of maintaining systems for optimum performance.

IN THIS CHAPTER, YOU WILL COVER:

A Selecting a hot water system
A1 Types of hot water systems

B Centralised storage hot water systems
B1 Centralised open-vented hot water storage systems
B2 Open-vented direct hot water storage systems
B3 Open-vented indirect hot water storage systems
B4 Combination centralised open-vented hot water storage systems
B5 Superduty hot water storage cylinders
B6 Immersion heater types and arrangements for use with open-vented hot water storage cylinders
B7 Gas-fired storage water heaters
B8 Storage cylinder insulation
B9 Grades of storage cylinders
B10 Storage cylinder sizes and capacities
B11 The purpose and use of secondary circulation
B12 Anodic corrosion protection of hot water storage cylinders
B13 Pipe sizes for open-vented hot water storage systems

C Centralised instantaneous hot water systems
C1 Thermal stores
C2 Combined primary storage units (CPSUs)
C3 Instantaneous gas-fired multi-point hot water heaters and combination boilers
C4 Pipe sizes for instantaneous hot water systems

D Localised hot water systems
D1 Instantaneous point-of-use water heaters
D2 Storage point-of-use water heaters

E General requirements for domestic hot water systems
E1 The installation of hot water pipework to BS 6700
E2 Temperature control of hot water systems
E3 The use of thermostatic mixing valves
E4 The insulation of hot water pipework
E5 Expansion of hot water pipework

F Installation of shower mixing valves and shower boosting pumps
F1 Cistern-fed supplies
F2 Cistern-fed supplies and booster pump
F3 Mains hot and cold supplies
F4 Unbalanced supply pressures

G Protection against backflow and back siphonage

H Testing and commissioning hot water systems

I Faults with open-vented water systems

J Decommissioning and maintenance of hot water systems

Test your knowledge

A SELECTING A HOT WATER SYSTEM

The first part of this chapter looks at the choice of system, from the many different types available. The type of system chosen will depend on:

- the size of the property and the distance to the outlets
- the number of occupants and the amount of hot water required
- the number of hot water outlets
- the type(s) of fuel to be used
- installation and maintenance costs
- running costs and fuel efficiency
- the pressure and flow rate of the incoming mains cold water supply.

A1 TYPES OF HOT WATER SYSTEM

There are many different types of hot water system that can be chosen for domestic applications.

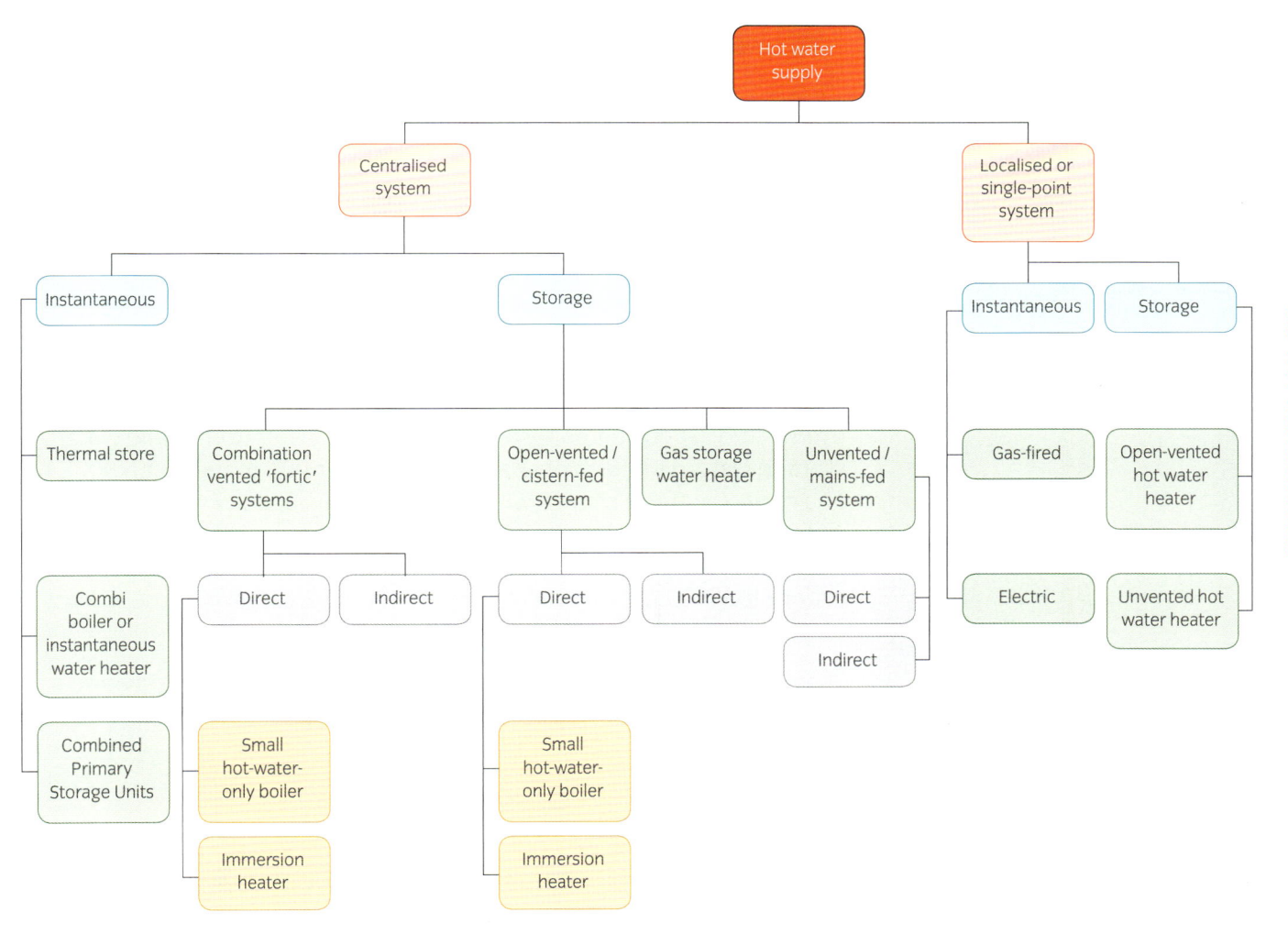

Types of hot water system

SmartScreen Unit 007 handout 2

As you can see from the diagram, hot water systems can be divided into two distinct types:

- Centralised hot water systems, where hot water is delivered from a central point to all hot water outlets in the dwelling. These can be further divided into centralised storage hot water systems (see Section B) or centralised instantaneous hot water systems (see Section C).

- Localised hot water systems, often called single-point or point-of-use systems. Hot water is delivered by a small water heater at the point where it is needed (see Section D).

B CENTRALISED STORAGE HOT WATER SYSTEMS

There are two main types of centralised storage hot water systems:

- **Open-vented systems** – fed from a cistern in the roof space.
- **Unvented systems** – fed directly from the cold water main (these will be covered at Level 3 and are not discussed here).

B1 CENTRALISED OPEN-VENTED HOT WATER SYSTEMS

In an open-vented storage hot water system, water is heated, generally by a boiler or an immersion heater, and stored in a hot water storage cylinder sited in a central location in relation to the hot water draw-off points in the property, often in an airing cupboard. Open-vented systems contain a vent pipe that remains open to the atmosphere, sited over the cold water storage cistern in the highest available location. The vent pipe allows air out on filling and in on draining down the system, and also acts as a safety relief in the unlikely event that the system becomes overheated. Open-vented systems are fed with water from a cold water storage cistern in the roof space (see Chapter 006, pages 300–302). The capacity of the cistern will depend on the capacity of the hot water storage vessel and whether any cold water outlets are fed from the same cistern. BS 6700 recommends that the capacity of the cistern feeding cold water to a hot water storage vessel must be at least equal to the capacity of the hot water storage vessel.

There are some important points to take into account when designing and installing open-vented hot water systems:

- The open vent pipe must not be smaller than 22mm and must terminate over the cold water storage cistern.

- The open-vent pipe must not be taken directly from the top of the hot water storage vessel, as this may result in parasitic (one-pipe) circulation.

KEY POINT

Parasitic circulation (or one-pipe circulation) is circulation that occurs within the same pipe. It generally occurs in open-vent pipes that rise vertically from the hot water storage cylinder. The hotter, central water rises up the vent pipe and the cooler water, near the walls of the pipe, falls back into the cylinder. This can be a major cause of heat loss from hot water storage cylinders.

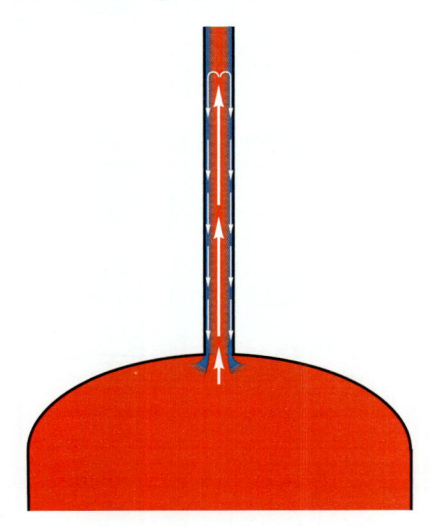

- The hot water draw-off pipe should rise slowly from the top of the cylinder to the open vent pipe and must incorporate at least 450mm of horizontal pipe between the storage cylinder and its connection point to the open vent. This is to prevent parasitic circulation from occurring.

- The cold feed pipe should be sized in accordance with BS 6700. The cold feed is the main path for expansion of water to take place within the cylinder when the water is heated. The heated water from the cylinder expands up the cold feed pipe, raising the water level in the cold water storage cistern.

- The cistern should be placed as high as possible to ensure good supply pressure. The higher the cistern, the greater the pressure at the taps. Poor pressure can be improved by raising the height of the cistern as far as Building Regulations will allow (see Chapter 006, page 302).

- All pipes should be laid with a slight fall (except the hot water draw-off) to prevent airlocks within the system.

- The cold feed pipe from the storage cistern must only feed the hot water storage cylinder.

- A drain-off valve should be fitted at the lowest point of the cold feed pipe.

- Centralised hot water storage systems may be either direct or indirect.

B2 OPEN-VENTED DIRECT HOT WATER STORAGE SYSTEMS

Direct systems use a direct-type cylinder that is heated either by a small hot-water-only boiler or an immersion heater. The boiler may either be a small gas-fired hot water heater (often called a gas circulator) designed to heat the water directly, or a small back boiler situated behind a solid fuel fire. Because the water in the boiler comes directly from the hot water storage cylinder, the boiler must be made of a material that does not rust (to prevent rusty water being drawn off at the taps). Suitable boiler materials are:

- copper
- stainless steel
- bronze.

The direct system with boiler

This type of hot water cylinder contains no form of heat exchanger and so is not suitable for use with central heating systems.

The hot water circulates from the boiler or circulator by the principle of convection. This is known as gravity circulation (see Chapter 004, page 177). The hot water rises in the primary flow pipe, directly heating the stored water in the cylinder before the cooler water returns to the boiler. The water in the cylinder does not heat uniformly. The water at the top

KEY POINT

Primary water is the water that is in the boiler, central heating system and the heat exchanger of an indirect-type hot water storage cylinder/vessel. It is called primary water because it is heated by the primary source of heat and hot water in the dwelling, namely the boiler. The pipes that connect the boiler to the heat exchanger are called the primary flow and the primary return. The secondary water is the stored water in the cylinder itself that is delivered to the hot water outlets and taps. The primary water heats the secondary water indirectly via the heat exchanger.

A direct cylinder with boiler connection

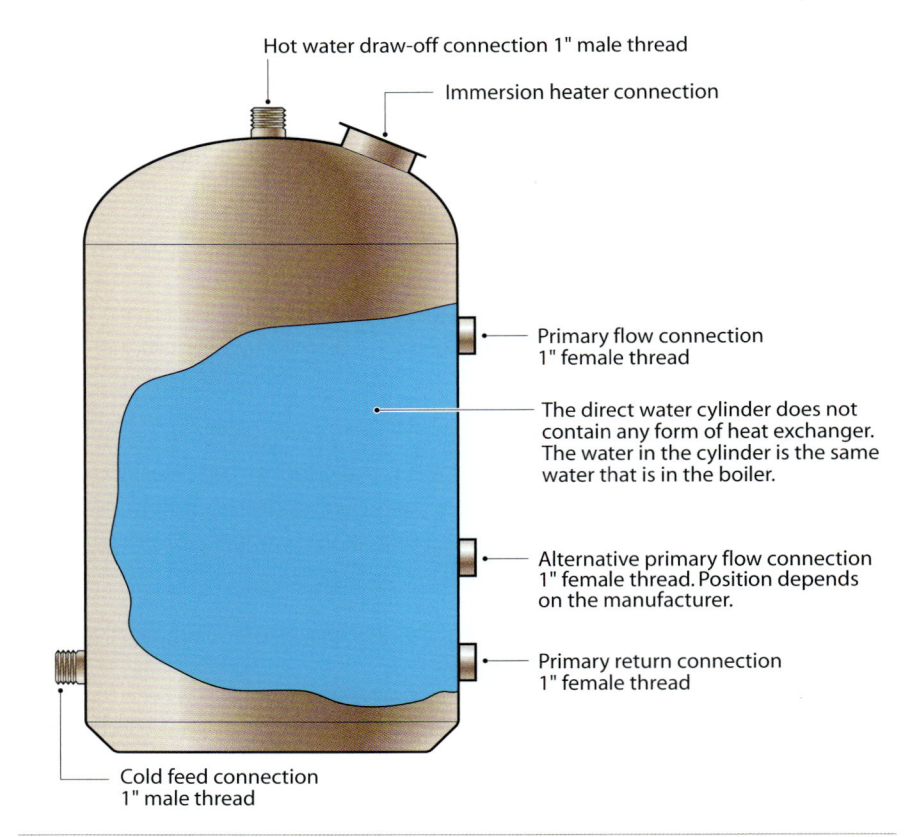

The direct cylinder

Labels on diagram:
- Hot water draw-off connection 1" male thread
- Immersion heater connection
- Primary flow connection 1" female thread
- The direct water cylinder does not contain any form of heat exchanger. The water in the cylinder is the same water that is in the boiler.
- Alternative primary flow connection 1" female thread. Position depends on the manufacturer.
- Primary return connection 1" female thread
- Cold feed connection 1" male thread

KEY POINT

The direct cylinders examined here are heated using either electricity via immersion heaters or special non-rusting solid fuel boilers and gas circulators. Direct cylinders are also available for solar hot-water-only systems but these are quite different internally from those that use more traditional fuels such as oil and gas. They are a rare type of installation; most installers will opt for the double-feed indirect solar systems previously mentioned (see page 118).

The direct cylinder with immersion heaters

of the cylinder is usually 10°C hotter than the bottom (generally 60°C at the top and 50°C at the bottom). This is known as stratification, and it is desirable in stored hot water systems.

The primary flow and return pipes to and from the boiler/circulator should be a minimum of 28mm regardless of pipe length, unless otherwise stated in the manufacturer's instructions.

Direct cylinders, when connected to solid fuel back boilers, are susceptible to boiling because there is no effective method of temperature control. This leads to a high risk of scalding, and the likelihood of scale build-up in hard water areas.

The direct system with immersion heater

As an alternative to direct systems with a circulator/back boiler, most modern direct systems use two 3kW immersion heaters placed in the side of the cylinder to heat the water. One immersion heater is placed at the bottom of the cylinder to heat all the contents, and a second immersion heater is placed halfway down the cylinder for daytime top-up. The immersion heaters are wired to a time controller for use with cheaper-rate overnight electricity. The temperature of the immersion heaters is best limited to 55°C to ensure lower running costs and minimise scale build-up.

The direct system of hot water

The direct system of hot water with immersion heaters

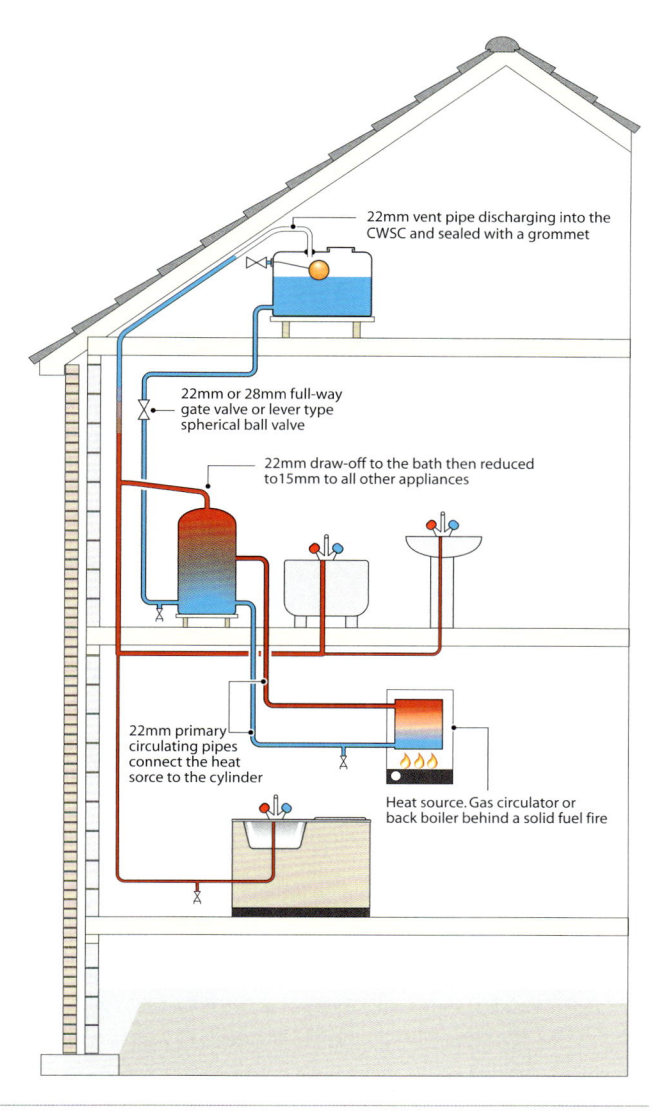

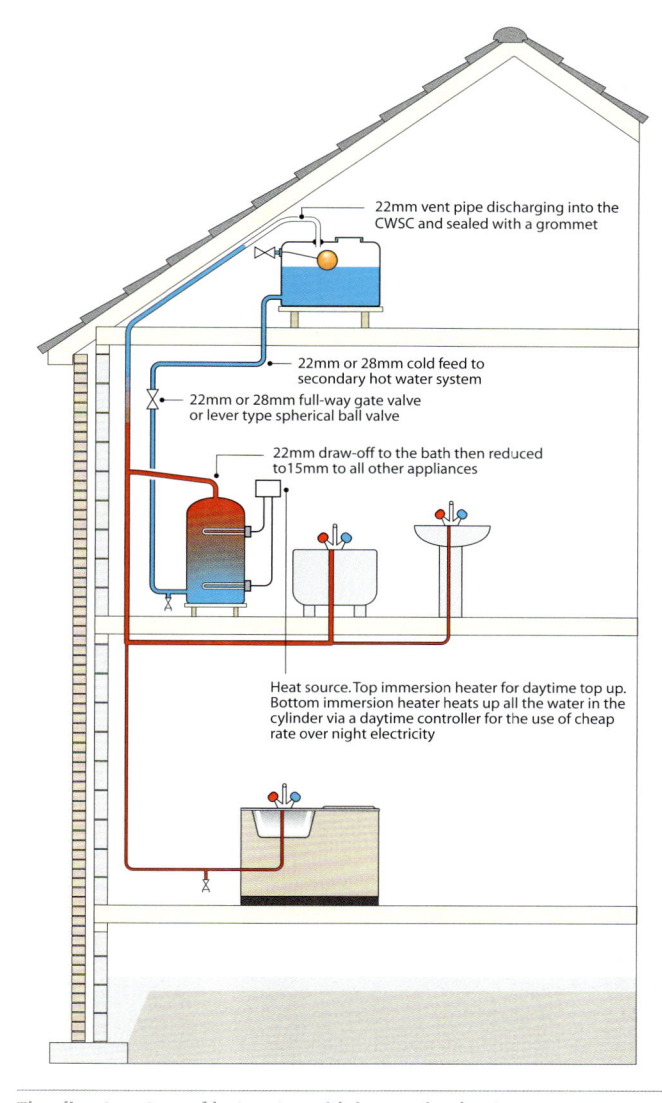

When to choose a direct system

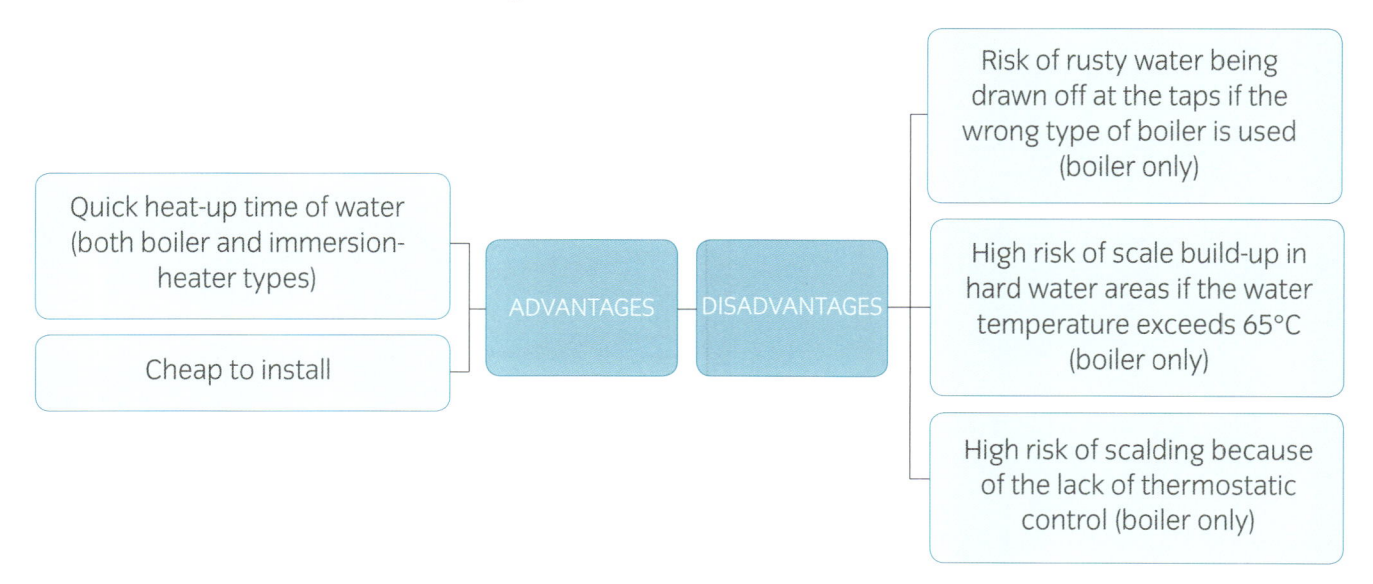

Advantages
- Quick heat-up time of water (both boiler and immersion-heater types)
- Cheap to install

Disadvantages
- Risk of rusty water being drawn off at the taps if the wrong type of boiler is used (boiler only)
- High risk of scale build-up in hard water areas if the water temperature exceeds 65°C (boiler only)
- High risk of scalding because of the lack of thermostatic control (boiler only)

007 HOT WATER SYSTEMS

Direct system considerations:

- **Property size** – suitable for most houses; generally only used in properties where there is no requirement for a wet central heating system (immersion heater only).
- **Storage capacity** – varies with occupancy. Generally, 210 litres for four people.
- **Fuel type** – mostly used with Economy 7 electricity but can also be used with some solid fuel boilers and gas circulators. Gravity hot water circulation only.
- **Installation cost** – the least expensive of all storage systems for houses when boilers and circulators are not fitted.
- **Fuel efficiency** – Economy 7 electricity is 100 per cent energy efficient. All the energy used goes into heating the water, but the price of the electricity can be high.

SmartScreen Unit 007 animation 1

B3 OPEN-VENTED INDIRECT HOT WATER STORAGE SYSTEMS

An indirect system uses an indirect-type hot water storage cylinder, of which there are two distinct types:

- double-feed indirect hot water storage cylinder
- single-feed self-venting indirect hot water storage cylinder.

These storage cylinders are different from the direct type in that they contain heat exchangers to heat the secondary water. The heat exchanger (sometimes referred to as a calorifier) contains primary water and so is classified as part of the dwelling's central heating system.

Double-feed indirect hot water storage systems

These are probably the most common of all hot water delivery systems installed in domestic properties in the UK. They use a double-feed

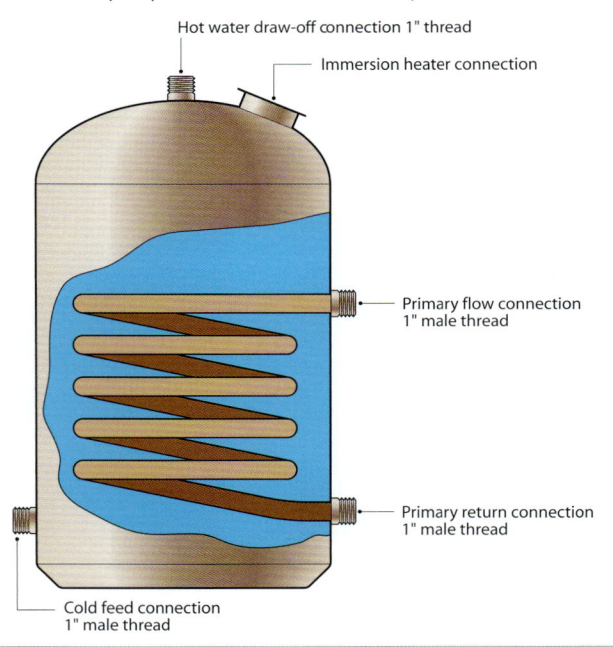

Heat exchanger
A device or vessel that allows heat to be transferred from one water system to another without the two water systems coming into contact with each other. The transfer of heat takes place via conduction (see Chapter 004, page 176).

An indirect cylinder

indirect hot water storage cylinder, which contains a heat exchanger, at the heart of the system. The heat exchanger within the cylinder is usually a copper coil but, in older-type cylinders, it can also take the form of a smaller cylinder called an annular. These systems are called indirect simply because the secondary water in the cylinder is heated indirectly by the primary water via the heat exchanger. The double-feed indirect cylinder contains a heat exchanger in the form of a coil and so is suitable for use with central heating systems.

In a double-feed indirect system, two cisterns are used: a larger cold water storage cistern for the domestic hot water and a smaller feed and expansion cistern for the heating. It is now general practice to install indirect cylinders on all new installations in preference to direct types, even if the indirect flow and return are capped off, ready to use at a later date.

SmartScreen Unit 007 animation 2

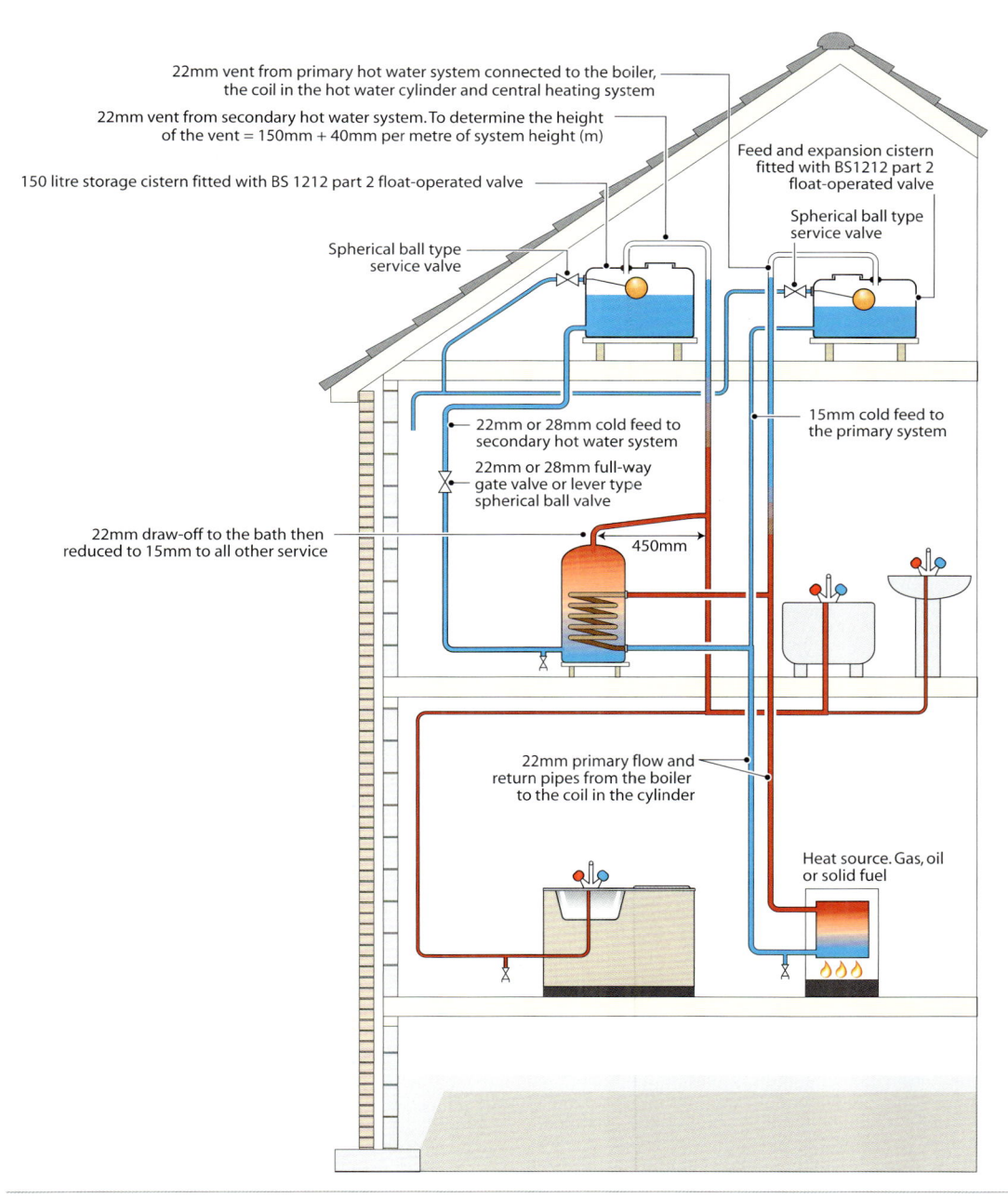

The indirect system of hot water with gravity primary circulation

The heat exchanger coil

The indirect cylinder showing coil connections

SmartScreen Unit 007 handout 7

The double-feed indirect hot water storage cylinder allows the use of boilers and central heating systems that contain a variety of metals, such as steel and aluminium, because the water in the cylinder is totally separate from the water in the heat exchanger. This means that there is no risk of dirty or rusty water being drawn off at the taps. The system is designed in such a way that the water in the boiler and primary pipework is hardly ever changed, with the only loss of water being in the feed and expansion cistern through evaporation.

The secondary water is drawn from the hot water storage cylinder to supply the hot taps. It is heated by conduction (see Chapter 004, page 176), as the water in the cylinder is in contact with the heat exchanger.

A feed and expansion cistern feeds the primary part of the system, and this must be large enough to accommodate the expansion of the water in the system when it is heated. The vent pipe from the primary system must terminate over the feed and expansion cistern. An alternative method is to use a sealed heating system, which is fed with water from the cold water main via a filling loop. Water expansion is accommodated in an expansion vessel.

Hot water storage cylinders must conform to BS 1566, which specifies the minimum heating surface area of the heat exchanger.

Existing double-feed indirect systems

Existing double-feed indirect systems use gravity circulation via 28mm gravity primary flow and return pipes to heat the water in the cylinder. This type of system can no longer be installed, as it extremely wasteful in terms of energy usage. Document L1b of the Building Regulations recommends that these systems should be replaced with fully pumped systems wherever possible (see Chapter 008, page 403) or must be updated to include a cylinder thermostat and a motorised zone valve arrangement as stated in the Domestic Heating Compliance Guide. This is to limit the amount of energy wastage.

New double-feed indirect systems

On new installations, double-feed indirect cylinders must incorporate pumped circulation in the heat exchanger. Document L1a of the Building Regulations dictates that all new installations must have pumped primary circulation with controls that prohibit energy wastage. This is achieved by installing thermostatic control over the hot water storage cylinder via a cylinder thermostat and a motorised zone valve arrangement as stated in the Domestic Heating Compliance Guide. Because the primary flow and return are pumped, the pipe size, in most cases, can be reduced to 22mm. This subject will be covered in greater detail in Chapter 008.

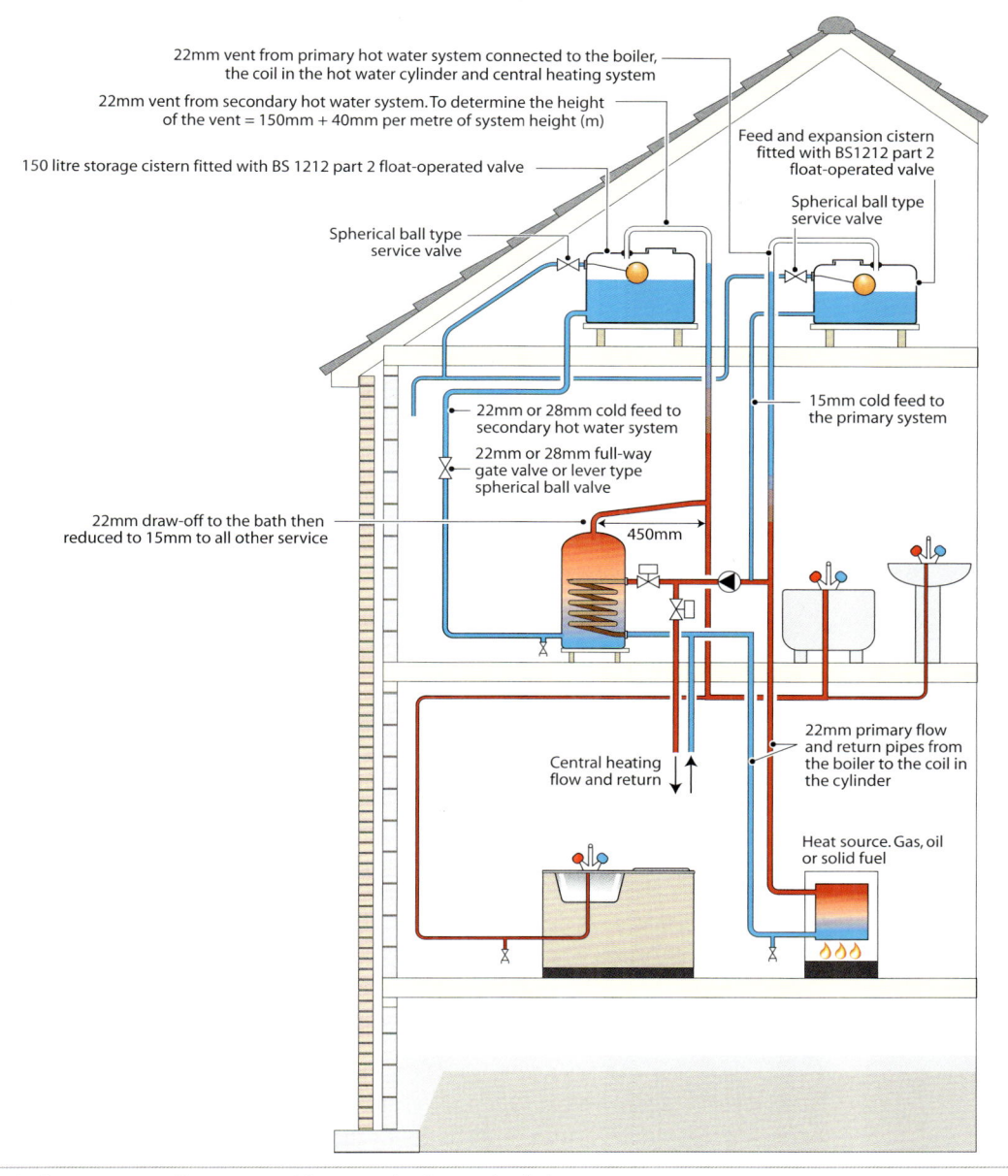

The indirect system of hot water with pumped primary circulation

Indirect cylinders for renewable energy hot water supply

Open-vented cylinders have been developed for installation onto renewable energy hot water supply systems, such as solar, geothermal and ground-source heat pumps. The cylinder contains two heat exchanger coils. The first coil is used with a conventional fuel source such as gas or oil and this accounts for 70 per cent of the cylinder's hot water volume. The second coil has 30 per cent volume dedicated to the renewable energy heat source and is usually situated in the top third of the cylinder. They are suitable for:

- modern fully pumped heating systems (see Chapter 008, page 403), both vented and sealed heating systems up to 3.5 bar pressure
- a double thickness of CFC-free polyurethane insulation
- capacities from 130 to 300 litres.

A renewable energy storage cylinder

When to choose a double-feed indirect system

ADVANTAGES
- Can be used with fully pumped heating systems and condensing boilers
- Wide range of capacities available
- Fully compliant with Document L of the Building Regulations

DISADVANTAGES
- Expensive to install
- Will require a second cistern and associated pipework or expansion vessel and filling loop

Double-feed indirect system considerations:

- **Property size** – suitable for all domestic properties when used with central heating systems.

- **Storage capacity** – varies with occupancy. Generally, 210 litres for four people.

- **Fuel type** – can be used with gas, oil, solid fuel, electricity and renewable energy fuel sources. Suitable for fully pumped heating systems. Conforms to Document L of the Building Regulations.

- **Installation cost** – more expensive than direct systems due to the extra pipework for the feed and expansion cistern.

- **Fuel efficiency** – gas and oil appliances must be energy-efficient condensing types. Can also be used with Economy 7 electricity.

Single-feed self-venting indirect hot water storage systems

This system uses a single-feed self-venting indirect cylinder. It contains a special heat exchanger that uses air entrapment to separate the primary water from the secondary water. These types of cylinders are no longer used on new installations, but you may come across them in existing systems.

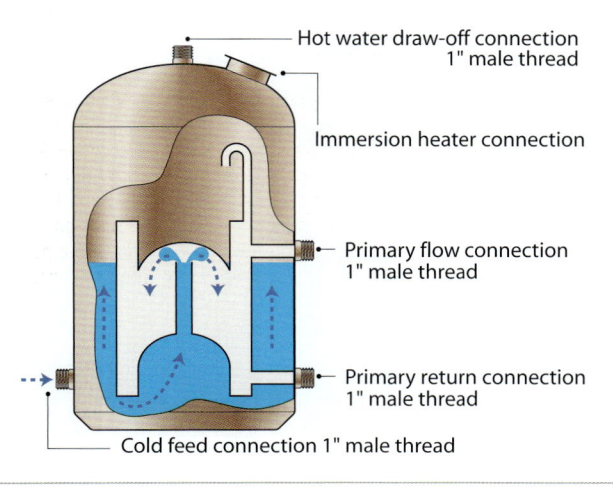

It is fitted in the same way as a direct system, with only one cold water storage cistern in the roof space but, unlike the direct system, it allows a boiler and central heating to be installed. It does not require a separate feed and expansion cistern. It has a special heat exchanger that works in such a way that the primary and secondary water are separated by a bubble of air that collects in the heat exchanger, preventing the waters from mixing.

A single-feed self-venting indirect cylinder

As the cylinder fills with water from the cold feed pipe, the water enters the heat exchanger through a tube in the bottom dome, passing up through the centre of the heat exchanger. As the heat exchanger fills, the water flows down the primary return, filling the heating system. When the heating system and the cylinder are full, a bubble of air forms on the underside of the top dome, effectively separating the primary and secondary water. As the water is heated, it expands. The expanded water forces the air from the top dome into the bottom dome, allowing the expansion of water to take place without breaking the air bubble.

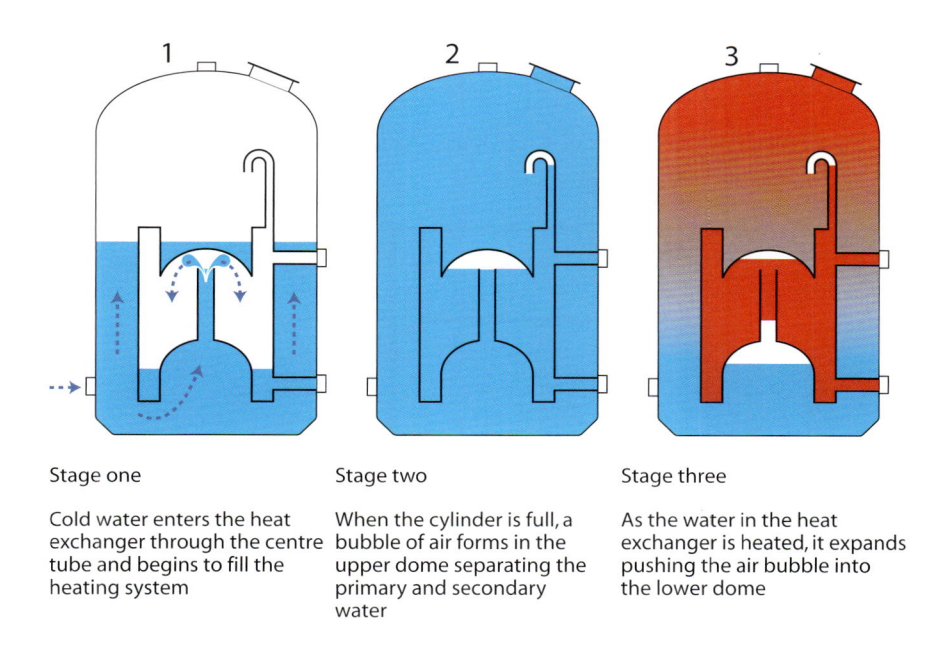

Stage one
Cold water enters the heat exchanger through the centre tube and begins to fill the heating system

Stage two
When the cylinder is full, a bubble of air forms in the upper dome separating the primary and secondary water

Stage three
As the water in the heat exchanger is heated, it expands pushing the air bubble into the lower dome

The operation of the air bubble and air entrapment

Periodically the air bubble may break, allowing the primary and secondary waters to mix. When this happens, discoloured water will be drawn off when a hot tap is opened. Should this occur, the boiler should be switched off and the system allowed to cool down. The air bubble should then re-form naturally. If it doesn't form, you would need to drain the whole system and refill.

One recurring problem that can occur with this system is that if the central heating system is too big for the heat exchanger, the air bubble ruptures almost every time the system is heated due to excessive water expansion. This causes rusty water to be drawn off at the hot taps and encourages the formation of red oxide sludge (rust sludge), which eventually blocks the primary flow to the heat exchanger, causing loss of circulation and hot water.

Installation of this type of cylinder is forbidden under Document L of the Building Regulations. The cylinder cannot be replaced with like-for-like on existing installations. Any system that contains this cylinder type must be upgraded to a double-feed indirect-type with full thermostatic control of the hot water on replacement installations.

SmartScreen Unit 007 handout 3

KEY POINT

The single-feed self-venting cylinder is often referred to as a Primatic cylinder (Primatic is a trade name of IMI Ltd). Another version of this cylinder was once available and may be found in some existing installations, known as an Aeromatic. This has an air-release valve on the side of the cylinder near the heat exchanger to bleed air from the heat exchanger.

SmartScreen Unit 007 animation 3

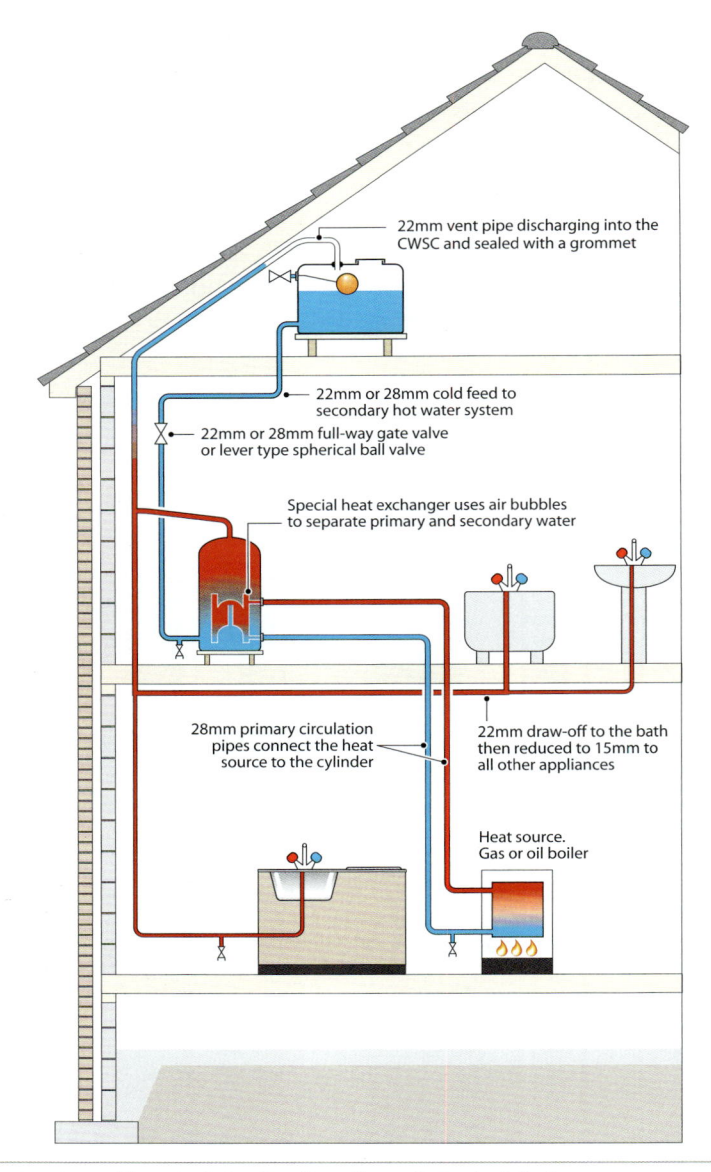

Diagram labels:
- 22mm vent pipe discharging into the CWSC and sealed with a grommet
- 22mm or 28mm cold feed to secondary hot water system
- 22mm or 28mm full-way gate valve or lever type spherical ball valve
- Special heat exchanger uses air bubbles to separate primary and secondary water
- 28mm primary circulation pipes connect the heat source to the cylinder
- 22mm draw-off to the bath then reduced to 15mm to all other appliances
- Heat source. Gas or oil boiler

An indirect system with a single-feed, self-venting cylinder installed

B4 COMBINATION CENTRALISED OPEN-VENTED HOT WATER STORAGE SYSTEMS

A combination cylinder has its own cold water storage cistern attached on the top of the cylinder. These are known as fortic cylinders and are available in both direct and double-feed indirect types. They come in a variety of sizes – 114 litres of hot water storage is the most common, with 40 litres of cold water storage above.

Integrating a cold water feed cistern and a hot water cylinder in a compact all-copper unit is an effective way of providing adequate supplies of hot water when storage space is limited, so is popular for use in small properties or flats.

This cylinder has a factory-fitted cold feed and vent pipes. Isolation of the hot water for maintenance should be via a full-way gate valve installed on the hot water draw-off pipe.

A combination open-vented hot water storage cylinder

The main problem with this type of cylinder is the lack of water pressure at the taps. The cold water storage is very close to the hot water cylinder and so the static head of pressure is very low. As a result of this, fortics must be installed as high as possible to improve the pressure at the outlets. Power-shower pumps should not be installed for use with this type of cylinder, because the cold water storage cistern will not be able to replenish itself quickly enough to feed the shower pump.

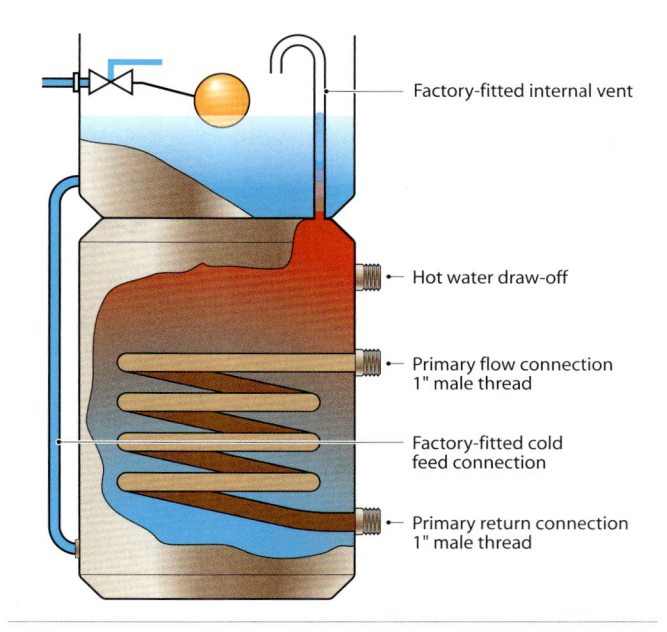

- Factory-fitted internal vent
- Hot water draw-off
- Primary flow connection 1" male thread
- Factory-fitted cold feed connection
- Primary return connection 1" male thread

The layout of a combination open-vented hot water storage cylinder

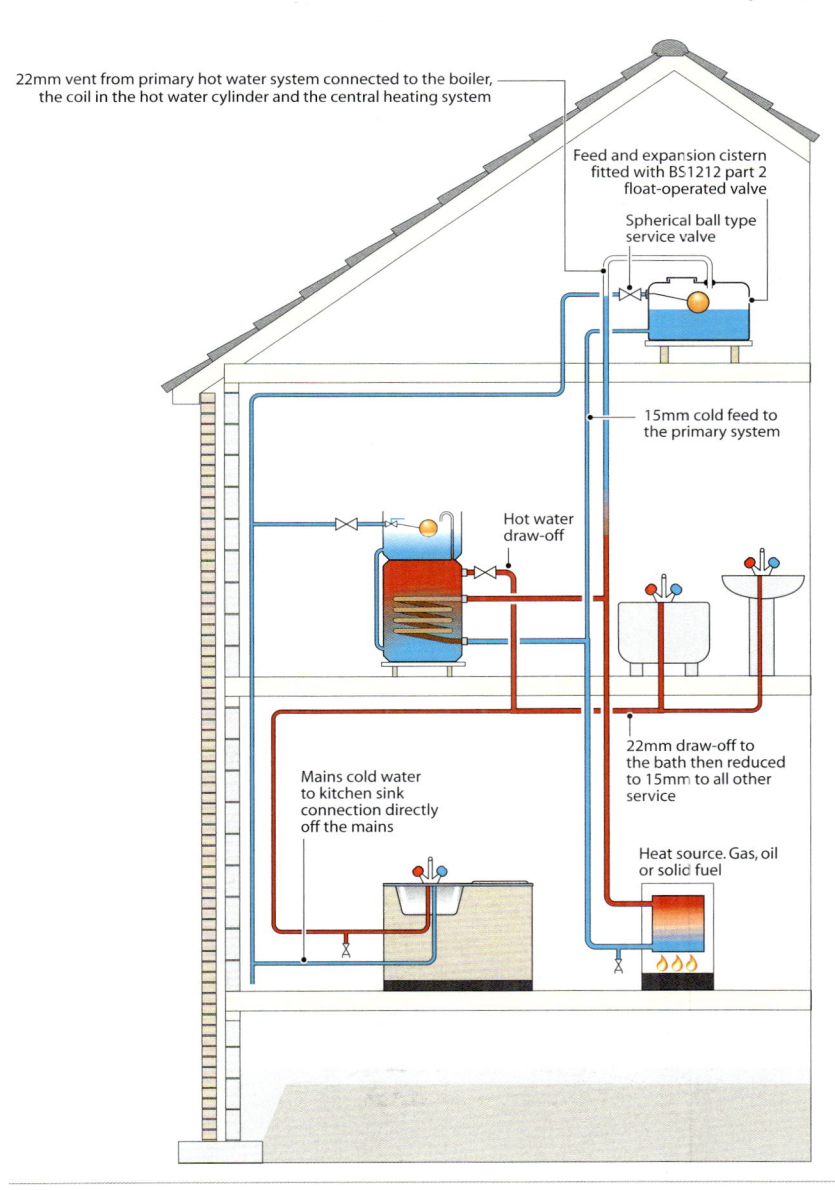

- 22mm vent from primary hot water system connected to the boiler, the coil in the hot water cylinder and the central heating system
- Feed and expansion cistern fitted with BS1212 part 2 float-operated valve
- Spherical ball type service valve
- 15mm cold feed to the primary system
- Hot water draw-off
- Mains cold water to kitchen sink connection directly off the mains
- 22mm draw-off to the bath then reduced to 15mm to all other service
- Heat source. Gas, oil or solid fuel

A combination open-vented hot water storage system

When to choose a combination cylinder system

ADVANTAGES
- A cheap solution for hot water systems, especially suited to flats and small houses
- Easy to install
- Fully compliant with Document L of the Building Regulations

DISADVANTAGES
- Suffers from lack of pressure unless installed at height
- Not suitable for pumped shower installations because of the lack of cold water storage

Combination cylinder system considerations:

- **Property size** – suitable for small properties and flats.
- **Storage capacity** – usually 114 litres for flats but larger capacities are available.
- **Fuel type** – best used with Economy 7 electricity but double-feed indirect cylinder type can also be used with gas and oil central heating.
- **Installation cost** – low installation and materials costs.
- **Fuel efficiency** – Economy 7 electricity is 100 per cent energy efficient. All the energy used goes into heating the water, but the price of the electricity can be high.

B5 SUPERDUTY HOT WATER STORAGE CYLINDERS

The superduty cylinder has a multi-coil heat exchanger, which is made up of several smaller-bore coils rather than one large-bore coil. This encourages a rapid recovery of the hot water, because the coil has a greater surface area of heat presented to the water in the cylinder. Superduty cylinders work in a similar way to instantaneous hot water cylinders and replenish the hot water even as it is being used. This reduces the amount of storage required and can save up to 40 per cent on fuel bills when compared to the standard cylinder type. When cheap-rate electricity is used for heating the water, it is most economical to heat the entire contents of the cylinder overnight. This reduces the need to use the immersion heater during the day when electricity is more expensive. Key points about them are:

- rapid heated water recovery – generally recovery times are 15 minutes for 45 litres, 45 minutes for 210 litres (assuming a boiler output of 9kW); 4 minutes for 45 litres, 19 minutes for 210 litres (assuming a boiler output of 30kW)
- multi-coil heat exchanger
- smaller storage cylinder means more space in the airing cupboard

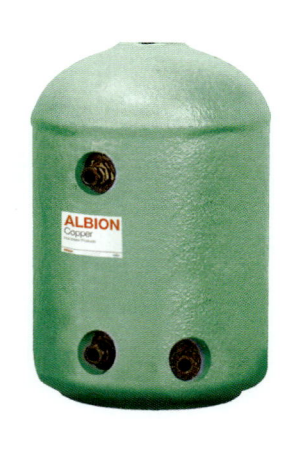

A superduty cylinder

- reduced boiler cycling
- saving on fuel bills
- can be used with conventional fully pumped wet central heating systems, as shown in the diagram.

Superduty cylinders are at their most efficient when installed alongside condensing boilers running at maximum temperature. This will ensure that recovery times are at their absolute minimum and the lower return temperature of the condensing boiler will maximise the time the boiler spends in condensing mode.

When to choose a superduty cylinder system

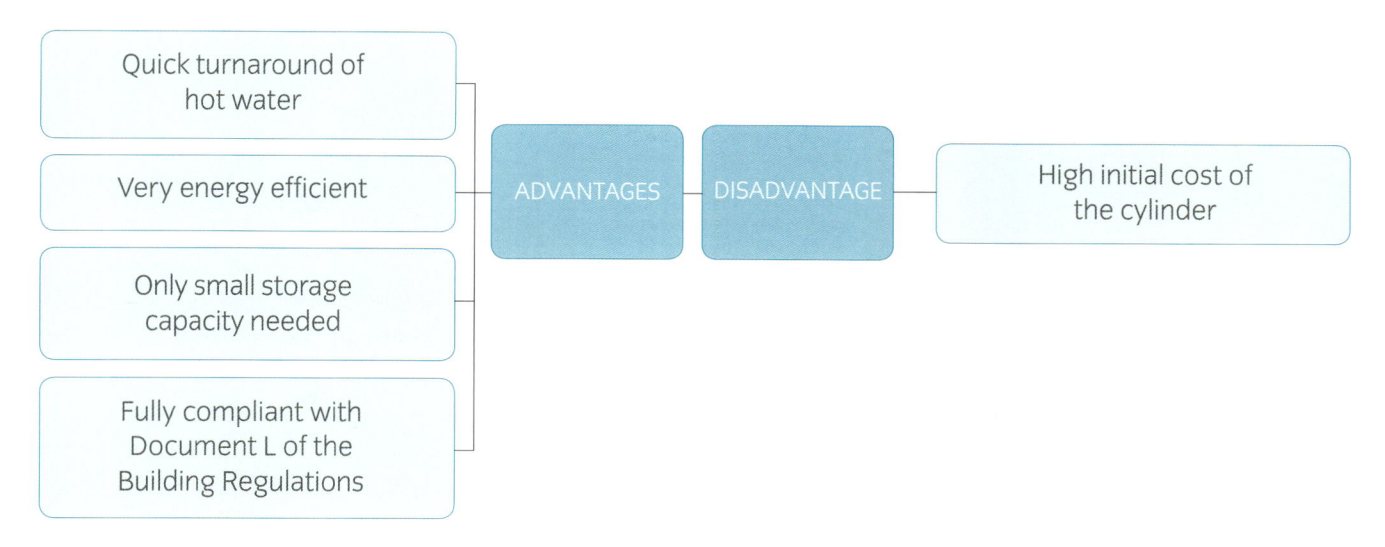

```
Quick turnaround of hot water  ┐
Very energy efficient          ├─ ADVANTAGES   DISADVANTAGE ─ High initial cost of the cylinder
Only small storage capacity needed ┤
Fully compliant with Document L of the Building Regulations ┘
```

Superduty cylinder system considerations:

- **Property size** – suitable for all domestic properties.
- **Storage capacity** – usually 80 litres.
- **Fuel type** – can be used with Economy 7 electricity and fully pumped heating systems.
- **Installation cost** – initial cost of the cylinder is high but is installed as a double-feed indirect cylinder with comparable running costs.
- **Fuel efficiency** – extremely energy efficient when used with condensing boilers due to quick heating of the water. Can cut fuel costs by up to 40 per cent.

B6 IMMERSION HEATER TYPES AND ARRANGEMENTS FOR USE WITH OPEN-VENTED HOT WATER STORAGE CYLINDERS

Almost all domestic open-vented hot water storage cylinders have an immersion heater boss to allow the installation of an immersion heater. There are several ways an immersion heater can be fitted to a cylinder, depending on the type of immersion heater and the type of cylinder it is to be installed into.

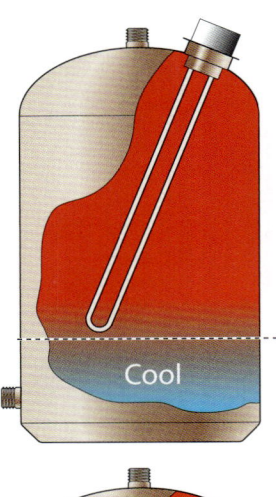

1. The simplest type of immersion heater installation. This single element immersion heater heats the water using a 3kW heating element. It does not heat the whole of the contents leaving the bottom 25% of only cool water.

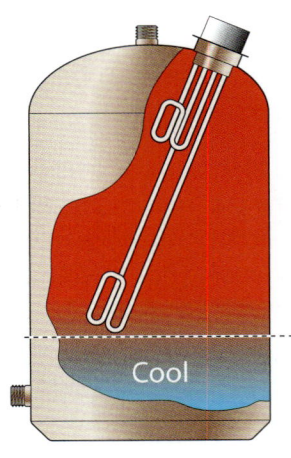

2. This installation uses a twin element immersion heater. The long element heats most of the cylinder contents whilst the top element is used for hot water top-up during the day. Each element has its own thermostat.

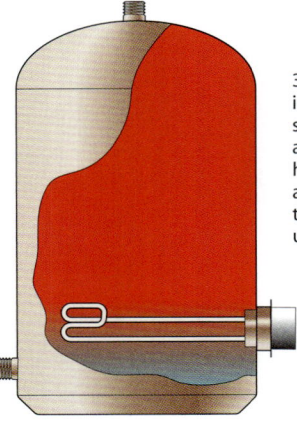

3. A bottom installed 3kW immersion heater on the side of the cylinder ensures a more even distribution of heat to the water so that almost all of the water in the cylinder is heated to a usable temperature

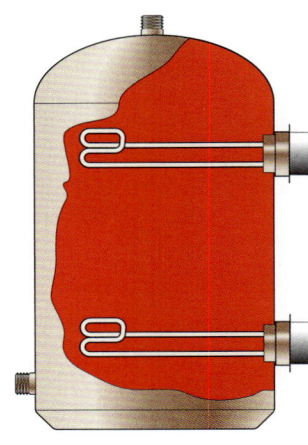

4. This is the best form of immersion heater installation that is best suited for use with cheap rate overnight electricity. The bottom immersion is used at night with the top heater used for top-up during the day. This arrangement is used with a time controller.

Immersion heater arrangements

Most immersion heaters are rated at 3kW. A cylinder with a capacity of 120 litres heats up from cold to around 65°C in approximately 2 hours and a 210-litre storage cylinder heats up in approximately 4 hours. Cylinders with either a dual element or twin immersion heater arrangement can supply 50 litres of fully heated water on a 1-hour boost during the day.

Immersion heaters are available with several different sheath metals. The sheath encases the heater element to protect it from the water. Sheath materials include:

- **tinned copper** – the least resistant to corrosion and scale formation
- **Incoloy 825** – a titanium alloy that offers greater resistance to scale formation
- **titanium** – especially made for use in areas where the water is aggressive, it offers the best overall resistance to corrosion and scale formation.

In hard water areas, the build-up of scale on the heater element can lead to overheating and eventual failure. Scale occurs quickly when water is heated to over 60°C and so temperatures should generally be kept a little lower than this to avoid scaling problems. The use of an immersion heater with a sheath that has a high resistance to scaling

A single-element immersion heater

A dual-element immersion heater

A side-entry immersion heater

is recommended in areas where the water is extremely soft or temporarily hard.

All immersion heaters must comply with BS EN 60335-2-73:2003 and have a resettable double thermostat (RDT) as standard. This enables problems with overheating to be recognised quickly. The thermostat has a non-self-resetting thermal cut-off device, which shuts off the electrical supply to the immersion heater if it overheats. The high limit thermostat is manually resettable.

There are a number of warning signs that could indicate that the water is being overheated by an immersion heater:

- excessively hot water coming out of the taps
- excessive noise or bubbling from the hot water cylinder
- hot or warm water coming from cold taps, which are served from the same cistern as the hot water cylinder
- steam and/or condensation in the roof space
- a warm cistern in the roof space.

Maxistore immersion heaters

Maxistore immersion heaters have been developed for use with overnight Economy 7 electricity. Maxistore immersion heaters are designed for 350mm side-entry hot water storage cylinders, and have:

- an Incoloy 825 nickel/titanium alloy sheath for high corrosion resistance
- a 3kW heater element
- a no-heat zone (often called a cold tail) to prevent overheating
- low watt-density for low-noise water heating
- an RDT combined thermostat and resettable safety cut-out.

Economy 7 electricity
A UK tariff that provides for seven hours of cheaper-rate electricity, usually between 1 am and 8 am in the summer and 12 am and 7 am in the winter (although times may vary between regions and suppliers).

SUGGESTED ACTIVITY...
To refresh your memory about the types and make-up of water, check out Chapter 004, page 166 and Chapter 006, page 282.

B7 GAS-FIRED STORAGE WATER HEATERS

The storage type of gas water heater is a self-contained unit. It is very similar in design to a domestic hot water cylinder except that the unit contains a gas burner to heat the water, with a thermostat to control the temperature. It requires a flue to vent the products of combustion to the atmosphere, either by an open flue or a balanced flue arrangement. Both vented units fed from a cistern and unvented types fed direct from the cold water main are available, with capacities ranging from 75 to 285 litres. Some units will also provide central heating.

A gas-fired storage water heater

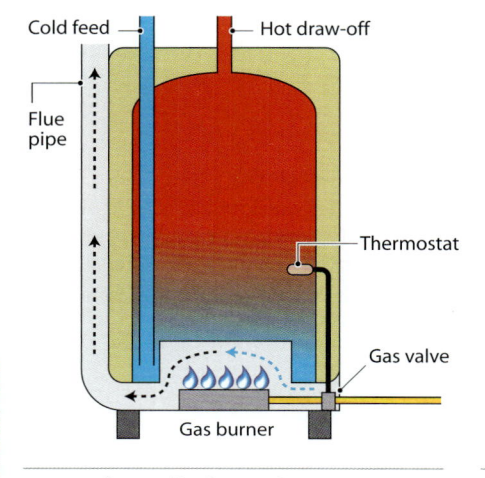

Layout of a gas-fired water heater

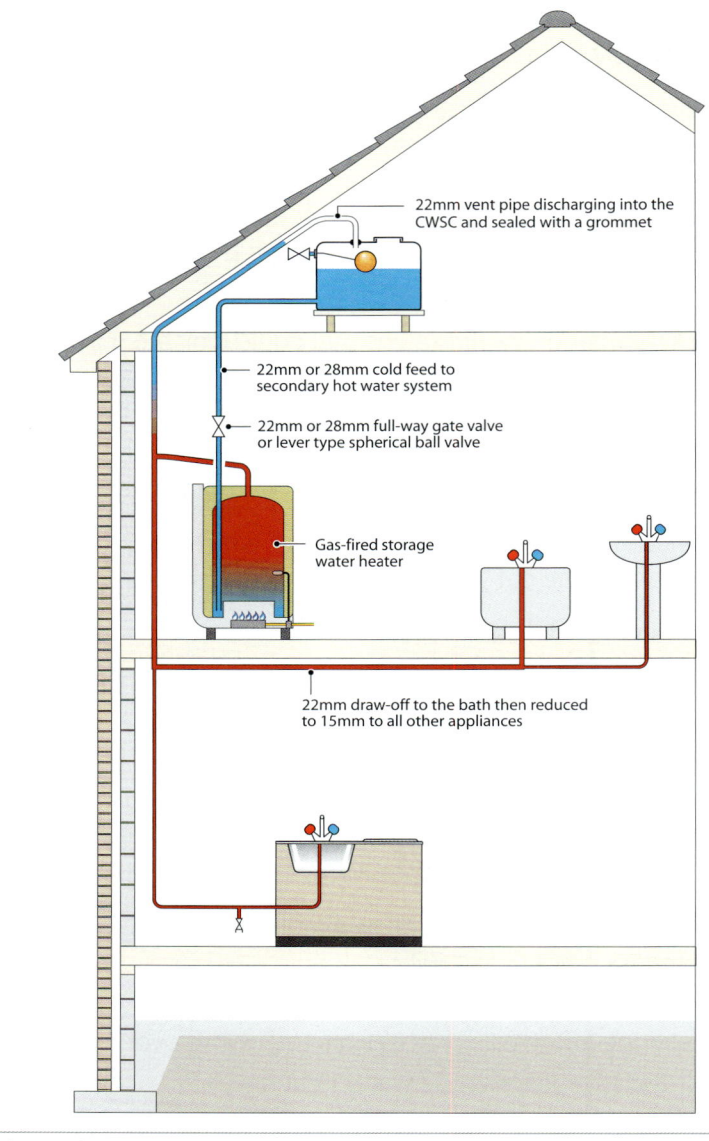

Gas-fired water heater installation

When to choose a gas-fired storage water heater system

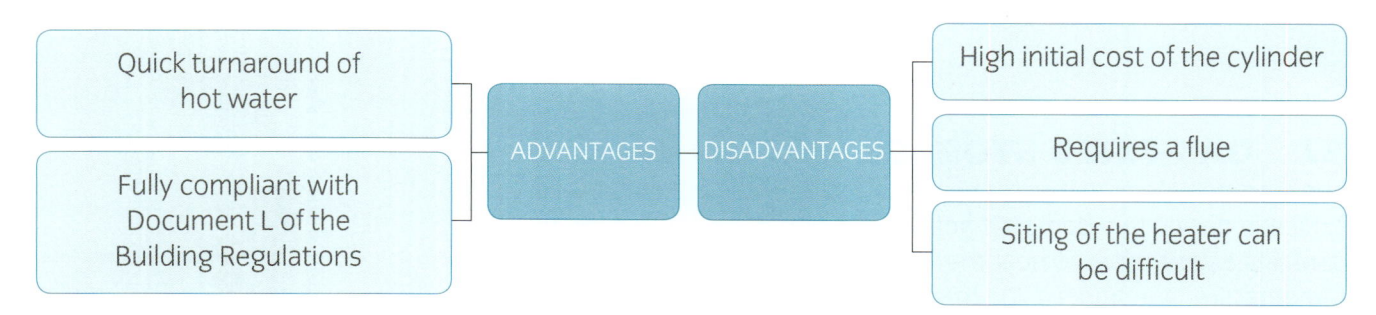

ADVANTAGES
- Quick turnaround of hot water
- Fully compliant with Document L of the Building Regulations

DISADVANTAGES
- High initial cost of the cylinder
- Requires a flue
- Siting of the heater can be difficult

Gas-fired storage water heater system considerations:

- **Property size** – suitable for all domestic properties.
- **Storage capacity** – up to 350 litres for domestic models.
- **Fuel type** – natural gas and LPG.

- **Installation cost** – initial cost of the cylinder is high, but some units can also supply central heating.
- **Fuel efficiency** – 85 per cent efficiency is typical for this type of water heater.

B8 STORAGE CYLINDER INSULATION

Cylinders are insulated with polyurethane foam, which is sprayed on to a predetermined thickness. The thickness of the insulation is covered by Building Regulations Document L: Conservation of fuel and power, which was updated in October 2010. The insulation thicknesses have been modified to deliver low standing heat loss and keep CO_2 emissions to a minimum in line with the Regulations:

- part L1a (new-build and replacement cylinders) have 50mm insulation
- part L1b (replacement-only cylinders) have 35mm insulation.

Cylinder insulation jackets are also available for uninsulated cylinders, made from fibreglass with a PVC outer layer. They are tied with a lace at the top and kept in place by either aluminium bands or plastic straps.

B9 GRADES OF STORAGE CYLINDERS

Open-vented hot water storage cylinders are manufactured to BS 1566-1:2002 – Copper indirect cylinders for domestic purposes. BS 1566 specifies three grades of cylinders; for each grade the pressure that the cylinder will withstand is indicated. The grades of cylinder are:

- grade 1: 25m head
- grade 2: 15m head
- grade 3: 10m head.

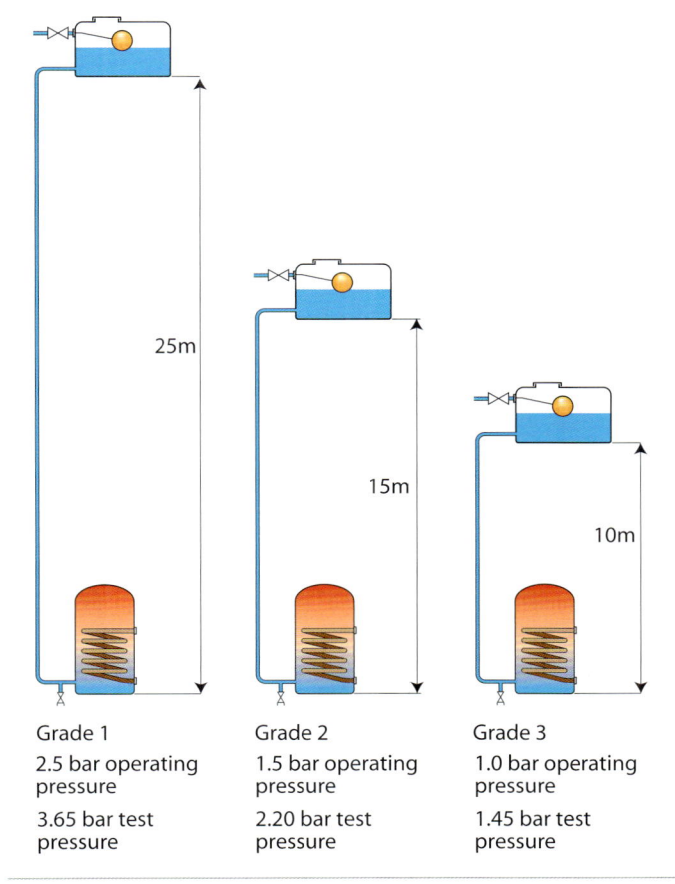

Grade 1
2.5 bar operating pressure
3.65 bar test pressure

Grade 2
1.5 bar operating pressure
2.20 bar test pressure

Grade 3
1.0 bar operating pressure
1.45 bar test pressure

Grades of storage cylinders

SmartScreen Unit 007 handout 4

B10 STORAGE CYLINDER SIZES AND CAPACITIES

Open-vented hot water storage cylinders are available in a wide range of sizes and capacities. The more common sizes are listed in the table below:

Size (mm)	Capacity (litres)
900 × 350	74
900 × 400	98
1050 × 400	116
900 × 450	120
1050 × 450	144
1200 × 450	166
1500 × 450	210

B11 THE PURPOSE AND USE OF SECONDARY CIRCULATION

If the hot water supply pipes have to run over long distances, cold water will be drawn off at the hot taps before the hot water arrives. This cold water is known as a dead leg, and its length is restricted by BS 6700:

Outside diameter of pipe (mm)	Maximum uninsulated length (m)
Up to 12	20
12–22	12
22–28	8
Over 28	3

If the hot water supply pipes exceed the recommended lengths, secondary circulation will need to be installed.

With secondary circulation, a return pipe runs from the furthest hot tap back to the cylinder, which it enters about one-third of the way down. A circulating pump placed on the return (pumping into the cylinder and made from bronze to ensure that corrosion does not pose a problem) aids efficient circulation to and from the cylinder. The system is shown in the diagram below.

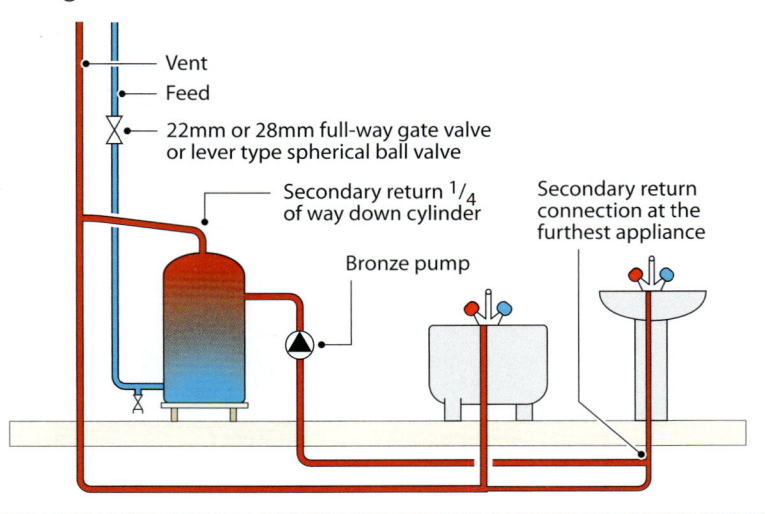

Secondary circulation

On larger installations, a clock can be wired to the pump so that circulation only occurs at certain times, for example in an office block where it is only needed at certain times of the day when the building is occupied.

B12 ANODIC CORROSION PROTECTION OF HOT WATER STORAGE CYLINDERS

Hot water storage cylinders can suffer from galvanic corrosion where there are two or more dissimilar metals, especially in areas where the water is soft (soft water is aggressive towards certain metals). This often occurs with certain types of brass fittings.

Hot water cylinders can be protected from galvanic corrosion by the use of a magnesium rod, which is either fastened to the bottom of the storage cylinder during manufacture or simply dropped in the draw-off connection during installation. This magnesium rod is known as the sacrificial anode. It works by encouraging the corrosion away from the weaker anodic metal in the installation, and is eaten away itself. If necessary, the magnesium rod can be replaced once it has been completely destroyed.

Magnesium sacrificial anode brazed to the bottom of the hot water storage cylinder

Sacrificial anode

B13 PIPE SIZES FOR OPEN-VENTED HOT WATER STORAGE SYSTEMS

Pipe sizes are critical if the correct flow rate is to be achieved at the outlets. For open-vented hot water systems fed from a cistern in the roof space, the size of the pipework would generally depend on the size of the system. A minimum 22mm cold feed pipe to the cylinder should be installed, with a full-flow ball type valve or a full-way gate valve to provide isolation of the hot water system. Occasionally, the cold feed may be 28mm if there is more than one bathroom in the property. The cold feed should be fitted with a drain-off valve at the lowest point to allow complete drain down of the hot water storage cylinder. The connection of the cold feed to the cold water storage cistern must be at least 25mm above any cold distribution pipework to ensure that, in the event of mains cold water failure, the hot water runs out first.

The hot water draw-off should have a gradual rise towards the vent and must have a minimum run of 450mm horizontally to prevent parasitic circulation occurring (see page 348). The vent pipe must rise vertically, terminate inside the cold water storage cistern and be sealed by means of a rubber grommet, and should have no valve installed anywhere along its length. The vent pipe and draw-off must be installed in minimum 22mm pipework.

The hot distribution pipework must be a minimum of 22mm to any large-volume appliances such as baths, but can be reduced in size to 15mm to supply kitchen sinks, wash basins and shower valves. It is good practice to install isolating valves at the appliances, although it is not a requirement of the Water Supply (Water Fittings) Regulations 1999. The pipework should have a gradual incline towards drain-off valves to permit total draining of the system for maintenance and repair.

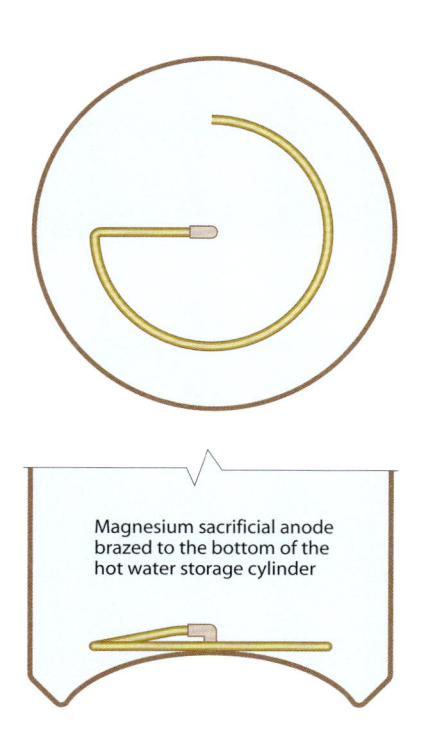

SUGGESTED ACTIVITY...

Calculation of the capacity of a cylinder is relatively simple, and involves the use of pi (π). Take π as being 3.142.

The formula for calculating the capacity of a cylinder is:

$\pi r^2 \times h \times 1000$

where:

π = 3.142
r = radius (m)
h = height (m)

Example:

A cylinder has a diameter of 500mm and a height of 1000mm; what is its capacity in litres?

Answer:

First, you will need to convert mm to m. Therefore, 500mm becomes 0.5m and 1000mm becomes 1m. The diameter is 0.5m so the radius will be half of that, therefore the calculation will read:

$3.142 \times (0.250 \times 0.250) \times 1 \times 1000 = 196.375$ litres

Now attempt the following calculations:

a A cylinder measures 300mm × 1050mm. What is its capacity?
b A cylinder measures 400mm × 850mm. What is its capacity?
c A cylinder measures 500mm × 1500mm. What is its capacity?

Answers: a. 74.22 litres, b. 106.82 litres, c. 294.56 litres

> **SUGGESTED ACTIVITY…**
> We covered the types of cisterns and their installation requirements in Chapter 006, page 301–302. Why not write down what you remember, and then check your notes against these pages?

> **SUGGESTED ACTIVITY…**
> To calculate the height of the vent pipe above the F&E cistern you must first determine the length from the overflow pipe to the lowest part of the cold feed pipe at the cylinder. If the distance between them is, say, 4m then the calculation is as follows:
>
> 4 × 40 + 150 = 310
>
> So the vent pipe must be taken above the overflow level of 310mm.
>
> Now try this for yourself.
>
> a There is a distance of 6m between the overflow level and the cold feed connection on the cylinder. What is the recommended height of the vent pipe?
>
> b There is a distance of 3m between the overflow level and the cold feed connection on the cylinder. What is the recommended height of the vent pipe?
>
> c The height of the vent pipe above the F&E cistern is 350mm. What is the distance between the overflow level and the cold feed connection to the cylinder?
>
> Answers: a. 390mm, b. 270mm, c. 5m

Where double-feed indirect cylinders are installed, the primary system must contain a separate feed and expansion cistern or an expansion vessel and disconnectable filling loop, which separates the primary water from the secondary water. The cold feed to the primary system from the feed and expansion (F&E) cistern can be installed in 15mm pipework and must not contain any form of isolation valve. The vent from the primary system must be installed in 22mm pipework; it should rise vertically and terminate over the F&E cistern. The height of the vent pipe above the F&E cistern should be not less than 150mm plus 40mm for every metre in height from the overflow level to the lowest point of the cold feed pipe.

The F&E cistern must be capable of accommodating an expansion of 4 per cent of the total amount of water contained in the primary system and any heating system installed.

C CENTRALISED INSTANTANEOUS HOT WATER SYSTEMS

Instantaneous hot water heaters take cold water direct from the mains cold water supply and pass it through a heat exchanger, where it is heated to a usable temperature before being discharged at the hot outlets. There is no storage of hot water in the units themselves. There are three basic types:

- thermal stores
- combined primary storage units
- instantaneous gas-fired multi-point hot water heaters/combination boilers.

C1 THERMAL STORES

Sometimes called water-jacketed tube heaters, thermal stores work by passing mains cold water through two heat exchangers that are encased in a large storage vessel of primary hot water fed from a boiler. They are very similar to an indirect system but they work in reverse.

Inside the unit are two heat exchangers, which the mains cold water passes through, and a small expansion chamber. The expansion chamber allows for the small amount of expansion of the secondary water. The primary water can reach temperatures of up to 82°C which could, potentially, be transferred into the secondary water. Because of this, an adjustable thermostatic mixing valve blends the secondary hot water with mains cold water so that the water does not exceed 60°C.

Thermal stores

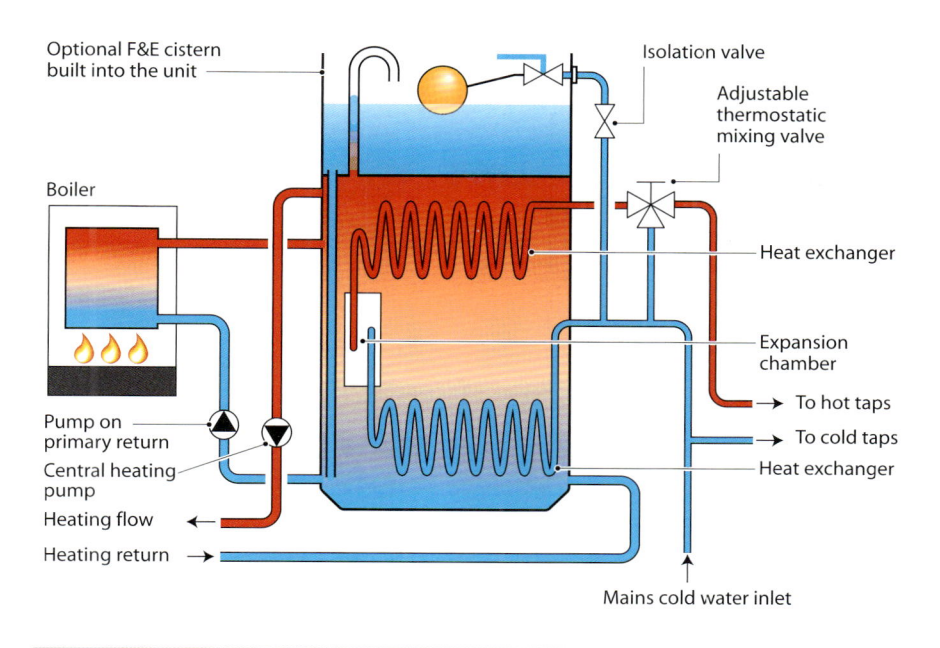

Layout of a thermal store

When to choose a thermal store system

Advantages:
- Instantaneous hot water
- Fully compliant with Document L of the Building Regulations
- Hot water pipework can be 15 mm, so cheaper installation costs

Disadvantages:
- High initial cost of the cylinder
- Requires a lot of space
- The thermal store needs to be kept hot at all times

Thermal store system considerations:

- **Property size** – suitable for all domestic properties.
- **Storage capacity** – none.
- **Fuel type** – used with condensing gas boilers.
- **Installation cost** – initial cost of the cylinder is high, but units can also supply central heating.
- **Fuel efficiency** – improves fuel efficiency by reducing boiler cycling; typical efficiencies comparable to condensing boilers.

C2 COMBINED PRIMARY STORAGE UNITS (CPSU)

These are very similar in design to thermal stores, in that cold water from the mains supply is passed through a heat exchanger. The difference is that the unit has its own heat source – a gas burner to heat the primary water – eliminating the need for a separate boiler.

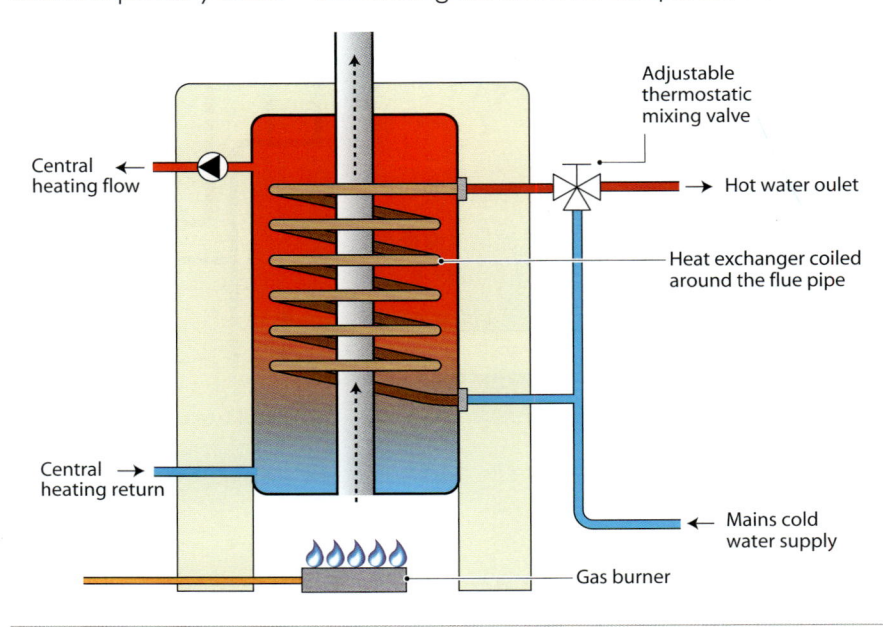

Combined storage unit

When to choose a CPSU system

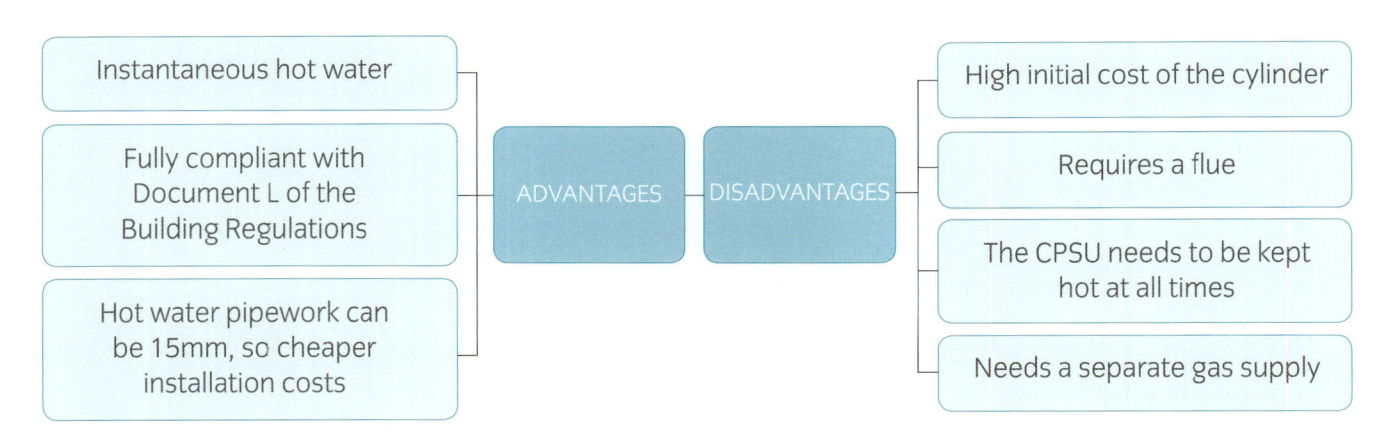

Advantages:
- Instantaneous hot water
- Fully compliant with Document L of the Building Regulations
- Hot water pipework can be 15mm, so cheaper installation costs

Disadvantages:
- High initial cost of the cylinder
- Requires a flue
- The CPSU needs to be kept hot at all times
- Needs a separate gas supply

CPSU system considerations:

- **Property size** – suitable for all domestic properties.
- **Storage capacity** – none.
- **Fuel type** – natural gas.
- **Installation cost** – initial cost of the cylinder is high, but units can also supply central heating.

- **Fuel efficiency** – efficiencies are less than for condensing boilers, at around 76 per cent.

C3 INSTANTANEOUS GAS-FIRED MULTI-POINT HOT WATER HEATERS AND COMBINATION BOILERS

Although these two appliances are quite separate, they both work in a similar manner in that they provide instantaneous hot water when the tap or outlet is turned on.

Instantaneous gas-fired multi-point hot water heaters

Instantaneous gas-fired multi-point water heaters work on the principle of pressure differential. A crucial part of this type of heater is the venturi tube, which works on Bernoulli's principle (see Chapter 004, page 185). It is the venturi tube that creates the pressure difference.

When the hot tap is opened, water passes through the heat exchanger. The movement of water causes the gas valve to open and the gas is ignited by a pilot light (also known as piezo ignitor). The gas valve opens because of a reduction in pressure on one side of a rubber diaphragm in the pressure differential valve as the water passes over the venturi tube. The movement of the diaphragm causes an attached push rod to open the gas valve, allowing gas to flow to the burner. When the tap is closed, the pressure within the pressure differential valve equalises and a spring closes the valve.

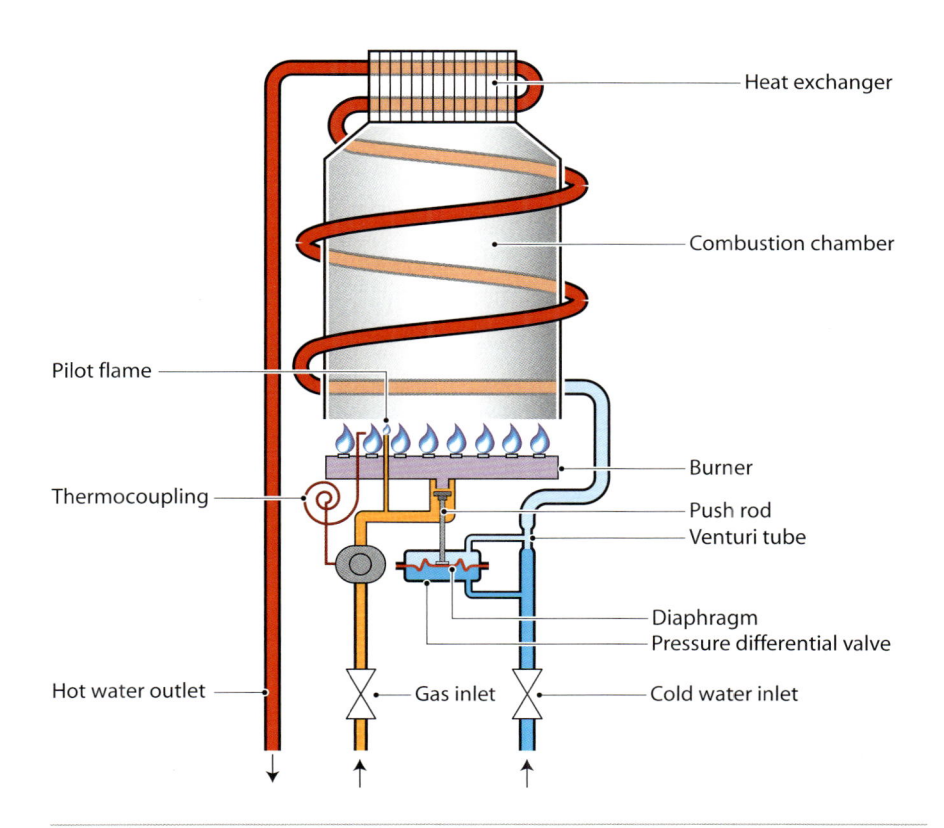

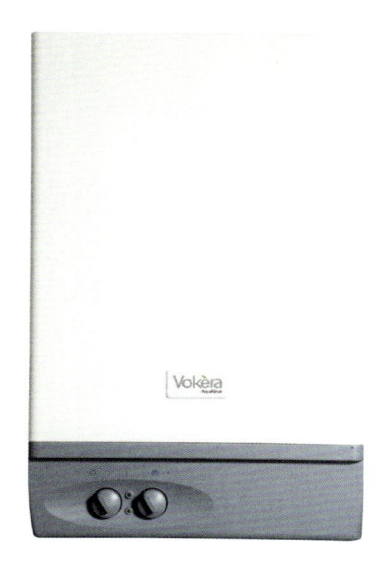

SmartScreen Unit 007 animation 5

The layout of an instantaneous hot water heater

An instantaneous multi-point hot water heater

These heaters can supply two or three wash basins individually in a single installation, but the hot taps should only be opened one at a time – the flow rate from this type of heater is insufficient to supply more than one open tap simultaneously.

When to choose an instantaneous gas-fired multi-point hot water heater system

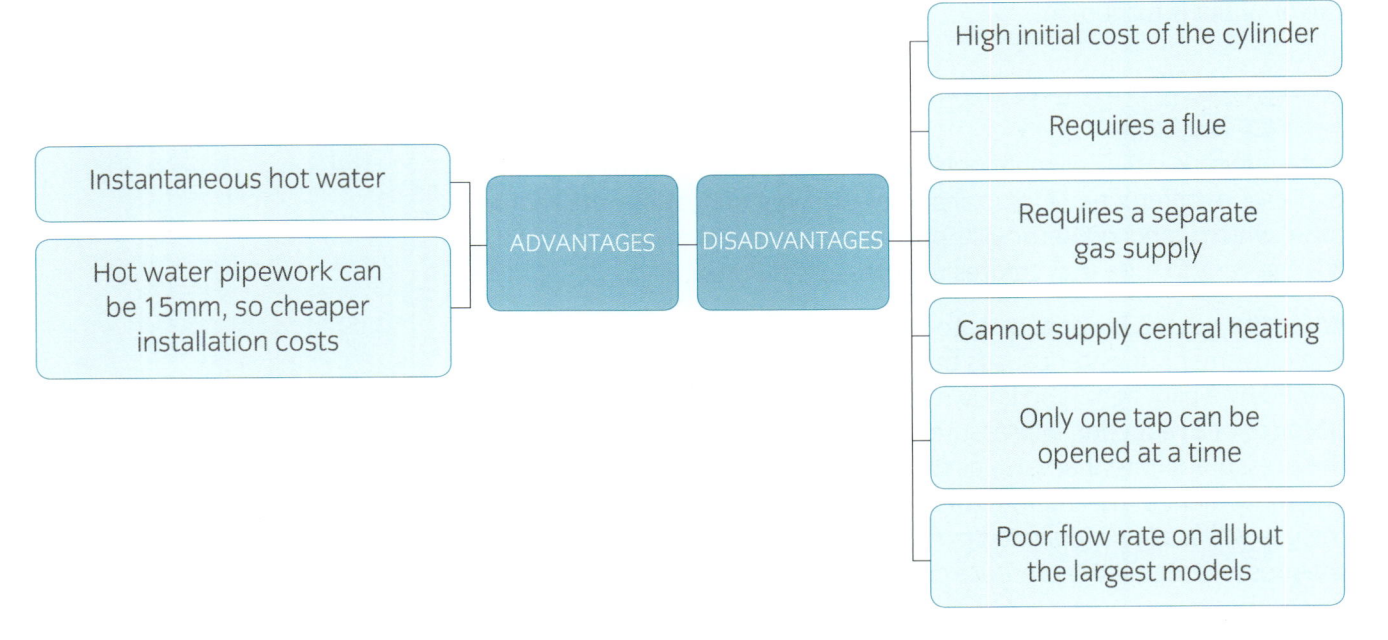

ADVANTAGES
- Instantaneous hot water
- Hot water pipework can be 15mm, so cheaper installation costs

DISADVANTAGES
- High initial cost of the cylinder
- Requires a flue
- Requires a separate gas supply
- Cannot supply central heating
- Only one tap can be opened at a time
- Poor flow rate on all but the largest models

Instantaneous gas-fired hot water heater considerations:

- **Property size** – small domestic properties and flats.
- **Storage capacity** – none.
- **Fuel type** – natural gas.
- **Installation cost** – initial cost of the cylinder is high, but the simple system layout makes the installation relatively cheap.
- **Fuel efficiency** – efficiencies are around 86 per cent.

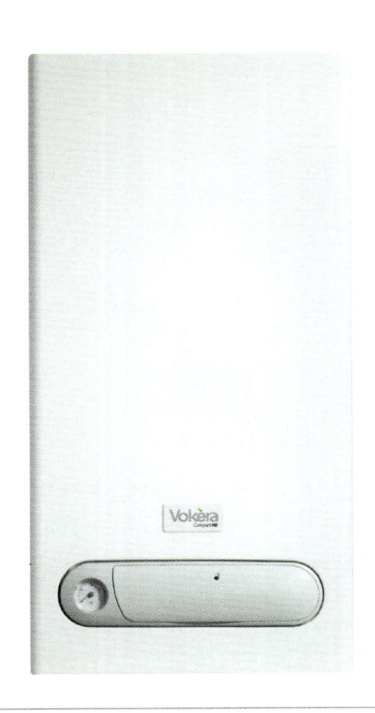

Combination boilers

Combination boilers (combis) provide instantaneous hot water and central heating from one centrally sited unit. These units use the same principle as thermal stores, heating the domestic water through a water-to-water heat exchanger.

Combis are the second most popular hot water supply system but they are not covered by BS 6700, other than to recommend that the water should be treated to reduce limescale, which can affect the plate heat exchanger. Key points of the combination boiler are:

- it provides instantaneous hot water, which means that water is only heated when it is needed

A typical condensing combination boiler

- cold water storage is not required, therefore the risk of damage through cisterns freezing is eliminated.

The installation of combination boilers will be covered in Chapter 008.

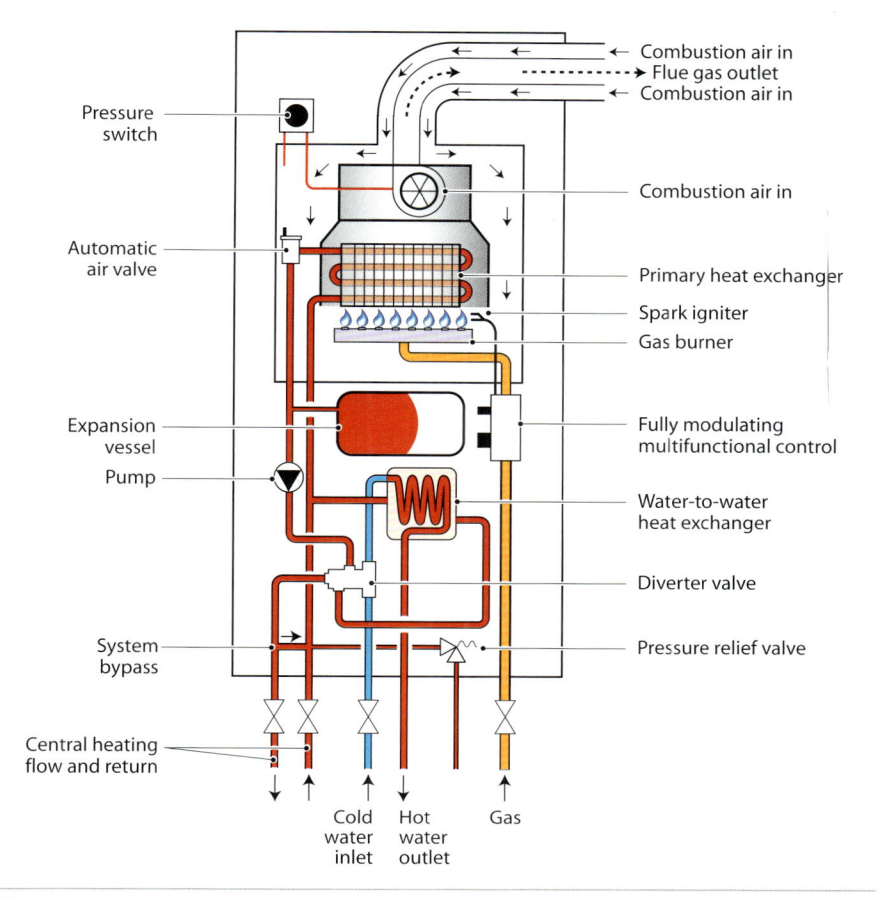

The layout of a combination boiler

When to choose a condensing combination boiler system

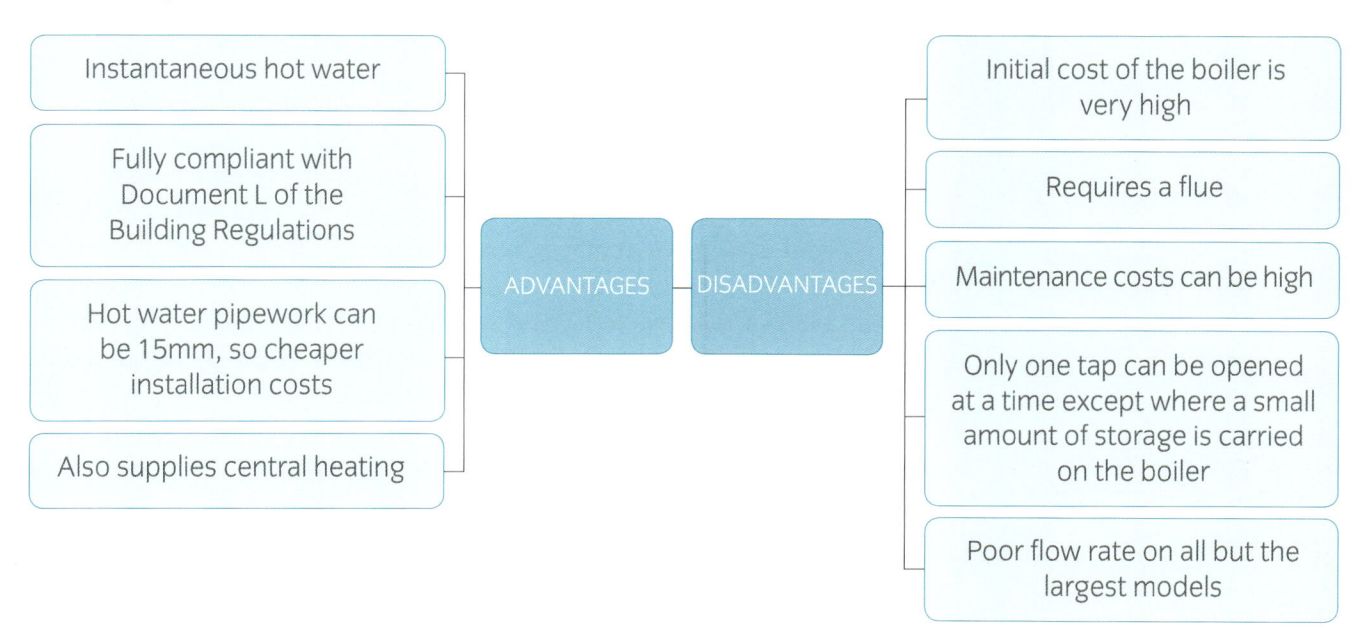

ADVANTAGES
- Instantaneous hot water
- Fully compliant with Document L of the Building Regulations
- Hot water pipework can be 15mm, so cheaper installation costs
- Also supplies central heating

DISADVANTAGES
- Initial cost of the boiler is very high
- Requires a flue
- Maintenance costs can be high
- Only one tap can be opened at a time except where a small amount of storage is carried on the boiler
- Poor flow rate on all but the largest models

Condensing combination boiler system considerations:

- **Property size** – small to medium domestic properties and flats. Not suitable for large properties.
- **Storage capacity** – none (except on some larger models).
- **Fuel type** – natural gas, LPG, oil.
- **Installation cost** – initial cost of the heater is expensive but the simple system layout makes the installation relatively cheap.
- **Fuel efficiency** – efficiencies are around 92 per cent for gas and 88 per cent for oil.

C4 PIPE SIZES FOR INSTANTANEOUS HOT WATER SYSTEMS

Since most instantaneous heaters, condensing combination boilers, thermal stores and CPSUs are supplied by a 15mm mains cold water connection, the installation pipework to the taps is relatively easy. You can take 15mm pipework to all appliances, including large-volume appliances such as baths. It should be remembered, however, that the flow rate from some instantaneous heaters such as gas instantaneous hot water heaters and combination boilers is only sufficient to for one tap at a time to be used. When two taps are open, the flow rate drops considerably.

D LOCALISED HOT WATER SYSTEMS

A localised system of hot water supply is always installed at the place where it is needed (such as a wash basin or a sink) where connection to a centralised system is impractical. For this reason, they are often called point-of-use or single-point systems. Like centralised systems, they are available in either storage or instantaneous versions.

D1 INSTANTANEOUS POINT-OF-USE WATER HEATERS

Like all instantaneous water heaters, this type of appliance heats the water straight from the water main and only operates as long as the water is flowing. It has no hot water storage. There are two different types: over sink and under sink. Over sink water heaters are sited over the sink or wash basin that they are serving and have a swivel spout to direct the water to where it is needed; under sink water heaters are fitted below the appliance. They can either be fuelled by gas (over sink only) or electricity and are usually inlet controlled, which means that the flow of water is controlled as it enters the appliance. Under sink models (electric) are outlet controlled.

Gas-fired over sink point-of-use water heaters work in a very similar way to their larger cousin, the gas-fired instantaneous multi-point water heater. This is a flueless-type water heater, which means that care

SmartScreen Unit 007 handout 10

should be taken with regard to its installation position. Restrictions apply where the room is too small in terms of volume (m³) and where ventilation of the room is difficult. They can only be installed by a gas-competent engineer, and you will not deal with them at this level.

Electric point-of-use water heaters are available in two forms:

- **Simple hand-wash heaters** – These are usually installed over a single wash basin (some models can be installed under sink). They are available in many different styles with outputs of about 3kW. They are either controlled via an external tap supplied with the unit or an integral tap/temperature selector. A feature of this type of unit is that the temperature of the water is dependent on the flow of water. The faster the water flows through the heater element, the cooler the water. Because of this, flow rates are poor but sufficient to allow hand washing.

- **Electric showers** – There are many different types on the market, with varying outputs from 8.5kW to 11kW. Many feature microchip technology, allowing the temperature of the water to be stabilised. Flow rates are dependent on the kW output of the unit; the higher the output, the better the overall flow rate at a showering temperature. All electric showers feature a low-pressure heater element cut-off so that the temperature of the water does not cause harm if the supply pressure/flow rate is low. Most will operate on 1 bar of water pressure and some models are available for cistern-fed cold water supplies. The flow of water is controlled by a simple solenoid valve, which only allows water to flow when the electricity supply is turned on.

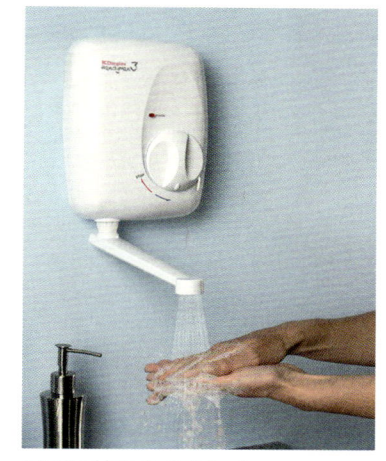

A typical hand wash heater

Solenoid valve

A solenoid valve operates with the aid of an electromagnet. When electricity is supplied to the electromagnet of the valve, the valve becomes magnetised and snaps open, allowing water to flow. Once the electricity has been switched off, the valve is no longer magnetised and a spring snaps the valve shut.

SmartScreen Unit 007 animation 4

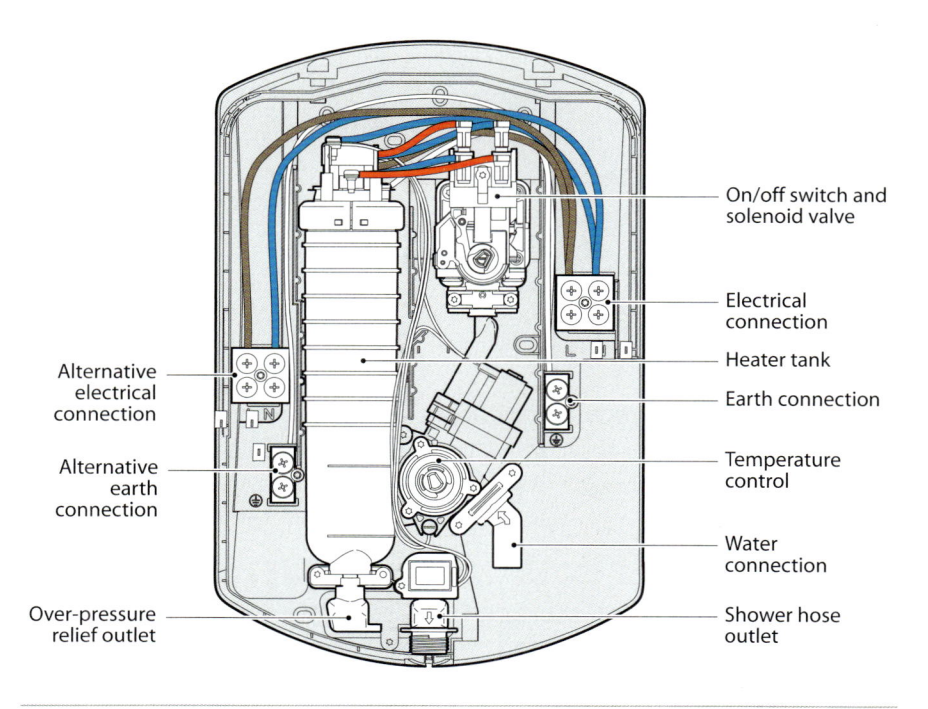

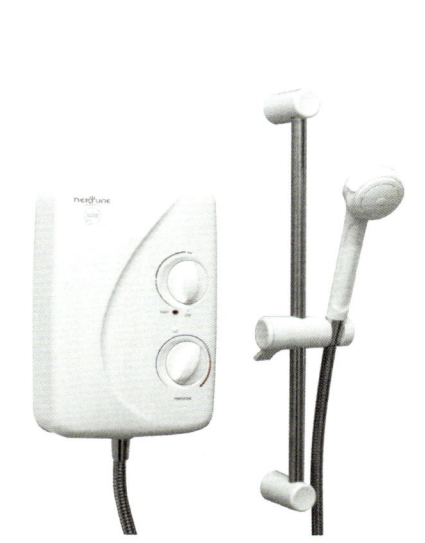

Inside an electric shower

Labels: Alternative electrical connection; Alternative earth connection; Over-pressure relief outlet; On/off switch and solenoid valve; Electrical connection; Heater tank; Earth connection; Temperature control; Water connection; Shower hose outlet

A typical electric shower

> **KEY POINT**
>
> You must not attempt to install any electrical appliance such as showers, immersion heaters or handwash heaters unless you are competent to do so and have the correct qualifications. Remember – electricity can kill!

D2 STORAGE POINT-OF-USE WATER HEATERS

Storage point-of-use heaters are small, usually inlet-controlled, heaters that have a small amount of storage, generally below 15 litres. There are three general types:

- **Over sink storage water heaters** – These, as the name suggests, are sited over the sink. They have a spout to discharge the water. One feature of this type of water heater is its tendency to drip when the water is being heated. This is because the spout acts as a vent, releasing the excess expanded water. This is a vital function that stops the heater from becoming over-pressurised due to the water expansion, and is the reason that these heaters must be inlet controlled. The cold water is usually supplied direct from the mains cold water supply via a spreader tee, which ensures even displacement of the hot water in the water heater.

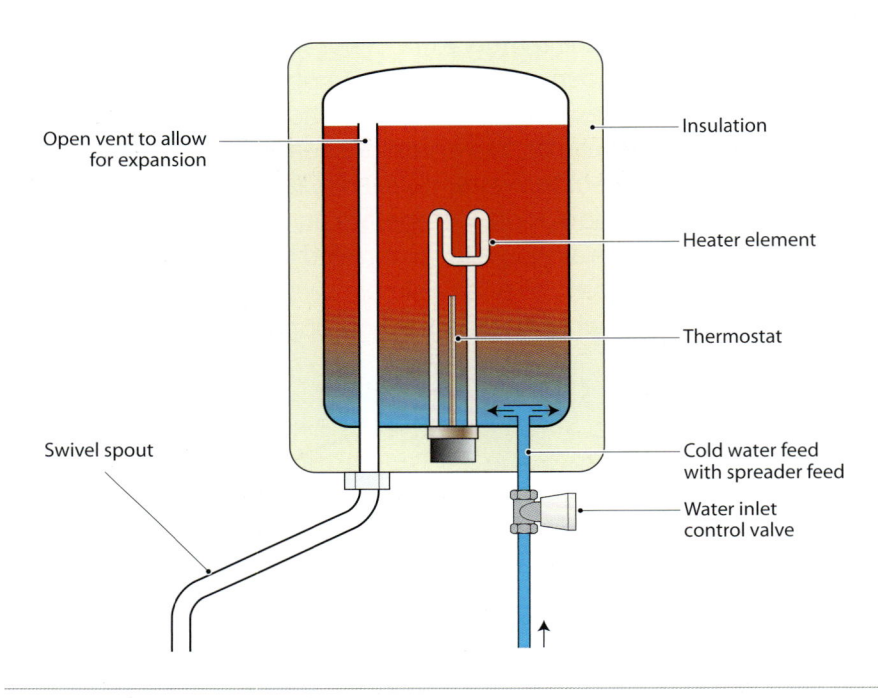

Inside an over sink storage water heater

- **Under sink storage water heaters** – These are fitted under the sink or wash basin rather than above it. They are similar in design to the over sink-type but they are installed with a special tap that allows inlet water control while still maintaining an open vent through the outlet of the tap (see diagram).

- **Unvented under sink storage heaters** – Unvented hot water storage heaters are connected direct to the mains cold water supply and deliver hot water at near-mains cold water pressure. Because they have less than 15 litres of storage, they are not subject to the stringent regulations that surround the installation of larger unvented hot water storage units.

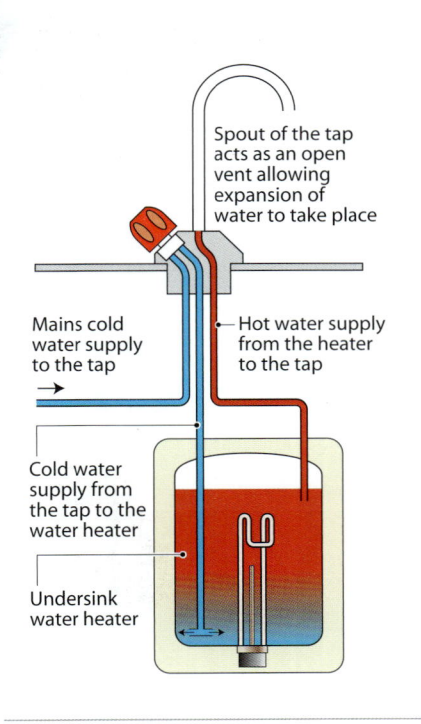

The special tap for use with under sink storage water heaters

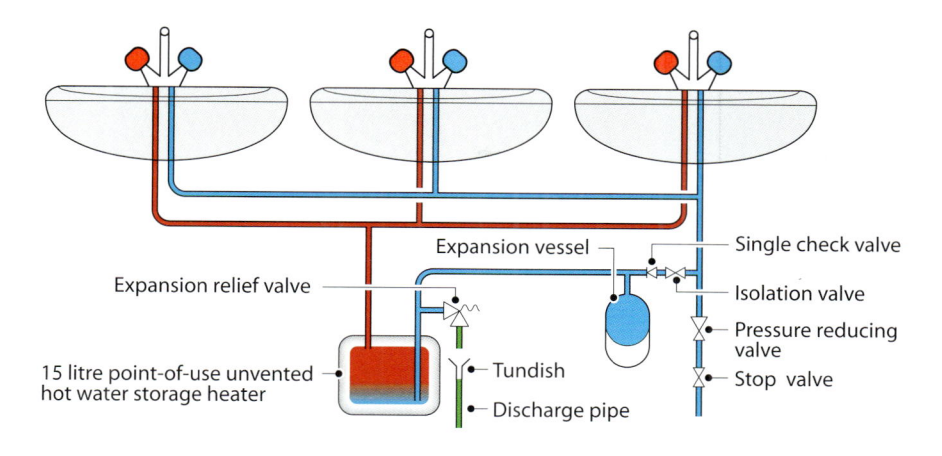

The pipework layout of unvented under sink storage water heaters

There are certain controls and fittings required with unvented hot water storage heaters and each has a specific job to do:

- **Pressure reducing valve** – This reduces the mains cold water to the correct operating pressure before it enters the heater. It is usually set to a specific pressure by the manufacturer.

- **Single check valve** – The positioning of the single check valve is critical as it prevents hot water from expanding and backflowing into the cold water main (see Chapter 006, page 312).

- **Expansion vessel** – This is required if the expansion of water cannot be taken up in the pipework without contaminating the mains cold water feed into the heater. It consists of a small vessel that is divided by a rubber membrane. One side of the membrane there is a charge of air at a set pressure; the other side is taken up by the water in the system. As the water is heated, the water expands into the expansion vessel, thereby maintaining the pressure within the heater at a safe level.

- **Expansion relief valve** – This (sometimes known as the pressure relief valve) is required in case of over-pressurisation due to failure of either the expansion vessel or the pressure-reducing valve. If the expansion of water cannot be accommodated within the system or the pressure reaches an unacceptable level, the expansion valve opens, relieving the excess pressure to a safe level.

- **Discharge pipework and tundish** – The discharge pipework is installed to remove any water that may be discharged from the expansion relief valve. It runs to a safe location outside the building and should be installed in copper tube, as the water that is discharged may be at high temperature. The tundish is installed as an air gap between the discharge pipework and the heater.

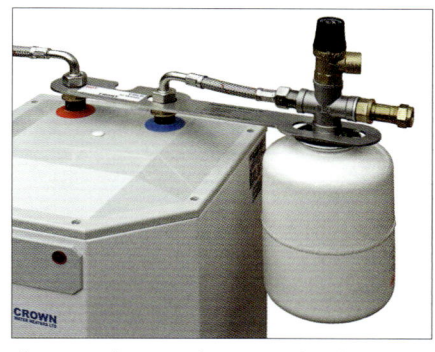

The expansion vessel connected to the water heater

The tundish

When to choose a localised hot water system

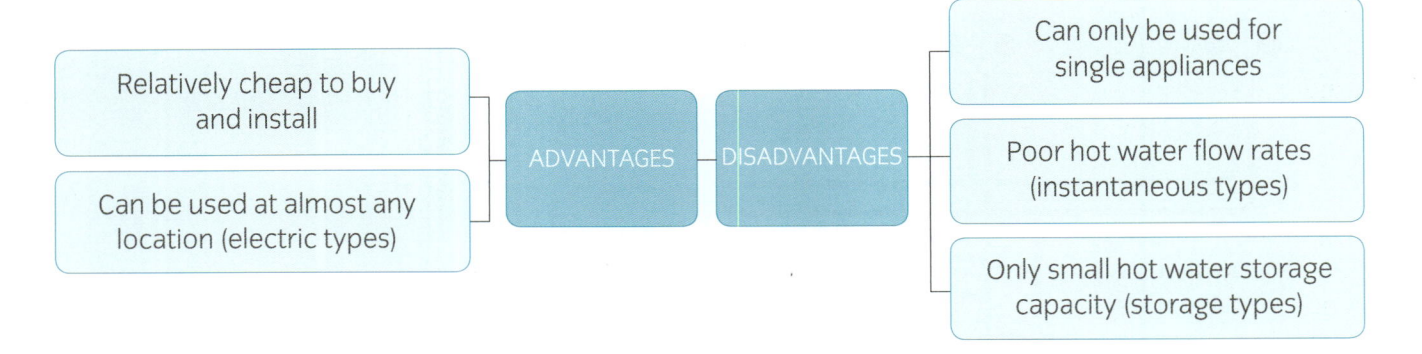

- Relatively cheap to buy and install
- Can be used at almost any location (electric types)

ADVANTAGES

DISADVANTAGES

- Can only be used for single appliances
- Poor hot water flow rates (instantaneous types)
- Only small hot water storage capacity (storage types)

Localised hot water system considerations:

- **Property size** – can be installed at almost any domestic or industrial property.
- **Storage capacity** – none (instantaneous); below 15 litres (storage types).
- **Fuel type** – natural gas, electricity.
- **Installation cost** – relatively cheap.
- **Fuel efficiency** – electric water heaters are 100 per cent efficient. Gas efficiencies are unknown.

E GENERAL REQUIREMENTS FOR DOMESTIC HOT WATER SYSTEMS

This section looks at the general requirements for planning and installing hot water systems within a dwelling, including the measures needed to regulate the temperature of water within the systems.

E1 THE INSTALLATION OF HOT WATER PIPEWORK TO BS 6700

The installation of hot water pipework is covered in BS 6700. The materials used are usually copper tubes to BS EN 1057 and polybutylene pipes and fittings. These are the only materials that do not cause contamination of the water and that can withstand the temperatures required. The pipework should be capable of withstanding at least one and a half times the normal operating pressure of the system and sustained temperatures of 95°C, with occasional increases up to 100°C to allow for malfunctions of hot water heating appliances. All systems must be capable of accommodating thermal expansion and movement within the pipework. Care should be taken when pressure testing open-vented cylinders to ensure that the maximum pressure that the cylinder can withstand is not exceeded. If necessary, the cylinder should be disconnected and the pipework capped before testing commences.

The installation methods for hot water systems are very similar to those for cold water systems. Care should be taken when installing hot and cold water pipework side by side so that any cold water installation is not adversely affected by the hot water pipework.

E2 TEMPERATURE CONTROL OF HOT WATER SYSTEMS

According to BS 6700, hot water systems must not be allowed to exceed 100°C at any time. A maximum normal operating temperature of 60°C is required to kill off legionella bacteria. There are several methods for maintaining and controlling the temperature of hot water systems and preventing them from exceeding the maximum temperature:

- A thermostat should be installed and set to the temperature required. A second thermostat, called a high limit thermostat, will start to operate should the maximum temperature be exceeded. This is known as a second tier level of temperature control.

- Some immersion heaters have a resettable double thermostat. One thermostat can be set to 50–70°C, the other is a resettable high limit thermostat designed to switch off the power to the unit when the maximum temperature is exceeded. It can be manually reset.

- Some immersion heaters have a non-resettable double thermostat. One thermostat can be set between 50–70°C; the other is a high limit thermostat designed to permanently switch off the power to the unit until the immersion heater is replaced and the fault rectified.

- Open-vented double-feed indirect cylinders with gravity or pumped primary circulation must be fitted with a minimum of a cylinder thermostat and a motorised zone valve that closes when the water in the cylinder reaches a preset level.

- Open-vented cylinders with no high limit thermostat can be fitted with a temperature relief valve which opens automatically at a specified temperature to discharge water via a tundish and discharge pipework safely to outside the property.

E3 THE USE OF THERMOSTATIC MIXING VALVES

As we have already seen, the maximum temperature of hot water in a dwelling should not exceed 60°C, but this is far too hot for bathing and showering. Water with a temperature as low as 52°C can cause serious burns to a child if it is exposed to the skin for two minutes or more.

In April 2010, new legislation under Document G of the Building Regulations required that all new-build properties and renovations have temperature control to baths not exceeding 48°C and all hot water storage cylinders where the stored water may exceed 80°C (usually solid-fuel heated cylinders). The Care Standards Act 2000 requires that all properties to which the public have access, such as schools, hospitals and nursing homes, have the temperature of water delivered to all hot

> **SUGGESTED ACTIVITY...**
> To refresh your memory about the tools, materials and installation requirements of pipework within dwellings, check out Chapter 005, Sections B and D (page 222 and 257) and Chapter 006, section E (page 328).

A thermostatic mixing valve

outlets (except where food preparation is carried out) limited to 43°C. This is done by the use of thermostatic mixing valves for appliances and inline blending valves for storage cylinders.

A thermostatic mixing valve mixes hot and cold water together and supplies it to an appliance at exactly the right temperature. It uses a temperature-sensitive element, usually a wax cartridge that expands and contracts, to maintain a specific temperature based on the temperatures of the hot and cold water entering the valve. The length of pipe from the mixing valve to the taps should be kept as short as possible.

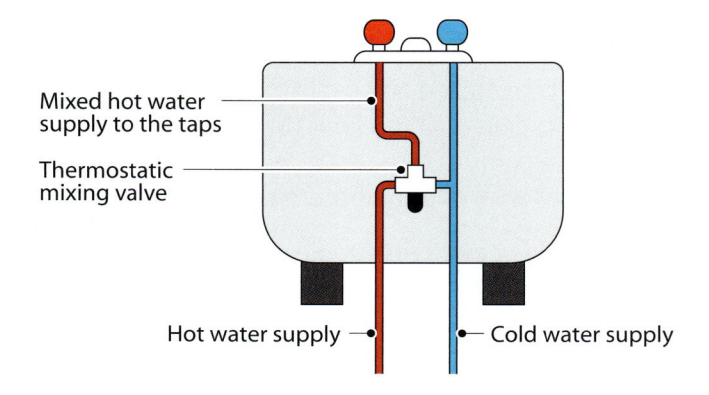

Installation of a thermostatic mixing valve

E4 THE INSULATION OF HOT WATER PIPEWORK

When installing new hot water installations in domestic properties, pipes should be wrapped with thermal insulation that complies with the Domestic Heating Compliance Guide. There are four main considerations:

- Primary circulation pipes for heating and hot water circuits should be insulated wherever they pass outside the heated living space, such as below ventilated suspended timber floors and through unheated roof spaces. This is for protection against freezing.

- Primary circulation pipes for domestic hot water circuits should be insulated throughout their entire length except where they pass through floorboards, joists and other structural obstructions.

- All pipes connected to hot water vessels, including the vent pipe, should be insulated for at least 1m from their points of connection to the cylinder or at least up to the point where they become concealed.

- If secondary circulation, such as a pumped circuit-feeding bath and basin taps in a large property, is installed, all pipes fed with hot water should be insulated to prevent excessive heat loss through the secondary circulation circuit.

SmartScreen Unit 007 handout 1

E5 EXPANSION OF HOT WATER PIPEWORK

When the pipework of the hot water system is filled with hot water, the heated pipework will expand. As the pipework cools down, it will contract. This expansion and contraction must be accommodated during the installation process. Pipes that pass through walls and floors where not enough room has been left for expansion will 'tick' and 'creak' as expansion and contraction takes place.

The rate of expansion (known as the coefficient of linear expansion) will depend on the material the pipe is made from. Generally, pipework made from plastic materials tends to expand more than that made from copper:

- the coefficient of linear expansion of plastic pipe is 0.00018 per metre per °C
- the coefficient of linear expansion of copper pipe is 0.000016 per metre per °C.

This means that for every degree rise in temperature, polybutylene pipe will expand 0.00018m in every metre and copper will expand 0.000016m in every metre.

F INSTALLATION OF SHOWER MIXING VALVES AND SHOWER BOOSTING PUMPS

Chapter 006, pages 322–25 looked at shower mixing valves and the various types of shower boosting pumps. This section investigates how to install these appliances within hot water systems.

There are a number of different shower valves available, ranging from bath/shower mixer taps and simple shower mixing valves to thermostatic and pressure-balancing shower valves. The method of installation is, in most cases, the same for each type of valve, with the requirement that equal pressure and flow rates exist on both the hot water and cold water installations. There are four main types of installation:

- **Cistern-fed supplies** – Simple installations from a storage cistern in a roof space that supply water to both hot and cold water systems, thus ensuring equal pressures across both systems.
- **Cistern-fed supplies and booster pump** – Installations that include an inlet twin impeller shower boosting pump or an outlet single impeller shower boosting pump (often called 'power showers').
- **Mains hot and cold supplies** – Installations that use mains cold and mains-fed hot water systems.
- **Unbalanced supply pressures** – Installations that use supplies where there is an imbalance in supply pressures (such as those systems that use a combination boiler/instantaneous hot water heater for the hot water supply).

SUGGESTED ACTIVITY...

To calculate the amount of expansion that takes place on a given length of pipe:

Length of pipe (m) × coefficient of linear expansion × temperature rise

What is the expansion on a 15mm copper pipe 6m in length, when the pipe is heated from 10°C to 60°C?

6 × 50 × 0.000016 = 0.0048m or 4.8mm

Now attempt these examples:

a What is the expansion on a 15mm polybutylene pipe 6m in length, when the pipe is heated from 10°C to 60°C?

b What is the expansion on a 15mm copper pipe 20m in length, when the pipe is heated from 15°C to 50°C?

c What is the expansion on a 15mm copper pipe 30m in length, when the pipe is heated from 12°C to 58°C?

Answers: a. 0.054m or 54mm, b. 0.0112m or 11.2mm, c. 0.02208m or 22.08mm

> **KEY POINT**
>
> In a hot water storage cylinder, water forms in layers of temperature from the top of the cylinder, where the water is at its hottest, to the base, where it is at its coolest. This is stratification, and it is necessary if the cylinder is to perform to its maximum efficiency. Manufacturers will purposely design storage vessels and cylinders with stratification in mind. They will design:
>
> - a vessel that is cylindrical in shape
> - a vessel that is designed to be installed upright rather than horizontally
> - a vessel with the cold feed entering the cylinder horizontally.

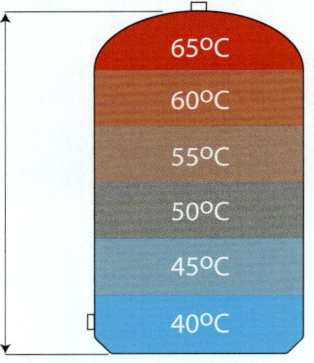

Hottest water at a max. temperature of 65°C is at the top of the cylinder

65°C
60°C
55°C
50°C
45°C
40°C

Hot water at a max. temperature of 40°C is at the bottom of the cylinder

SmartScreen Unit 007 handout 8

F1 CISTERN-FED SUPPLIES

Shower mixing valves fed from a storage cistern require equal pressures on both the hot and cold supplies to maintain the correct mixing ratio of hot and cold water. The safest type of valve to use is the thermostatic type, which maintains a constant temperature irrespective of the temperature of the incoming hot and cold supplies to the valve. Ordinary mixing valves also work well with cistern-fed supplies. Remember that ordinary mixing valves are not thermostatically controlled, so the water will eventually become cooler the longer the shower is used. This is because of stratification within the cylinder.

To create enough pressure to give a reasonable shower, there has to be a minimum of 1m from the bottom of the cistern to the shower head at its highest position.

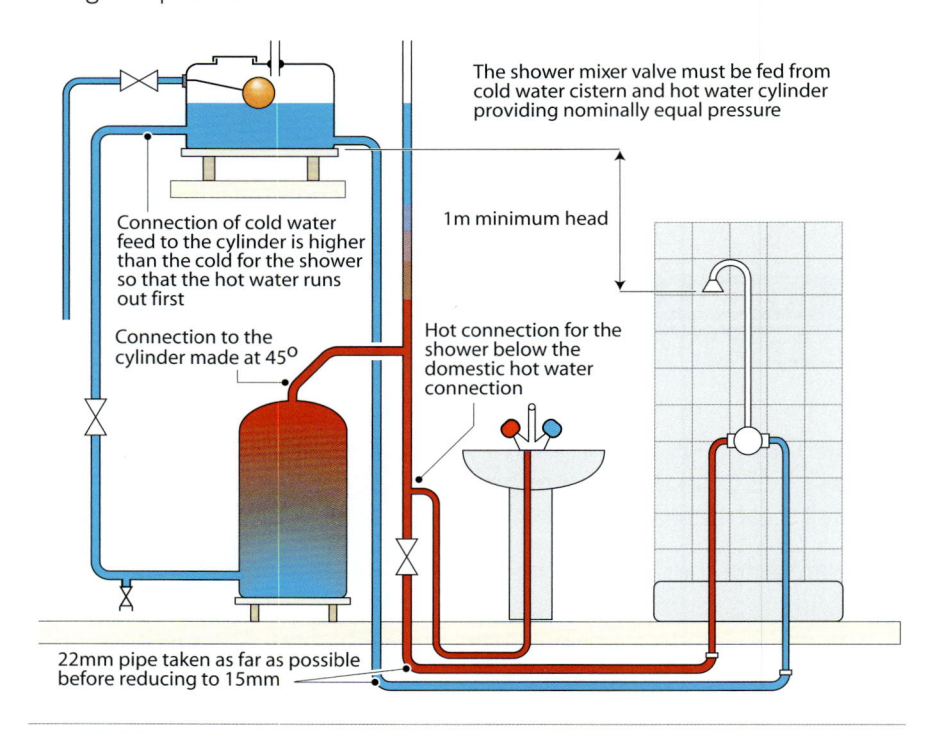

The shower mixer valve must be fed from cold water cistern and hot water cylinder providing nominally equal pressure

1m minimum head

Connection of cold water feed to the cylinder is higher than the cold for the shower so that the hot water runs out first

Connection to the cylinder made at 45°

Hot connection for the shower below the domestic hot water connection

22mm pipe taken as far as possible before reducing to 15mm

Gravity-fed shower installation

F2 CISTERN-FED SUPPLIES AND BOOSTER PUMP

There are two systems that use a shower booster pump:

- **Systems that use a twin impeller pump on the inlet to the mixer valve** – The pump increases the pressure of the hot and cold water supplies to the mixer valve independently. The water is then mixed to the correct temperature in the valve before flowing to the shower head (see diagram). There are two ways to make the hot connection to the cylinder. The first involves installing the hot water draw-off from the cylinder at an angle of between 30° and 60°, with the hot shower pump connection being made at an angle of 90° with a tee piece. This allows any air in the system to filter up to the vent and away from the hot shower pump inlet. The second method involves making a direct connection to the cylinder using a special fitting: either an Essex flange or a Surrey flange. The hot water is taken directly from the hot water storage vessel, avoiding any air problems that may occur.

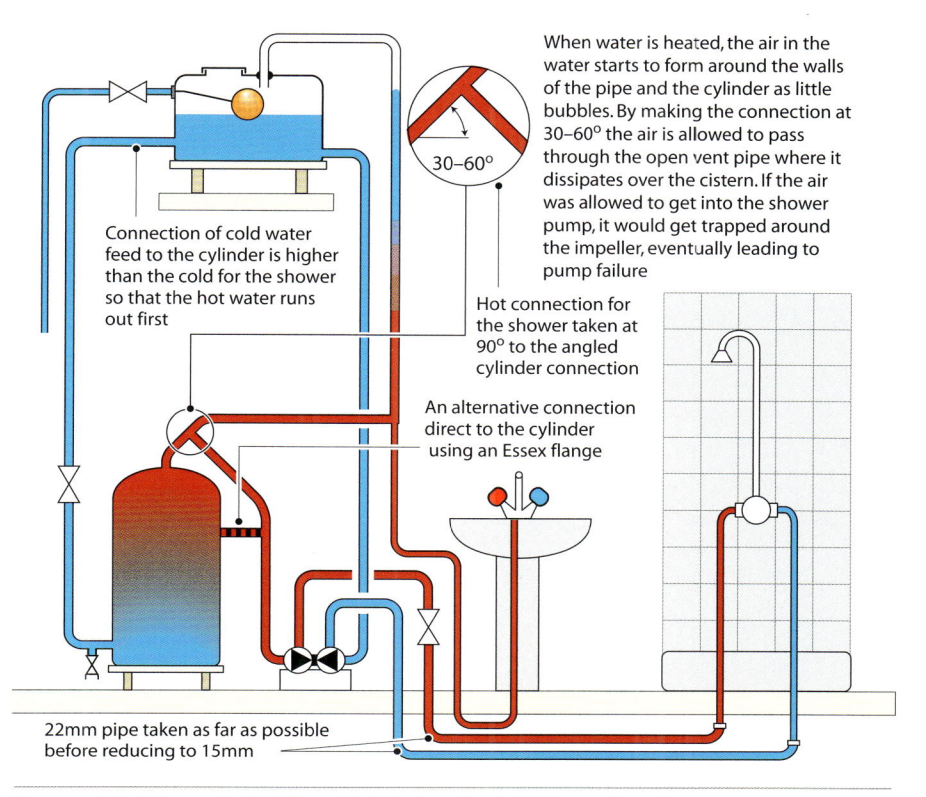

Connection of cold water feed to the cylinder is higher than the cold for the shower so that the hot water runs out first

When water is heated, the air in the water starts to form around the walls of the pipe and the cylinder as little bubbles. By making the connection at 30–60° the air is allowed to pass through the open vent pipe where it dissipates over the cistern. If the air was allowed to get into the shower pump, it would get trapped around the impeller, eventually leading to pump failure

Hot connection for the shower taken at 90° to the angled cylinder connection

An alternative connection direct to the cylinder using an Essex flange

22mm pipe taken as far as possible before reducing to 15mm

Pump-assisted shower installation with twin impeller, inlet shower booster pump

An Essex flange

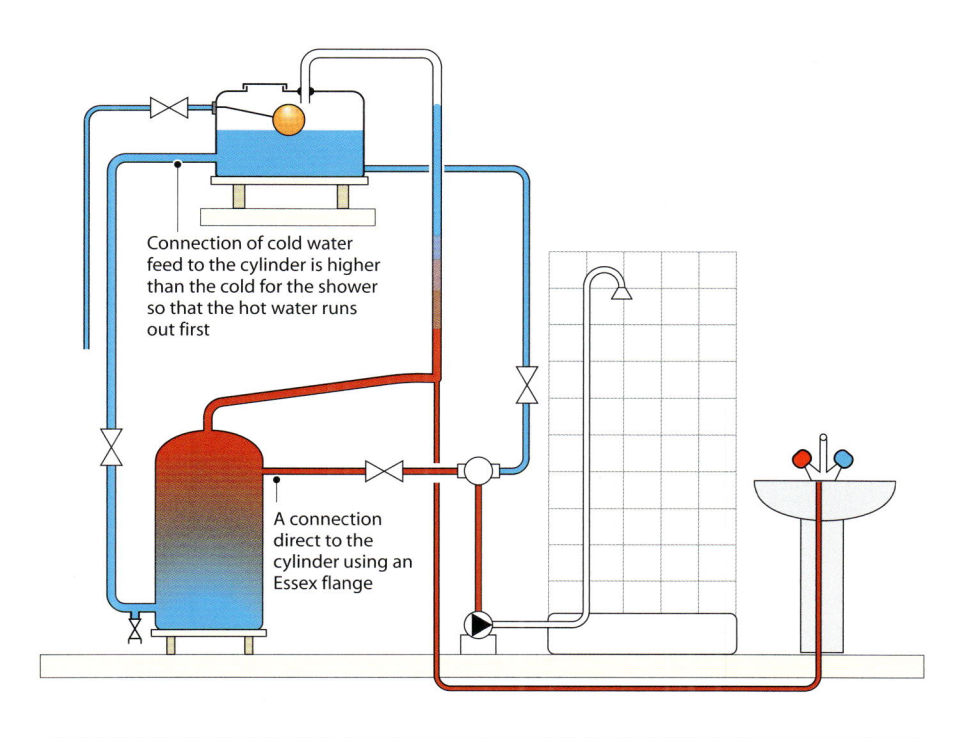

Connection of cold water feed to the cylinder is higher than the cold for the shower so that the hot water runs out first

A connection direct to the cylinder using an Essex flange

Pump-assisted shower installation with single impeller, outlet shower booster pump

383

A large water-volume 'deluge' shower head

- **Systems that use a single impeller pump off the outlet from the mixer valve** – These boost the water after it has left the mixer valve. They are usually used with concealed shower valves and fixed 'deluge' type, large water-volume shower heads.

In both of these installations, the pump increases the pressure of the water, which means that the minimum 1m head is not necessary. However, a minimum head of 150mm is required to lift the flow switches that turn the pump on. It may be possible to install the pump with a negative head, where the cistern is lower than the pump, provided a means of starting the pump is in place, such as a pull-cord switch.

F3 MAINS HOT AND COLD SUPPLIES

The installation of unvented hot water storage cylinders is covered at Level 3, but you may be required to install or maintain shower mixing valves that are installed on this type of system.

Shower pumps are not required, as the hot and cold supplies are fed directly from the mains cold water supply via a pressure-reducing valve that reduces the pressure of the water to the operating pressure of the unvented hot water storage cylinder. The obvious advantages of this are:

- the large amount of water that can be delivered to the shower head
- the force of the water leaving the shower head, giving a powerful Continental-type shower.

Because the unvented hot water cylinder usually operates at a slightly lower pressure than the mains cold water supply, the cold water to the shower must be at the same pressure as the hot water supply. This means that the cold supply must be connected after the pressure-reducing valve but before the single check valve on the unit (see diagram) to ensure equal hot and cold pressures.

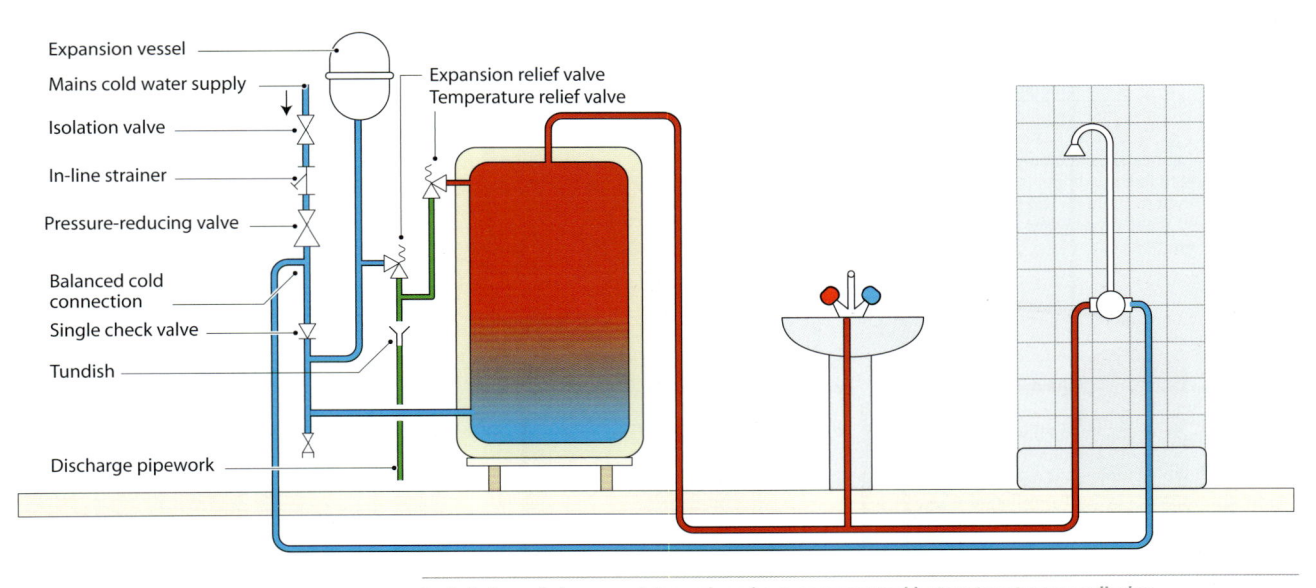

Installation of shower mixing valves from an unvented hot water storage cylinder

F4 UNBALANCED SUPPLY PRESSURES

Showers installed on instantaneous water heaters and combination boilers require a shower valve that is pressure compensating, because as the cold water passes through the hot water heater/combi boiler it loses pressure and flow rate, so an imbalance of pressure/flow rate between the mains cold water and the hot water from the heater occurs. The pressure-compensating shower mixer valve adjusts both pressure and flow rate within the shower valve body to give a reasonably powerful shower.

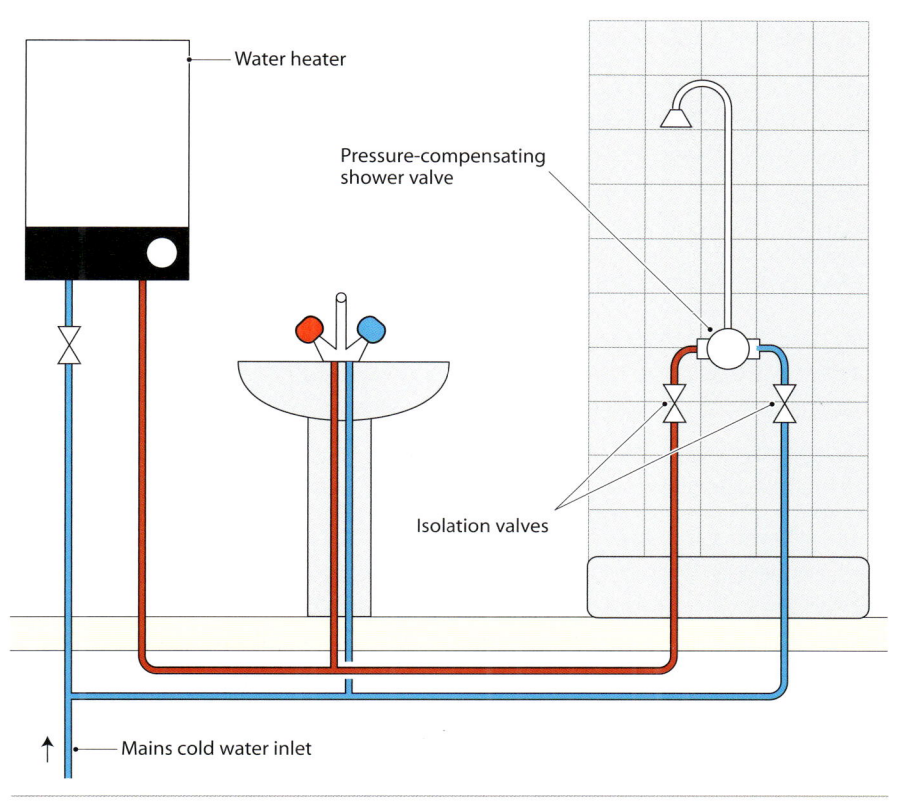

SmartScreen Unit 007 worksheet 3

Installation of a pressure-compensating shower mixing valve

G PROTECTION AGAINST BACKFLOW AND BACK SIPHONAGE

Chapter 006 looked at the different types of basic backflow prevention devices and air gaps. That theme is continued in this chapter – hot water is itself categorised as fluid category 2 simply because heat has been added to the cold, wholesome water. Other considerations here are that many of the bathroom appliances that are connected to the hot and cold supply are also at risk from fluid categories 3 and 5 (see page 309).

SUGGESTED ACTIVITY...
To refresh your memory about the types of fluid categories, backflow prevention devices and air gaps, check out Chapter 006, pages 309–14.

Appliances that may be at risk from backflow

Wash basins: fluid categories 2 and 3 risk	Taps for use with wash basins should discharge at least 20mm above the spill-over level of the appliance (AUK2 air gap). Mixer taps should be protected by the use of single check valves on the hot and cold supplies. Twin-flow mixer taps do not require any backflow protection, as the water mixes on exit from the tap.
Kitchen sinks: fluid category 5 risk	No backflow protection is required, as the height of the outlet is well above the spill-over level of the appliance. This is classified as an AUK3 air gap. If a mixer tap where both hot and cold water mix in the tap body is installed, single check valves must fitted on both hot and cold supplies. Twin-flow mixer taps do not require any backflow protection as the water mixes on exit from the tap.
Baths: fluid categories 2, 3 and 5 risk	As for wash basins, except that the air gap should be 25mm. Bath/shower mixer taps, where the water is fed from the mains cold water supply and there is a risk of the shower head being below the water level in the bath, should be protected by double check valves, or a shower hose retaining ring, which maintains an AUK2 air gap above the spill-over level of the bath.
Bidets: fluid categories 2, 3 and 5 risk	There are two types of bidet that are at risk from backflow: • Over-rim bidets using taps are protected from backflow by having an AUK2 air gap, as the taps discharge over the spill-over level of the appliance. These would be installed in the same way as wash basins. • Ascending spray bidets and over-rim bidets using taps with a hose connection must be installed as shown in the diagram below to prevent contamination.
Shower valves: fluid categories 2 and 3 risk	When both hot and cold supplies are fed from a cistern, no backflow protection is required. However, when both are fed from mains-fed supplies, single check valves are required with a hose-retaining ring to prevent the hose entering the water. If no retaining ring is fitted then both hot and cold supplies should have a double check valve installed.
Electric shower units: fluid categories 2 and 3 risk	A double check valve is required where a hose retaining ring is not fitted.

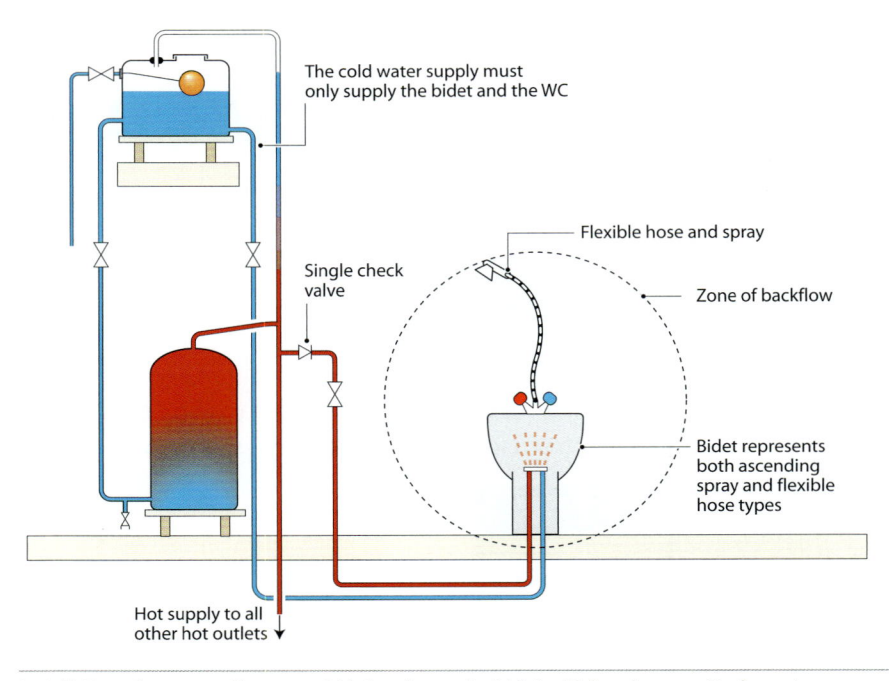

Installation of an ascending spray bidet and over-rim bidet with hand spray attachment

H TESTING AND COMMISSIONING HOT WATER SYSTEMS

The testing and commissioning of hot water systems are very similar to the processes for cold water installations (see Chapter 005, page 269–72 and Chapter 006, page 331–33). Hot water systems should be tested in accordance with the methods given in BS 6700 depending on the materials that have been installed. Commissioning also follows similar procedures. Hot water systems should be filled and flushed with cold water from the mains cold water supply before the system is filled to full capacity and run to full operating temperatures. A visual check for leaks should be maintained during this process, as leaks could appear when the hot water is pulled through the installation to the taps and the pipework expands.

Flow rates and temperatures should be checked at taps and shower mixing valves to ensure that they are in line with the manufacturer's installation instructions and any relevant regulations. Flow rates can be checked using a flow meter or weir gauge and temperatures checked using suitable thermometers. Thermostats on immersion heaters and boilers should be checked to ensure safe and correct operation. Benchmarking of hot water appliances should be completed and the benchmark certificates signed by the commissioning plumber.

Taps and terminal fittings should be checked for correct operation and any faults rectified before handover to the customer.

> **SUGGESTED ACTIVITY...**
> We covered the different types of taps and terminal fittings in relation to cold water systems in Chapter 006, Section D (page 314–22). The same principles apply to hot water systems.

SmartScreen Unit 007 handout 11

I FAULTS WITH OPEN-VENTED HOT WATER SYSTEMS

There are many faults that can occur with open-vented hot water systems. Some of these may be due to poor system design but most occur with use. Some of the more common faults are:

- **Loss of hot water** – This may be due to evaporation of the water in the feed and expansion (F&E) cistern installed on double-feed indirect cylinders, with gravity circulation to the heat exchanger. This is usually due to a sticking float-operated valve (FOV) that fails to top the water up as evaporation occurs. Because the FOV is stuck in the off position, the water evaporates down to the primary flow pipe and this stops circulation to the heat exchanger and prevents the cylinder getting hot. To rectify the fault, the FOV in the F&E cistern should be removed and repaired/replaced.

- **Immersion heater element failure** – This is usually due to corrosion of the immersion heater element sheath, allowing water to penetrate the heater element. This causes a short circuit, which usually blows the fuse. The immersion heater will need to be replaced.

- **Cylinder thermostat failure** – A very rare fault. The thermostat should first be tested to confirm that it has failed before replacing it.

- **Motorised valve failure** – This is a common occurrence with fully pumped systems. The valve should be tested to confirm whether it is the valve itself that has failed or just the motor in the actuator head.

- **Boiler failure** – This is a more serious fault that may mean specialist diagnosis and repair by an experienced plumber.

- **Airlocks** – These can usually be traced to long horizontal runs in the cold feed to the cylinder as it leaves the cistern. The closer the horizontal run is to the cistern, the less head of pressure there is on the cold feed. This can create an airlock before the cold feed drops vertically to the cylinder. Low-pressure systems always work better when the pipework exits the roof space quickly. Long horizontal runs create problems with flow rate when the head of pressure is low.

- **The cold feed has a backfall towards the cistern** – Air collects in the high point in the pipework. The pipework leaving the cistern should fall away from the cistern to ensure a good flow rate.

- **Noise in the system** – This can be due to oscillation of the float-operated valve. This may be because of a faulty float-operated valve or a missing cistern wall-strengthening plate, which prevents the cistern wall from vibrating. Vibration may also come from the immersion heater when the electricity is turned on. The heater element vibrates quickly, making a humming sound. The only action here is to replace the heater.

- **Overheating of the water** – This causes the water to boil and is a problem found in some older direct systems with a coal-fired back boiler.

- **Expansion of the pipework** – This causes ticking and creaking noises when not enough room has been allowed for expansion of the pipework. On new properties, this type of noise is not allowed and must be traced and rectified.

- **Excessively hot water** – This is usually caused by immersion heater thermostat failure. This will need testing and replacing with a thermostat that has a high limit stat cut-out.

- **Uncontrolled heat from a solid fuel appliance** – This may occur in direct systems.

- **Cylinder collapse** – Due to the creation of a vacuum in the cylinder caused by the hot water dropping as soon as it leaves the cylinder before it enters the vent pipe; having no vent pipe installed; a blocked vent pipe; or an isolation valve installed on the vent pipe which is turned off.

J DECOMMISSIONING AND MAINTENANCE OF HOT WATER SYSTEMS

Decommissioning hot water systems for maintenance and replacement of components can be a delicate task. It is important to ensure that the heat source is totally isolated before work on the system begins. A notice should be placed next to the heat source informing that the system is decommissioned and must not be turned on. Fuses to electric heaters, thermostats and motorised valves should be removed and retained. If appliances are removed, any open pipes should be capped off. The customer should be informed when the system is turned off.

The main components of hot water systems that require periodic maintenance are:

- **The hot water storage vessel** – This should be periodically checked for any signs of corrosion. Diminishing flow rates could indicate scale build-up in either the cold feed connection or the hot water draw-off connection. These can be removed and descaled as necessary. When replacing hot water cylinders, the cylinder should be preassembled as much as possible before installation begins to reduce the period for which the hot water supply is turned off.

- **The hot water appliances** – These should be serviced annually in line with the manufacturer's instructions.

- **The cistern (for open-vented systems)** – This should be checked periodically for sediment build-up on the bottom of the cistern. If a cistern is to be replaced, the replacement cistern should be pre-assembled before the system is decommissioned. This will reduce the decommissioning time.

- **Taps and terminal fittings such as float-operated valves** – These should be checked for correct shut-off, and water levels checked and adjusted as necessary.

- **Isolation valves such as full-way gate valves and service valves** – These should be checked to ensure that they shut off the flow of water fully.

- **Thermostats** – Systems, such as immersion heaters and boilers, should be run to operating temperature to ensure the correct operation of any thermostats. They should be checked using digital thermometers.

- **Shower mixing valves and pumps** – These should be inspected to ensure that they are functioning in accordance with the manufacturer's specifications. Flow rates can be confirmed by using a weir gauge. Filters can be removed and cleaned. The operation of the flow switch on shower boosting pumps should be checked as these turn the shower pump on. Shower heads should be cleaned of any scale build-up as this can significantly reduce the flow of water.

> **SUGGESTED ACTIVITY...**
> How much do you remember about basic maintenance tasks? Why not write notes from memory and check them against Chapter 006, Section F (page 334–41).

CONCLUSION

Hot water is a necessity. How we deliver it is a matter of choice. There are so many systems to choose from and each one has its advantages and disadvantages. In this chapter, we have investigated a sample of the most popular systems, from simple point-of-use heaters to Building-Regulations-compliant storage and non-storage systems for whole-house hot water distribution for a variety of property types and sizes. These systems should be considered carefully to give the best possible combination of initial cost, efficiency, hot water control, maintenance costs and eventual replacement.

007 TEST YOUR KNOWLEDGE

SmartScreen Unit 007 revision sample questions

1. Which seven factors should be considered in design selection of hot water systems?

2. Briefly describe a centralised system.

3. Name three main types of centralised storage system.

4. In a vented system of hot water,
   a What should the maximum temperature of hot water be?
   b State two reasons why this is so.

5. A 450mm offset prevents which type of circulation at the hot water draw-off connection?

6. Which type of heat exchanger does a double-feed indirect cylinder contain?

7. What does the single-feed indirect system use to separate primary and secondary water?

8. What is the purpose of the secondary return?

9. What type of pump should be used on pumped secondary return?

10. What is the purpose of a sacrificial anode?

11. Identify these systems A, B and C:

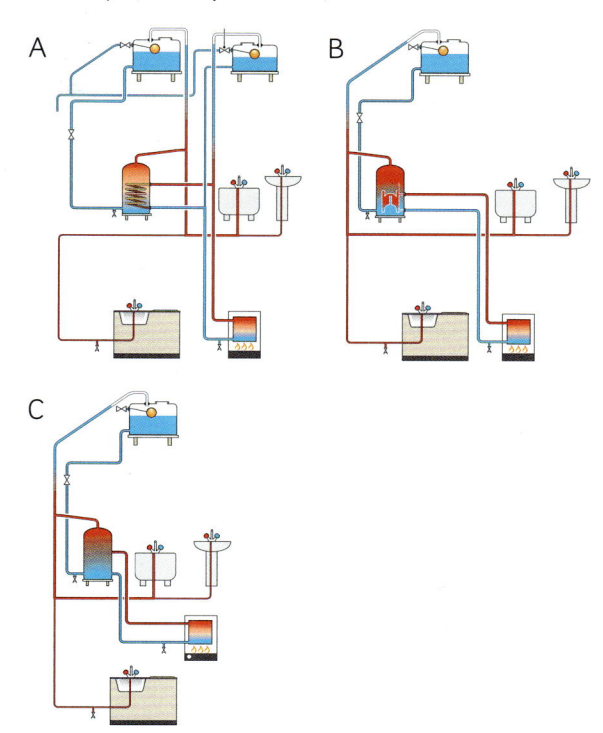

12. Give another phrase for the term **point of use system**.

13. What is meant by the term **multipoint**?

14. What is the purpose of the thermostatic mixing valve on a water-jacketed tube heater and where should it be positioned?

15. Why do electric showers require a check valve?

16. For a low-pressure shower to be effective, how high must the base of the cold water cistern be from the shower head?

17. Name two types of shower pump.

18. When installing a shower valve, the hot and cold supplies must be balanced. What does this mean?

19. Who should be informed before carrying out any maintenance tasks in a domestic property?

008
UNDERSTAND AND APPLY DOMESTIC CENTRAL HEATING INSTALLATION AND MAINTENANCE TECHNIQUES

Some 97 per cent of homes in the UK have central heating and most of this is in the traditional form of a boiler and radiators. In the past ten years, central heating has developed into a sophisticated home heating system that incorporates energy-saving appliances and controls designed to heat the dwelling quickly and efficiently using as little fuel as possible. This saves thousands of tonnes of carbon dioxide from being released into the atmosphere. This chapter looks at the subject of wet central heating from a domestic perspective. It investigates existing and modern systems, their pipework layouts, methods of control, the various types of appliances and the fuels they use.

IN THIS CHAPTER, YOU WILL COVER:

A Factors affecting central heating design
- A1 The purpose of central heating
- A2 Internal design temperatures and air change rates

B Central heating systems
- B1 Pumped central heating only systems
- B2 Semi-gravity heating systems
- B3 Fully pumped systems
- B4 The open vent, cold feed and circulating pump position for fully pumped systems
- B5 The feed and expansion cistern
- B6 Primary open safety vent
- B7 Sealed (pressurised) heating systems
- B8 Alternative central heating designs

C Heat-producing appliances
- C1 Solid fuel appliances
- C2 Gas central heating boilers
- C3 Oil-fired central heating appliances

D Typical flue systems for central heating appliances
- D1 Open flues
- D2 Room-sealed (balanced) flues

E Heat emitters
- E1 Panel radiators
- E2 Column radiators
- E3 Low surface temperature radiators
- E4 Fan convectors
- E5 Tubular towel warmers
- E6 Towel warmers with integral panel radiators
- E7 Skirting heating
- E8 Underfloor heating

F Mechanical central heating controls
- F1 Radiator valves
- F2 Automatic air vents
- F3 Automatic bypass valves
- F4 Thermo-mechanical cylinder control valves
- F5 Anti-gravity valves
- F6 Drain valves

G Electrical central heating controls
- G1 Time clocks and programmers
- G2 Room thermostats
- G3 Cylinder thermostats
- G4 Frost thermostats and pipe thermostats
- G5 Motorised valves
- G6 System design and control
- G7 Boiler interlock

H Installation of central heating systems
- H1 Pipework materials
- H2 Installation methods
- H3 Testing and filling
- H4 Commissioning and balancing
- H5 Decommissioning
- H6 Corrosion protection

I Maintenance of systems
- I1 Pump replacement
- I2 Radiator replacement
- I3 Tasks that may require system drain-down
- I4 Power flushing a system
- I5 Routine maintenance tasks
- I6 Dealing with simple system faults

Test your knowledge

A   FACTORS AFFECTING CENTRAL HEATING

The recommendations for good central heating installations are set out in the following documents:

- **BS EN 14336:2004 Heating systems in buildings** – Installation and commissioning of water-based heating systems.
- **BS EN 12831:2003 Heating systems in buildings** – Method for calculation of the design heat load.
- **BS EN 12828:2003 Heating systems in building**s – Design for water-based heating systems.
- **Various other documents** – such as the Domestic Heating Compliance Guide and Building Regulations Approved Document L1.

The first part of this chapter looks at the purpose of central heating and explores the criteria for efficient central heating design.

> **KEY POINT**
> Check out Chapter 003, pages 106–10 for more information about these important documents.

A1   THE PURPOSE OF CENTRAL HEATING

The main purpose of central heating is to provide thermal comfort conditions within a building or dwelling. Central heating is preferable to open fires as it heats the whole property. Other factors to consider when discussing comfort are:

- **Humidity** – The amount of moisture there is within the environment. Ideal conditions for humans require 40 to 60 per cent humidity. Anything below 40 per cent can make the eyes and throat very dry. Above 60 per cent makes the atmosphere feel damp and uncomfortable.
- **Air changes** – The amount of air movement (not air velocity) within the building. Air movement is important because it replaces used air with fresh air, which we need for breathing. BS EN 12831 gives ideal air changes and temperatures for specific rooms within a dwelling.
- **Air temperature** – Air temperatures should range between 16 and 22°C, dependent upon the type of activity being carried out, the age of the occupants and the level and quality of their clothing. Air temperature at ground level should not be greater than 3°C below that at head height, and room surface temperatures not above the air temperatures.
- **Air velocity** – This is the speed at which the air travels within the building. If it travels too fast, a draught will be felt; if there is no movement, the air changes will not be satisfied. Airflow past the body should be horizontal and at a velocity of between 0.2m and 0.25m per second. A variable air velocity is preferable to a constant one.

- **Activity within the building** – This applies to the type of work that is being carried out within the building. The more physical activity that is carried out by the occupants, the lower the temperature required, and temperatures may have to be adjusted to suit.
- **Clothing** – This relates to the type of clothing worn by the occupants of the building.
- **Age and health** – These are major factors in heating design. The age of the occupants will have a direct effect on the type of the systems installed. Older people feel the cold more than younger people, which may mean that design calculations will have to be modified, especially when it comes to the temperatures of key rooms such as the lounge and bedroom.

Thermal comfort is achieved when a desirable heat balance between the body and surroundings is met. How we achieve this balance is down to the design of the central heating system and the way it is installed. There are three definitions of central heating systems within dwellings:

Definition	Interpretation
Full central heating – The simultaneous heating of all spaces in a dwelling to maintain specified temperatures based on calculated heat losses.	This is a system where the heat losses for the whole dwelling have been correctly calculated (either longhand or using computer heat-loss software) to an agreed comfortable temperature.
Selective central heating – The simultaneous heating of some of the spaces in a dwelling to maintain specified temperatures based on calculated heat losses.	This is where only some of the rooms, such as the bathroom and master bedroom, are heated with heat emitters. The heat losses for those rooms, however, are calculated.
Background central heating – The simultaneous heating of all or some of the spaces in a dwelling to temperatures below those specified based on calculated heat losses.	This type of heating is mainly installed on cost grounds. Smaller boilers and heat emitters are cheaper but only provide enough heat to take the chill off the rooms at a lower temperature. Full or selective heating systems must be able to achieve comfort levels when the external temperature is at −1°C. If this cannot be achieved, it would be classified as background heating.

Central heating system types can be classified by the sizes of tube they use, for example:

- **minibore** – often called small bore, pipework sizes 15mm to 35mm
- **microbore** – pipework sizes 8mm to 10mm.

A2 INTERNAL DESIGN TEMPERATURES AND AIR CHANGE RATES

Internal design temperatures and air change rates for domestic central heating installations are recommended by BS EN 12831 for full and part central heating systems. They are based on providing the customer with an acceptable level of thermal comfort throughout the dwelling. The minimum design temperature and air change rates required by BS EN 12831 are as follows:

Typical design temperature and ventilation rates		
Room	Design temperature	Design air changes per hour
Living room	21°C	1.5
Dining room	21°C	1.5
Bedsit room	21°C	1.5
Hall	18°C	1.5
Bedroom	18°C	1
Study	21°C	1.5
Bathroom	22°C	2
Toilet	18°C	2
Kitchen	18°C	2

> **KEY POINT**
>
> The phrase 'design temperatures' simply means the temperatures that the designer has used to work out the heat losses for the building. These can vary with the occupants' age or health. Although British Standards tell us the most appropriate temperatures to use, some people may require warmer or cooler temperatures than those specified. For instance, when calculating the design temperatures for a nursing home or complex for the elderly, 23°C should be used throughout the building instead of the normal temperatures specified.

B CENTRAL HEATING SYSTEMS

Domestic central heating systems fall into two categories. These are based on the way the system is filled with water and the pressure at which it operates. The two categories are:

- Low-pressure, open-vented central heating systems, fed from a feed and expansion (F&E) cistern in the roof space. These can be both modern fully pumped systems and existing gravity hot water/pumped heating installations.

- Sealed, pressurised central heating systems, fed direct from the mains cold water supply and incorporating an expansion vessel to take up the expansion of water due to the water being heated. These are generally more modern fully pumped and combination boiler systems.

The water in low-pressure open-vented central heating systems is kept below 100°C. For existing systems the flow water from the boiler is usually about 80°C and the return water temperature is usually 12–15°C lower.

Circulation of the water can be either by gravity circulation to the heat exchanger in the hot water cylinder and pumped heating to the heat emitters or by means of a fully pumped system where both the hot water heat exchanger and heat emitters are heated using a circulating pump. Fully pumped systems have the advantage that system resistance created by the pipework, fittings and heat emitters can be overcome more easily and this enables the system to heat up more quickly, giving the occupants a much more controllable system.

Sealed heating systems operate at a higher pressure, with modern systems incorporating condensing boilers operating at the slightly higher temperature of 82°C for the flow temperature with a return temperature 20°C lower (62°C) – this increases the heat flow rate and thereby the efficiency of the system.

SmartScreen Unit 008 handout 1

In both cases the difference between the flow and return temperatures represents the amount of heat lost to the heated areas.

The chart below shows the development of central heating, from the open-vented one-pipe system through to the more modern sealed combination boiler systems and fully pumped systems using system boilers.

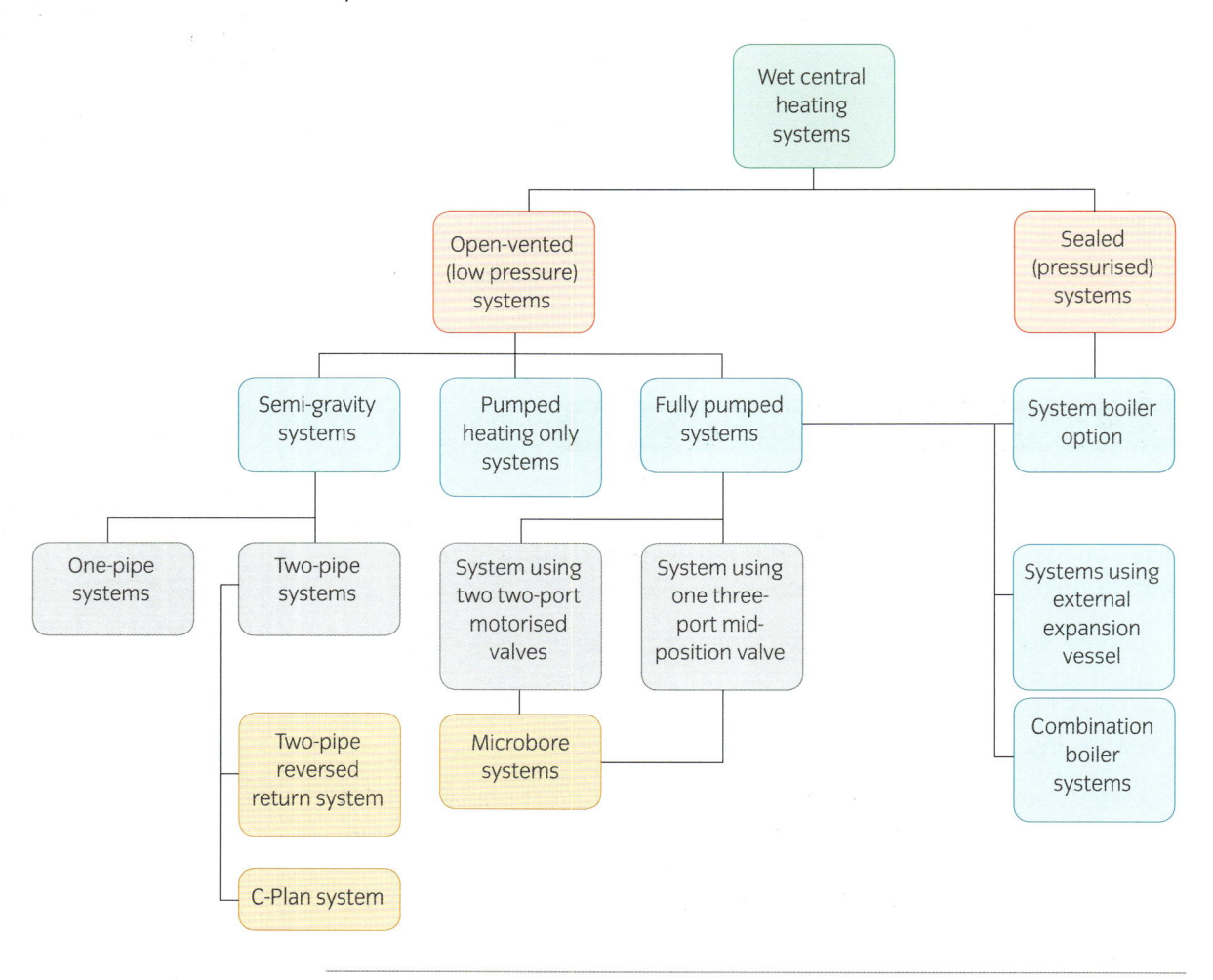

```
                    Wet central
                      heating
                      systems
                         |
         ┌───────────────┴───────────────┐
    Open-vented                       Sealed
   (low pressure)                  (pressurised)
      systems                         systems
         |                               |
   ┌─────┼─────────┐                     |
 Semi-    Pumped   Fully pumped      System boiler
gravity  heating    systems             option
systems  only                              |
         systems                           |
   |        |         |                    |
┌──┴──┐     |    ┌────┴────┐         ┌─────┴─────┐
One-  Two-  |   System    System     Systems using
pipe  pipe  |   using two using one   external
systems sys |  two-port  three-port  expansion
        |   |  motorised  mid-        vessel
        |   |  valves    position
        |   |            valve       Combination
        |   └────┬───────┘           boiler
        |    Microbore               systems
        |    systems
   Two-pipe
   reversed
   return system
        |
   C-Plan system
```

The development of pumped central heating

B1 PUMPED CENTRAL HEATING ONLY SYSTEMS

Pumped central heating only systems do not contain any provision for heating domestic hot water. They serve only the heat emitters, usually radiators/convectors, for domestic installations. The cold feed and the vent pipe can either be taken direct from the boiler or direct from the heating pipework. They are generally two-pipe systems, with the central heating circulating pump installed on the flow pipe.

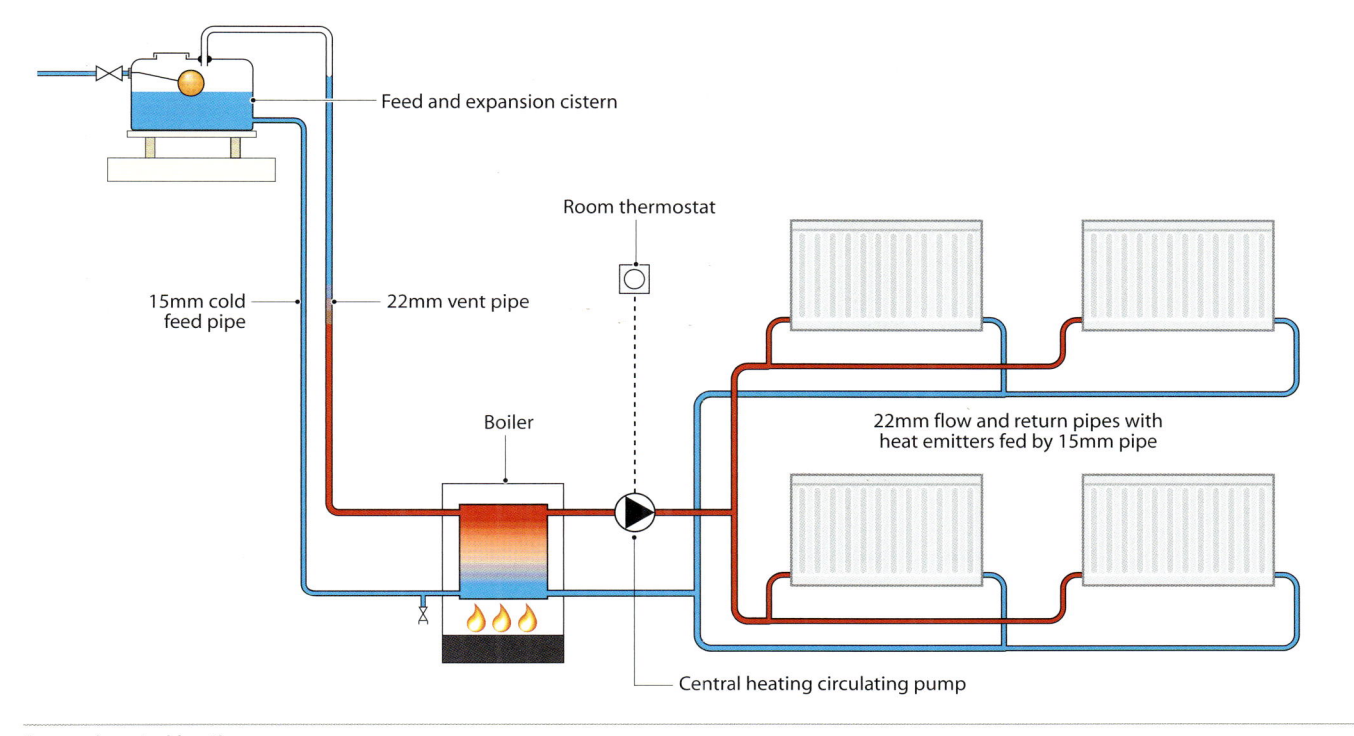

Pumped central heating

B2 SEMI-GRAVITY HEATING SYSTEMS

Semi-gravity heating systems use gravity circulation to heat the domestic secondary water and pumped central heating circulation. The heat exchanger within the hot water storage cylinder is connected to the boiler by the primary flow and return pipes, usually 28mm in diameter with a 22mm vent pipe branched from the primary flow and a 15mm cold feed pipe connected to the primary return. They may still be found as existing systems in older properties. There are four basic systems, and each is an advance on the previous one:

- the one-pipe semi-gravity system
- the two-pipe semi-gravity system
- the C-Plan semi-gravity system
- the two-pipe semi-gravity system with heat sink for solid fuel boilers.

The one-pipe semi-gravity system

This is a simple ring circuit of pipework to and from the boiler and as such, there are no separate flow and return pipes. The main 'ring' is pumped and the water circulates through the radiators by gravity. The size of the radiators is calculated from the temperature drop at each successive one, with the last radiator always being around 15°C cooler than the first. Balancing the flow of water to each radiator is a simple process with the use of radiator valves but this increases system resistance and slows the heating process.

> **KEY POINT**
>
> These systems are often referred to as 'gravity' or 'pumped' systems because they incorporate old heating technology in the form of gravity circulation and forced circulation via a central heating circulating pump.

SmartScreen Unit 008 handout 2 (page 1)

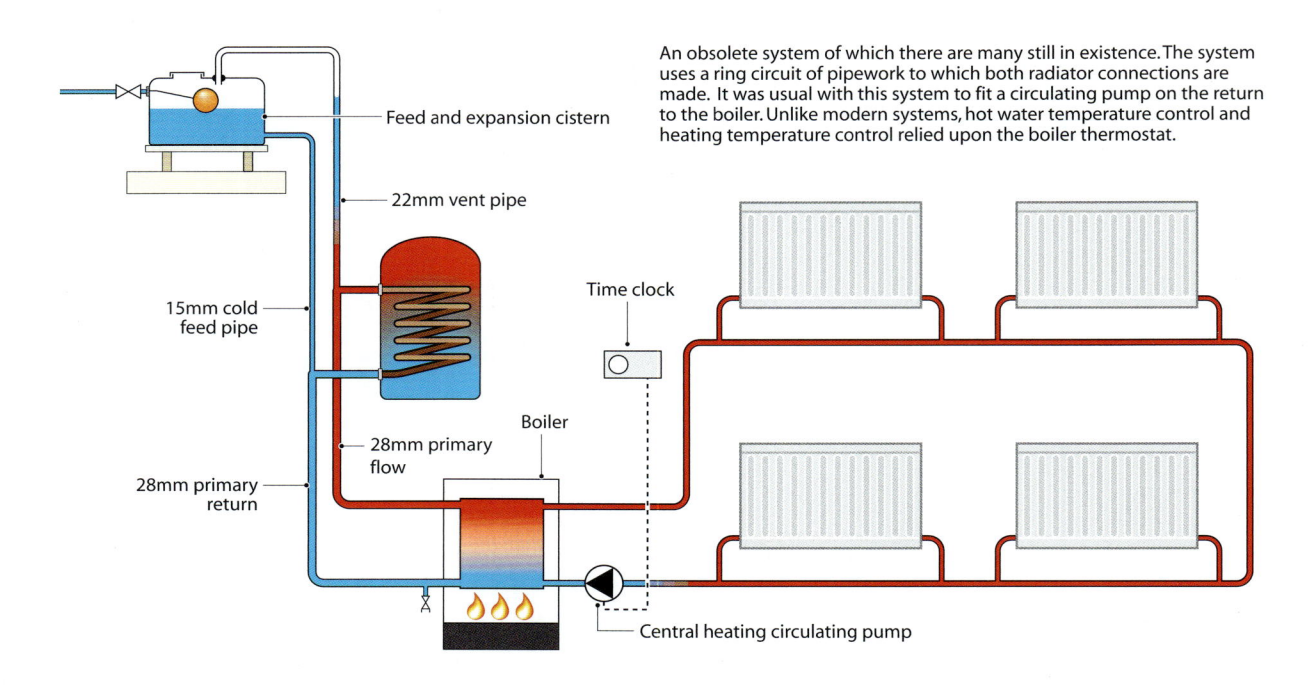

An obsolete system of which there are many still in existence. The system uses a ring circuit of pipework to which both radiator connections are made. It was usual with this system to fit a circulating pump on the return to the boiler. Unlike modern systems, hot water temperature control and heating temperature control relied upon the boiler thermostat.

The one-pipe system

ADVANTAGES	DISADVANTAGES
▪ Cheap to install because there is less pipework involved in the installation as compared to other heating systems. **Boiler cycling** This happens when a heating system has reached temperature, and the boiler shuts down. A few minutes later the boiler will fire up again to top the temperature up as the system loses heat, and after a few seconds shuts down again. This constant firing up and shutting down as the system water cools slightly wastes a lot of fuel energy.	▪ The water in the system cools as it travels from one heat emitter to the next. If the same heat output is to be obtained, larger heat emitters are needed the further from the boiler they are situated and/or the greater the number of heat emitters that the water travels through. ▪ The system tends to circulate water within the main pipework ring. Circulation within the heat emitters can only be induced by a difference in the density of the water entering and leaving the system (gravity circulation). ▪ Uncontrolled heating of the primary circuit by the boiler leading to overheating of the domestic secondary hot water by the coil heat exchanger. ▪ Constant **boiler cycling** even when the hot water and heating are up to temperature leads to wastage of fuel energy. ▪ The system is not compliant with Building Regulations Document L and should be updated. ▪ The boilers fitted to this type of system are only about 78 per cent efficient or lower. ▪ Condensing-type boilers cannot be fitted to this type of installation because of the gravity circulation required by the hot water storage cylinder.

The two-pipe semi-gravity system

Like the one-pipe system, this has gravity circulation to the hot water circuit and pumped circulation to the central heating circuit. The system differs considerably from the one-pipe system in having two pipes, a flow and a return, which are connected to the boiler. The heat emitters are connected to separate branches of the main flow and return pipes so, in effect, each heat emitter has its own flow and return pipework to the boiler. This means that all of the heat emitters achieve the same temperature and this negates the need to increase heat emitter size due to temperature loss. The temperature difference across each flow and return is usually 12–15°C with a flow temperature of around 80°C.

One of the biggest problems with older central heating systems was the lack of temperature control on both the hot water and heating circuits, which meant that the hot water and the radiators became as hot as the water in the boiler. The two-pipe semi-gravity system went some way to addressing this problem with the inclusion of a room thermostat, which simply switched off the pump when the desired room temperature was reached. The secondary water, however, was still uncontrolled and was often too hot. Because of this, the two-pipe semi-gravity system is no longer used as a new installation as it does not comply with Building Regulations Document L. Systems of this type must be updated to include full thermostatic control over both hot water temperature and room temperature by the inclusion of separate controls. The updated system is known as the C-Plan.

SmartScreen Unit 008 handout 2 (page 2)

SmartScreen Unit 008 worksheet 1

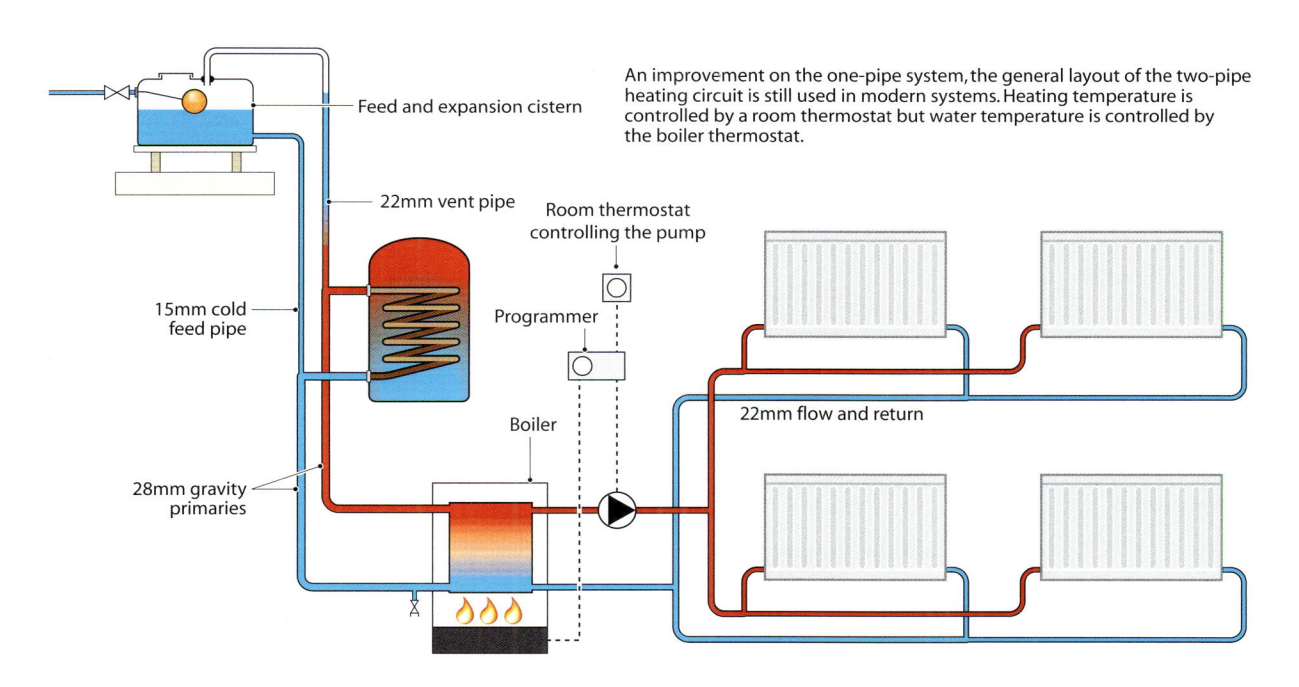

An improvement on the one-pipe system, the general layout of the two-pipe heating circuit is still used in modern systems. Heating temperature is controlled by a room thermostat but water temperature is controlled by the boiler thermostat.

Two-pipe semi-gravity system

ADVANTAGES	DISADVANTAGES
- All of the heat emitters reach the same temperature. - The two-pipe system is much quicker at heating up than the one-pipe system. This saves on fuel usage.	- Uncontrolled heating of the primary circuit leads to overheating of the domestic secondary hot water. - Constant boiler cycling even when the hot water and heating are up to temperature leads to wastage of fuel energy. - The system is not compliant with Building Regulations Document L and must be updated to a C-Plan system as a minimum standard. - The boilers fitted to this type of system are only about 78 per cent efficient or lower. - Condensing type boilers cannot be fitted to this type of installation because of the gravity circulation needed by the hot water storage cylinder.

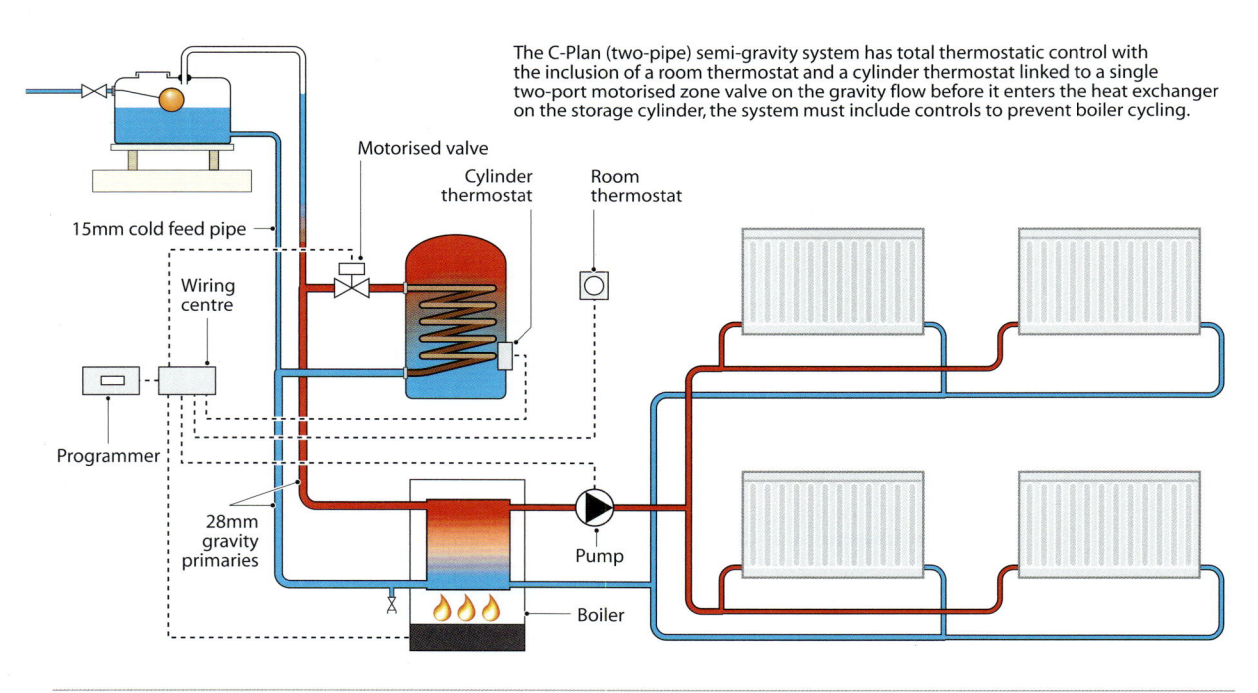

SmartScreen Unit 008 handout 3 (page 1)

The C-Plan (two-pipe) semi-gravity system

The C-Plan is an updated version of the semi-gravity two-pipe system that incorporates full thermostatic control of both heating and hot water circuits. Room temperatures are controlled by a room thermostat and thermostatic radiator valves, while the hot water temperature is controlled by a cylinder thermostat linked to a single two-port motorised zone valve. This is installed on the gravity flow before it enters the heat exchanger at the hot water storage cylinder.

The C-Plan (two-pipe) semi-gravity system has total thermostatic control with the inclusion of a room thermostat and a cylinder thermostat linked to a single two-port motorised zone valve on the gravity flow before it enters the heat exchanger on the storage cylinder, the system must include controls to prevent boiler cycling.

C-Plan semi-gravity system

The C-Plan system is accepted as compliant with Building Regulations Document L1b for the updating of existing systems.

ADVANTAGES	DISADVANTAGES
■ All of the heat emitters reach the same temperature. ■ The two-pipe system is much quicker at heating up that the one-pipe system. This saves on fuel usage. ■ The system is compliant with Building Regulations Document L. ■ Full control over both heating and hot water circuits is possible.	■ The system is not as controllable as more modern fully pumped systems. ■ Condensing-type boilers cannot be fitted to this type of installation because of the gravity circulation required by the hot water storage cylinder. ■ The boilers fitted to this type of system are only about 78 per cent efficient or lower.

The two-pipe semi-gravity system with heat sink for solid fuel boilers

This system is a variation on a theme developed for solid fuel appliances. It is very similar to the C-Plan system but contains a heat sink radiator (often called a heat leak) connected to the gravity primary flow and return pipes.

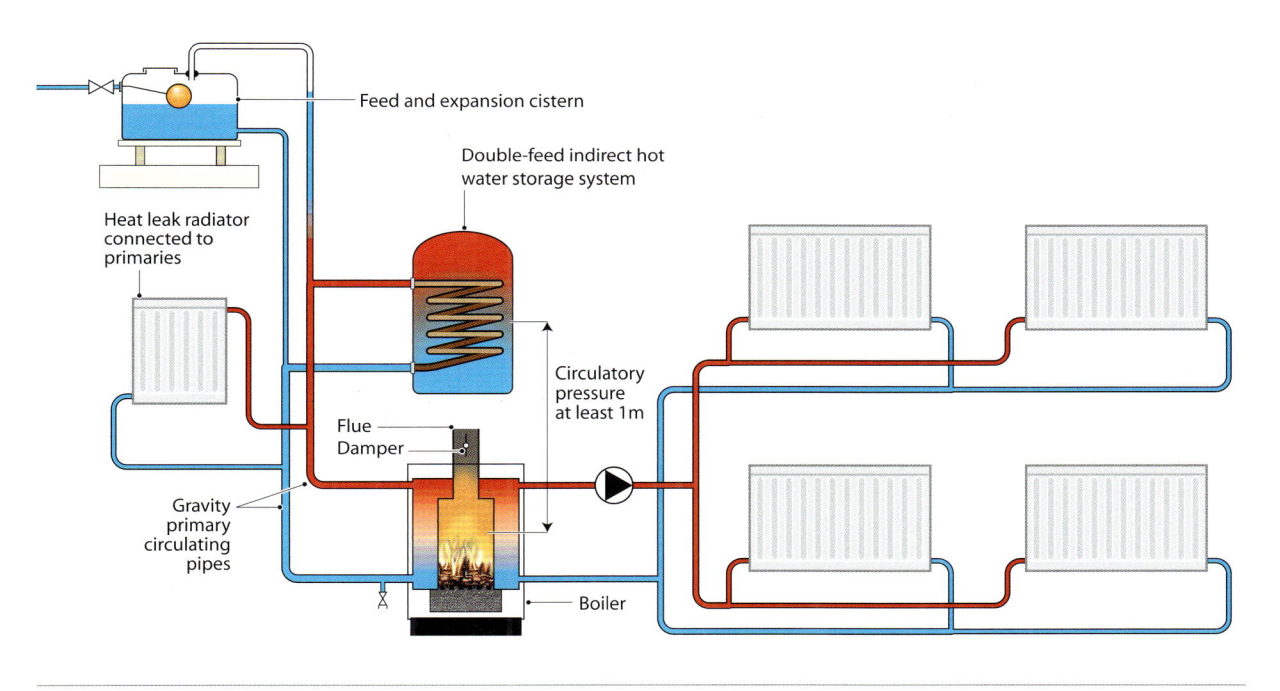

Two-pipe semi-gravity system with heat sink for solid fuel boilers

Solid fuel appliances are not as controllable as other fuels such as gas and oil and it is very hard to stop the boiler from generating heat. Gravity circulation relies on there being a temperature difference between the primary flow and return. In other words, the flow temperature must be higher than the return temperature for circulation to occur. When the water in the cylinder gets hot, the primary return pipe temperature starts to rise to the point where it is almost at the same temperature as the flow. When that happens, gravity circulation nearly stops. The boiler, however, continues to generate heat and if this is not dispersed in some way, it will lead to the boiler overheating. The heat sink radiator allows the excess heat to be dissipated to the surrounding atmosphere, which ensures that a minimum amount of circulation occurs through the primary flow and return, thus preventing the boiler from overheating. A heat sink radiator must be controlled by a lockshield radiator valve at either end so that there is no householder adjustment. The cap must be removed and the valve adjusted with a spanner.

As you can see from the illustration, the basic pipework layout is the same as other semi-gravity systems covered here, and the system has similar time and heat control functions, with two subtle differences:

- Solid fuel appliances control water temperature using mechanical, not electrical, thermostats, so electrical thermostatic control has to be added to the pipework system.
- It is essential that some means of dissipating the excess heat is permanently connected to the system, such as the heat sink radiator. For this reason, only an open-vented system can be used with solid fuel systems.

Generally, solid fuel systems will include:

- A room thermostat and a single time clock to control the circulating pump.
- A pipe thermostat on the gravity flow set to 45°C to prevent the pump from running when the temperature drops below this temperature. This will prevent the stored hot water from being dumped if the fire dies down, and cool water from being circulated through the boiler. This is a frequent cause of condensation and corrosion.
- A second pipe thermostat set to 90°C to automatically run the pump and dissipate excess heat through the central heating system, irrespective of how any other controls are set.
- A large capacity hot water storage vessel.

All radiators except the heat sink can be fitted with thermostatic radiator valves. The heat sink radiator must be fitted with lockshield-type valves to prevent tampering. An 'anti-gravity' valve should also be included on the upstairs central heating circuit to prevent unwanted gravity circulation to the bedroom radiators when the central heating pump is off.

ADVANTAGES	DISADVANTAGES
- This system is compliant with Building Regulations Document L.	- This system, because it is solid fuel, is not as controllable as other similar systems. - Only open-vented systems can be used. - Requires extra external controls to prevent overheating. - Can be very expensive to install.

B3 FULLY PUMPED SYSTEMS

Modern heating systems utilise pumped primary circuits as well as pumped heating circuits. By installing two two-port zone valves or a three-port mid-position valve the user can have hot water only, heating only or a combination of both. There are two basic types:

- fully pumped system with mid-position valve (commonly referred to by the Honeywell trade name Y-Plan)
- fully pumped system with two two-port valves (commonly referred to by the Honeywell trade name S-Plan).

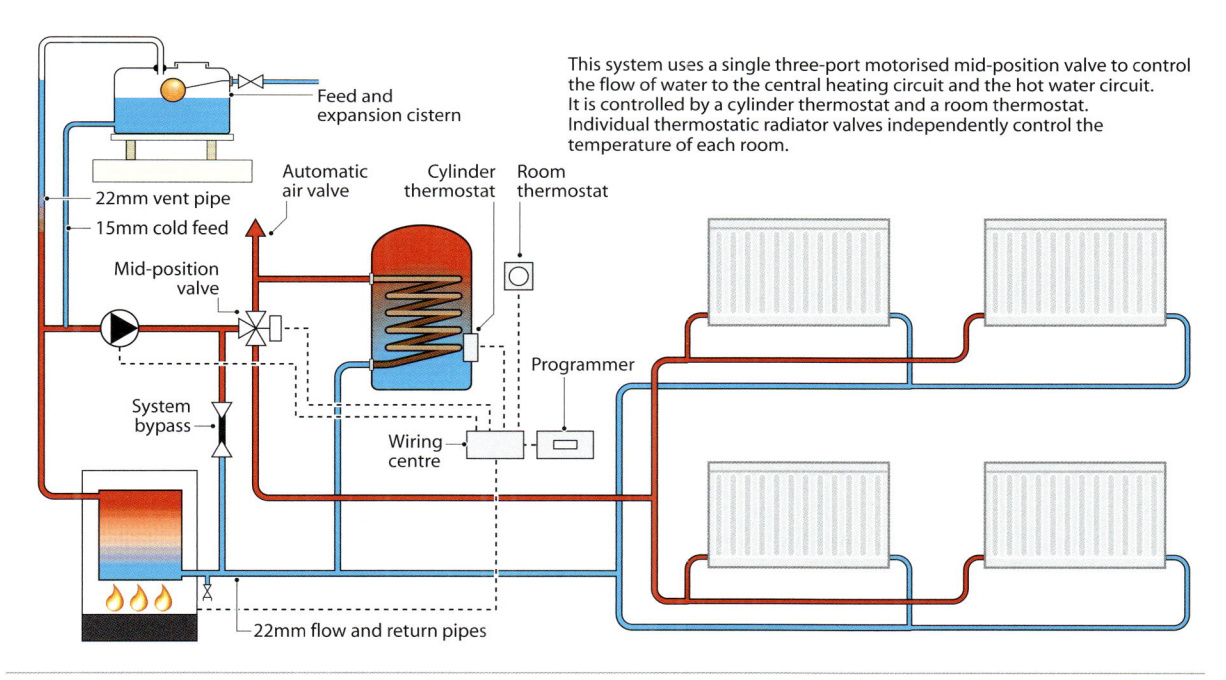

Fully pumped systems offer a better choice both of system design and boiler and, because the need for gravity circulation has been eliminated, they also give a much greater scope for installation options, especially when positioning the boiler. The boiler no longer needs to be lower than the storage cylinder.

Full thermostatic control is available to both hot water and heating circuits by means of a cylinder thermostat, a room thermostat and thermostatic radiator valves. Fully pumped systems heat up much more quickly than semi-gravity systems, offering savings on fuel and operating costs, and both Y-Plan and S-Plan systems can be used with natural gas, LPG and oil appliances.

Fully pumped system with mid-position valve

This system uses a single three-port motorised mid-position valve to control the flow of water to the central heating circuit and the hot water circuit. It is controlled by a cylinder thermostat and a room thermostat. Individual thermostatic radiator valves independently control the temperature of each room.

Y-Plan fully pumped system

The three-port mid-position valve controls the flow of water to the primary (cylinder) circuit and the heating circuit. The valve reacts to the demands of the cylinder thermostat or the room thermostat in the following way:

- At a set time, the programmer activates the system, calling for both hot water and heating.

Three-port mid-position valve

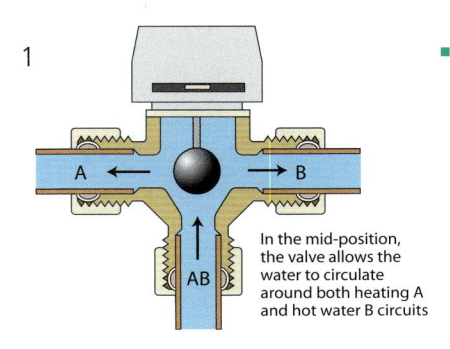

1

In the mid-position, the valve allows the water to circulate around both heating A and hot water B circuits

The mid-position valve in the mid position serving heating and hot water

- With the motorised valve in the mid position, water from the boiler circulates around both the primary and heating circuits. The boiler fires up and the circulating pump begins to circulate the water.

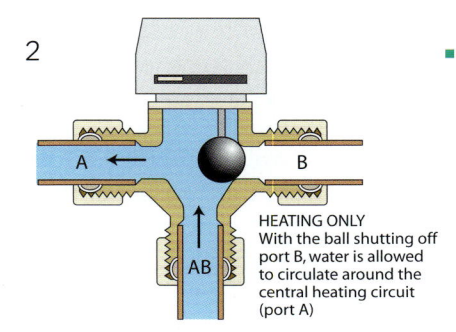

2

HEATING ONLY
With the ball shutting off port B, water is allowed to circulate around the central heating circuit (port A)

The mid-position valve with the hot water port closed

- When the cylinder reaches temperature the valve is energised by the cylinder thermostat, which closes the hot water port and prevents water flowing to the hot water cylinder heat exchanger.

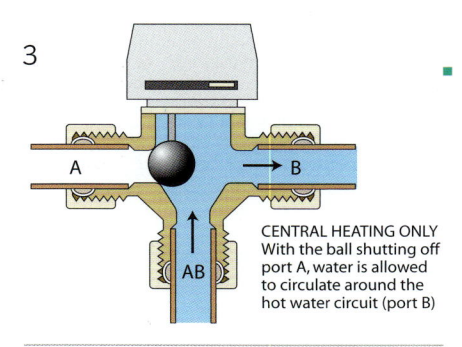

3

CENTRAL HEATING ONLY
With the ball shutting off port A, water is allowed to circulate around the hot water circuit (port B)

The mid-position valve with the heating port closed

- When the room reaches its set temperature, the valve is energised by the room thermostat, which closes the heating port and prevents water flowing to the heating circuit.

- With both the room thermostat and the boiler thermostat satisfied, the pump and the boiler shut down and the valve returns to the mid position. In this condition the system will only operate should either the room thermostat or cylinder thermostat call for heat. This is known as 'boiler interlock'.

The system contains a system bypass fitted with an automatic bypass valve, which simply connects the flow pipe to the return pipe. The bypass is required when all circuits are closed either by the motorised valve or the thermostatic radiator valves as the rooms reach their desired temperatures. The bypass valve opens automatically as the circuits close, to protect the boiler from overheating by allowing water to circulate through the boiler and thus keeping it below its maximum temperature. This prevents the boiler from 'locking out' on the overheat energy cut-out.

Locking out
A process by which a thermostat protects the boiler from overheating by shutting it down when a temperature of around 85°C is reached. High limit thermostats are manually resettable by pushing a small button on the boiler itself.

Fully pumped system with two two-port valves

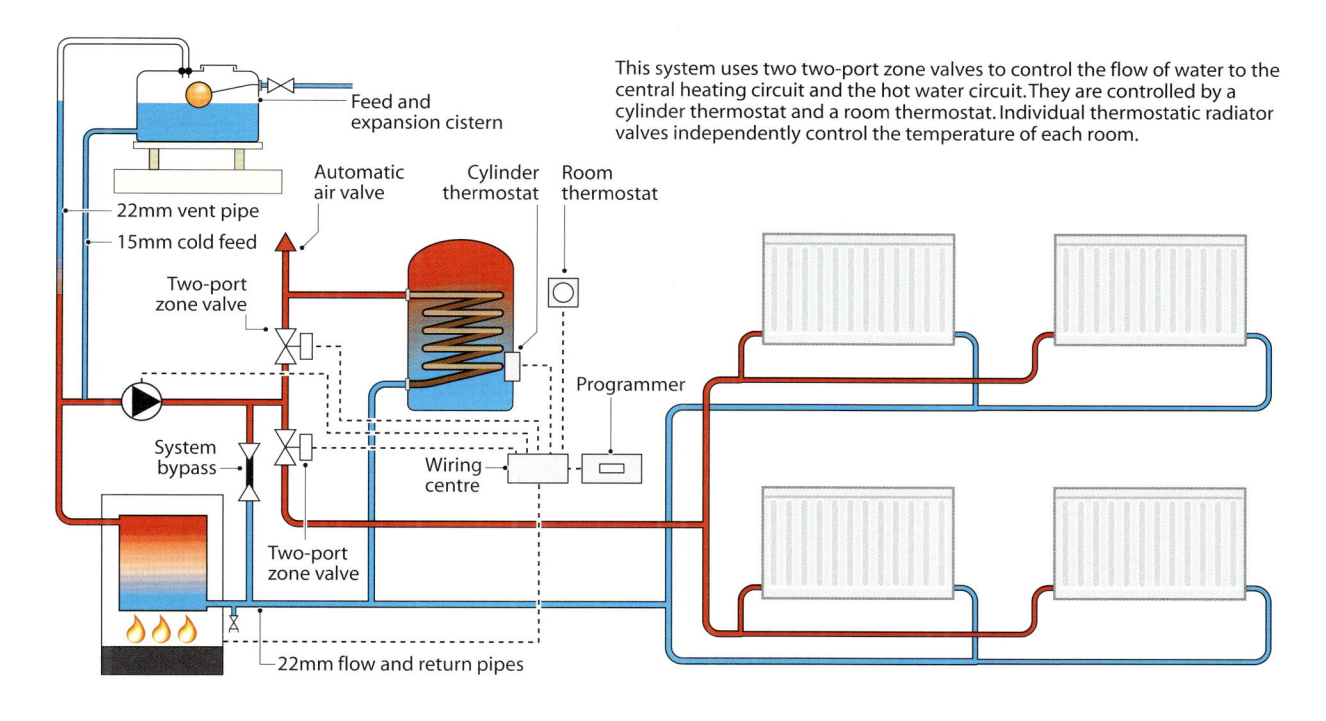

This system uses two two-port zone valves to control the flow of water to the central heating circuit and the hot water circuit. They are controlled by a cylinder thermostat and a room thermostat. Individual thermostatic radiator valves independently control the temperature of each room.

S-Plan fully pumped system

This has two two-port motorised zone valves to control the primary and heating circuits separately, by the cylinder and room thermostats respectively. This system is recommended for dwellings with a floor area greater than 150 square metres because it allows the installation of additional two-port zone valves to zone the upstairs heating circuit separately from the downstairs circuit. A separate room thermostat and perhaps a second time clock/programmer would also be required for upstairs zoning. The system operates as follows:

- At a set time, the programmer activates the system, calling for both hot water and heating.

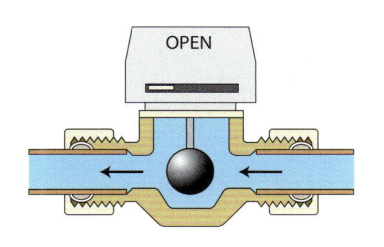

Zone valve open

- Both of the two-port motorised zone valves open and water from the boiler circulates around both the primary and heating circuits. The boiler fires up and the circulating pump begins to circulate the water.

- When the cylinder reaches temperature the two-port zone valve is energised by the cylinder thermostat, which closes the hot water zone valve and prevents water flowing to the hot water cylinder heat exchanger.

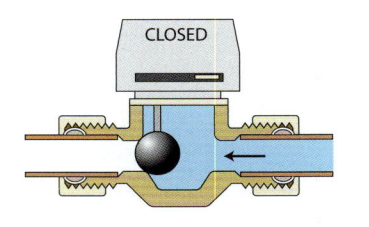

When the room reaches its set temperature, the two-port zone valve is energised by the room thermostat, which closes the valve and prevents water flowing to the heating circuit.

Zone valve closed

With both the room thermostat and the boiler thermostat satisfied, the pump and the boiler shut down. In this condition, the system will only operate should either the room thermostat or cylinder thermostat call for heat. This is known as 'boiler interlock'.

As with the Y-Plan, a system bypass is required for overheat protection of the boiler. The following table compares the S-Plan and Y-Plan systems:

	Full thermostatic control	Compliant with Building Regulations Document L	Recommended for larger properties	Can be used with sealed (pressurised) systems	Can be used with system boilers	Can be zoned	Anti-cycling boiler interlock
Y-Plan system	✓	✓	✗	✓	✓	✗	✓
S-Plan system	✓	✓	✓	✓	✓	✓	✓

B4 THE OPEN VENT, COLD FEED AND CIRCULATING PUMP POSITION FOR FULLY PUMPED SYSTEMS

The position of the open vent pipe, the cold feed pipe and the circulating pump in relation to a fully pumped system is an important part of the system design. If these elements are positioned incorrectly, the system will not work properly and may even induce system corrosion due to constant aeration of the system water.

The open vent and the cold feed should be positioned on the flow from the boiler on the suction side of the circulating pump with a maximum distance of 150mm between them. This is called the neutral point, as the circulating pump acts on both the feed pipe and the open vent pipe with equal suction. If they are any further apart, the neutral point becomes weak and the pump will act on the feed pipe with a greater force than the open vent pipe. This creates an imbalance, which leads to a lowering of the water in the F&E cistern. When the pump switches off, the water returns to its original position. This constant see-sawing motion aerates the water, creating corrosion within the system.

SmartScreen Unit 008 handout 3 (pages 2–3)

KEY POINT

Central heating systems do not like air. Aeration of the water is one of the biggest causes of corrosion in heating systems because the air in the water contributes to rust occurring throughout the system and the formation of red oxide sludge. Water alone will not cause corrosion, even with ferrous metals present, such as radiators and convectors. It is the air present in the water that causes metals to rust, and constant water movement at the F&E cistern will aerate the water in the system enough for corrosion to take place.

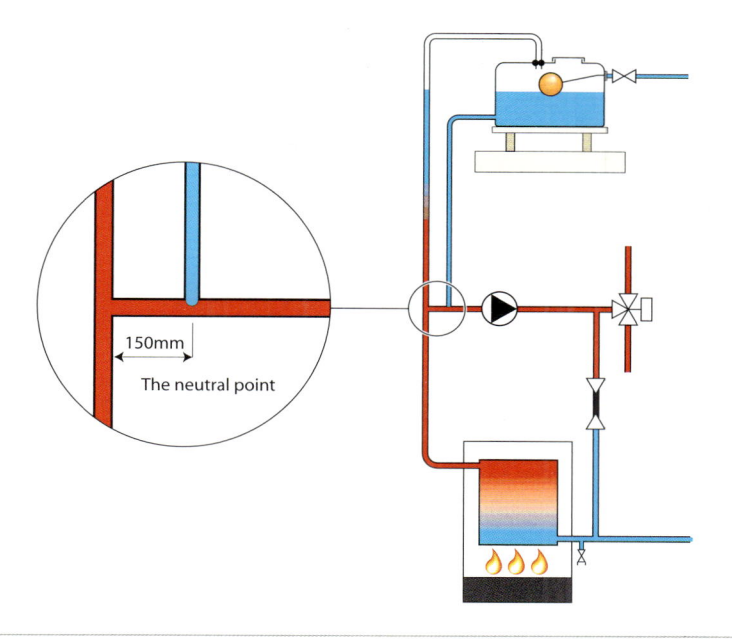

The position of the cold feed and open vent pipes

The circulator (pump)

The circulator (or pump) must be positioned with care to avoid design faults that could lead to problems with corrosion by aeration of the water due to water movement in the F&E cistern. This occurs when water is either pushed up the cold feed pipe and the open vent pipe or is circulated between the cold feed pipe and the open vent pipe.

> **KEY POINT**
> Remember! The relationship between the vent pipe, the cold feed and the pump is an important one. The order in which they should be connected to the system is easily remembered by the acronym **VCP** – **V**ery **C**orrect **P**rocedure, or **V**ent, **C**old Feed, **P**ump.

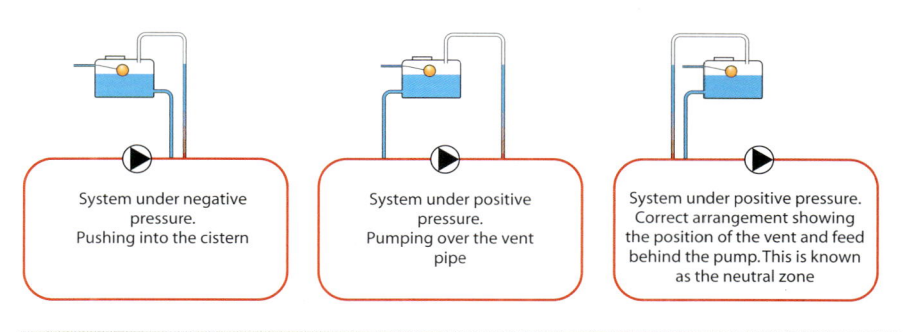

System under negative pressure. Pushing into the cistern

System under positive pressure. Pumping over the vent pipe

System under positive pressure. Correct arrangement showing the position of the vent and feed behind the pump. This is known as the neutral zone

The position of the circulator (pump)

Central heating circulator

SmartScreen Unit 008 handout 4

SmartScreen Unit 008 worksheet 2

The circulator, or hydronic central heating circulator, is a simple electric motor with a fluted waterwheel-like impeller that circulates the water around the system by centrifugal force. The faster the impeller rotates, the greater the circulation that occurs in the system. For quiet operation of the system, the flow rate should not exceed 1 litre per second (1.5 litres per second for microbore systems, see page 416). Most domestic circulating pumps have three speeds, which correspond to varying circulatory pressures, or heads. Domestic circulating pumps have either a 6m or a 10m head.

The use of air separators

The use of an air separator helps in the positioning of the feed and vent by ensuring that the neutral point is built into the system. The positioning of the pipework on an air separator creates a turbulent water flow in the separator body and this helps to remove air from the system, which makes the system quieter in operation and significantly reduces the risk of corrosion.

SmartScreen Unit 008 interactive activity 2

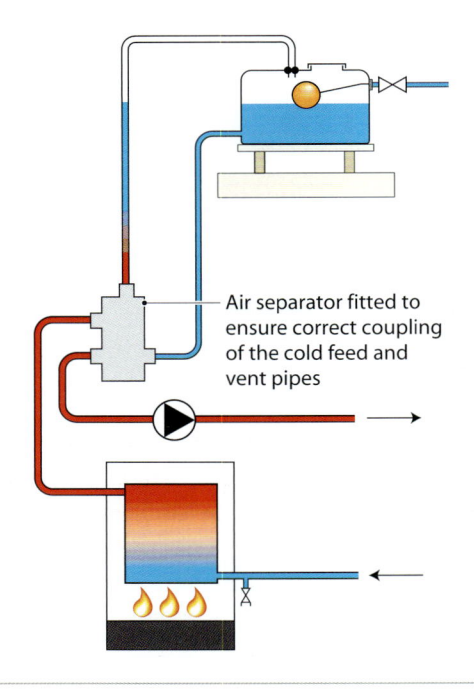

The use of an air separator

B5   THE FEED AND EXPANSION CISTERN

Open-vented systems contain a feed and expansion (F&E) cistern, which fulfils two important functions:

- it is the means by which water enters the system for filling and top-up
- it allows space for the system water to expand when it is heated.

Generally, the size of the F&E cistern will depend on the size of the system, but for most domestic systems an 18-litre cistern is recommended. The bigger the system, the more water it will contain and so the water expansion will be greater. The water level in the cistern should, therefore, be set at a low level.

The cistern must be located at the highest part of the central heating system and must not be affected by the operation of the circulating pump. For fully pumped systems, the cistern must be at least 1 m above the highest part of the pumped primary flow to the heat exchanger in the hot water storage cylinder. For gravity systems, the minimum height of the cistern can be calculated by taking the maximum operating head of the pump and dividing it by three.

The cold feed for the system for most domestic properties is 15mm. The cold feed pipe should not contain any service or isolation valves.

This is to ensure that there is a supply of cold water in the event of overheating and leakage, preventing the system from boiling. Should the valve be inadvertently closed, a dangerous situation could develop, especially if the vent was also blocked as the pressure would build up in the system, raising the boiling point of the water to dangerous levels. Both the cistern and any float-operated valve it may contain must be capable of withstanding hot water at a temperature close to 100°C.

B6 PRIMARY OPEN SAFETY VENT

The purpose of the open vent pipe is one of safety. The open vent is installed to:

- ensure that the system always remains at atmospheric pressure, limiting the boiling point to 100°C
- provide a safety outlet should the system overheat due to a component failure.

In a fully pumped system, the height of the open vent should be a minimum of 450mm from the water level in the cistern to the top of the open vent pipe. This is to allow for any pressure surges created by the circulating pump. The minimum size of pipe for the open vent is 22mm and this, like the cold feed pipe, should not be fitted with any valves.

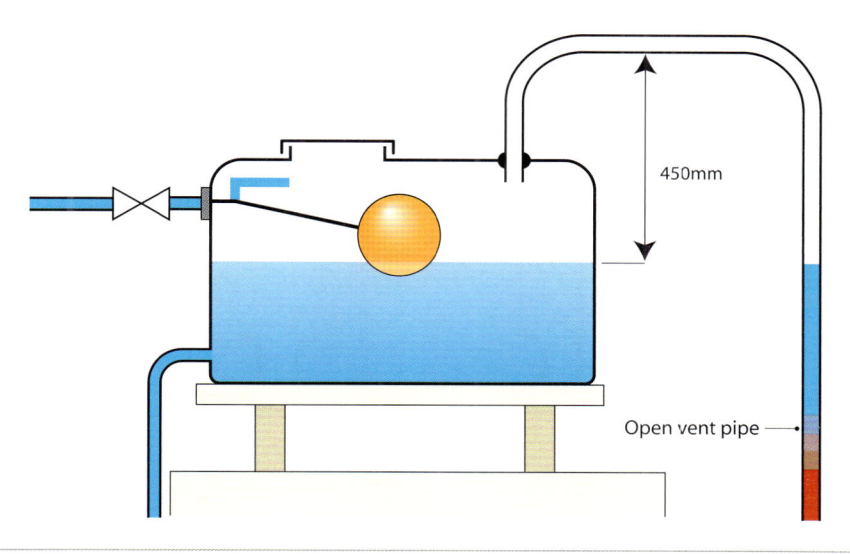

SmartScreen Unit 008 worksheet 3

Height of the open vent pipe

B7 SEALED (PRESSURISED) HEATING SYSTEMS

Sealed heating systems are those that do not contain an F&E cistern but are filled with water directly from the mains cold water supply via a temporary filling loop. The expansion of water is taken up by the use of an expansion vessel and the open vent is replaced by a pressure relief valve, which is designed to relieve the excess pressure by releasing the system water and discharging safely to a drain point outside of the dwelling. This is vital, as the water may be in excess of 80°C. A pressure gauge is also included so that the pressure can be set when the system is filled and periodically checked for rises and falls – these could indicate a potential component malfunction. The system is usually pressurised to around 1 bar. There are several types:

- sealed systems with an external pressure vessel
- system boilers that contain all necessary safety controls
- combination boilers.

All fully pumped systems can be installed as sealed systems or they can be purpose designed 'heating only' systems as described in Section B1 of this chapter (pages 396–97).

Sealed systems with an external pressure vessel

This system uses an expansion vessel in the place of the feed and expansion cistern. It is filled directly from the main cold water supply via a filling loop. A pressure relief valve safeguards the system from over pressurisation.
The system shown here is a typical Y-Plan system complete with all the same controls found on an open-vented Y-Plan.

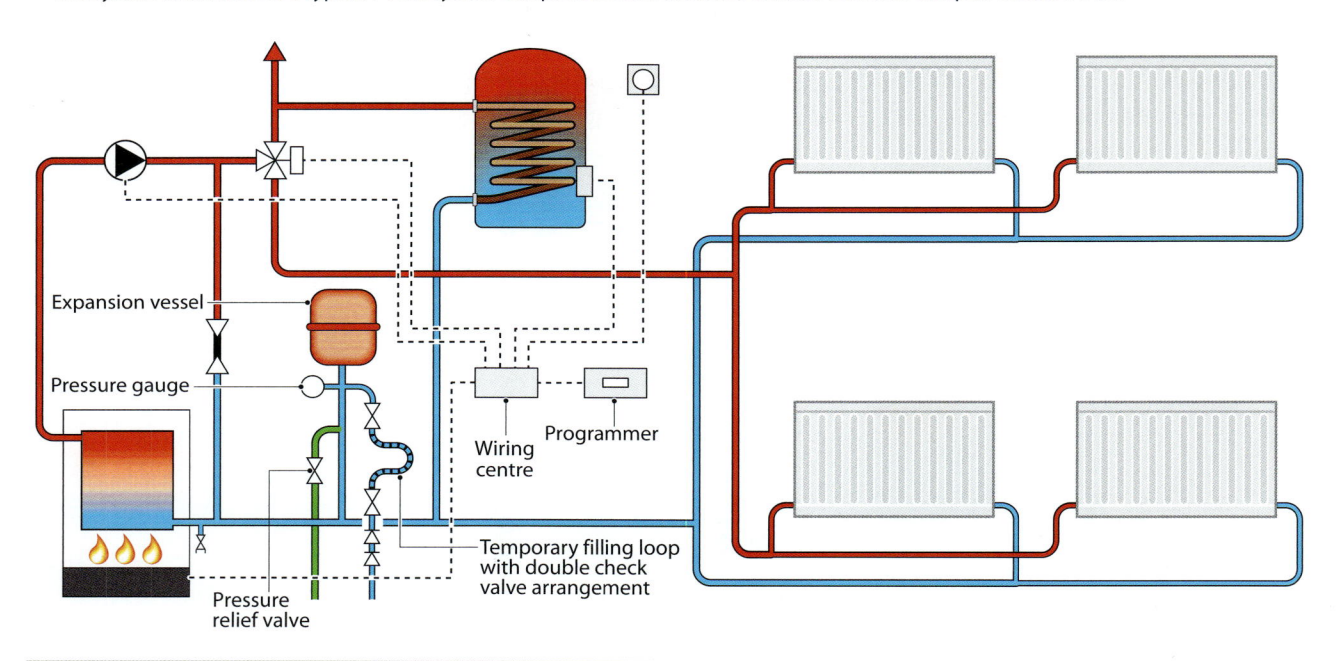

Sealed system with external pressure vessel

This system, as you can see from the illustration, is identical to the open-vented Y-Plan system except that it does not use an F&E cistern, cold feed pipe or open vent pipe. Instead, it has the following components:

- an external expansion vessel fitted to the system return
- a pressure relief valve

- a temporary filling loop
- a pressure gauge.

The expansion vessel

The expansion vessel is a key component of the system. It replaces the F&E cistern on the vented system and allows the expansion of water to take place safely. It comprises a steel cylinder, which is divided in two by a neoprene rubber diaphragm.

The vessel is installed on the return, because the return water is generally 20°C cooler than the flow water and this does not place as much temperature stress on the expansion vessel's internal diaphragm as the hotter flow water would. If installing the vessel on the flow is unavoidable, it should be placed on the suction side of the circulating pump in the same way as the cold feed and open vent pipe on the open-vented system.

On one end of the expansion vessel is a Schrader air pressure valve where air is pumped into the vessel to 1 bar pressure; this forces the neoprene diaphragm to virtually fill the whole of the vessel.

On the other end is a ½-in male BSP thread and this is the connection point to the system. When mains-pressure cold water enters the heating system via the filling loop and the system is filled to a pressure of around 1 bar, the water forces the diaphragm backwards away from the vessel walls, compressing the air slightly as the water enters the vessel. At this point, the pressure on both sides of the diaphragm is 1 bar.

As the water is heated, expansion of about 4 per cent takes place. The expanded water forces the diaphragm backwards, compressing the air behind it still further and, since water cannot be compressed, the system pressure increases.

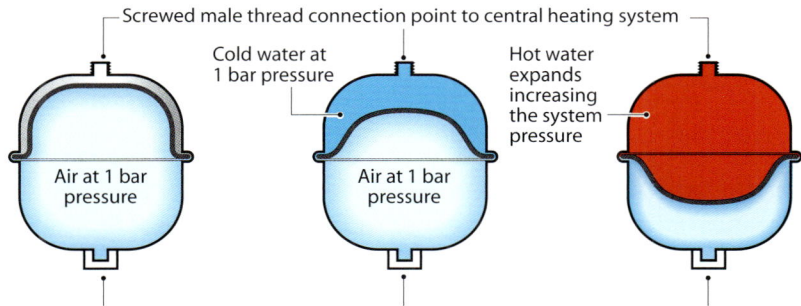

Expansion vessel with filling loop, pressure relief valve and pressure gauge

Screwed male thread connection point to central heating system

Cold water at 1 bar pressure

Hot water expands increasing the system pressure

Air at 1 bar pressure

Air at 1 bar pressure

Schrader-type pressure valve for checking and topping up the air charge

1. When the system has no water in it, the diaphragm is pressed against the wall of the expansion vessel by the charge of air in it.

2. When the system is filled with cold water, the water forces its way in slightly compressing the diaphragm. The water pressure is topped up to 1 bar

1. As the system heats up, the water expands and forces its way further into the expansion vessel increasing the pressure to around 1.5 bar and compressing the air charge behind the diaphragm to 1.5 bar.

Operation of an expansion vessel

ACTIVITY

Exactly how much expansion takes place?

The amount of expansion that takes place will depend on how many litres of water the heating system contains. As we have found in previous chapters, water expands at atmospheric pressure by 4 per cent when it is heated but in this case, the water is under pressure, so by how much does pressurised water expand?

To answer this question, we must first calculate the expansion factor, which can be used to calculate water expansion for a given volume and pressure. If the density of the cold water and the density of the water at maximum operating temperature are known, this is a fairly simple exercise. The calculation is as follows:

$$\frac{d_1 - d_2}{d_2}$$

Where:

d_1 = density of water at filling temperature (kg/m^3)

d_2 = density of water at maximum operating temperature (kg/m^3)

If the system has 250 litres of water and the system is filled with water at 4°C and the maximum temperature is 85°C, what is the expansion factor?

Water @ 4°C has a density of 1000kg/m^3

Water @ 85°C has a density of 968kg/m^3

The equation is therefore:

$$\frac{1000 - 968}{968} = 0.0330$$

So, the expansion factor (e) = 0.0330

Now, we must use this in another equation.

To find the amount of expansion of water in a system containing 250 litres of water operating at a maximum temperature of 85°C, the equation is:

$$V = 1 - \frac{p_1}{p_2}$$

Where:

V = The total volume of the expansion vessel
C = The total volume of water in the system in litres (250 litres)
p_1 = The fill pressure in bar pressure (1 bar)
p_2 = The setting of the pressure relief valve in bar pressure (3 bar)
e = The expansion factor (0.0330)

If these are entered into the equation, the equation becomes:

$$\frac{0.0330 \times 250}{1 - \frac{1}{3}} = 12.36$$

As a percentage of 250, 12.36 is:

$$\frac{12.36 \times 100}{250} = 4.94 \text{ per cent}$$

Therefore:

Water under a pressure of 1 bar when cold, expands by 4.94 per cent when heated to 85°C.

On cooling, the water contracts, the air in the expansion vessel forces the water back into the system and the pressure reduces to its original pressure of 1 bar.

Periodically, the pressure in the vessel may require topping up. This can be done by removing the cap on the Schrader valve and pumping the vessel up to its original pressure with a foot pump.

The pressure relief valve

The pressure relief valve (also known as the expansion valve) is installed on the system to protect against over-pressurisation of the water. Pressure relief valves are usually set to 3 bar pressure. If the water pressure rises above the maximum pressure that the valve is set to, the valve opens and discharges the excess water pressure safely to the outside of the property through the discharge pipework.

Pressure relief valves are most likely to open because of lack of room in the system for expansion due to a malfunction with the expansion vessel:

- the diaphragm in the expansion vessel has ruptured allowing water both sides of the diaphragm
- the vessel has lost its charge of air.

Pressure relief valve

The filling loop

The filling loop is an essential part of any sealed system and should contain an isolation valve at either end and a double check valve on the mains cold water supply side of the loop. The filling loop is the means by which sealed central heating systems are filled with water. Unlike open-vented systems, sealed systems are filled directly from the mains cold water via a filling loop. The connection of a heating system to the mains cold water supply constitutes a cross connection between the cold main (fluid category 1) and the heating system (fluid category 3), which is not allowed under the Water Supply (Water Fittings) Regulations 1999. The filling loop must protect the cold water main from backflow, which is done in two ways:

> **SUGGESTED ACTIVITY...**
> Air gaps and backflow prevention devices were covered in Chapter 6, page 310–14. See if you can draw the different layouts from memory, and then check your answers against these pages.

- it has a type-EC verifiable double check valve included in the filling loop arrangement
- it must be disconnected (or isolated if integral) after filling, creating an AUK3-type air gap for protection against backflow.

The filling loop is generally fitted to the return pipe close to the expansion vessel and may even be supplied as part of the expansion vessel assembly (see image at the top of page 411).

The pressure gauge
This is to allow the correct water pressure to be set within the system. It also acts as a warning of component failure or an undetected leak should the pressure begin to rise or fall inexplicably.

Filling loop

Condensing system boilers
A system boiler is an appliance where all necessary safety and operational controls are included and fitted directly to the boiler. There is no need for a separate expansion vessel, pressure relief valve or filling loop and this makes the installation much simpler.

The system boiler has all the components for a sealed system contained within the boiler unit. It is filled directly from the mains cold water via a filling loop which is often fitted by the boiler manufacturer.

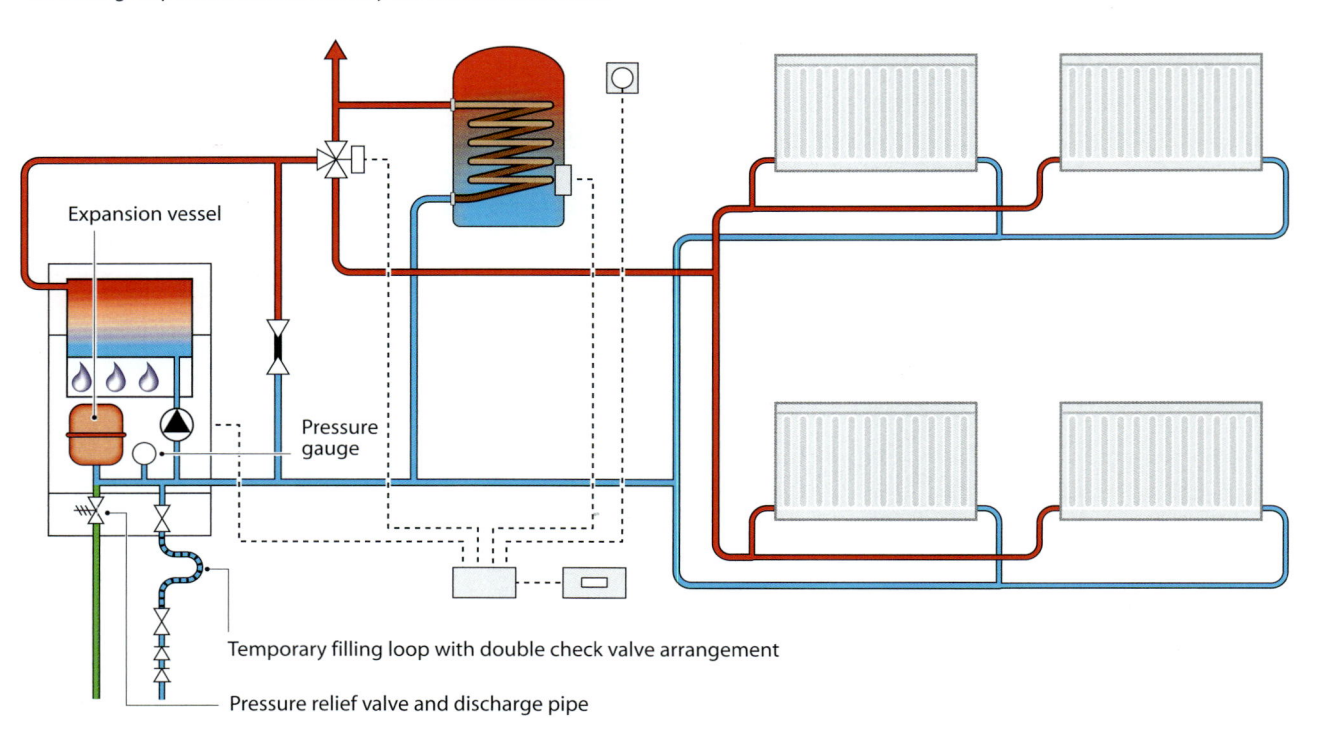

Sealed system with a system boiler

Condensing combination boilers
In recent years, combination boilers have become one of the most popular forms of central heating in the UK. A combination boiler provides central heating and instantaneous hot water supply from a single appliance. Modern combination boilers are very efficient and they contain all the safety controls, ie expansion vessel and pressure relief valve, of a sealed system. Most combis also have an integral filling loop.

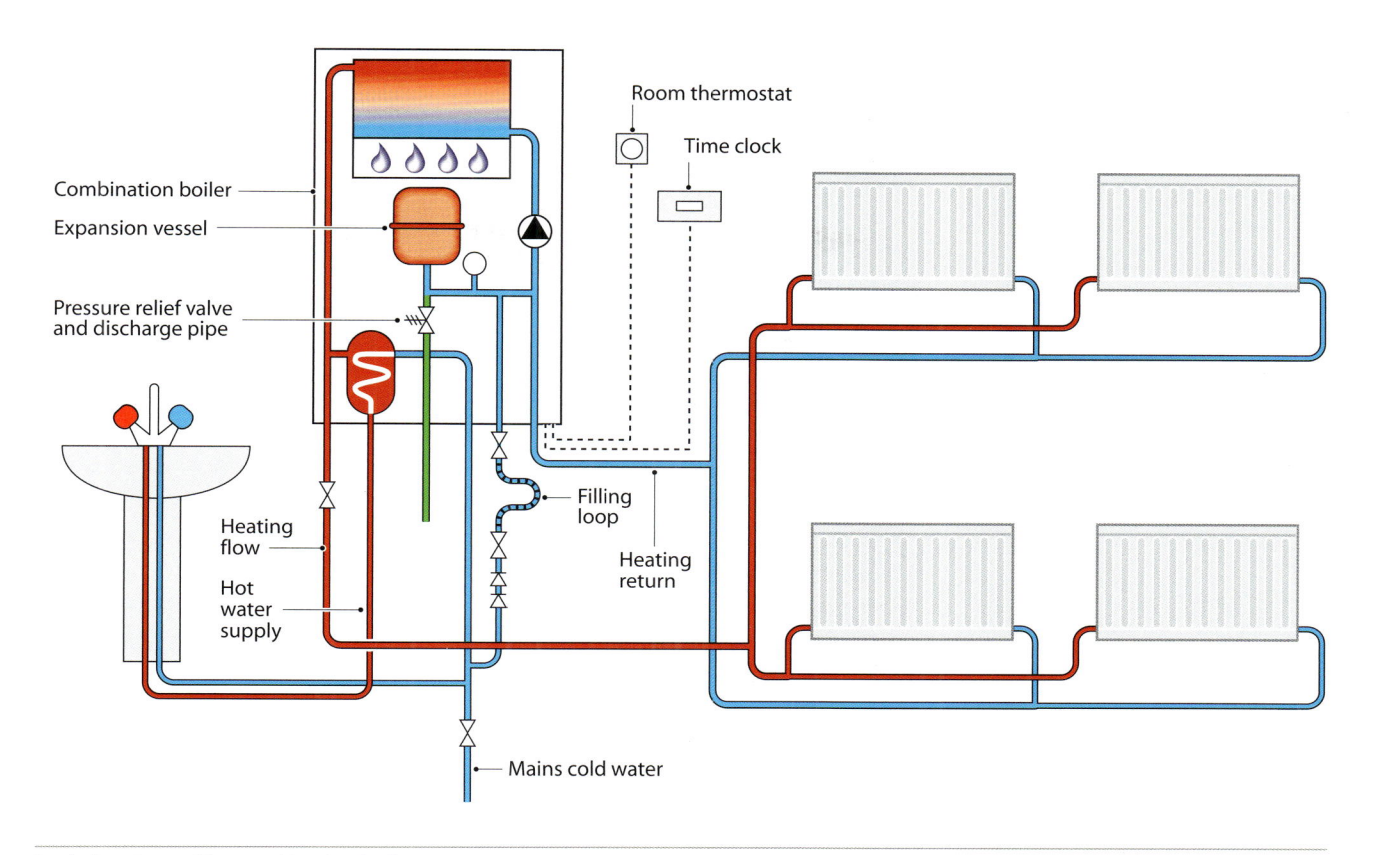

Sealed system with a combination boiler

Table comparing different types of sealed system:

	Full thermostatic control	Compliant with Building Regulations Document L	Recommended for larger properties	Instantaneous hot water supply	Can be used for large dwellings	Anti-cycling boiler interlock
Sealed system with external expansion vessel	✓	✓	✓	✗	✓	✓
Sealed system with system boiler	✓	✓	✓	✗	✓	✓
Sealed system with combination boiler	✓	✓	✓	✓	✗	✓

B8 ALTERNATIVE CENTRAL HEATING DESIGNS

Apart from the central heating systems we have already looked at, there are two other pipework arrangements that can be installed in domestic premises. These are:

- the microbore system
- the reversed return system.

The microbore system

The microbore system is a form of two-pipe system that uses a very small-bore pipe to feed the heat emitters. The system uses a multi-connection fitting, known as a manifold, fitted to the flow and return pipes and, depending on the size of the system, these are either 22mm or 28mm in size. All of the flow pipes to the heat emitters are taken from the flow manifold and all of the returns to the return manifold. The heat emitters are supplied through microbore pipework, generally 8mm or 10mm in diameter. Manifolds are fitted in pairs with the flow and return manifolds beside each other.

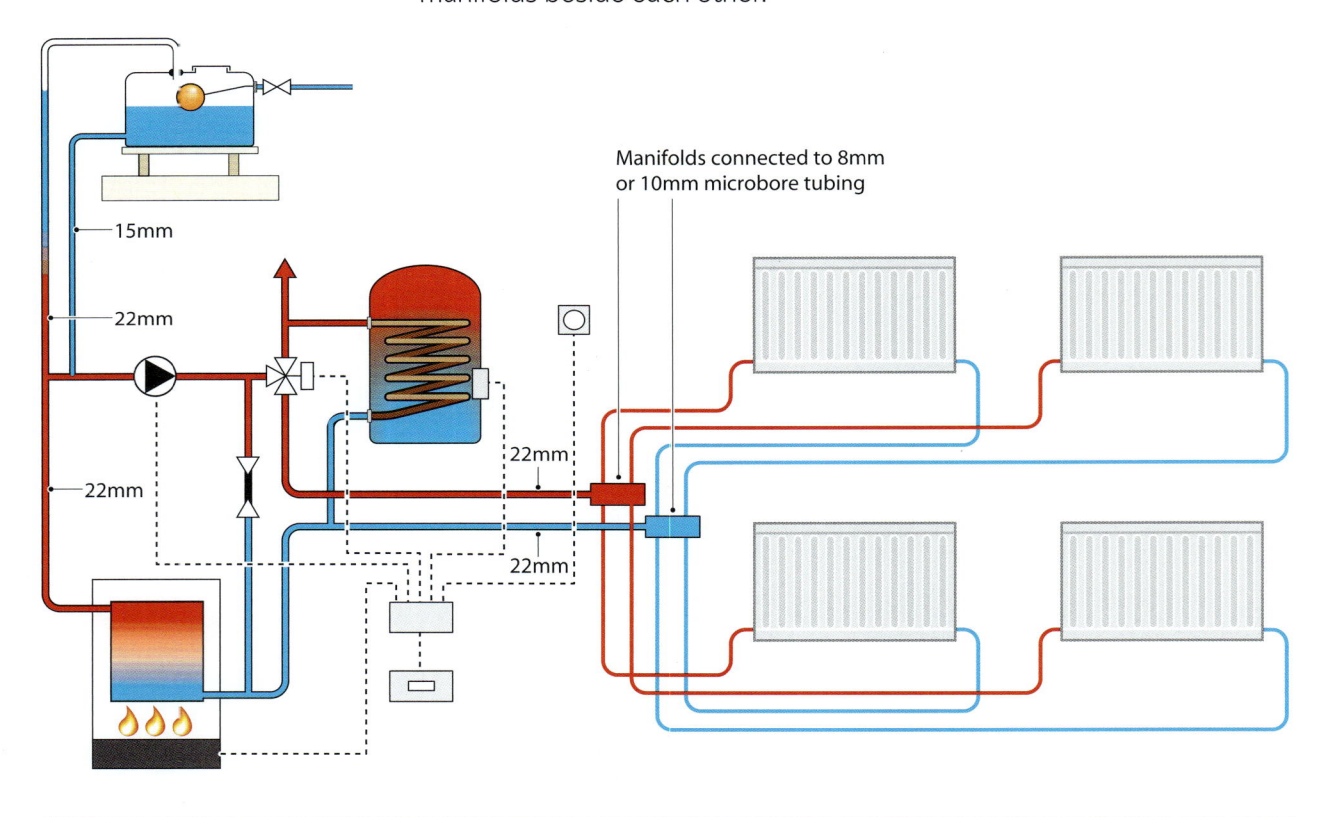

Microbore system

In small dwellings all the radiators may be taken from one pair of manifolds, which can accommodate up to eight radiators. It is usual, however, to fit a separate pair of manifolds on each floor in a house, and larger properties may have two pairs on each floor. The pipework loops that serve the largest radiators should not be longer than 9m.

ADVANTAGES	DISADVANTAGES
■ Contains only a small amount of water, so is quickly heated. ■ Microbore tubing comes in fully annealed coils and is easily bent by hand and easily hidden. ■ It can sometimes be a cheaper form of installation. ■ Long lengths of tubing means fewer joints. ■ Can be used with sealed and open-vented systems, Y-Plan or S-Plan. ■ The system is compliant with Building Regulations Document L.	■ Microbore piping is easily damaged and not very resistant to knocks. ■ Microbore tubes can get easily blocked with sludge if the system is installed poorly. ■ The system can suffer from scale build-up in areas where temporary hard water exists. ■ Because of smaller pipework, frictional resistance is greater, which can put increased strain on the circulator.

Microbore manifolds

The reversed return system

The reversed return system is designed for larger systems and is a variation of the two-pipe system. In the reversed return system, the return travels away from the boiler in the same direction as the flow before looping around to be connected to the return at the boiler. By doing this, the amount of pipe used on both the flow and the return is almost equal and this has the effect of ensuring that all of the heat emitters reach full temperature at about the same time. Reversing the return makes balancing the system much quicker and easier and in some cases, balancing is eliminated completely.

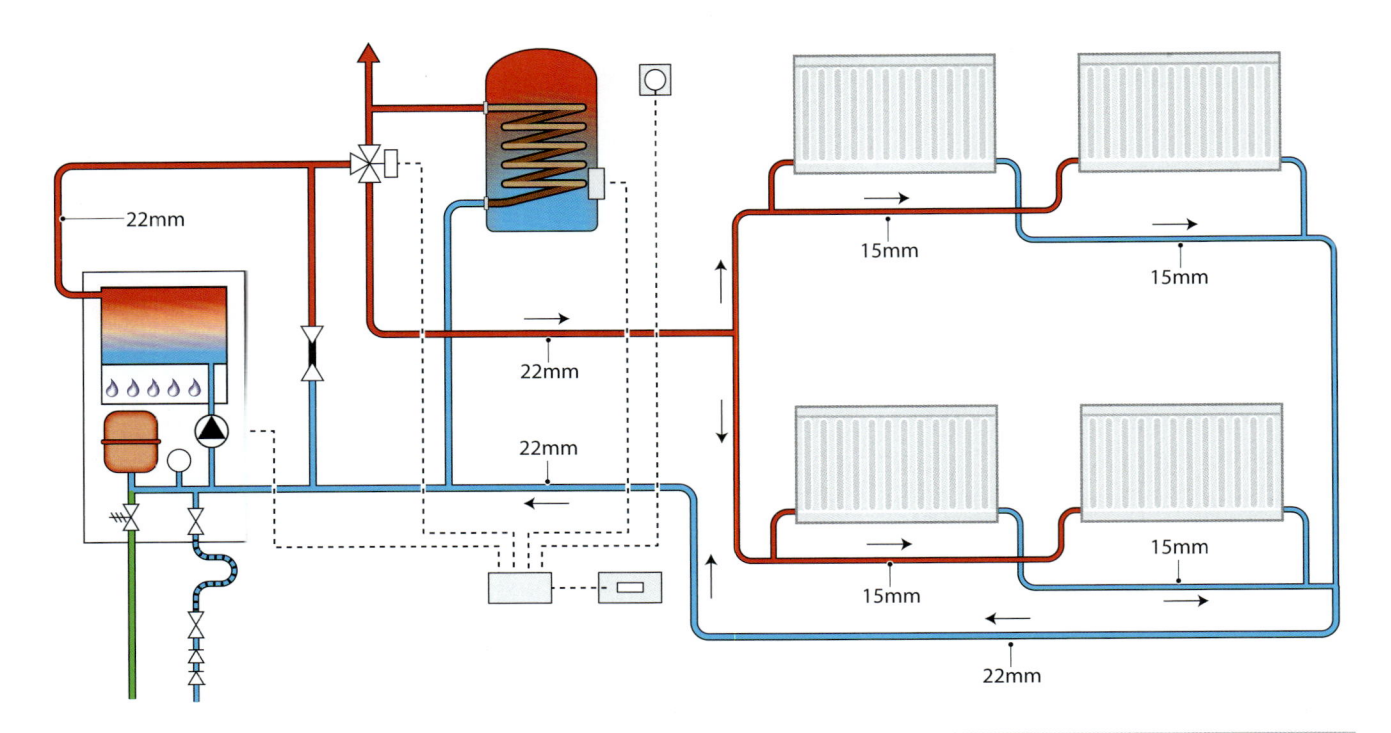

Reversed return system

ADVANTAGES	DISADVANTAGES
■ Eliminates the need for complex boiling procedures.	■ It is difficult to install.
■ Can be used with sealed and open-vented systems, Y-Plan or S-Plan.	■ It is a more expensive system due to the extra time taken on installation and the extra materials required.
■ The system is compliant with Building Regulations Document L.	■ The system installation requires careful planning and design.

KEY POINT

The legal requirements for the installation of solid fuel and oil heat-producing appliances, such as boilers, cookers and room heaters, are covered in Building Regulations Document J (Heat producing appliances). The legal requirements for the installation of gas appliances are given in the Gas Safety (Installation and Use) Regulations 1998. In all cases, manufacturers' instructions must always be followed when installing heat-producing appliances of any kind. The governing bodies for the different fuels used with heating appliances are:

Gas: Gas Safe http://www.gassaferegister.co.uk/

Oil: Oftec http://www.oftec.org/

Solid fuel: Hetas http://www.hetas.co.uk/

C   HEAT-PRODUCING APPLIANCES

This section investigates the different appliances that can generate the heat required to warm the systems and the different fuels they use. Boilers used for central heating systems are generally heated by one of four different fuels:

- solid fuel
- gas
- oil
- electricity in the form of an electric boiler – these are available for domestic systems but will not be mentioned here as they are specialist appliances.

The following table compares different types of fuel for heating appliances:

	Open-flued	Room-sealed (natural draught)	Room-sealed (fan assisted)	Freestanding / independent	Wall-mounted	Condensing	Non-condensing (traditional)	System boiler	Cookers	Open fire with high-output back boiler	Room heaters
Solid fuel	✓	✗	✗	✓	✗	✗	✓	✗	✓	✓	✓
Gas	✓	✓	✓	✓	✓	✓	✓	✓	✓	✗	✗
Oil	✓	✗	✓	✓	✓	✓	✓	✓	✓	✗	✗

C1 SOLID FUEL APPLIANCES

Solid fuel appliances are still used in rural areas of the UK where access to piped fuel supply is difficult. Solid fuel is available in many different forms, including:

- coal
- coke
- anthracite
- biomass wood pellets (carbon-neutral).

In this section we will look at the most common types of solid fuel appliances.

Open fires with a high-output back boiler

High-output back boilers are installed behind a real open coal fire. These appliances give their heat output in two forms:

- radiation from the open fire for direct room heating
- hot water from the boiler which is available for domestic hot water supply and central heating.

These appliances work on an open flue or chimney and contain a manual flue damper to regulate the amount of updraught through the chimney. By regulating the updraught, a certain amount of control can be administered over the heat of the fire. Typically, with the damper open, a fire of this type will give around 6.8–10kW of hot water heating output. With the damper closed, outputs vary from 5.3kW to 8.4kW. Radiated heat outputs from the coal fire directly into the room peak at around 2.6kW.

Solid fuel high-output back boiler

> **SUGGESTED ACTIVITY...**
> The different types of solid fuel were discussed in Chapter 003, page 111 and page 116. Have a look at that chapter to remind yourself about their characteristics.

Room heaters

A solid-fuel room heater is an enclosed appliance, usually with a glass door so that the fire can be viewed. It is installed directly into a chimney or open flue capable of accepting solid fuel and can either be stand-alone or fitted into a chimney breast with a high-output back boiler capable of serving up to ten heat emitters. Room heaters provide radiant heat for direct warmth and a constant circulation of convected heat.

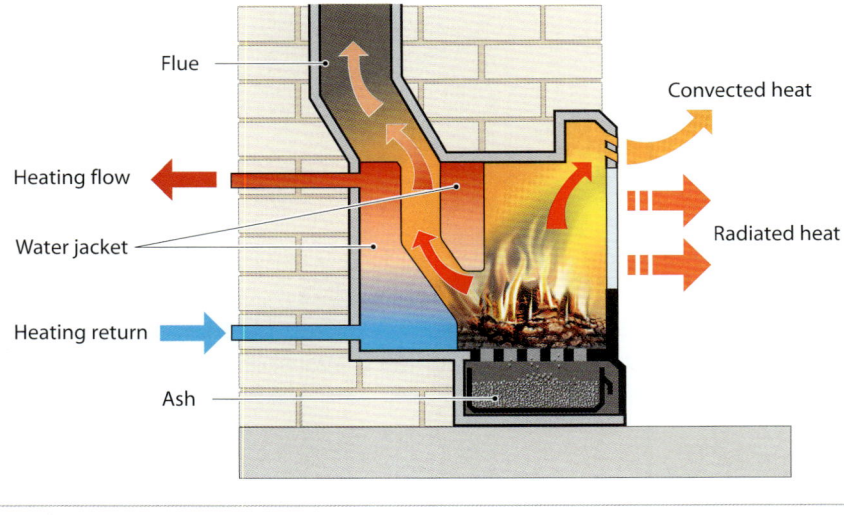

Solid-fuel room heater

Room heater cutaway

Solid fuel cookers (Aga type)

Open-flued solid fuel cookers have been around for many years. The concept of the solid fuel cooker is very simple: a controllable fire, burning continuously, inside a well-insulated cast iron shell, which retains the heat. When cooking is required, the heat is transferred to the ovens. The hotplates, because they are always hot, are covered with insulated cast iron covers, which lift up when hotplate cooking is required. Many models provide hot water and central heating as well as radiated heat in the room where they are fitted.

Solid fuel cookers burn a wide variety of solid fuels, including wood, and all have easy-to-empty ash pans so that the fire never goes out.

Independent boilers (freestanding)

Domestic open-flued independent solid fuel boilers are designed to provide both domestic hot water and central heating in a whole range of domestic premises from the very large to the very small.

There are two main types of independent boilers for domestic use:

- **Gravity feed boilers** – Often called hopper-fed boilers, these appliances incorporate a large hopper, positioned above the firebox, which can hold two or three days' supply of small-sized anthracite. The fuel is fed automatically to the fire bed as required, and an in-built, thermostatically controlled fan aids combustion. This provides a rapid response to an increase in demand. They are available in a wide range

Aga-type solid fuel cooker

KEY POINT

As we shall see later on in the chapter, Aga-type cookers are also available as oil-fired appliances.

of sizes and outputs. The main danger with this type of boiler is the risk of fire in the hopper. The fuel fed to the fire bed needs to be regulated with care.

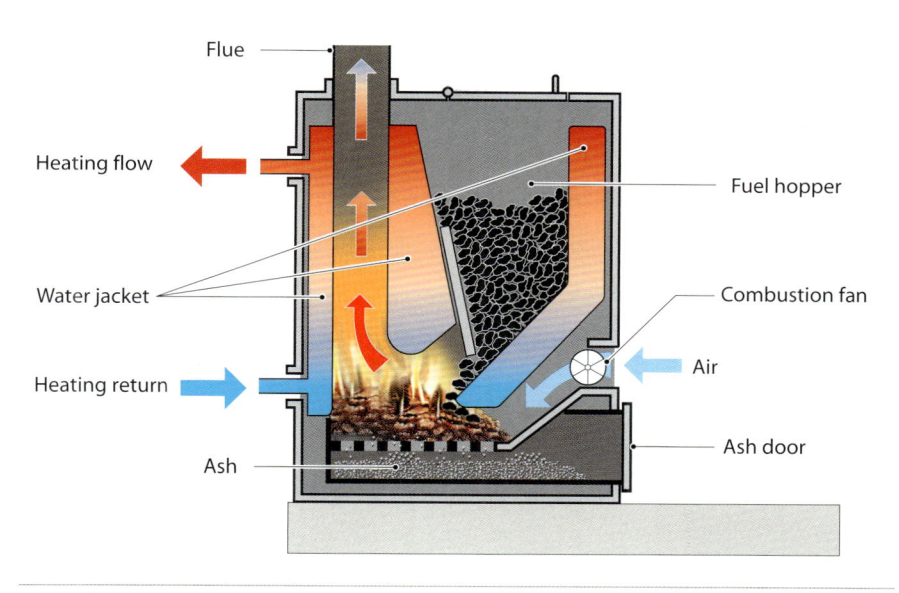

Gravity feed boiler

- **Batch feed boilers** – These are 'hand-fired' appliances requiring manual stoking. They require much more refuelling than hopper-fed boilers. They can, however, be less expensive to run in some cases and will often operate without the need for an electrical supply, thereby providing hot water and central heating during power failure.

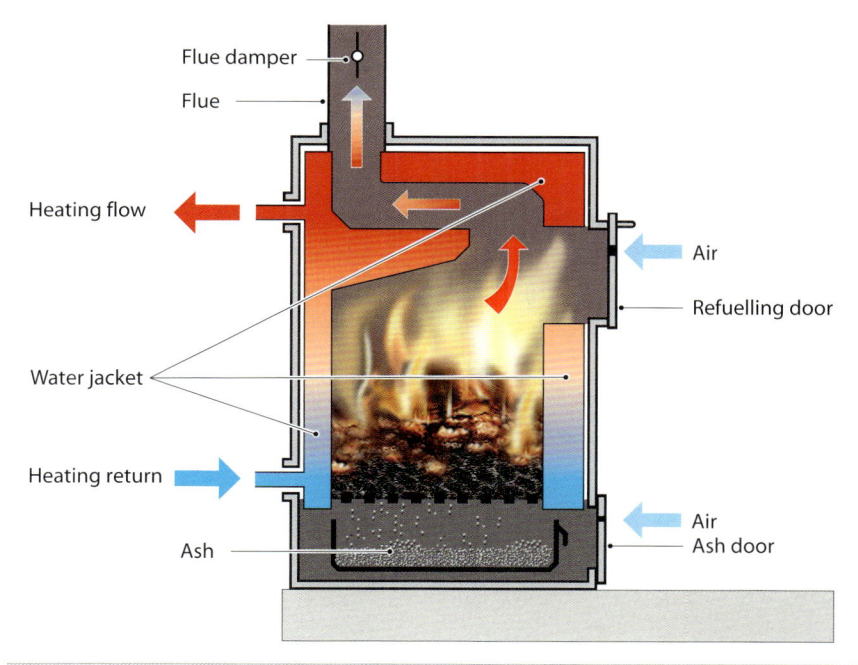

Batch feed boiler

C2 GAS CENTRAL HEATING BOILERS

Gas central heating boilers are the most popular of all central heating appliances. Over the years there have been many different types, from

> **SUGGESTED ACTIVITY...**
> The different types of commercially available gas were dealt with in Chapter 003, pages 113–14. Look through that chapter to re-familiarise yourself with them.

> **SUGGESTED ACTIVITY...**
> Condensing combination boilers and condensing system boilers were looked at on pages 414–15 of this chapter. Read that section again if you need to.

large multi-sectional cast iron domestic boilers to small, low-water-content condensing types. Both natural gas (those that burn a methane-based gas) and LPG (those that burn propane) types are available.

Central heating boilers can be categorised as:

- traditional boilers (non-condensing)
  - cast iron heat exchangers
  - low-water-content heat exchangers
  - combination boilers (non-condensing)
- condensing boilers
- condensing system boilers
- condensing combination boilers.

The following table compares different boilers and flue arrangements:

	Energy efficient	Cast iron heat exchanger	Low water content	Open-vented system	Sealed (pressurised) system	Open-flue	Room-sealed (natural draught)	Room-sealed (fan-assisted)	Wall-mounted	Freestanding
Traditional boilers	✗	✓	✓	✓	✓	✓	✓	✓	✓	✓
Condensing boilers	✓	✗	✓	✗	✓	✗	✗	✓	✓	✓
Condensing system boilers	✓	✗	✓	✗	✓	✗	✗	✓	✓	✗
Condensing combination boilers	✓	✗	✓	✗	✓	✗	✓	✓	✓	✓

Traditional boilers (non-condensing)

Traditional non-condensing boilers have been around for many years and in many different forms. In this part of the chapter we will look at some of the boilers you may come across when working on the many existing systems there are installed.

Boilers with cast iron heat exchangers

For many years, boilers were made with cast iron heat exchangers. They were often very large and heavy, even for small domestic systems. Some heat exchangers were made from iron cast in a single block, while older types were made up of cast iron sections bolted together. The more sections a boiler had, the bigger the heat output.

Fuel efficiency was typically 55–78 per cent, with much wasted heat escaping through the flue. Most traditional boilers were fitted on open-vented systems but sealed (pressurised) systems could also be installed with the inclusion of an external expansion vessel and associated controls (see page 411–14).

Cast iron boilers can be found either freestanding (floor mounted) or wall mounted using a variety of flue types:

- open
- room-sealed (natural draught)
- fan-assisted room-sealed (forced draught).

ADVANTAGES	DISADVANTAGES
- Long-lasting, typically 20 to 30 years. - Very robust.	- Heavy. - Not energy efficient. - Do not comply with Building Regulations Document L. - Noisy. - Very basic boiler controls.

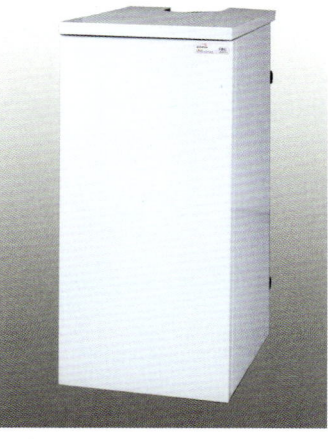

Traditional open-flue gas boiler

Boilers with low-water-content heat exchangers

Low-water-content heat exchangers were usually made from copper tube with aluminium fins, or lightweight cast iron. They were an attempt to reduce the water content of the heating system, thus speeding up heating times and improving efficiency. Typical efficiencies for this type of boiler were around 82 per cent.

The boilers were mostly wall mounted, very light in weight and, as a consequence, often quite small in size, designed for fully pumped S- and Y-Plan heating systems only. They were the first generation of central heating boilers to use a high temperature limiting thermostat (or energy cut-out) to guard against overheating and often used a basic printed circuit board to initiate a pump overrun, which kept the pump running for a short period after the boiler had shut down. This was required to dissipate any latent heat build-up in the water in the heat exchanger which could trip the energy cut-out and result in boiler lock-out.

Low-water-content boilers can be found with a variety of flue types:

- open
- room-sealed (natural draught)
- fan-assisted room-sealed (forced draught).

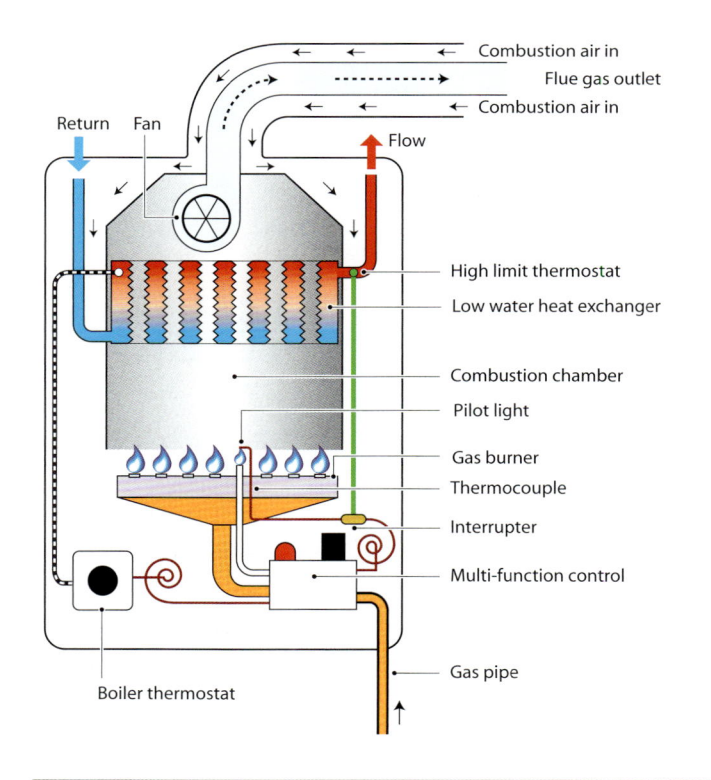

Fan-assisted low-water-content boiler

ADVANTAGES	DISADVANTAGES
■ Light in weight. ■ Often a cheaper appliance. ■ Relatively quick water heating times.	■ Can be very noisy. ■ Not energy efficient. ■ Do not comply with Building Regulations Document L. ■ Relatively short working life. ■ High maintenance compared with other boilers.

Combination boilers (non-condensing)

Combination boilers that supply instantaneous hot water as well as central heating have been around for many years. Early models, although wall mounted, were very large. Most had a sealed (pressurised) heating system but some were the low-pressure, open-vented type. Hot water flow rates were often poor by comparison with modern condensing types.

Early combination boilers can be found with a variety of flue types:

- open
- room-sealed (natural draught)
- fan-assisted room-sealed (forced draught).

ADVANTAGES	DISADVANTAGES
■ Instantaneous hot water supply. ■ Sealed system means no F&E cistern required in the roof space.	■ Not energy efficient. ■ Do not comply with Building Regulations Document L. ■ Often delivered poor hot water flow rates compared to other hot water delivery systems.

Condensing boilers

The latest addition to the gas central heating family is the condensing boiler. This works in a very different way from the traditional boiler.

Natural gas, when it is combusted, contains carbon dioxide, nitrogen and water vapour. As the flue gases cool, the water vapour condenses to form water droplets. It is this process that condensing boilers use. The flue gases first pass over the primary heat exchanger, which extracts about 80 per cent of the heat. The flue gases, which still contain 20 per cent of latent heat, are then passed over a secondary heat exchanger where a further 12–14 per cent of the heat is extracted. When this happens, the gases cool to their dew point, condensing the water vapour inside the boiler as water droplets, which are then collected in the condensate trap before being allowed to drain via the condensate pipe. The process gives condensing boilers their distinctive 'plume' of water vapour during operation, which is often mistaken for steam.

Dew point
This is the temperature at which the moisture within a gas is released to form water droplets. When a gas reaches its dew point, the temperature has been cooled to the point where the gas can no longer hold the water and it is released in the form of water droplets.

Modern condensing boilers are around 93 per cent efficient, releasing only 7 per cent of wasted heat in the cooler flue gases to the atmosphere.

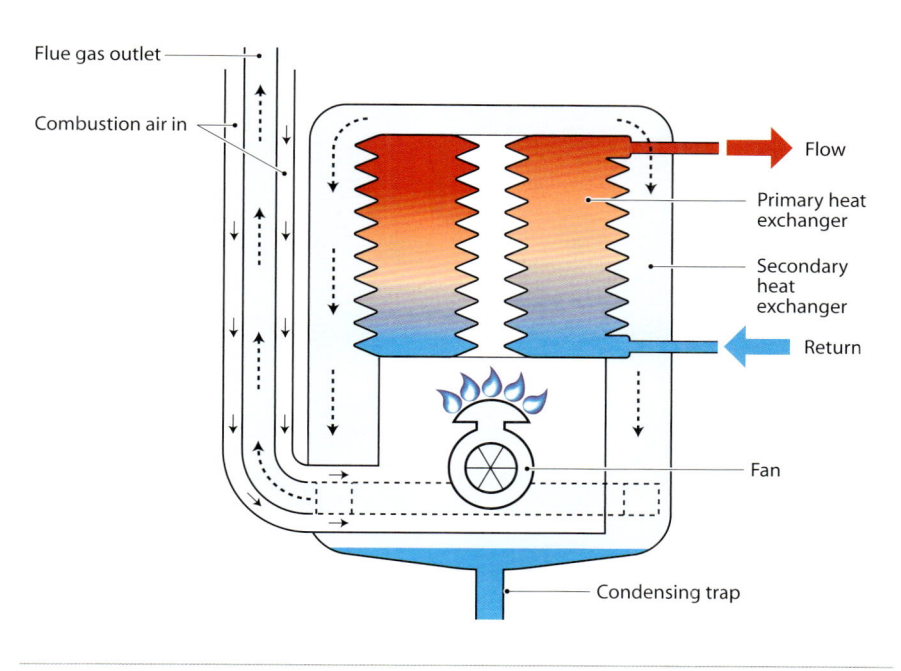

How a condensing boiler works

ADVANTAGES	DISADVANTAGES
- Compliant with Building Regulations Document L. - Very high efficiency. - Sealed (pressurised) system gives better heating flow rates. - System corrosion can be reduced. - Very quiet in operation. - Can be used with all modern fully pumped heating systems (system boilers). - No F&E cistern required in the roof space. - Very good flow rate on hot water supply (condensing combination boilers).	- High maintenance compared with other boilers. - Siting of the condensing pipework can often prove difficult. - Do not work if the condensing lines freeze during cold weather. - Use more gas when not in condensing mode.

> **SUGGESTED ACTIVITY...**
> To refresh your knowledge of fuel oil for domestic appliances, check out Chapter 003, page 112–13.

C3 OIL-FIRED CENTRAL HEATING APPLIANCES

Oil-fired appliances are popular where access to mains gas is difficult. They offer a viable alternative to gas appliances. Most oil-fired appliances use C2-grade 28-second viscosity oil (kerosene).

The following table compares different oil-fired appliances:

	Energy efficient	Cast iron heat exchanger	Open-vented system	Sealed (pressurised) system	Forced-draught open flue	Natural-draught open flue	Room-sealed (fan-assisted)	Wall mounted	Freestanding	Pressure jet burners	Vaporising burners
Traditional boilers	✗	✓	✓	✓	✓	✗	✓	✓	✓	✓	✗
Condensing boilers	✓	✓	✗	✓	✓	✗	✓	✓	✗	✓	✗
System boilers	✓	✓	✗	✓	✓	✗	✓	✓	✗	✓	✗
Combination boilers	✓	✓	✗	✓	✓	✗	✓	✓	✓	✓	✗
Cookers	✗	✓	✓	✗	✗	✓	✗	✗	✓	✗	✓

As can be seen from the table above, oil-fired appliances are available in a variety of different types, which generally use two different firing methods:

- pressure-jet or atomising burners
- vaporising burners.

Pressure-jet or atomising burners

Pressure-jet burners use an oil burner that mixes air and fuel. An electric motor drives a fuel pump and an air fan. The fuel pump forces the fuel through a fine nozzle, breaking the oil down into a mist. This is then mixed with air from the fan and ignited by a spark electrode. Once it is lit, the burner will continue to burn as long as there is a supply of air and fuel in the correct ratio.

Oil pressure-jet type boilers are installed on all modern oil-fired central heating systems including condensing system boilers, condensing combi boilers and wall-mounted types.

Typical oil pressure-jet burner

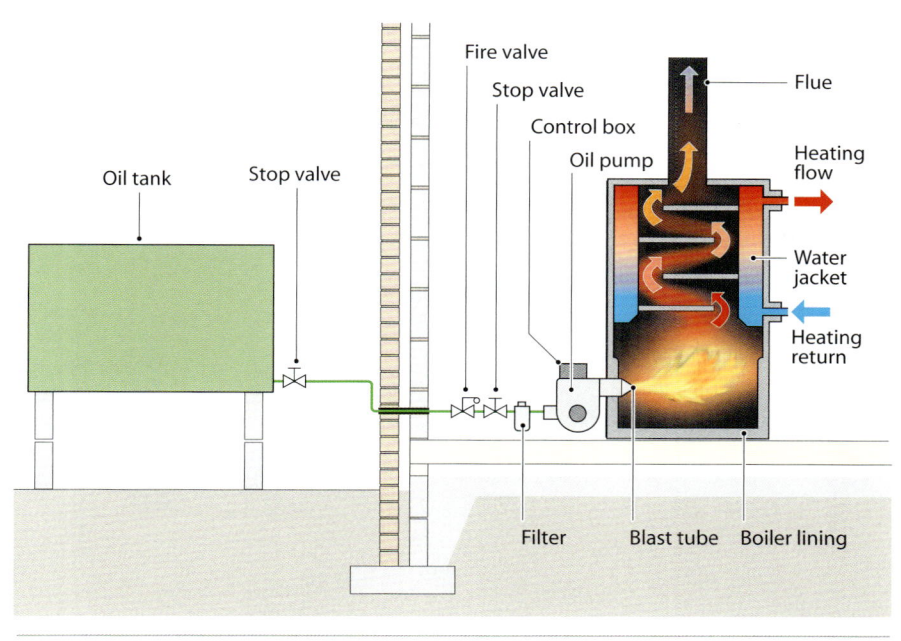

Oil pressure-jet burner installation

ADVANTAGES	DISADVANTAGES
• Compliant with Building Regulations Document L. • Very high efficiency. • Sealed (pressurised) system gives better heating flow rates. • Can be used with all modern fully pumped heating systems (system boilers). • No F&E cistern required in the roof space (system and combination types). • Very good flow rate on hot water supply (condensing combination boilers).	• High maintenance compared with gas boilers. • Noisy in operation. • Require oil tank for fuel storage.

Vaporising burners

Vaporising burners work on gravity oil feed; there is no pump. The oil flows to the burner where a small oil heater warms the oil until vapour is given off, and it is the vapour that is then ignited by a small electrode. As the oil burns, vapour is continuously produced, which keeps the burner alight.

They are generally only used in oil-fired cookers.

ADVANTAGES	DISADVANTAGES
• Very quiet in operation.	• Very limited use (cookers only).

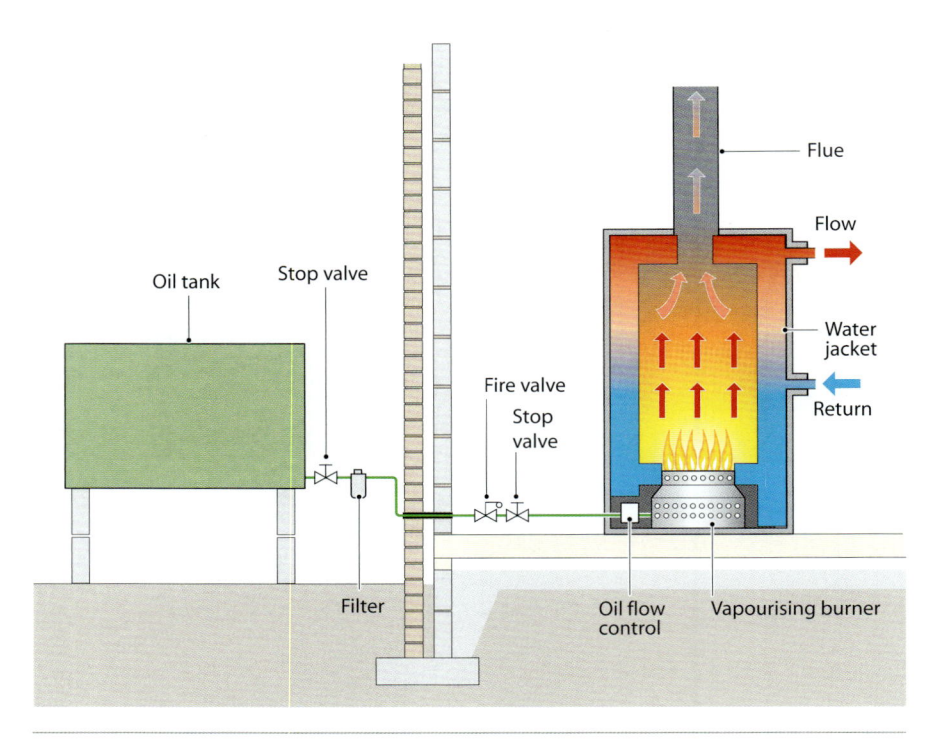

Vaporising oil burner installation

D TYPICAL FLUE SYSTEMS FOR CENTRAL HEATING APPLIANCES

All central heating appliances need a flue to remove the products of combustion safely to the outside. The basic concept is to produce an updraught, whether by natural means or by the use of a fan, to eject the fumes away from the building. There are two main flue concepts:

- open flues
- room-sealed (balanced) flues.

The following table compares different flue types for different appliances:

	Open-flue (natural draught)	Open-flue (forced draught)	Room-sealed (natural draught)	Room-sealed (fan-assisted)
Solid fuel boilers	✓	✓	✗	✗
Gas boilers	✓	✓	✓	✓
Pressure-jet oil burners	✗	✓	✗	✓
Vaporising oil burners	✓	✗	✗	✗

D1 OPEN FLUES

The open flue is the simplest of all flues. Because heat rises, it relies on the heat of the flue gases to create an updraught. There are two different types:

- natural draught
- forced draught.

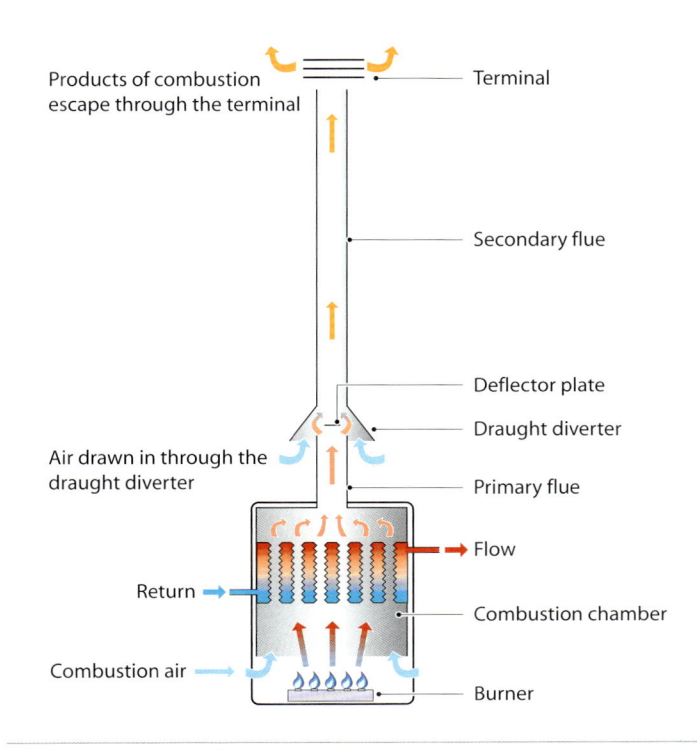

Operation of an open flue

With a boiler having this type of flue, air for combustion is taken from the room in which the boiler is located. The products of combustion are removed by natural draught vertically to the atmosphere, through a suitable terminal. The room in which the appliance is installed must have a permanently opened vent that allows fresh air to enter the boiler, permitting combustion to take place. This is usually supplied through an airbrick on an outside wall. All natural-draught open-flue appliances work in this way. The material from which the flue is made, however, will differ depending on the type of fuel used.

Occasionally, an open flue may have a forced draught. This is where a purpose-designed fan is positioned either before the combustion chamber or close to the primary flue. The fan helps to create a positive updraught by blowing the products of combustion up the flue. Forced-draught open flues are not suitable for all open-flue types and it will depend upon the boiler manufacturer and the boiler/flue design.

D2 ROOM-SEALED (BALANCED) FLUES

This boiler type draws its air for combustion directly from the outside through the flue assembly used to discharge the flue products. It is inherently safer than an open-flue type, since there is no direct route for flue products to spill back into the room. There are two basic types:

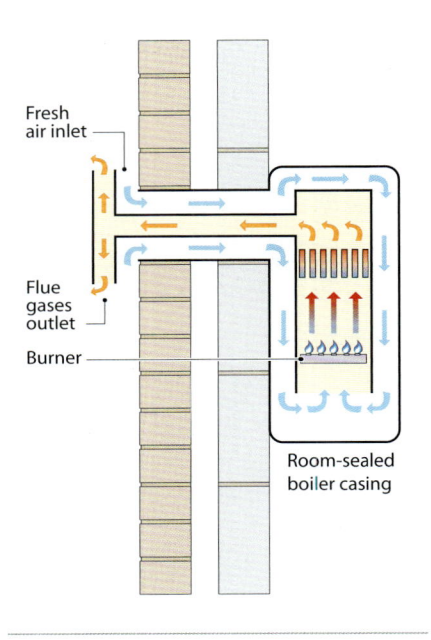

- natural draught
- fan-assisted (forced draught).

Natural draught

Natural draught room-sealed appliances have been around for many years and there are still many thousands in existence. The basic principle is very simple – both the combustion air (fresh air in) and the products of combustion (flue gases out) are situated in the same position outside the building. The products of combustion are evacuated from the boiler through a duct that runs through the combustion air duct – one inside the other.

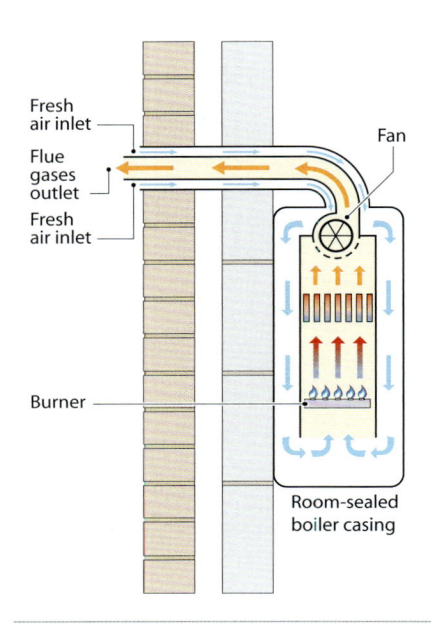

The boiler terminals are either square or rectangular and quite large in size. Terminal position is critical to avoid fumes going back into the building through windows and doors.

Fan-assisted (forced draught)

Fan-assisted room-sealed appliances work in the same way as their natural draught cousins, with the products of the combustion outlet being positioned in the same place (generally) as the combustion air intake, but there are two distinct differences:

- The process is aided by a fan, which ensures the positive and safe evacuation of all combustion products and any unburnt gas that may escape.
- The flue terminal is circular, much smaller and can be positioned in many more places than its predecessors.

There are two very different versions of the fan-assisted room-sealed boiler:

- **The fan positioned on the combustion products outlet from the heat exchanger** – This creates a desired negative pressure within the casing.
- **The fan positioned on the fresh air inlet, blowing a mixture of gas and air to the burner** – This creates a positive pressure within the boiler casing. Nearly all condensing boilers use this principle.

E HEAT EMITTERS

So far, this chapter has looked at the heating systems and the appliances that drive them. Here, we will look at the methods of getting the heat into the room or dwelling. For this, we need to look at the many different types of heat emitters that are available.

E1 PANEL RADIATORS

Modern panel convectors/radiators are designed to emit heat by convection and radiation. Seventy per cent of the heat is convected. They have fins welded to the back that warm the cold air that passes through them, creating warm air currents that flow into the room. This dramatically improves the efficiency of the radiator. Steel radiators that do not have fins rely more on radiant heat and this can lead to cold spots in the room. Positioning of the radiator is, therefore, critical. Radiators should be sited on a clear wall with no obstructions, such as windowsills, above them. If this is not possible, enough space should be left between the top of the radiator and the obstruction to allow the warm air to circulate. It is recommended by radiator manufacturers that radiators should be fitted at least 150mm from finished floor level to the bottom of the radiator (depending on the height of the skirting board) to allow air circulation.

The most common types of radiators are shown below.

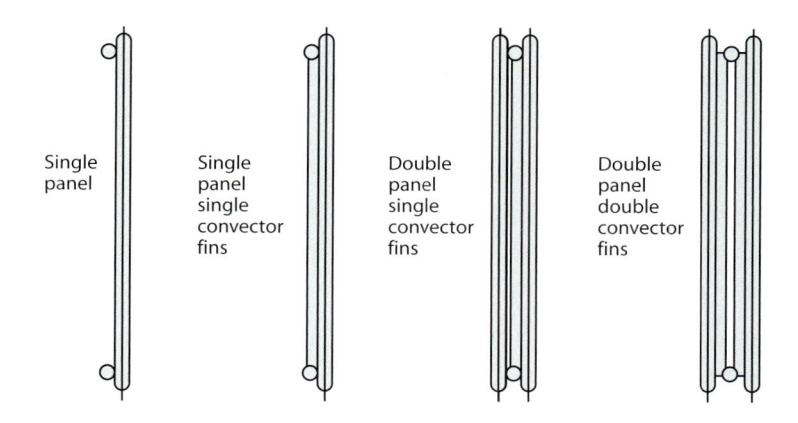

Single panel
Single panel single convector fins
Double panel single convector fins
Double panel double convector fins

Types of panel radiator

Manufacturers provide a wide range of heights – from 300mm through to 900mm, and lengths – from 400mm increasing by 100/200mm increments through to 3m.

It is important that radiators are fixed according to the manufacturer's instructions if the best output performance is to be achieved. Outputs vary from manufacturer to manufacturer. Each of these different radiator types is produced in three different styles:

- **Seamed top** – A very common style of radiator that was the market leader for many years. Top grilles and side panels are available for this radiator style.

- **Compact** – These have factory-fitted top grilles and side panels, making them a more attractive radiator style. Currently the most popular radiator style available.

- **Rolled top** – The least popular of all radiator styles. They are somewhat old-fashioned-looking with exposed welded seams either side.

Seamed top radiator   Compact radiator   Rolled top radiator

> **KEY POINT**
>
> Radiator connections are classified by their abbreviations. For example:
>
> **TBOE** means Top, Bottom, Opposite End (used on heat sink radiators with solid fuel systems and one-pipe systems). This is probably the best combination for energy efficiency as the hot water enters the radiator at the point where it is most needed but it is not used for aesthetic reasons. Pipework that enters at the top of a radiator can look unsightly and is more easily damaged.
>
> **BBOE** means Bottom, Bottom, Opposite End (the usual method of radiator connection for aesthetic reasons: it looks better than TBOE).
>
> **TBSE** means Top, Bottom, Same End (used with some one-pipe systems).

Domestic panel radiators have ½-in BSP female threads at either side, top and bottom and these will accept a variety of radiator valves. One end of the radiator has an air-release valve, with the other end being blanked by the use of a plug. These, together with hanging brackets, are usually supplied by the radiator manufacturer.

Dressing a radiator

Dressing a radiator involves getting the radiator ready for hanging by putting in the valves, the air release valve and the plug. The process is as follows:

1 Carefully remove the radiator from its packing. Inside the packing you will find the hanging brackets, the air-release valve and the plug and, often, small U-shaped pieces of plastic that are to be placed on the brackets where the radiator fits. These are designed to prevent the radiator from rattling.

2 Take out the factory-fitted plugs. Be careful here, especially if you are working in a furnished property, as the radiator often contains a small amount of water from being tested at the factory.

3 Split the valves at the valve unions and wrap PTFE tape around the valve tail. About 10–15 wraps will ensure that the joint between tail and radiator does not leak. Make the tail into the radiator using a radiator spanner.

4 Insert and tighten the air-release valve and plug using an adjustable spanner.

Radiator spanner

Hanging a radiator

1 Before hanging the radiator, you must decide how close you want it to be to the wall. Radiator brackets have two options – near and far – therefore, select the one that is best for the installation and the customer. Maximising the space between the radiator and the wall increases convection.

2 Mark the centre of the radiator and the position of the radiator brackets.

3 Place a radiator bracket into position on the radiator and measure from the bottom of the bracket to the bottom of the radiator. This is usually (depending on the manufacturer) 50mm. This is measurement A.

4 Mark the centre of the position of the radiator on the wall where the radiator is to be hung.

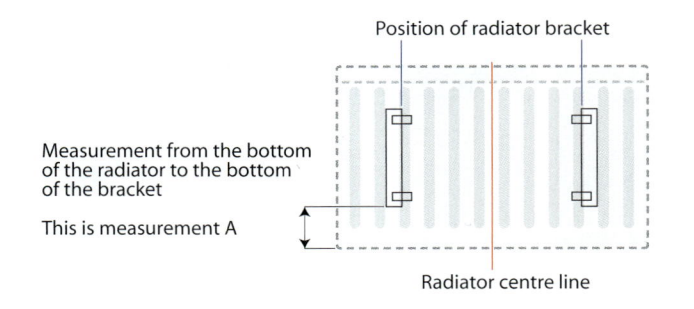

Marking bracket positions on radiator

5. Place the radiator against the wall on the centre line and mark the position of the brackets on the wall. Using a spirit level, draw two vertical lines where the brackets are to be fixed.

6. Radiators are best hung at 150mm from the floor (depending on the skirting board height) to allow air circulation through the fins, so add measurement A to 150mm and mark across the two bracket marks on the wall, using a spirit level.

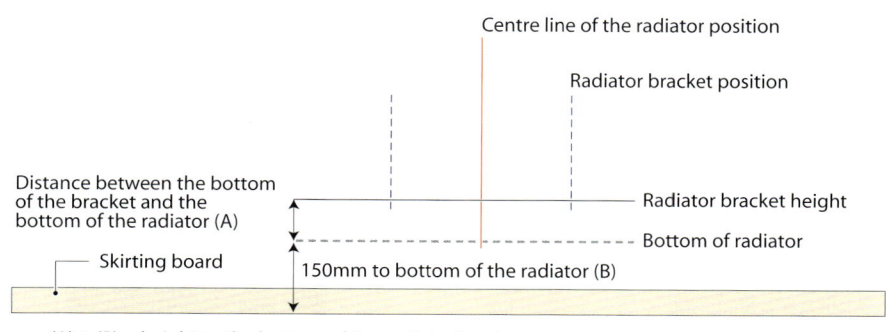

(A) + (B) = height to the bottom of the radiator brackets

Marking heights on wall

7. Radiator brackets can usually be hung either with the radiator close to the wall, or with a larger gap. Decide which way the brackets are to be fixed, then place the bracket against the marked position on the wall, making sure that the bottom of the bracket is sitting on the bottom bracket mark. Mark the fixing position.

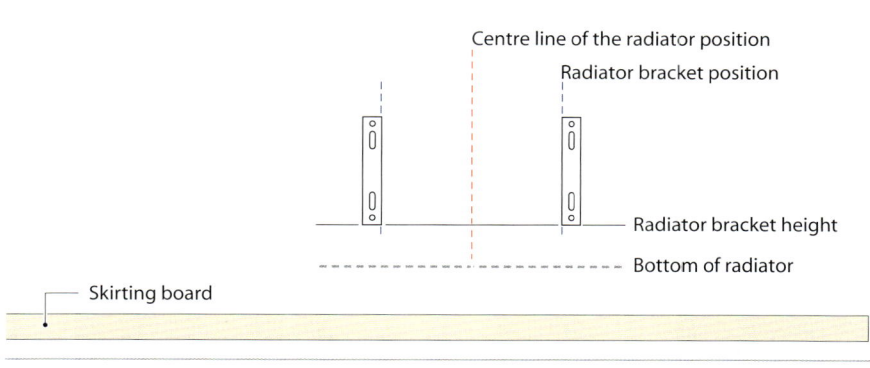

Marking bracket positions on wall

8. For masonry walls: using a suitable masonry drill bit, drill the four bracket holes (a 7-mm masonry drill bit and brown wall plugs are usually suitable). Screw the brackets to the wall using 50mm × 10 screws.

9. For timber-studded walls: use plasterboard fixings that are capable of carrying the weight of the radiator plus the water inside.

10. Hang the radiator onto the brackets. Check that it is level using a spirit level and that it is 150mm from the finished floor level.

E2 COLUMN RADIATORS

Column radiators have been available for many years. As the name suggests, they are made up of columns; the more columns the radiator has, the better the heat output. They are increasingly being used with modern heating systems, especially on period refurbishments.

Column radiators can be made from three different metals – cast iron, steel and aluminium – with many modern column radiator designs now being produced by a variety of manufacturers.

Modern column radiator

Traditional column radiator

E3 LOW SURFACE TEMPERATURE RADIATORS

Low surface temperature radiators (LSTs) were specifically designed to conform to the NHS Estates guidance note 'Safe hot water and surface temperature' which stated that:

> 'Heating devices should not exceed 43°C when the system is running at maximum design output…'

This has been adopted not only by the NHS but also by Local Authorities and commercial buildings installations to which the general public may have access, including residential care homes and schools. LSTs are also becoming popular in domestic installations, especially in children's bedrooms and nurseries and where the elderly, infirm or disabled are likely to come into contact with radiators.

SmartScreen Unit 008 handout 5

Low surface temperature radiator

E4 FAN CONVECTORS

Fan convectors work on the same basic principle as traditional finned radiators. A finned copper heat exchanger is housed in a casing, which also contains a low-volume electrically operated fan. As the heat exchanger becomes hot, a thermostat operates the fan and the warm air is blown into the room. Because the warm air is forced into the room, more heat can be extracted from the hot, circulating water. Once the desired temperature has been reached, the fan is switched off again by the thermostat.

Fan convectors tend to be larger than traditional radiators and they also require a mains electric connection, usually via a switched fuse spur. There are two separate types of fan convector:

- **Wall mounted** – These tend to be quite large in size. The manufacturer's data should be consulted to allow the correct heat output to be selected.

Operation of a fan convector

A wall-mounted fan convector

- **Kick space heaters** – These are specifically designed for kitchen use where space to mount a radiator is limited. They are installed under a kitchen unit and blow warm air via a grille mounted on the kick plinth.

Kick-space fan convector

A kick-space fan convector

E5 TUBULAR TOWEL WARMERS

These are available in a range of different designs and colours and are often referred to as designer towel rails. They can be supplied for use with wet central heating systems with an electrical element option, for use during the summer when the heating system is not required. They are usually mounted vertically on the wall and can be installed in bathrooms and kitchens.

Tubular towel warmers

Towel warmer with integral radiator

E6 TOWEL WARMERS WITH INTEGRAL PANEL RADIATORS

Less popular than tubular towel rails, these heat emitters combine a towel rail and radiator in one unit. It allows a towel to be warmed without affecting the convection current from the radiator. They are generally only installed in bathrooms.

E7 SKIRTING HEATING

Skirting heating consists of a finned copper tubular heat exchanger in a metal casing, which replaces the skirting boards in a room. It is usually used where unobtrusive heat emitters are required. Skirting heating can be used as perimeter heating below glazing or for background heat in some areas.

The heat output, at 450 watts per metre, is quite low which means that, to be effective, the skirting heating would need to be at floor level on all walls of the room to offset the room heat losses – although the heat coverage is very similar to that experienced with specialist underfloor heating.

One disadvantage is that efficiency is reduced by dust collecting in the fins.

E8 UNDERFLOOR HEATING

Underfloor heating, sometimes referred to as embedded pipe coils, is becoming more popular, especially in buildings that are environmentally friendly and in refurbishments of bespoke dwellings such as barn conversions. Underfloor heating consists of pipework laid on a thermally insulated bed in a loop or coil formation around the floor of the room. This is then screeded over with concrete.

The insulated bed prevents heat loss downwards and ensures that as much of the heat as possible is projected upwards into the room. The pipes are usually installed 150–225mm apart across the floor of the room.

The pipework loops are fed back to a central pair of manifolds, one for the flow and one for the return. A special mixing valve regulates the water temperature to ensure that the floor does not reach too high a temperature. Each room is thermostatically regulated by its own room thermostat.

F MECHANICAL CENTRAL HEATING CONTROLS

Mechanical central heating controls do not use electricity but still play a vital role in helping to ensure the correct and efficient operation of the system. This section looks at the most common mechanical controls used on domestic central heating systems.

F1 RADIATOR VALVES

There is a wide selection of radiator valves available from many different manufacturers. However, there are three basic types:

- **Thermostatic radiator valves (TRVs)** – These control the temperature of the room by controlling the flow of water through the radiator. They react to air temperature. TRVs have a heat-sensitive head that contains a cartridge, which is filled with a liquid, a gas or a wax, and this expands and contracts with heat. As the room heats up, the cartridge expands and pushes down on a pin on the valve body. The pin closes and opens the valve as the room heats up or cools down. The valve head has a number of temperature settings to allow a range of room temperatures to be selected. Document L1 of the Building Regulations requires that thermostatic radiator valves are installed on new installations to control individual room temperatures and on all radiators except the radiator where the room thermostat is fitted. Most TRVs are bidirectional. This means that they can be fitted on either the flow or the return.

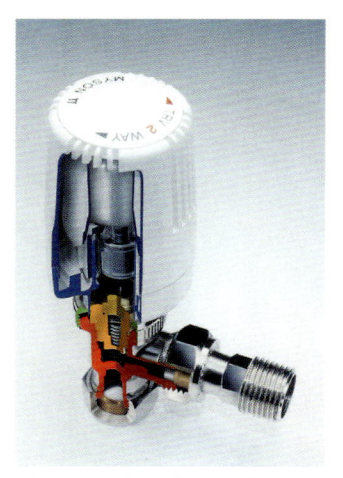

Thermostatic radiator valve

- **Wheel head valves** – These allow manual control of the radiator by being turned on or off. The valve is turned on by rotating the wheel head anticlockwise and turned off by rotating it clockwise.

- **Lockshield valves** – These are designed to be operated only by a plumber and not by the householder. They are adjusted during system balancing to regulate the flow of water through the radiator. The lockshield head covers the valve mechanism. They can be turned off for radiator removal.

Wheel head radiator valve with lockshield cover

F2 AUTOMATIC AIR VALVES

Automatic air valves are fitted where air is expected to collect in the system, usually at high points. They allow the collected air to escape from the system but seal themselves when water arrives at the valve.

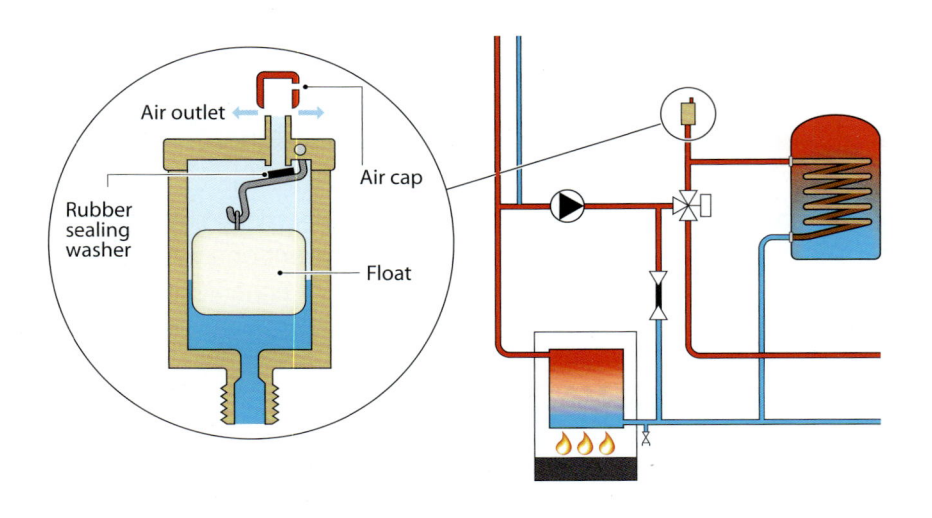

Automatic air valve

When water reaches the valve the float arm rises, closing the valve. As more air reaches the valve the float momentarily drops, allowing the air out of the system. These valves are often used with a check valve that prevents air from being drawn into the system backwards through the valve.

F3 AUTOMATIC BYPASS VALVES

The automatic bypass valve controls the flow of water across the flow-and-return circuit of fully pumped heating systems by opening automatically as other paths for the water close, such as circuits with motorised valves and radiator circuits with thermostatic radiator valves. This occurs as the hot water circuit and heating circuit/thermostatic radiator valves begin to reach their full temperature. As the circuits close, the bypass will gradually open, maintaining circulation through the boiler and reducing noise in the system due to water velocity. Most boiler manufacturers require a bypass to be fitted to maintain a minimum flow rate through the boiler and prevent overheating.

Automatic bypass valve

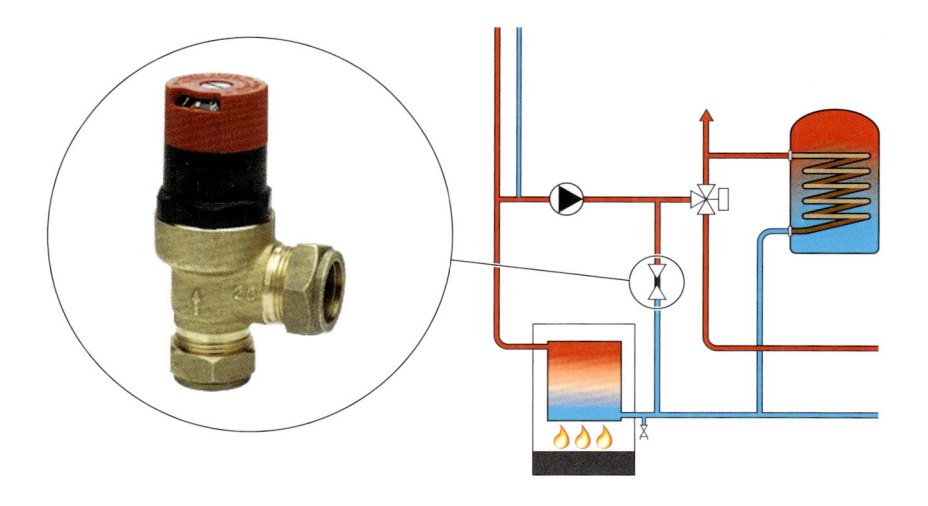

Automatic bypass valves are much better than fixed bypass valves, as these, being permanently open, take the flow of hot water away from the critical parts of the system, which increases the heating time for both hot water and heating circuits. This reduces the efficiency of the system and increases fuel usage.

F4 THERMO-MECHANICAL CYLINDER CONTROL VALVES

These are non-electrical valves used to control the temperature of a hot water cylinder. They are mainly used with gravity primary circulation as part of an upgrade to give some control over secondary hot water temperature. According to the Domestic Heating Compliance Guide, if the hot water cylinder only is being replaced and no control over the hot water temperature exists, a thermo-mechanical thermostat is the minimum standard of hot water control required to comply with Document L of the Building Regulations.

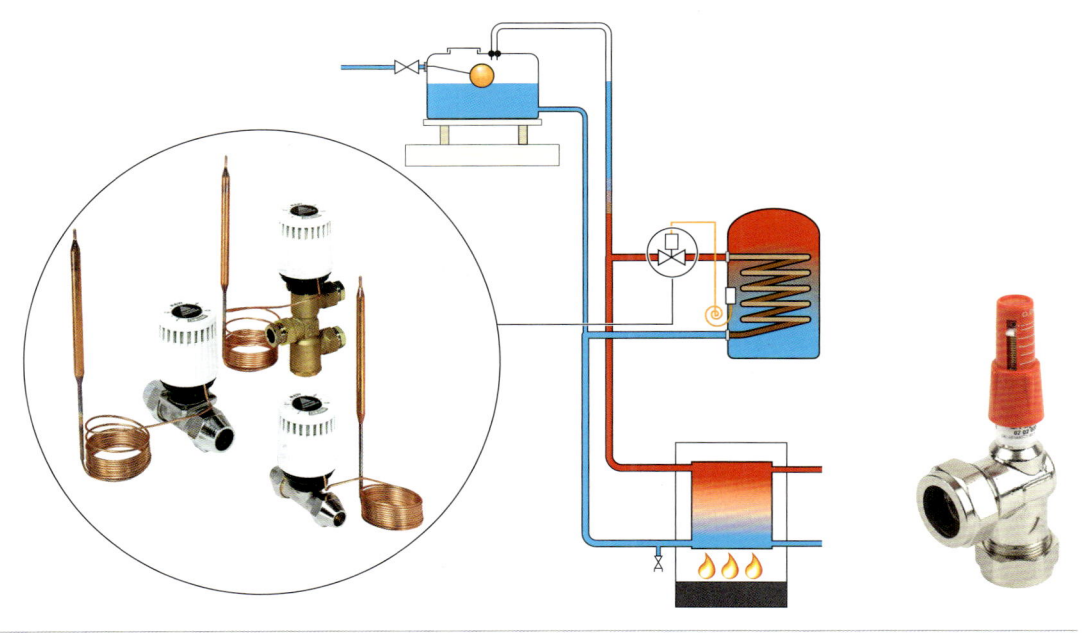

Thermo-mechanical thermostat

Thermo-mechanical thermostats work on the principle of thermal expansion of a liquid or gas in much the same way as thermostatic radiator valves, except that with this valve, the temperature of the water is sensed by a remote sensor. The sensor should be placed about one-third up from the bottom of the cylinder

F5 ANTI-GRAVITY VALVES

Anti-gravity valves prevent unwanted gravity circulation to the upstairs radiators on semi-gravity systems when only the hot water is being heated. They are essential on all semi-gravity systems, especially those fuelled by solid fuel. Anti-gravity valves should be positioned on the vertical flow to the upstairs heating circuit.

Anti-gravity valves are very similar in design to the single check valves mentioned in earlier chapters. They only allow water flow in one direction and, when the heating system is off, they are in the closed position. In this position, gravity circulation cannot take place. As soon as the central heating circulation pump switches on, the flow of the water opens the valve to allow heating circulation.

F6 DRAIN VALVES

Drain valves should be fitted at the lowest points in the heating installation to allow complete draining of the water in the system and this includes all radiators, especially if the flows and returns to the radiators are on vertical drops from above. For this purpose, radiator valves with built-in drain valves are available.

Radiator valve with built-in drain valve

SUGGESTED ACTIVITY...
To refresh your knowledge of drain valves, check out Chapter 005, page 316. Don't forget, when soldering drain valves, the valve must be taken apart before soldering commences otherwise you might melt the rubber drain valve washer onto the valve seating.

G ELECTRICAL CENTRAL HEATING CONTROLS

Modern central heating systems cannot function without electrical controls. They are required at every stage of operation from switching the system on to shutting it down when the required temperature has been reached. They provide both functional operation and safety and are a requirement of Building Regulations Document L (Conservation of fuel and power).

Before looking at the various controls, consider the implications of Document L, which was updated in October 2010. The main points are listed in the table, opposite.

(Table and summary of general requirements reproduced by permission of TACMA. TACMA is the association of UK manufacturers and suppliers of controls used in heating and hot water systems and the internal environment of buildings: www.tacma.org.uk.)

Areas covered by the regulations	Requirements in the regulations	Suitable installations for compliance	Exemptions/ recommendations
Heating – temperature control	1. The dwelling must be divided into at least two heating zones.	A zone valve on the pipework to control the flow to each zone.	Single-storey open-plan dwellings in which the living area is more than 70 per cent of the total floor area can be controlled as one zone.
	2. Each zone must have temperature control with both a thermostat and individual radiator controls such as TRVs.	Either: A room thermostat or programmable room thermostat in each zone, plus thermostatic radiator valves (TRVs) on all radiators except bathrooms and rooms with a thermostat. A centrally controlled system of TRVs on each radiator that can be demonstrated to provide interlock when no heat is required.	Where only the boiler is replaced, compliance with zone requirements can be achieved by a single room thermostat or programmable room thermostat plus TRVs on all radiators except the one in the room with the thermostat. Installation of TRVs when the system is drained down should always be done except where the radiators or pipework make this impractical
Heating and hot water – time control	The dwelling must have automatic time control so that the heating and hot water system is turned on and off at set times that can be adjusted by the occupant(s). Time control must be provided as follows: Dwellings up to 150m² where the hot water is produced instantaneously only require a single time-control circuit. Dwellings up to 150m² where there is a hot water cylinder require separate time control for the heating and for the hot water. Dwellings over 150m² require separate time control for the hot water (unless instantaneous) and separate time control for each heating zone.	All required timed circuits must have independent time control. This can be achieved through either: A separate timer or programmer on each circuit. A full programmer with separate timing of heating and hot water. A programmable room thermostat on each timed heating circuit plus a timer on the hot water circuit. A multi-channel programmer providing full control of each timed circuit from a central point.	Where only a hot water cylinder is being replaced it is acceptable to have a single timing control for both space and water heating. However, if separate time control for hot water is present in such a situation then the new installation must retain this level of control.

Hot water – temperature control	The dwelling should have control of the temperature of any stored hot water. Dwellings over 150m² should have more than one hot water circuit with separate time and temperature control for each circuit.	A cylinder thermostat with a zone valve or three-port valve. In some circumstances such as thermal stores a second pump should be substituted for the zone valve.	Where only a hot water cylinder is being replaced in an emergency situation (i.e. non-planned) either a wireless or thermo-mechanical hot water cylinder thermostat should be installed as a minimum.
Hot water supply – temperature control	The hot water supply to any fixed bath must be designed and installed so as to incorporate measures to ensure that the temperature of the water that can be delivered to that bath does not exceed 48°C.	A thermostatic mixing valve should be installed and it is recommended that this be set to deliver hot water to the bath at 43°C.	Only applies in new dwellings or through a material change of use of an area within an existing dwelling, e.g. a new bathroom.

Summary of general requirements for all heating and hot water systems with a gas or oil boiler

(These requirements apply for both the installation of new systems and when replacing a boiler unless specifically mentioned as an exemption.)

1 Install the system with fully pumped circulation. When replacing a boiler in an existing system with semi-gravity circulation convert the system to fully pumped circulation.

2 Install an automatic bypass valve where manufacturer's instructions advise installation of a bypass. TACMA does not recommend the installation of an automatic bypass valve if the boiler is of a fully modulating type.

3 Install a 'boiler interlock' so that the boiler and pump are switched off when there is no demand for heating or hot water. This is achieved by correct wiring of the room thermostats or programmable room thermostats, the cylinder thermostat and zone valves in conjunction with the timing device(s). The use of traditional TRVs alone does not provide interlock though some systems of programmable TRVs can do so – check manufacturer's data for information.

Additional requirements for installations

1 On completion of the installation all equipment should be commissioned in accordance with the manufacturer's instructions. The operation of all controls should be tested and the distribution system should be fully balanced to ensure correct operation of the thermostatic radiator valves.

2 The installer must also give a full explanation of the system and its operation to the user. This will include a description of how to use all of the controls and the relevant User Instructions must be left with the user. For new systems in existing homes the Part L approved document states that 'a way of complying would be to provide a suitable set of operating and maintenance instructions aimed at achieving economy in the use of fuel and power in terms that householders can understand in a durable format that can be kept and referred to over the service life of the system(s)'.

Controls upgrades in existing homes

While upgrades to controls in existing heating systems, other than at times of boiler replacement, are not specifically required under the Building Regulations it is good practice for all homes to have a set of controls that at least complies with the minimum standards in the Building Regulations – a boiler interlock, room thermostat, programmer, thermostatic radiator valves and a hot water cylinder thermostat. These will ensure that the existing heating system is not operating inefficiently and allow the occupants to make further reductions in their energy costs through behaviour change.

Heating installers should recommend controls upgrades as required to meet these standards when visiting homes for maintenance and repairs. UK Government is committed to reducing energy use in homes and householders should be reminded that 84 per cent of energy use in homes is from heating and hot water.

Wireless controls are a convenient choice for controls upgrades, allowing ease of installation and minimal disruption by taking away the requirement for wiring runs.

To comply with the requirements, the correct electrical controls must be fitted. For more information on central heating controls, you can visit:

http://www.planningportal.gov.uk/buildingregulations/approveddocuments/partl/

and

http://www.sap-online.co.uk/

G1 TIME CLOCKS AND PROGRAMMERS

Time clocks are the simplest of all central heating timing devices. They are only suitable for switching on one circuit such as the heating circuit and so are ideally suited for combination boiler installations. Both mechanical and digital time clocks are available.

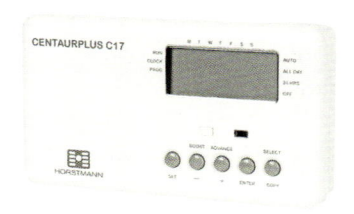

Programmers are similar to time clocks but have two functions: to switch on the hot water and the central heating. There are three basic types:

- **Mini-programmer** – This allows the heating and hot water circuits to be on together, or hot water alone, but not heating alone. Ideally suited to C-Plan systems.

- **Standard programmer** – This uses the same time settings for space heating and hot water.

- **Full programmer** – This allows the time settings for space heating and hot water to be fully independent. Some will allow seven-day programming of both heating and hot water so that the two circuits can be used individually or both together.

Programmers are often fitted to the front fascia of the boiler and integrated into the boiler design. This, however, is not always convenient, especially if the boiler is sited in a garage or roof space.

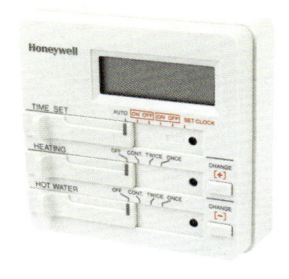

Central heating time clock

G2 ROOM THERMOSTATS

A room thermostat senses air temperature. It is simply a temperature-controlled switch that connects or breaks an electrical circuit when either calling for heat or shutting the circuit down as the correct temperature has been reached. Most room thermostats contain a very small heater element called an accelerator, which tops up the heat to the room thermostat by 1–2°C, smoothing out the temperature cycle and preventing the boiler from cycling when it isn't required.

Programmable room thermostats allow different temperatures to be set for different days of the week. They also provide a night setback feature where a minimum temperature can be maintained at night. Some units also allow time control of the hot water cycle.

Digital programmer

Room thermostat

Cylinder thermostat

SmartScreen Unit 008 handout 6

Frost thermostat

Pipe thermostat

SmartScreen Unit 008 worksheet 4

G3 CYLINDER THERMOSTATS

A simple control of stored hot water temperature, usually strapped to the side of the hot water cylinder about one-third of the way up from the bottom. It is used with a motorised valve to provide close control of water temperature and should be set to 55°C.

G4 FROST THERMOSTATS AND PIPE THERMOSTATS

The purpose of the frost thermostat is to stop the boiler and any other vulnerable parts of the system from freezing in extremely cold weather. It is wired into the system to override all other programmers and thermostats. They should be set to between 3°C and 5°C and should be placed close to the vulnerable parts of the system, especially if they are fitted in unheated garages and roof spaces.

Frost thermostats are much more effective when installed alongside a pipe thermostat, which is strapped to vulnerable pipework and senses water temperature. It is designed to override all other controls when the temperature of the water is close to 0°C and works in conjunction with the frost thermostat. The pipe thermostat and frost thermostat should be wired in series (see Chapter 004, page 201).

G5 MOTORISED VALVES

We have already seen that both the two-port zone valve and the three-port mid-position valve are key controls for the S-Plan and Y-Plan fully pumped systems and the C-Plan semi-gravity system. To recap the key points of these valves:

- **Three-port diverter valve** – Very similar in appearance to the three-port mid-position valve, this valve is designed to control the flow of water on fully pumped central heating/hot-water systems, where hot water priority is required.

- **Three-port mid-position valve** – Used on fully pumped central heating/hot water systems to provide full temperature control of both the hot water and heating circuits when linked to cylinder and room thermostats. The circuits can operate together or independently of each other.

- **Two-port motorised zone valve** – Valves of this type can be found on both C-Plan-plus systems where a single valve linked to a cylinder thermostat controls the hot water temperature, and S-Plan fully pumped systems where two two-port zone valves control the heating and hot water circuits via room and cylinder thermostats. They can also be used to zone different parts of the heating circuit.

G6 SYSTEM DESIGN AND CONTROL

Now that we have seen the controls and the system layouts, we must look at how the controls work together to ensure efficient operation of the systems. This section concentrates on fully pumped systems, as these are the systems that must be used for new installations. The tables opposite compare the Y-Plan and S-Plan systems. For more about how they work, see pages 403–406.

The Y-Plan system	
The three-port valve	The flow from the boiler must be connected to the AB port, which is marked on the valve. The A port must be connected to the heating circuit. The B port must be connected to the hot water circuit. The valve must not be installed upside down as leakage of water could penetrate the electric actuator.
Time control	This must be provided by a programmer that allows individual use of hot water and heating circuits.
Heating circuit	Must have a room thermostat positioned in the coolest room (normally the hallway) away from heat sources and cold draughts. It should be wall-mounted at 1.5m from floor level. The room thermostat controls the three-port valve. All radiators must have thermostatic radiator valves fitted apart from the radiator in the room where the room thermostat is located.
Hot water circuit	The hot water temperature must be controlled by a cylinder thermostat placed one-third up from the base of the cylinder. The cylinder thermostat controls the three-port vale.
Bypass	An automatic bypass valve is required.
Frost/pipe thermostat	Must be provided where parts of the system are in vulnerable positions.

The S-Plan system	
The two-port zone valves	A single zone valve must be installed on the hot water circuit controlled by a cylinder thermostat. The heating circuit must contain one or more (if the system is to be zoned) two-port zone valves. These are controlled by individual room thermostats.
Time control	This must be provided by a programmer that allows individual use of hot water and heating circuits. A second time clock may be required if the system is zoned.
Heating circuit	One or more room thermostats controlling downstairs and upstairs heating circuits. These should be installed at 1.5m from floor level.
Hot water circuit	The hot water temperature must be controlled by a cylinder thermostat placed one-third up from the base of the cylinder.
Bypass	An automatic bypass valve is required.
Frost/pipe thermostat	Must be provided where parts of the system are in vulnerable positions.

G7 BOILER INTERLOCK

The boiler interlock is not a single control device but the interconnection of all of the controls on the system, such as room thermostats, cylinder thermostats and motorised valves. The idea behind the boiler interlock is to prevent the boiler firing up when it is not required, a problem with older systems. A boiler interlock can also be achieved by the use of advanced controls, such as a building management system (BMS), usually reserved for the larger systems but now available for domestic properties.

H INSTALLATION OF CENTRAL HEATING SYSTEMS

In this part of the chapter, we will consider the materials that can be used to install domestic central heating systems and the installation methods for both new-build properties and existing installations.

H1 PIPEWORK MATERIALS

Most domestic systems use one of three types of pipe materials:

- **Copper tubes and fittings** – Grades R220 and R250 are generally used for domestic central heating installations. Grade R250 in sizes 15mm, 22mm and 28mm is used for minibore installations, while R220 is used for microbore systems. Most modern microbore system use 10mm pipe.

- **Low carbon steel pipes and fittings** – These are very rarely used for domestic installations but are used extensively on commercial and industrial systems.

- **Polybutylene pipes and fittings** – These are fast becoming the material of choice for new-build installations because of the ease of installation. Remember, though, that the connections to any heat-producing appliances must be made using copper for the metre nearest to the appliance, although a combination of polybutylene and copper is acceptable at the heat emitter where copper is used to connect to the emitter and the polybutylene barrier pipe is used underfloor.

H2 INSTALLATION METHODS

The installation of tubes and fittings has been covered extensively in earlier chapters of this book but central heating systems demand careful consideration because of the temperature that the systems run at. With water at 80°C for the flow and 60°C for the return, the pipework – regardless of the material used – will expand and contract as the pipe heats up and cools down. Obviously, not all of the materials expand at the same rate but provision should be made at the installation stage to allow for expansion and contraction if problems with noise are to be avoided. Here are some points to consider:

> **SUGGESTED ACTIVITY...**
> Pipework materials were extensively covered in Chapter 005, pages 222–53. Read through that section to remind yourself about these.

- Polybutylene pipe expands more than copper tube but copper is much more rigid than polybutylene. When installing pipes in wooden floors, enough room should be allowed in any notches made. If the pipes are too tight in the joist, they will tick as they expand and contract. This is very pronounced with central heating systems installed using copper tubes because the water reaches a higher temperature.

- Clipping and securing pipework becomes very important. The clipping distances for the various pipes and tubes become critical where polybutylene pipe is concerned, especially when used with central heating installations. As the pipe becomes hot, it starts to soften and this leads to the pipe sagging between joists and clips. This not only looks unsightly but can put excessive strain on the joints.

- On new-build installations, it is common practice to install microbore pipework behind the dry lining plasterboard. In this instance, if the pipework is made from copper, it should be clipped well and wrapped to avoid noise and corrosion. Polybutylene pipe should also be wrapped because the expansion of pipe on a hard surface could cause undue abrasion on the soft plastic. A metallic tape should be placed at the back of the polybutylene to allow the pipe to be found by metal detecting tools when it is covered.

- Pipes placed in chases should be wrapped against corrosion and insulated where required.

General installation requirements are:

- F&E cisterns must be fitted in accordance with the Water Supply (Water Fittings) Regulations 1999. Cistern requirements are mentioned in Chapter 006, page 300.

- Filling loops, expansion vessels and associated equipment should be installed where they do not create an eyesore but are accessible. The installation of expansion vessels should always be in accordance with the manufacturer's installation instructions. With system boilers and combination boilers this does not present a problem, as they are an integral part of the appliance.

- Radiator positions should be considered with care. It is generally accepted that radiators should be placed under windows but this is not always the best position if an even circulation of warm air is to be achieved. On new-builds and refurbishments, the radiator positions are usually marked on the detailed building plans.

- Pipework must be insulated in places where there is a risk of freezing, such as under a suspended timber floor or in an unheated garage. The Building Regulations also advise that pipework in airing cupboards must be insulated to prevent unwanted heat loss.

- All metallic pipework and metal parts within the system must be electrically bonded to earth.

> **SUGGESTED ACTIVITY...**
> In previous chapters, we learnt how we can calculate the head of pressure in a system. You will remember that 10m head is equivalent to 1 bar pressure. Also, you will remember that the test pressure is 1.5 times normal operating pressure. With these two facts in mind, calculate the test pressure required from the following information:
>
> a A system has a head of 4m. What is the test pressure measured in bar?
>
> b A system has a head of 6m. What is the test pressure measured in bar?
>
> c A system has a head of 10m. What is the test pressure measured in bar?
>
> Answers: a. 0.6 bar, b. 0.9 bar, c. 1.5 bar.

H3 TESTING AND FILLING

Testing

Before initial testing takes place, the system should be visually checked to make sure that it is correct, that all visible joints are tight and that all clipping is in accordance with the British Standards distances.

The testing procedure is very similar for both hot and cold water installations but the test pressure will depend on the type of system installed. As with other systems, the test pressure is 1.5 times normal operating pressure, and that pressure will vary depending on the type of system installed. For instance:

- for sealed (pressurised) systems working at 1 bar pressure, the test pressure is 1.5 bar
- for open-vented systems, where the head of pressure is, say, 8m, the test pressure is 12 m or 1.2 bar.

Test timing should be in accordance with BS 6700:2006+A1:2009 and will depend on the material used in the installation. Testing should be conducted using a hydraulic test pump.

Filling

The procedure for filling central heating systems will depend on the type of system that is installed.

Open-vented systems

Filling open-vented systems is a fairly simple procedure. Having conducted a pressure test at the installation stage, there should be no surprises when it comes to system filling:

1. Ensure that all radiator valves and radiator air-release points are closed.

2. Check the F&E cistern to ensure that all joints are tight.

3. Temporarily replace the pump with a short piece of tubing; this will ensure that no debris enters the pump.

4. Ensure that all motorised valves are manually set to the open position for initial system filling.

5. Turn on the service valve to the F&E cistern and allow the system to fill.

6. Starting with the furthest away radiator on the downstairs circuit, open the radiator valves and fill and bleed the air from each radiator. Work backwards towards the boiler – the downstairs circuit first, then the upstairs circuit. This will ensure that air is not trapped in pockets around the system.

7. Once the system is full, allow it to stand for a short while. Visually check for leaks at each radiator and all exposed pipework and controls/valves etc.

8. Check the water level in the F&E cistern.

> **SUGGESTED ACTIVITY...**
> Pipework testing was extensively covered in Chapter 005, pages 270–71 – take a look at those pages to refresh your memory.

9 Drain down the system. This will flush the system through, removing any flux residues, steel wool, etc.

10 Refit the pump and turn on the pump valves.

11 Refill the system as before.

Sealed systems

The main difference when compared to open-vented systems is that there is no F&E cistern, so the system will have to be filled in short bursts via the filling loop. In other words, turn on the filling loop, fill the system up to operating pressure, turn off the filling loop, bleed the air from the radiators until the pressure has depleted and then restart the process until the system is full. All other points are the same as above.

H4 COMMISSIONING AND BALANCING

After the system has been filled it should be commissioned by a competent person, who will complete all the necessary forms and certificates. The system should be run to full operating temperature and then switched off and drained down again while the system is still hot. This is known as the hot flush and ensures that the system is clean before corrosion inhibitors are added to the system. It is also considered good practice to use a system cleanser at this stage to ensure that all flux and swarf residues are removed from the system. When refilling, corrosion inhibitors can be added.

Commissioning a system will involve:

- checking that a 3 amp fuse is present in the electrical switch fused spur for the system

- running the system up to full operating temperature and, again, checking for any leaks

- checking that all radiators are working correctly and balancing the system to ensure that all radiators reach the same temperature in about the same time frame

- checking the system temperatures of both hot water and heating circuits with a digital thermometer

- checking the operation of all electrical thermostats, motorised valves, time clocks and programmers

- benchmarking the system: this is the formal recording of flow rates, temperatures, controls fitted, boiler gas rates and any other vital information that proves the system is installed in compliance with both the manufacturer's instructions and any relevant standards

- setting any time clocks/programmers for the customer and instructing the customer on the use of the system. Tell them what they can and, more importantly, cannot adjust

- ensuring that the customer is provided with all the manufacturer's installation and servicing literature. It is a good idea to remind them to keep all data in a safe place as these will be needed by the engineer when the appliance is serviced.

> **KEY POINT**
>
> The method of balancing a system is covered at Level 3 but it is important that you have an idea of what balancing is at this level even if you don't actually practise it.
>
> Balancing a system is ensuring an even heat distribution to all of the radiators on the system. Water will always take the least line of resistance and this means that the radiators nearer to the circulating pump will always get more than their fair share of heat. This is detrimental to the furthest radiators on the system and has the effect of slowing down the heating effect.
>
> Balancing is simply evening out the heat distribution by restricting the flow of water to the quickest-heating radiators by adjusting the flow rate through the lockshield valves. This has the effect of slowing the heat circulation through the quicker radiators and increasing the flow rate to the slower (furthest away) radiators, thus balancing the heat distribution throughout the system.

H5 DECOMMISSIONING

Decommissioning central heating systems follows much the same process as with other systems we have looked at. There are a number of scenarios where systems would need to be decommissioned:

- where the system is being completely stripped out prior to a new system installation or where the building is being demolished
- where the boiler is being replaced and the F&E cistern is being taken out
- where the system is being added to or altered
- where system components such as radiators are being permanently taken out
- general maintenance activities, such as a pump replacement, a radiator replacement or replacement of valves and other controls.

Always remember to:

- keep the customer and/or other trades informed of the work being carried out, ie when the system is being isolated and the expected length of time it will be out of service
- ensure that any services, such as electricity, gas etc, are safely isolated and pipework capped
- use warning notices, such as 'do not use' or 'system drained' on any taps, valves, appliances, electrical components, etc.

H6 CORROSION PROTECTION

Corrosion is probably the biggest problem that takes place within wet central heating systems and it commonly occurs in two forms:

- the formation of red oxide sludge (rust) because of constant air infiltration
- the formation of black oxide sludge and sediment because of galvanic corrosion.

Corrosion can attack a system very quickly. As soon as the system is filled with water, corrosion begins to work to break down certain elements within it.

Air infiltration

This is a constant problem with some systems, especially those that are open vented. Central heating systems last longer once the water in the system has lost all of its oxygen. Without oxygen, rust cannot occur. Air infiltration happens for a number of reasons:

- Micro leaks let air in but do not show as a water leak. These are extremely hard to trace and usually occur around the packing glands of lockshield radiator valves and air-release valves. They always occur on the negative pressure side of any system.

- Air is sucked down the vent pipe due to poor system design.
- A constant see-sawing of water within the F&E cistern aerates the water.
- Small leaks introduce fresh aerated water into the system.

Electrolytic corrosion

Within central heating systems, there are a number of metals used: steel radiators, brass valves (brass contains zinc), copper tubes and stainless steel heat exchangers. On older systems there may also be cast iron boilers or parts containing aluminium. All of these metals lie at different points on the electromotive series of metals (see Chapter 004, page 162) and once they are connected via water (an electrolyte), corrosion begins immediately. This problem is accelerated when the water becomes hot. The net result of this reaction is that the steel of the radiators begins to be eaten away, with the fine particles of steel falling to the bottom of the radiator as a sediment that forms a magnetic black sludge. As a by-product, the radiator may also fill with hydrogen, which requires constant venting. The sludge not only blocks pipework and finds its way into all of the low points of the system, but it also causes boiler noise and creates pitting corrosion in the radiators. The spider chart below shows some of the problems that can result from system sludging.

Black oxide sludge

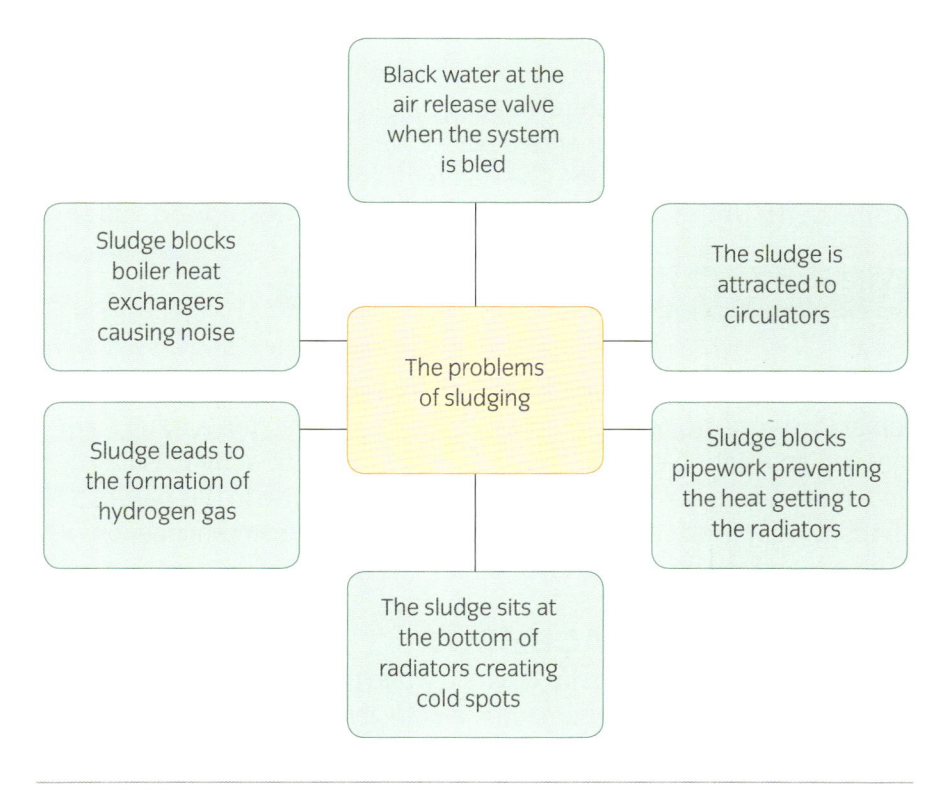

The problems of sludging

- Black water at the air release valve when the system is bled
- The sludge is attracted to circulators
- Sludge blocks pipework preventing the heat getting to the radiators
- The sludge sits at the bottom of radiators creating cold spots
- Sludge leads to the formation of hydrogen gas
- Sludge blocks boiler heat exchangers causing noise

System sludging

Corrosion inhibitor

Corrosion inhibitor must be added to the system to comply with the manufacturer's warranty. Corrosion inhibitor slows down the process of corrosion and black sludge forming, and helps to lubricate pump bearings and valves. Once added, corrosion inhibitor does not need to be replaced except when the system is drained down. Corrosion inhibitor:

- stops a build-up of black oxide sludge, the major cause of central heating problems
- helps to reduce fuel costs
- helps prevent the formation of hydrogen gas
- has a non-acidic neutral formation and so is harmless to the environment
- prevents pinholing of radiators and pipework
- prevents scale formation.

Corrosion inhibitor must not be added to systems that contain a single-feed self-venting cylinder, as these use air entrapment to separate the primary and secondary systems. Should the air bubbles within the cylinder break, it would lead to the inhibitor chemicals mixing with the domestic hot water supply, causing contamination.

The use of magnetic filters

Black oxide sludge is made up of minute particles of steel that have been robbed by galvanic corrosion. This sludge is attracted to components such as circulating pumps, causing pump failure and damage to the system.

Magnetic filters protect central heating systems by using very powerful magnets to attract the suspended black oxide steel particles in the central heating system water. This can remove almost 100 per cent of suspended particles, preventing further build-up of black oxide sludge.

Magnetic filter

MAINTENANCE OF SYSTEMS

Maintenance of central heating systems takes many forms, from replacing valves to replacing boilers. It can also include adding to or altering an existing system. Whatever maintenance activity is being undertaken, safe isolation of the system is paramount. In this part of the chapter, we will look at some of the more common maintenance activities and the processes involved.

I1 PUMP REPLACEMENT

The system should not require draining when replacing a pump. Before attempting to remove the pump, the electricity should be isolated at the switched fused spur and the fuse retained to prevent accidental switching on of the circuit. You should only attempt this task under the supervision of a qualified plumber.

1. Check that the electrical circuit is dead by using a multimeter or some other effective electrical testing device.

2. Make a simple drawing of the live/neutral/earth connections on the pump and disconnect the cable.

3. Turn off the isolating valves either side of the pump.

4. Carefully loosen the unions on the pump by turning them anticlockwise using water pump pliers. Have some old towels handy to catch any water.

5. Once both unions have been disconnected, remove the pump. The pump unions should have the old washers removed and the union faces cleaned. The new pump will come with replacement sealing washers (typically made of rubber).

6. Position the new pump, with the sealing washers in place between the valves, and hand tighten the unions. Take care to ensure that the pump is facing in the right direction for the system.

7. Fully tighten the unions with the water pump pliers. If the pump is installed horizontally, make sure that the bleed point is slightly above horizontal, as this will help to remove any air in the pump.

8. Turn on the pump valves and check for leaks.

9. Carefully reconnect the electrics to the pump – live to the L point, neutral to the N point and earth to the E point. Make sure that all electrical connections are tight.

10. Remove the centre bleed point on the pump and remove any air.

11. Reinstate the fuse in the consumer unit. Switch on and test for correct operation.

12. Check the F&E cistern in the roof space to ensure that the pump is not pumping water over the cistern through the vent pipe.

12 RADIATOR REPLACEMENT

If the new radiator is the same size as the one being replaced, the pipework should fit without too many problems. If the new radiator is either larger or smaller, the pipework will either have to be altered or a radiator valve extension will need to be fitted. It is desirable, when replacing a radiator, to replace the valves as well – they will probably be as old as the radiator you are replacing. If this is the case, then all or part of the system will need to be drained. The following description assumes that the radiator is downstairs and will require complete system drain-down.

Radiator valve extension

Before attempting to remove the radiator, the electricity should be isolated at the switched fuse spur to the system and the fuse retained to prevent accidental switching on of the circuit. The system should also be cold. It may be a good idea to ask the customer to turn the central heating off before you get to the job. Before you begin, make sure that you have protected carpets and furnishings with lots of dust sheets.

1. Isolate the F&E cistern at the service valve.

2. Locate a suitable drain valve, attach a hose and drain the system. Take care that the system contents are disposed safely to a drain as it will probably be very dirty, especially if the system is an old one. The black water will stain all it comes into contact with.

3. As the system drains, open the air-release valves on all radiators, starting upstairs and working to the downstairs.

4. When the system is drained, carefully loosen the two radiator valve compression nuts and remove the radiator. It is a good idea to leave the valves on the radiator and to turn them off before removal. This will help in preventing any residual dirty water leaking from the radiator. If possible, turn the radiator upside down (turn the air-release valve off first!) as this will further prevent accidental spillage.

5. The new radiator should be dressed as described on page 432 and hung as described on pages 432–33.

6. Reconnect the pipework, ensuring that the old compression nuts and olives are removed first. If the old olives have crushed the pipe too much, it may have to be replaced.

7. Ensure that all radiator unions and compression nuts are fully tight.

8. Turn off the drain valve. It may be a good idea to replace the drain-off valve washer at this stage. Drain valve washers quite often go stiff and brittle with the heat from the water.

9. Turn off all air-release valves.

10. Turn on the service valve to the F&E cistern or (if applicable) reconnect the filling loop and refill the system.

11. Bleed the air from all the radiators, starting downstairs and working to the upstairs. Leave the new radiator isolated at this stage. This will be the last radiator to be filled.

12. Open the valves to the new radiator and bleed the air from it. Check for leaks.

13. Replace the fuse in the fuse spur and run the system to full temperature to ensure that the new radiator is working perfectly.

14. If corrosion inhibitor had been added to the system in the past, this will need to be replaced. It must be replaced like-for-like. If this is not possible, the system should be flushed several times to ensure removal of all previous inhibitors.

13 TASKS THAT MAY REQUIRE SYSTEM DRAIN-DOWN

There are many situations where draining of the system is required, such as:

- replacing the hot water storage cylinder
- boiler replacement
- decommissioning of components such as radiators (the radiator, brackets and pipework should be removed and the pipes capped off at the branch to the flow and return pipes)
- replacement of motorised valves
- cutting into an existing system to alter or extend it – drain-down should be conducted when all other installation work has been carried out
- power flushing.

14 POWER FLUSHING A SYSTEM

It may become apparent that the system contains a lot of black water and even sludge. If this is the case, the system may be in need of a power flush. When replacing boilers, a power flush is required to remove any sludge within the system as part of the warranty. Manufacturers' warranties are void if this is not carried out.

Power flushing involves using a special high-powered pump to circulate cleaning chemicals and de-sludging agents through the system. These powerful chemicals strip the old corrosion residue from the system, ensuring that the system does not contain sediment, which may be harmful to new boilers, controls and valves. Power flushing is a lengthy process, often taking all day, which involves opening and closing individual radiators and circuits to ensure a thorough cleaning.

After power flushing is complete, the system may have an inhibitor added to the water to keep it free from corrosion.

Power flushing kit

15 ROUTINE MAINTENANCE TASKS

Routine maintenance should be conducted on a twelve-monthly basis. Routine maintenance includes:

- checking the pressure charge in expansion vessels on sealed systems, system boilers and combination boilers
- checking the operation of pressure relief valves on sealed systems, system boilers and combination boilers
- checking and topping up (if required) the pressure on system boilers and combination boilers
- visually checking for any signs of leakage on pipework, controls and appliances
- boiler servicing

- checking the correct operation of thermostats, motorised valves and thermostatic radiator valves
- checking the water level in F&E cisterns and adjusting as necessary
- ensuring that the system is reaching full temperature.

16 DEALING WITH SIMPLE SYSTEM FAULTS

It is impossible to cover all scenarios when dealing with system faults. Often, the reason for a fault developing is clear, and stems from poor design when the system was installed; other faults require rather more investigative work. Sometimes the system itself will lead you to the problem by the way it behaves or the noises it makes and so diagnosis becomes an easy task.

The following table looks only at some of the more common, simple system faults and the signs to watch out for – not appliance faults, which are covered at Level 3.

Symptom	Fault	Rectification
Discoloured water appearing at hot water taps. System has a double-feed indirect cylinder fitted.	The cylinder heat exchanger coil has pinhole corrosion, allowing primary water to mix with secondary water.	Drain down both hot water and heating systems and replace the hot water storage cylinder.
The overflow of the F&E cistern runs constantly, even when the heating system is off, but the float-operated valve is working correctly.		
A radiator is cold at the top but works once the air has been bled. It then works for about four weeks before filling with air again.	It is not air that is filling the radiator, but hydrogen; this is a clear sign of galvanic corrosion in the radiator.	A very common occurrence with systems that contain single-feed (primatic) cylinders. Because inhibitor cannot be used here, the only action is to replace the cylinder with a double-feed type, power flush the system and add corrosion inhibitor.
The hot water via the primary circulation pipes on a semi-gravity system is working correctly. However, the radiators on the system are lukewarm upstairs and cold.	Pump failure.	Replace the pump.

The radiators on a semi-gravity system work correctly, but there is no hot water. The gravity primary circulation pipes are cold.	This is unlikely to be an airlock. The biggest cause of this problem is evaporation of water in the F&E cistern linked to the float-operated valve, with the cistern sticking in the up position.	Re-washer or replace the float-operated valve and refill the F&E cistern.
A radiator is cold in the middle.	Black oxide sludge is blocking some of the radiator's water sections.	A temporary solution would be to take the radiator off and flush it out with cold water but unless the problem is identified, it will reoccur. The system requires a power flush and corrosion inhibitor adding to the system water.
A number of radiators on a downstairs heating circuit only reach lukewarm temperature. All other radiators are working correctly.	Black oxide sludge is blocking the circuit pipework leading to poor water circulation.	See above.
A boiler is noisy when the water begins to reach temperature.	This is known as kettling because the noise resembles a kettle just before boiling. Its correct name is localised boiling and occurs because the waterways of the boiler are partially blocked with either black oxide sludge or calcium deposits (limescale). As the water heats up, it momentarily boils before being moved away by the pump.	The system requires a power flush with sludge remover and descaler and corrosion inhibitor adding to the system water. It is also a good idea to do a litmus paper test to see if the water is acid or alkali. Alkali water tells us that the likely cause is calcium deposits and a scale preventer can then be added to the system to stop the problem recurring.

There are other possible faults, such as motorised valve and thermostat failure but these require a knowledge of electrical systems and testing with an electrical multimeter.

CONCLUSION

This has perhaps been the most challenging chapter in the book so far. The myriad of systems, layouts, appliances, components and fuels can be confusing but each one has telltale signs that make it unique. The art to good system recognition is looking, just as the key to good system fault diagnosis is listening. This chapter gives you the foundation for doing both.

A good central heating system is one that is efficient in use, warms the home to the right temperature, is quiet in operation and is installed to the highest possible standards. This can only be achieved with the knowledge that allows us to recognise the possibilities of efficiency, to design with the customer in mind and to ensure that installation meets best practice standards whenever practicable.

008 TEST YOUR KNOWLEDGE

SmartScreen Unit 008 revision sample questions

1. What is meant by the term **whole house central heating**?

2. Which system is being described here?
A simple ring circuit where the last radiator on the system will always be cooler that the first.

3. What are the main differences between the two-pipe semi-gravity system and the C-Plan semi-gravity system?

4. What are the names of the two fully pumped vented systems that can be installed?

5. With reference to question 4, which of these systems is shown in the diagram below?

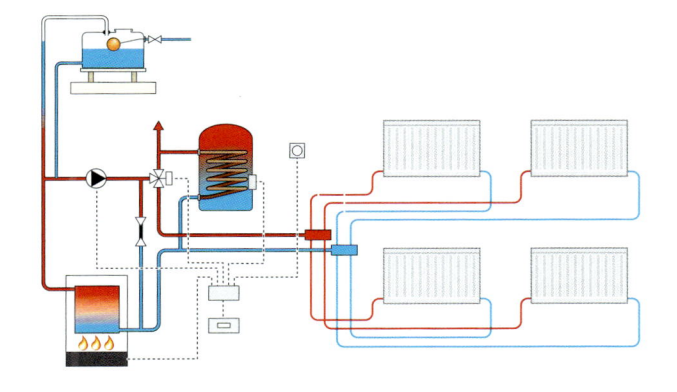

6. Which component takes the place of the feed and expansion cistern on a sealed heating system?

7. How is a sealed heating system filled with water?

8. Describe what a manifold is and where it is fitted.

9. What is the main purpose of the feed and expansion cistern?

10. What is the purpose of an air separator?

11. Name four types of heat emitter.

12. Identify the following radiator types:

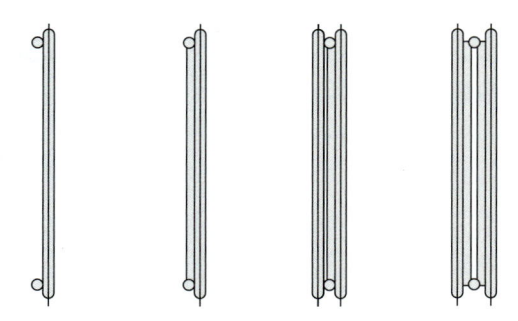

13. Briefly describe the use of an automatic bypass valve.

14. On the Y-Plan system, which port on the three-port motorised valve is the flow from the boiler connected to?

15. Where on a system would an automatic air valve be fitted?

16. Where should a frost thermostat be fitted and which other thermostat should it be used in conjunction with?

17. What is a **boiler interlock**?

18. Before filling a central heating system, why is it a good idea to remove the pump and replace it with a piece of pipe?

19. At what pressure should a central heating system be pressure tested?

20. What is the cause of the black sludge that can appear in a central heating system, and what are the main problems it can cause?

009
UNDERSTAND AND APPLY DOMESTIC RAINWATER SYSTEM INSTALLATION AND MAINTENANCE TECHNIQUES

IN THIS CHAPTER, YOU WILL COVER:

A The purpose of guttering and rainwater systems
A1 The working principles of gravity rainwater systems used on dwellings
A2 The sources of information required when carrying out work on gravity rainwater systems
A3 Factors determining the type of gutter system

B Gutter profiles and materials
B1 PVCu guttering systems
B2 Cast iron guttering systems
B3 Extruded aluminium guttering systems
B4 Jointing guttering of different materials and profiles

C Installation of PVCu gutters and rainwater pipes

D Safe working practices
D1 Working at height
D2 Protecting the customer's property

E Handling and storage of materials

F Maintenance of gutter systems
F1 Visual inspections and fault finding
F2 Leakage repairs
F3 Replacement of defective gutters and fittings
F4 Cleaning and clearing blockages

Test your knowledge

The United Kingdom has more than its fair share of rain. Rainfall varies greatly in the different regions. On average, the south-east of the UK has around 0.5m of rainfall a year compared to around 1.8m for the north-west. Rain penetrating a building can do a vast amount of damage. Without guttering systems the rainfall will run off a roof and erode the ground around a dwelling; it will penetrate the structure and may even affect the building foundations.

This chapter investigates the function and design of guttering systems, the materials they are made from and their methods of jointing and installation. As installing guttering invariably involves working at high level, we will also review previous learning on working safely at height.

A THE PURPOSE OF GUTTERING AND RAINWATER SYSTEMS

All dwellings have some form of collection system to carry rainfall that falls onto the building structure away from the building. This is achieved by the use of an eaves-level, usually fascia-board-mounted guttering system, which collects the water that runs off the roof and discharges it harmlessly away. The main purposes of a guttering system are:

- to protect the building's foundations
- to reduce ground erosion
- to prevent water penetration and damp in the building structure
- to provide a means for collecting rainwater for later use, ie rainwater harvesting.

A1 THE WORKING PRINCIPLES OF GRAVITY RAINWATER SYSTEMS USED ON DWELLINGS

The principle of any guttering and rainwater system is to remove the rainfall in such a way that it does not:

- constitute a nuisance for the occupiers of the dwelling
- damage the building structure or the building foundations or those of any adjacent building.

Domestic gutter and rainwater systems work by removing the rainwater that runs off roofs in channels known as gutters and discharging the water, via rainwater pipework, safely away from the building structure by gravity. The water may be discharged into:

- **A surface water drain** – Used where the dwelling has a separate system of drainage for foul water and for surface (rain) water.
- **A combined sewer** – A combined system of drainage where both foul and surface water discharge into a common drainage system.
- **A water course (stream, river etc)** – Where the water discharges directly into a flowing, nearby water source.
- **A soakaway drain** – A specially designed and located pit, sited away from the dwelling, which allows the water to soak away naturally to the water table.
- **A rainwater harvesting system for further use within the dwelling** – These were discussed in Chapter 003 (pages 148–49). They are specifically designed to serve WCs.

A2 THE SOURCES OF INFORMATION REQUIRED WHEN CARRYING OUT WORK ON GRAVITY RAINWATER SYSTEMS

There are a number of documents that we must consult when designing and installing rainwater systems. Like all other aspects of the building process, gutters and rainwater systems are subject to various legislative restrictions to ensure that the systems we design and install collect the rainwater from the roof structure and dispose of it safely. To ensure the correct design and installation of rainwater systems, we must, therefore, refer to:

- **Building Regulations Document H3:2002 Rainwater drainage** – This document states that adequate provision shall be made for rainwater to be carried from the roof of a building. It contains important information regarding design and installation of rainwater systems. It makes reference to BS EN 12056-3:2000.

- **BS EN 12056-3:2000: Gravity drainage systems inside buildings** – Roof drainage, layout and calculation. Like all British Standards, the document takes the form of recommendations. It relays the more technical aspects of rainwater system design, such as rainfall intensity calculations and outlet provision. It should be used in conjunction with the Building Regulations.

- **Manufacturers' instructions** – The manufacturers of gutters and rainwater pipework will have designed their systems to accommodate both the Building Regulations and British Standards. Wherever possible, the manufacturers' recommendations must be followed.

A3 FACTORS DETERMINING THE TYPE OF GUTTER SYSTEM

A guttering system should have sufficient capacity to carry the expected flow of water at any point on the system. The actual flow in the system depends upon the area of the roof to be drained, the rainfall intensity and the position of the rainwater outlets. The system's efficiency will also be affected by the fall of the gutter and any changes in direction in the gutter run.

Rainfall intensity

In England, the county of Cumbria (in the north-west) has the greatest total rainfall, at around 1.8m per year, with Essex and Kent (in the south-east) having considerably less, at around 500mm.

Average rainfall, however, is only half of the story. While it may rain much more in Cumbria than in Essex over a 12-month period, the number of litres discharged in a single 2-minute rainstorm is greater in Essex, at 0.022 l/s/m² (litres per second per square metre), compared with Cumbria, at 0.014 l/s/m². This is called rainfall intensity and must be factored into any guttering system design – it must be able to cope with a sudden, intense downpour.

BS EN 12056-3:2000 gives rainfall intensity in l/s/m² for a 2-minute storm event. The British Standard contains various maps that show the intensity for periods from one year to 500 years. It divides the rainfall intensity into four categories, which are used depending on the type of building. Domestic dwellings are category 1 in the following table:

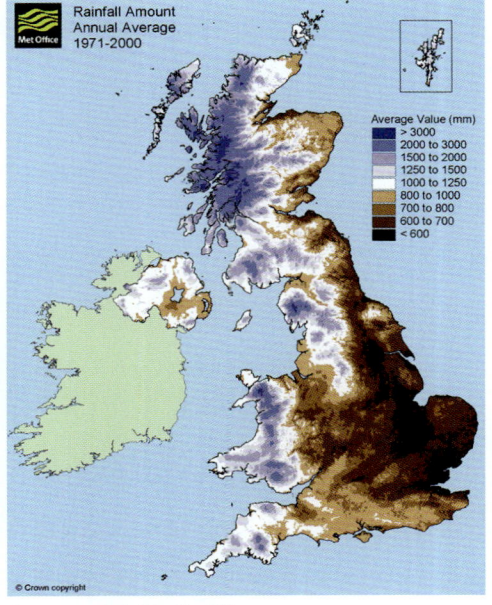

Average rainfall in the UK

Category 1	Eaves gutters and flat roofs
Category 2	Valley and parapet gutters for normal buildings
Category 3	Valley and parapet gutters for higher risk buildings
Category 4	Highest risk buildings

Roof area

The angle and area of the roof is a key consideration in any guttering system design. Take a look at the diagram below:

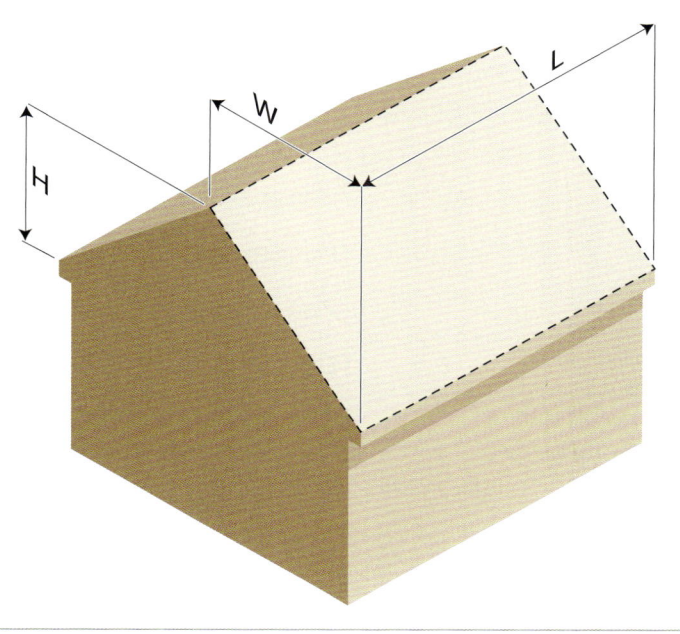

SUGGESTED ACTIVITY...
Using the formula given in the text, calculate the following effective roof areas:

a A roof has a length of 12m, a width of 7m and a height of 3m.

b A roof has a length of 8m, a width of 8m and a height of 4m.

c A roof has a length of 10m, a width of 8m and a height of 4m.

Answers: a. 102m^2 b. 80m^2 c. 100m^2

Roof angle and area

The drawing shows the roof of a dwelling. If the area of the roof increases, the amount of water collected and discharged from it also increases. Similarly, if the angle of the roof increases the area will increase, the amount of water will increase and the velocity at which the water enters the gutter will also increase.

The area of a roof can be calculated by using the following formula in accordance with BS EN 12056-3:2000:

Effective max roof area (allowance for wind)

$(W + \frac{H}{2}) \times L$ (length of roof) = area in m^2

Where:

W = Horizontal span of slope

H = Height of roof pitch

L = Length of roof

> **SUGGESTED ACTIVITY...**
> Using the pitch factors given in the text, calculate the following effective roof areas:
> a A roof has a length of 12m, a width of 7m and pitch of 45°.
> b A roof has a length of 8m, a width of 8m and pitch of 60°.
> c A roof has a length of 10m, a width of 8m and pitch of 30°.
>
> Answers: a. 126m², b. 119.68m², c. 103.2m²

Example 1

A roof has a length of 10m, a width of 6m and a height of 3m. Calculate the effective area of the roof:

$$\frac{(6 + 3) \times 10}{2} = 75m^2$$

The area of a flat roof should be regarded as the total plan area. If the roof has a complex layout with different spans and pitches, each area should be calculated separately.

Building Regulations Document H3 gives an acceptable alternative for the calculation of roof area where the area of the roof is multiplied by a pitch factor. These calculations are detailed in the table below. For this calculation, only the length of the roof and the span are required.

Type of surface	Design (m²)
Flat roof	Plan area of relevant portion
Pitched roof at 30°	Plan area of portion × 1.29
Pitched roof at 45°	Plan area of portion × 1.50
Pitched roof at 60°	Plan area of portion × 1.87
Pitched roof over 70° or any wall	Elevational area* × 0.5
*The elevational area of a roof is the width of the roof x the height of the roof. See drawing below:	If the width of the roof is 10m and the height of the roof is 3m, then: 10 × 3 = 30m Now multiply by 0.5. 30 × 0.5 = 15m

In this instance, if the angle of the pitch of the roof is known, the calculation is simplified. For example, if we use the data from the previous example, we arrive at the following:

Example 2

A roof has a length of 10m and a width of 6m. Calculate the effective area of the roof if the pitch of the roof is 30°:

Length of roof	=	10m
Width of roof	=	6m
The pitch factor from the table	=	1.29
10 × 6 × 1.29	=	77.4m²

We can now calculate the amount of rainwater to be expected on any given roof area in a sudden storm deluge. This is generally accepted to be 75mm rainfall per hour but will vary from area to area. To convert the area to litres per second (l/s), multiply the roof area (m²) by 0.0208.

Example 3
The area of the roof in Example 1 is 75m²; what is the expected rainfall in l/s?

75 × 0.0208 = 1.56 l/s

Running outlet position

The image on the right shows a running (rainwater) outlet. It is the connection between the guttering and the rainwater pipe, and is designed to cope with rainwater entering from two directions.

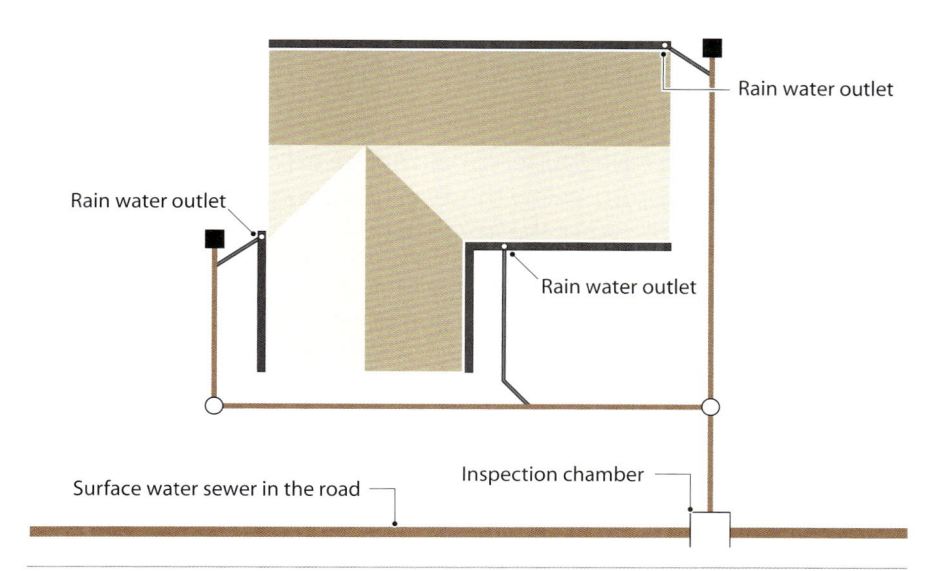

> **SUGGESTED ACTIVITY...**
> Using the effective areas calculated in the activity on page 464, calculate the rainfall rate in litres per second.
>
> Answers: a. 2.62 l/s b. 2.48 l/s c. 2.14 l/s

Building layout drawing

Running outlet

The position of the running outlets is usually based on the position of the gullies for the surface water sewer/drain to the property. These can be found on the building layout drawing (above).

The more outlets there are on a gutter system, the shorter the distance the water has to travel within the gutter and the more effective the system is at discharging the rainwater. Consider the drawing, right:

The outlet has to be able to cope with the total rainwater run-off from the whole roof area. The outlet in this situation could be positioned at either end of the roof but the total flow rate would be the same, as the outlet is receiving water from only one direction. This means that only half the capacity of the outlet can effectively be used. Placing the outlet centrally would increase the total area of roof that the gutter can serve.

The outlet position in drawing 2 is more effective than in drawing 1 simply because there are now two outlets and each outlet is coping with half the expected rainwater run-off. Again, an alternative but equally effective layout would be one outlet placed in the centre of the gutter run (position 3).

With outlets placed as in drawing 3, each half of the outlet has only a quarter of the flow rate to cope with; this layout is much more effective

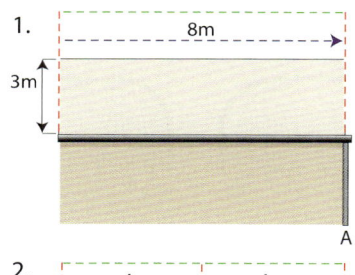

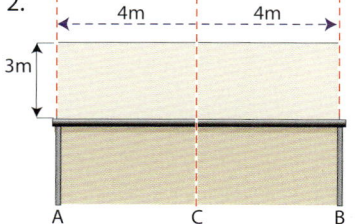

Alternative position C: Here the single outlet is equal to two outlets either end because of the outlet design

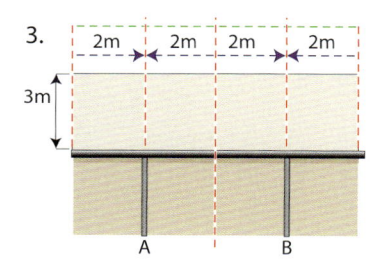

Outlet positions

009 RAINWATER SYSTEMS

465

KEY POINT

This is the least effective outlet design. The corners are sharp-edged, which restricts the flow of water down the outlet by causing a clash of water streams at the shaded area. This creates turbulent water flow. Some water will travel across the outlet and against the flow on the opposite side of the outlet.

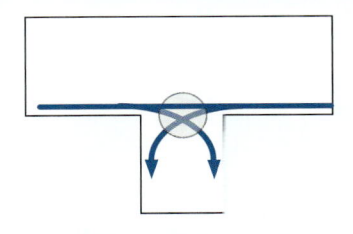

Here the corners are slightly rounded, which assists the flow of water down the outlet. However, the two water streams are likely to clash, creating some turbulence.

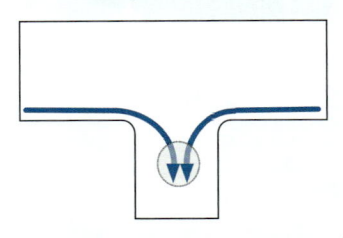

Fully rounded corners give a much better flow of water down the outlet. The two streams are kept more or less separate, which assists gravity flow down the rainwater pipe. This is known as hydraulic efficiency.

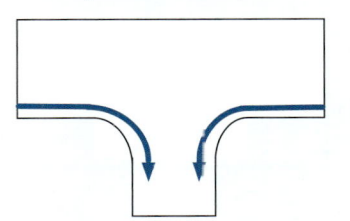

---

at discharging the rainfall without the risk of flooding because both outlets are being used to their full flow rate capacity.

Each manufacturer will have different rainwater flow rates for their running outlet designs. It should not be assumed that all manufacturers' flow rates will be equal. Therefore, the particular manufacturer's data should be considered before the installation begins.

To find out how many outlets are required on a rainwater system design, simply divide the expected flow rate of the roof area by the flow rate for the outlet, given in the manufacturer's technical literature.

The fall of the gutter

BS EN 12056-3:2000, section 7.2.1 and NE.2.1 states that:

- Gutters should be laid to a fall of between 1mm/m and 3mm/m wherever possible.
- The gradient of an eaves gutter should not be excessive to the point where the gutter drops below the level of the roof and the water running off the roof misses the front edge of the gutter.

In most cases, manufacturers interpret these two points as a slight fall of 1:600 (25mm in 15m). Laying a gutter with a fall greatly increases the flow capacity and, therefore, the area of roof that can be drained. It also ensures that silting of the gutter is reduced. However, manufacturers design guttering systems in such a way that gutters will still remove the roof water if it is laid level with little or no fall. A fall of 1:600 ensures that the gutter will not fall so low as to be below the discharge point of the roof.

Changes of direction in the gutter run

In most domestic gutter systems, changes of direction cannot be avoided. Where changes in direction greater than 10° occur within a guttering system, they restrict the flow of water through the system. A 90° gutter angle reduces the effectiveness of the run of gutter where the angle is situated by 15 per cent, effectively reducing the roof area that the gutter can usefully serve. Each subsequent change of direction reduces the gutter's effectiveness still further. A gutter angle that is placed near an outlet will also reduce the effectiveness of the outlet.

General design requirements are as follows:

- Straight gutter runs give maximum flow rate. Gutters should be installed so as to avoid unnecessary changes of direction.
- Where changes of direction cannot be avoided, outlets should not be installed close to them.
- In systems where there are considerable changes of direction, the maximum fall ratio should be used to avoid the gutter collecting water.

- A larger-capacity gutter should be considered where many changes in direction occur.

B GUTTER PROFILES AND MATERIALS

Over the years, gutters have been manufactured from many different materials and in many different profiles (shapes). In the past, the gutter profile was designed in line with the house style at the time. For example, the ornamental gutter profile (OG, or ogee) was designed during the Victorian era in the mid to late 1800s. As we shall see, a modern ogee profile is still available today to give a 'period' feel to a dwelling's exterior.

In this part of the chapter, we will take a look at modern materials and profiles. We will look at the different types of fittings for guttering and rainwater pipework and the typical methods of jointing.

B1 PVCU GUTTERING SYSTEMS

PVCu guttering systems are manufactured to the following British Standards:

- **BS EN 607:2004** – Eaves gutters and fittings made of PVCu.
- **BS EN 122001:2000** – Plastic rainwater piping systems for above-ground external use.

Most of the guttering systems used on domestic dwellings today are made from unplasticised polyvinyl chloride (PVCu), the characteristics of which were studied in Chapter 005. The advantages and disadvantages of PVCu as material for guttering systems are as follows:

ADVANTAGES	DISADVANTAGES
- It is easy to install. - It is lightweight and easy to handle. - Minimal maintenance is required. - It requires no painting. - It is economical. - It is corrosion free. - It has a smooth internal bore. - It has a life expectancy of 50 years.	- It is adversely affected by wood preservatives. - It has a greater coefficient of thermal expansion (0.06mm/m°c) compared to other materials; this can lead to the gutter pushing itself out of the joint if the expansion gap in the gutter fittings are not observed. - It goes brittle in cold temperatures and softens at a relatively low temperature.

PVCu gutter profiles

There are four main gutter profiles manufactured from PVCu. These are:

- **Half round** – The standard gutter profile, used on many domestic properties throughout the UK.

- **High capacity (often called deep half round or storm flow)** – A deeper version of the half-round profile. It is slightly elliptical in shape and is generally used on larger or steeper angled roofs where the velocity and volume of the water entering the gutter is high.

- **Square section** – Very popular in the 1980s and 1990s. Used with square-section rainwater pipes. Square section has a very good rainwater capacity.

- **Ogee (or OG, for 'ornamental gutter')** – A modern redesign of a Victorian gutter profile. It is used where a period look is important on new-builds and on many Victorian refurbishments.

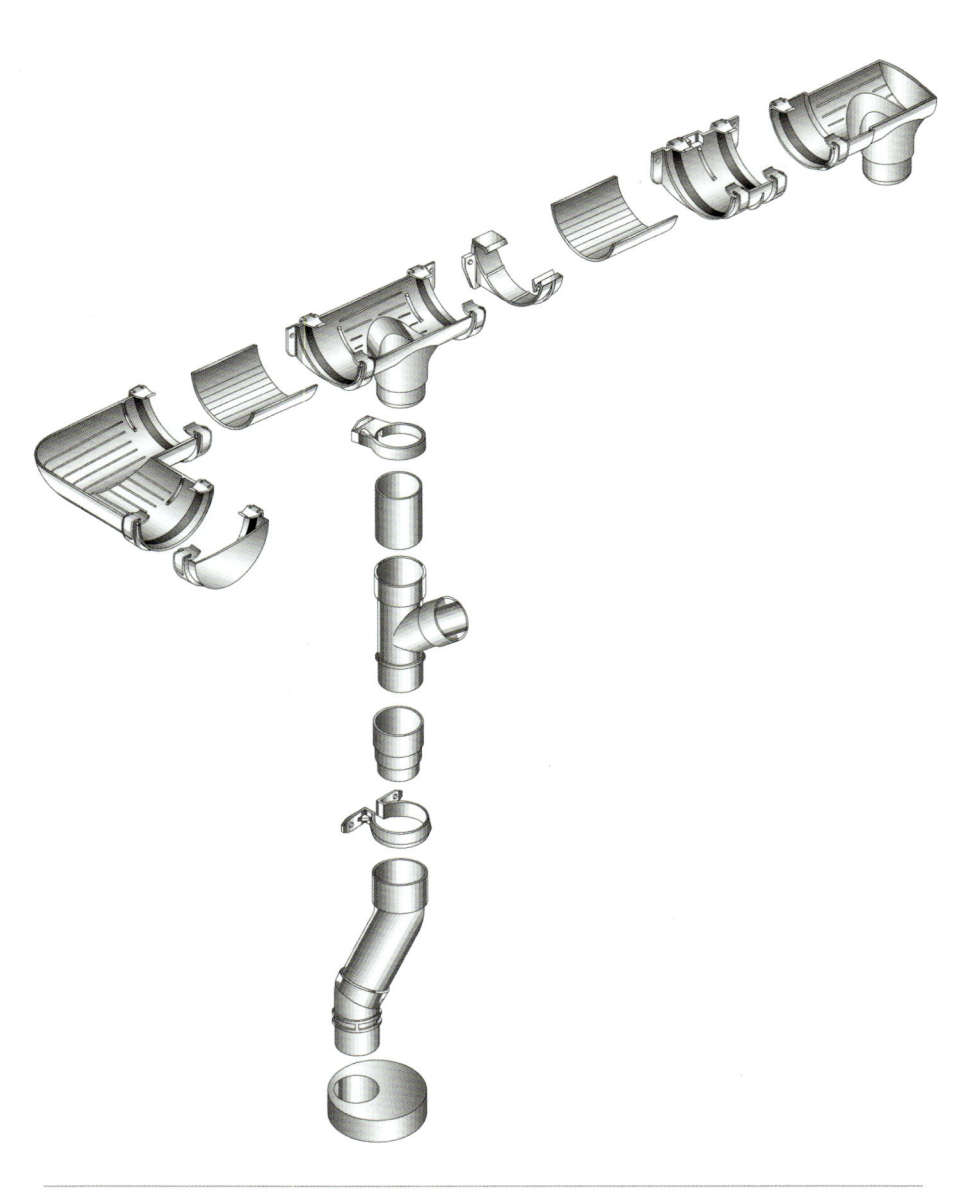

A typical PVCu gutter system

PVCu gutter fittings and jointing method

The image opposite shows the fittings in a typical 112mm half-round guttering system. The common fittings are shown below in all profile styles:

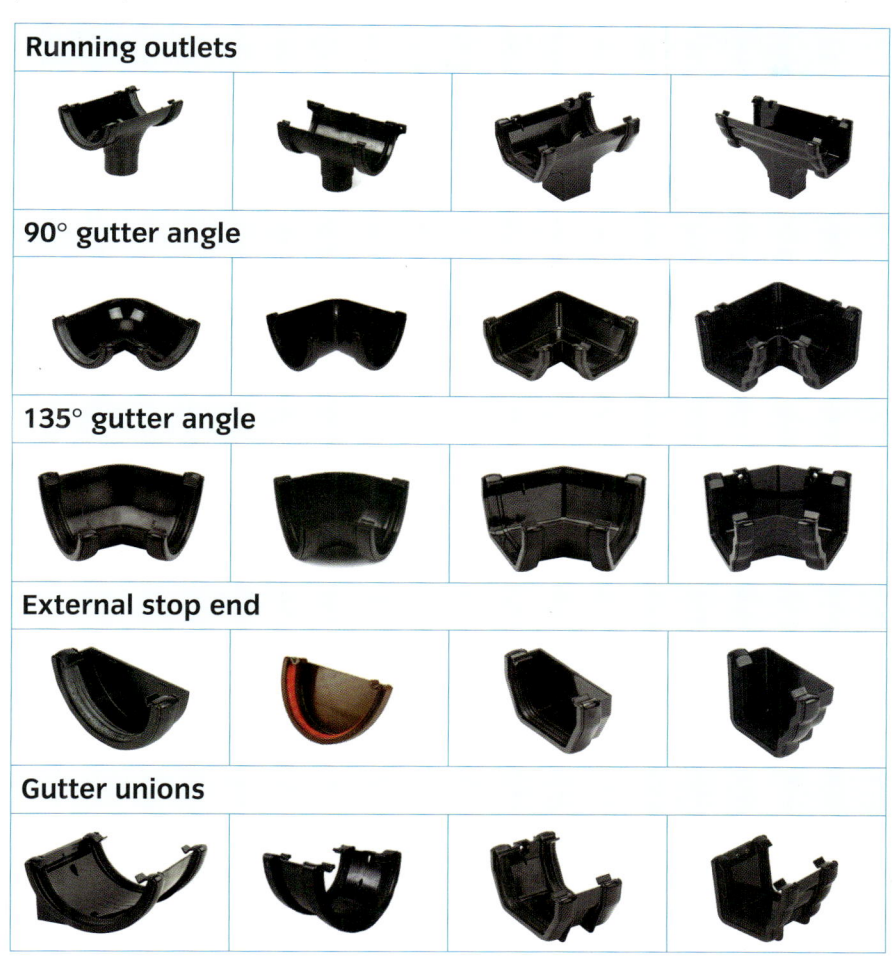

Running outlets

90° gutter angle

135° gutter angle

External stop end

Gutter unions

Rainwater pipes are available as either 68mm round or 62mm square section. Their common fittings are as follows:

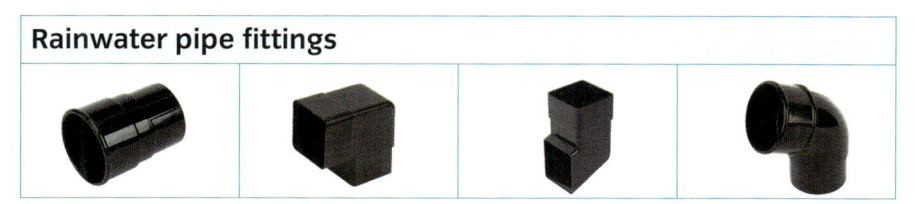

Rainwater pipe fittings

Thermal expansion of PVCu gutters and fittings

One of the problems with PVCu gutters is the large expansion rate. This can cause the gutters to creak as they get warmed by the sun and, in extreme cases, it can cause joint failure. PVCu has a coefficient of linear expansion of 0.06mm/m/°C. This means that for every metre of gutter, PVCu expands by 0.06mm for every degree rise in temperature. For example:

If a 1m length of gutter is subjected to a rise in temperature of 10°C, it will expand by the following amount:

$1 \times 0.06 \times 10 = 0.6$mm

This might not seem a lot but let us look at this in more detail.

> **SUGGESTED ACTIVITY...**
> Using the method shown in the worked example, calculate the following:
>
> a A south-facing gutter 10m long is subjected to a 15°C temperature rise. What is the expansion of the gutter when the coefficient of linear expansion of the gutter is 0.06mm/m/°C?
>
> b A south-facing gutter 20m long is subjected to a 30°C temperature rise. What is the expansion of the gutter when the coefficient of linear expansion of the gutter is 0.06mm/m/°C?
>
> c A south-facing gutter 5m long is subjected to a 20°C temperature rise. What is the expansion of the gutter when the coefficient of linear expansion of the gutter is 0.06mm/m/°C?
>
> Answers: a. 9mm b. 36mm c. 6mm

Example 4

A south-facing gutter 15 m long is subjected to a 25°C temperature rise. What is the expansion of the gutter when the coefficient of linear expansion of the gutter is 0.06mm/m/°C?

All the information we need to be able to calculate this is in the question:

Length of gutter	=	15m
Temperature diff. (Δt)	=	25°C
Coefficient of linear expansion	=	0.06mm/m/°C
15 × 25 × 0.06	=	22.5mm

To counteract the expansion, every manufacturer builds into their fittings a 10mm expansion gap. This must be observed when installing PVCu gutters if problems with thermal expansion are to be avoided.

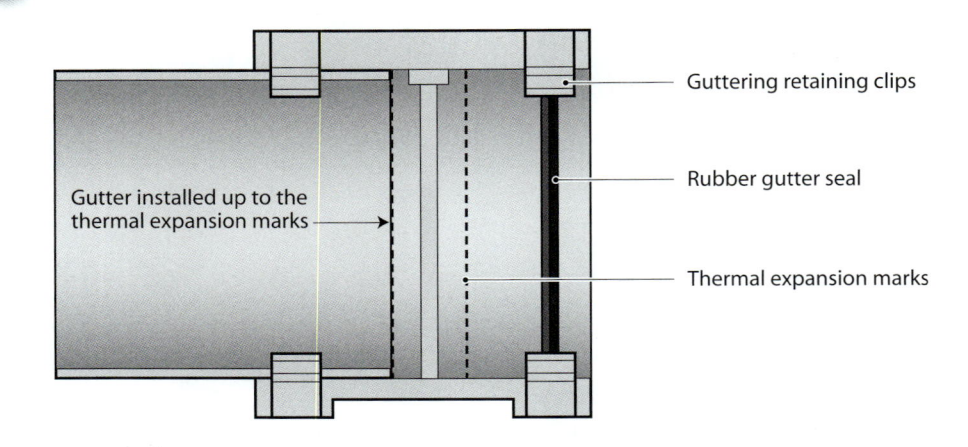

Expansion gap on PVCu gutter fittings

B2 CAST IRON GUTTERING SYSTEMS

Until the advent of PVCu guttering, cast iron was probably the most common material for gutters and rainwater pipework. It can still be seen on many older houses. It is strong and durable but can be difficult to maintain, as it requires regular painting to stop corrosion.

Cast iron may still be specified by the local authority, English Heritage or the National Trust if the building is listed or in a conservation area. The most common profiles for cast iron are:

- **Half-round section** – Visually very similar in shape to PVCu half-round profile.

- **Ogee section** – There are several variations of the ogee profile manufactured in cast iron; some are specific to a particular area, such as Notts ogee which can only be found in the Nottinghamshire area.

- **Deep half round** – Found on larger buildings.

ADVANTAGES OF CAST IRON	DISADVANTAGES OF CAST IRON
▪ It is strong and durable.	▪ Installation is expensive and time consuming. ▪ It is expensive. ▪ It requires regular painting and maintenance to prevent corrosion. ▪ It is heavy. ▪ Jointing is time consuming and messy.

Cast iron-type fittings and guttering are also available in cast aluminium, which offers certain advantages over its cast iron equivalent:

ADVANTAGES OF CAST ALUMINIUM	DISADVANTAGES OF CAST ALUMINIUM
▪ It is strong and durable. ▪ It does not rust. ▪ It is much lighter in weight than iron.	▪ Installation is expensive and time consuming. ▪ It is expensive.

Jointing cast iron (and cast aluminium) guttering systems

As you can see from the photograph, cast iron guttering has a socket on one end. The other end is a plain gutter. A successful joint involves fitting the end of one length of gutter into the socket of another with a jointing material in between. The two lengths of gutter are then bolted together using special zinc-plated gutter bolts. The jointing material can be one of the following:

Cast iron gutter

- **Paint and putty joint** – The traditional method of jointing cast iron guttering systems. The method of jointing is as follows:

1. The inside of the socket and outside of the spigot are first painted with black bitumen paint.
2. Linseed oil putty is then placed into the socket before mating the socket and spigot together.
3. A zinc gutter bolt is inserted through the holes on the socket and spigot and the two sections bolted together. Care should be taken not to over-tighten the bolt or the gutter will crack.
4. After the excess putty is cleaned off, the inside and outside of the joint can be painted to finish it.

- **A special silicone sealant** – The silicone is placed inside the joint and the two sections are bolted together (this is normally only used on new cast iron guttering installations).

- **A rubber grommet** – This method is not generic and is usually only available on specific manufacturers' gutters and fittings.

B3 EXTRUDED ALUMINIUM GUTTERING SYSTEMS

This type of guttering system is usually installed by a specialist company. Extruded seamless aluminium guttering systems are a modern innovation – they are light in weight and corrosion-resistant. This type of guttering is manufactured on site from a roll of coloured aluminium sheet by a special machine that is carried in the back of a van. The aluminium sheet is passed through the machine and this presses it into the shape required. As the gutter exits the former, strengtheners are fitted at regular intervals to give the gutter added rigidity.

Lengths of up to 30m can be manufactured in one continuous length without the need for an expansion joint, which reduces the amount of joints and, therefore, potential leaks. The gutter is installed with internal brackets spaced at 400mm and this means it is able to withstand shock-load from equipment such as ladders.

Most companies offer a variety of profiles, including half round and ogee, in a variety of colours.

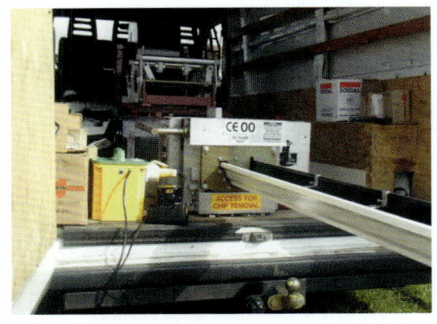

How seamless gutters are made

ADVANTAGES	DISADVANTAGES
- It is strong and durable. - It is lightweight. - Long lengths can be installed. - Fewer leaks than cast iron. - A variety of profiles and colours. - Minimal thermal expansion.	- It is an expensive system. - Does not suit all properties, especially mid-terrace and townhouses where there are gutters either side.

B4 JOINTING GUTTERING OF DIFFERENT MATERIALS AND PROFILES

Occasionally, it may be necessary to make joints between systems of guttering that use different materials or have different profiles. This can be done easily using specific adapter fittings. Below is a selection of gutter-to-gutter adapters:

Half-round PVCu to half-round cast iron	Half-round PVCu to ogee PVCu	Half-round PVCu to ogee cast iron	Half-round PVCu to square-section PVCu

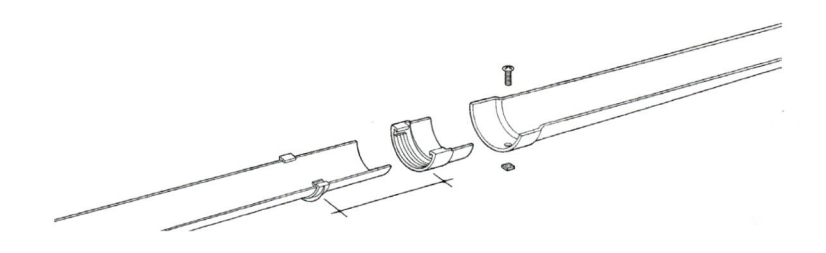

Connections to existing cast iron gutters

The diagram above shows how cast iron guttering can be joined to modern PVCu guttering systems. The PVCu gutter is made using a standard clip-in gutter joint. The cast iron joint should be made in accordance with the jointing technique mentioned on page 471. The adapter is made from PVCu and so care must be taken when bolting the cast iron gutter to the adapter. Too much tightening will result in the adapter splitting.

C INSTALLATION OF PVCu GUTTERS AND RAINWATER PIPES

Before starting the installation, fascia boards must be checked to ensure that they are straight and level and do not need replacing. Fascia boards that are not level or straight can give the gutter a crooked or wavy appearance, and rotted fascias will not hold the gutters properly. Occasionally, it may be necessary for fascia boards to be painted before the gutter is installed. The underfelt drip, which is a strip of felt positioned under the front row of tiles, should also be checked to ensure that it has not ripped or rotted. The felt stops rainwater from leaking behind the gutter.

The following table shows the correct tools to use when installing gutters and rainwater pipes:

Hand tools	Power tools
Pozidriv screwdrivers	110 V SDS power drill
Hacksaw	24 V battery-powered cordless drill
Claw hammer	
String/plumb line	
Bradawl	
File/rasp	

Installing the gutter
1 Establish the position of the outlets.

2 Establish the high point of the gutter and fix a fascia bracket at the high point on to the fascia board using 25mm × 10 zinc-plated roundhead screws.

3 Using a plumb line, centre the outlet over the gully or drain.

4 The distance between the high point and the outlet should be measured and a fall of 1:600 determined. Using this fall, fix the outlet at the low point on the fascia board.

5 A line can now be strung between the high fascia bracket and the outlet. For gutters that are to be fixed level, a spirit level should be used against the string.

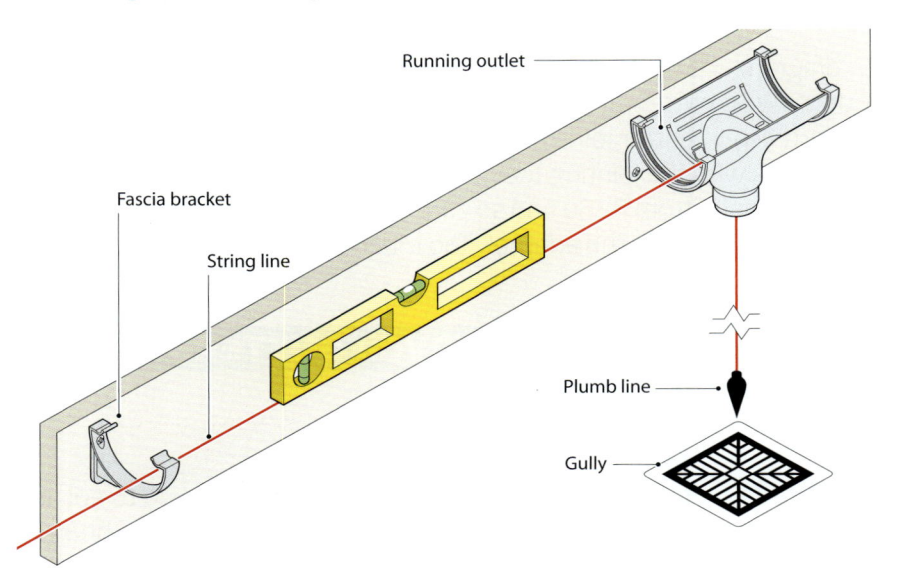

Setting the gutter fall

6 Screw further fascia brackets on to the fascia board, working away from the running outlet. The brackets should just touch the line but not distort it. Most manufacturers recommend a distance between the fascia brackets of 1m (750mm in areas that suffer heavy snowfall) but the manufacturer's instructions should be checked beforehand. There is no need to fix a bracket close to the running outlet as it is secured using screws and therefore acts as a bracket.

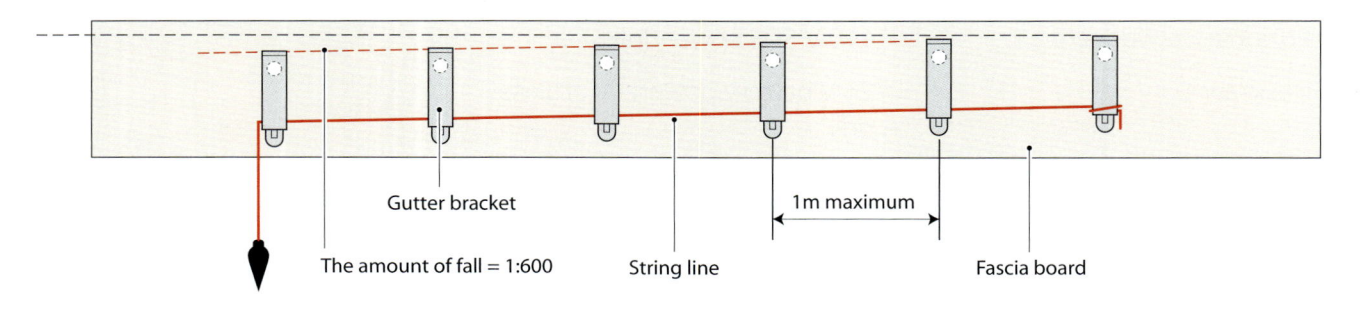

Installing fascia brackets

There are two methods of fixing guttering to dwellings without fascia boards. These are:

- **The use of top- or side-fitted rafter brackets** – These are galvanised steel brackets that are screwed to the top or the side of the roof rafters. The fascia brackets are then bolted to the rafter brackets. It is often necessary to replace sections of rafters that have been exposed to the elements. Check rafters before installation.

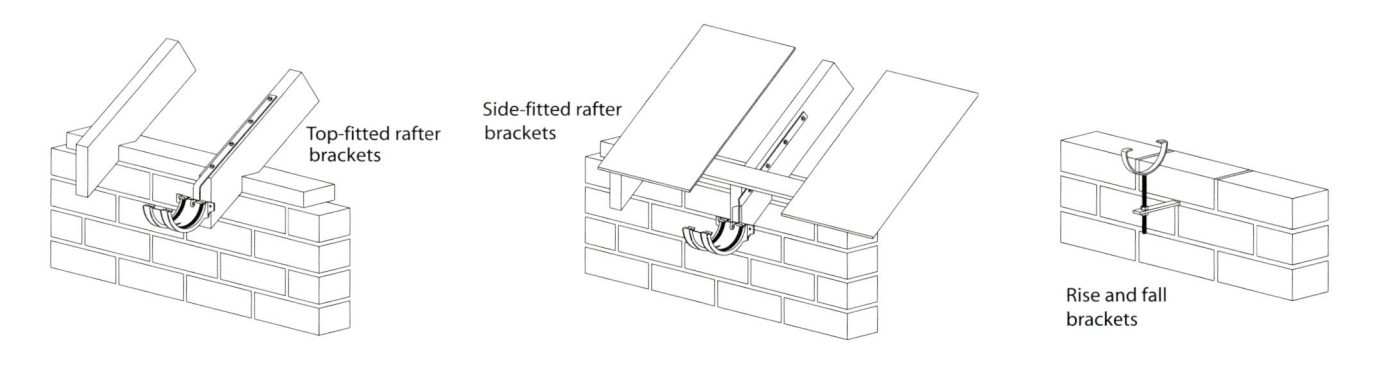

Top- and side-fitted rafter and rise and fall brackets

- **The use of drive-in rise and fall brackets (also known as rise and fall irons)** – These are flat, pointed strips of galvanized steel that are built into the brickwork joints. Threaded rod is then fitted, with a gutter bracket attached, which can be adjusted up or down to give a fall.

When installing gutter angle fittings, stop ends and gutter unions that are unsupported, fascia brackets should be fitted no more than 150mm away from either side of the fitting or end of the gutter.

Cast iron gutter fitted to rafter brackets

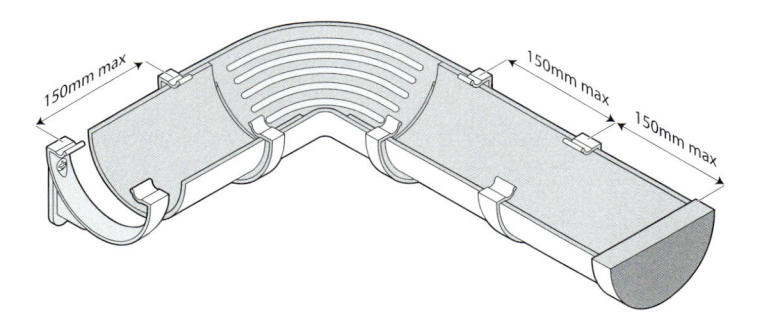

Installing gutter angles

Rise and fall brackets

Once all of the fascia brackets have been fixed, the gutter can be fitted. It is advisable to work away from the outlet towards the high point; this will save time on installation, as fewer cuts will be needed.

Cutting the gutter

Manufacturers recommend that gutter and rainwater pipes are cut using a fine-tooth saw or a hacksaw with a 24 teeth/in blade.

1. Measure and mark the gutter to the required length.

2. Cut the gutter or rainwater pipe carefully using a fine-tooth saw and deburr the cut using a file or rasp.

Making the guttering into the fittings

PVCu guttering systems use a snap-fit jointing system. To make a watertight joint, simply insert the gutter into the fitting, up to the expansion mark. Push the gutter up into the back of the gutter-fitting clip. Pull the front of the gutter down and clip the gutter in with the front gutter-fitting clip using the thumb.

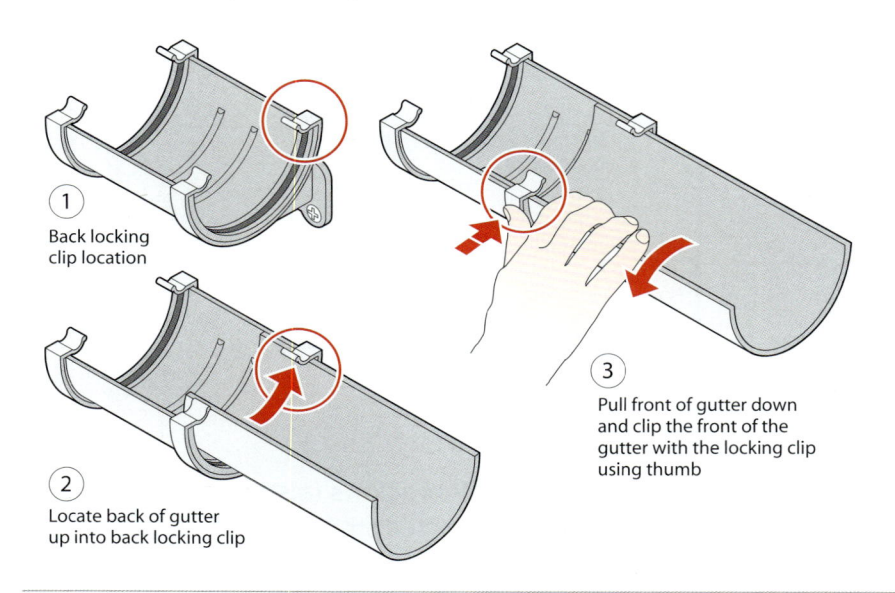

① Back locking clip location

② Locate back of gutter up into back locking clip

③ Pull front of gutter down and clip the front of the gutter with the locking clip using thumb

Installing the gutter

Installing the rainwater pipe

Before installing the rainwater pipe (RWP), it is advisable to fabricate the offset bend at the top of the pipe where it connects to the running outlet:

1 Measure the distance between the two 112.5° bends marked 'L' on the diagram.

2 Cut the length of RWP, deburr the pipe and, using solvent-weld adhesive, glue the offset bend together. Leave it for 5 minutes to set.

3 Install the offset onto the outlet and measure the distance to the shoe at the base of the RWP (if the pipe is to be fitted directly to the drain, measure the distance to the drain connection).

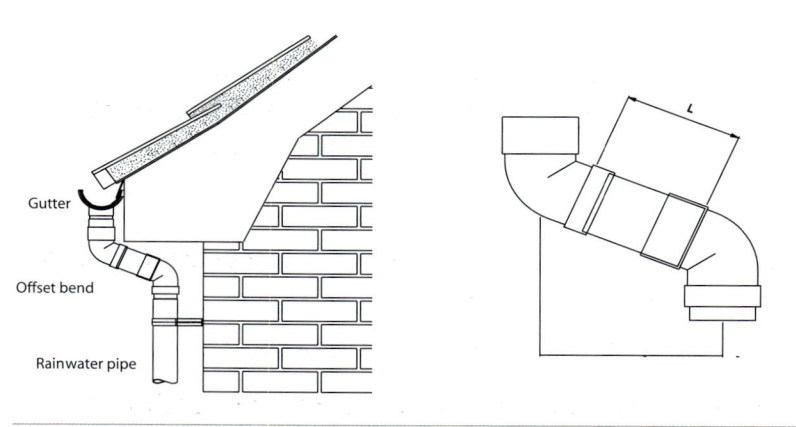

Gutter
Offset bend
Rainwater pipe

Making the offset bend

Rainwater pipe to drain connections

4 Cut the length of pipe required and deburr. Install the pipe onto the bottom of the offset and, using a spirit level, mark and drill the bottom RWP clip and screw the clip and pipe against the wall using wall plugs and 50mm × 10 alloy or stainless steel screws.

There is no restriction to the number of bends that can be installed on rainwater pipes. Where two RWPs converge, it is possible to take both pipes into a hopper head.

Before installing the RWP clips, always check the clip distances in the manufacturer's instructions. The distance between the RWP clips is shown in the table below:

Pipe size	Rainwater clip support centres: vertical (m)	Rainwater clip support centres: horizontal (m)
55	1.2	0.6
62	2.0	1.2
68	2.0	1.2
70	2.0	1.2
82	2.0	1.2
110	2.0	1.2

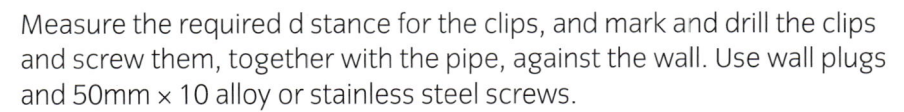

Measure the required distance for the clips, and mark and drill the clips and screw them, together with the pipe, against the wall. Use wall plugs and 50mm × 10 alloy or stainless steel screws.

Rainwater pipe clips

Hopper head

Once the system installation is complete, testing can be carried out by discharging water from a hosepipe at all high points in the system, and checking to make sure that the water discharges down the outlets and through the rainwater pipes without leakage or pooling of water in the gutter.

D SAFE WORKING PRACTICES

In Chapter 001, we looked at the dangers of working at height and, since guttering installation takes place at heights above head level, it is relevant to revisit some of the more important aspects of these procedures.

We will also investigate how we can protect customers' property while working above ground level.

D1 WORKING AT HEIGHT

The safest way to install gutters and rainwater pipes is from a correctly erected and secured scaffold, and on new-build housing, this is usually the case. Unfortunately, erecting a scaffold for the purpose of replacing existing gutters and rainwater pipes is often uneconomical and so most of this type of work is performed using ladders. It should always be remembered that a ladder is not a safe working platform and extreme care should be taken when working from a ladder. Here are some points to remember:

- Always assess the work before any working at height is undertaken. A risk assessment should be performed.

- Never attempt the job alone. PVCu gutter is very light but it can catch the wind.

- There is no height threshold but if you are high enough to become injured from a fall, you must adhere to the Working at Height Regulations 2007.

- Always select the most appropriate equipment for the task, such as mobile scaffold towers or elevated working platforms. If working from a ladder is unavoidable, a ladder stand-off should be used, especially when performing gutter maintenance tasks, as this allows better access to the gutter for cleaning without leaning the ladder against the gutter itself.

- Ensure that you are properly trained in the use of ladders and mobile scaffolds.

- Always check ladders to ensure that they are in good order and free from defects.

- Always use the appropriate fall restraints and harnesses when working at height.

- Always be aware of what or who is below you when working at height. Never drop tools, equipment or materials.

- Always make sure that the ladder is secure before attempting the work. If securing the ladder is not possible, then a second person should foot the ladder.

Using a ladder stand-off

KEY POINT
More information about working at height can be found in Chapter 001, pages 63–75, and at www.hse.gov.uk. Check these sources to make sure that you understand what precautions to take. Be safe when working at height – don't take risks.

D2 PROTECTING THE CUSTOMER'S PROPERTY

In previous chapters, we have seen how we should protect the customer's property when working inside the dwelling. The same care and attention should extend to outside the property.

It is important that the outside of the property is checked for any existing damage before work begins, and any you find should be pointed out to the customer. Precautions that can be taken to protect the customers' property include:

- When using a ladder, a ladder stand-off should be fitted to prevent the scraping damage that can be caused by ladders to the brickwork or masonry.
- If a ladder is to be erected on a lawn, first cover the lawn with a plywood sheet to prevent damage to the grass and flowerbeds.
- Lawns should have walk boards placed on them to prevent damage.
- Take care where vehicles are parked on the customer's drive below where you are working. To prevent possible damage, ask the customer to move them.
- Take care not to erect ladders on soft ground as they could sink, causing slippage of the ladder and possible damage to the building. If this is unavoidable, ensure that the ground is supported beforehand.
- Place barriers around where work is being carried out to prevent people from being injured when walking nearby.

E   HANDLING AND STORAGE OF MATERIALS

Care should be taken when handling PVCu guttering and rainwater pipes. Excessive scratching can ruin the aesthetic appearance of the gutter and affect joint sealing. Cold weather reduces the impact strength of PVCu and extra care is required in wintry conditions.

When pipe is delivered to the site, it is recommended that loading and unloading of pipe and gutter lengths is performed by hand, without the use of mechanical lifting aids. Always store pipes and gutters on flat surfaces, ensuring that the surface is free from sharp protrusions. Bundles of pipes and gutters can be stored up to three high without support. Loose gutter and pipe requires supports every 2m. Fittings should remain in their packaging until needed to reduce damage by scratching.

F   MAINTENANCE OF GUTTERING SYSTEMS

Maintenance of guttering systems is an essential activity to keep them working correctly. This section looks at essential tasks that are carried out during planned preventative maintenance or fault rectification. Different gutter materials require different methods of working and repair and it is important to have knowledge of the basic repair techniques required.

F1   VISUAL INSPECTIONS AND FAULT FINDING

Visual inspections are the first part of maintenance and repair. Visual inspections help in establishing the overall condition of the gutter and rainwater pipe installation, joints and fittings, and in pinpointing specific problems.

Fault	Remedy
Leaking joints	Carry out rectification operations (see below)
Cracked and broken gutters and rainwater pipes	Carry out rectification operations (see below)
Bad falls and bowing gutters	These will require realigning with the correct fall or the installation of extra fascia brackets
Blocked gutters and rainwater pipes causing water to overflow at the outlets	These require cleaning and clearing (see below)
Incorrectly spaced fascia and rainwater pipe brackets	Fascia brackets at 1m distance, vertical pipework brackets at 2m and horizontal brackets at 1.2m
Water overflowing from the gutter during periods of heavy rain after a major extension to the gutter system due to increase in roof area	Generally a sign of inadequate gutter choice for the size of roof area: • install more rainwater pipe outlets, or • replace the gutter with high-capacity gutter

F2 LEAKAGE REPAIRS

There are different visual signs for leaking joints, depending on the material the gutter is made from. With PVCu gutters the leak may not be obvious until water is discharged down the gutter, especially if the gutter is black in colour. In some instances, leaks may show on the surface of the gutter as a black/green moss growth. A joint that is leaking, usually because the rubber seal has either shrunk or become misaligned, is generally an easy problem to fix by replacing the defective fitting. Remember that some gutter manufacturers use different fitting dimensions, and one type of gutter may not fit another. Most manufacturers produce compatibility charts to show which gutters fit together.

With leaking PVCu fittings, remember the following:

- Always try to replace like-for-like. This is sometimes not possible; there are many manufacturers that no longer exist, or that have changed their specifications and improved/updated their fittings.

- Do not be tempted to repair leaking joints with silicone sealant. While the joint may be sealed initially, as soon as the gutter expands and contracts, the joint will break again and begin to leak again.

Leaking cast iron fittings are generally visible from the ground without the need to pour water down the gutter. Leaking cast iron joints have visual telltale signs, such as:

- rust staining on the mouth of the joint
- moss and lichen growth on the mouth of the joint
- water staining in the joint area
- rust around the gutter bolt.

Repairing a leaking cast iron joint is a reasonably easy task that involves removing the gutter bolt, breaking (parting) the joint, cleaning out the old jointing medium (usually paint and putty) and repainting and re-puttying the joint before remaking the joint with a fresh gutter bolt. Care should be taken, however, as movement of the gutter can break further joints down the gutter run. Again, silicone sealant is not a satisfactory jointing medium in this situation. The area must be dry before jointing is attempted.

Signs of corrosion on cast iron gutters

F3 REPLACEMENT OF DEFECTIVE GUTTERS AND FITTINGS

Perhaps the most obvious of all gutter defects are cracked and broken gutters and rainwater pipes.

PVCu gutters and rainwater pipes are under constant attack from ultraviolet rays from sunlight. This can often lead to gutters becoming brittle, causing them to shatter or crack. Placing ladders directly against PVCu gutters, when undertaking maintenance and cleaning, can also damage the guttering further.

The main problem here, especially where replacement is necessary, is compatibility. Most manufacturers now use generic gutter and rainwater pipe sizes, but older guttering systems are often smaller in size with no adapters available. In this case, replacement of the entire system is the only option.

Where the gutter is compatible with other systems, the replacement of gutter is a fairly simple process:

1. Visually inspect the job and assess the risks. A risk assessment should be carried out. Guttering is a two-person job if working from a ladder.

2. The correct PPE should be worn when attempting this task. Eye protection is essential.

3. If it is possible to remove the cracked section between two fittings, this will be the simpler option. It is advisable to replace the fittings either side as well as the length of gutter, as the rubber seals may not create a seal when the new gutter is installed.

4. Unclip the gutter from the fittings and begin to remove the gutter from the fascia brackets by pulling the gutter and bracket towards you and down. Unclip the gutter by lifting the front edge of the fascia bracket and clicking it over the gutter. Be careful here. The brackets may be as brittle as the gutter itself.

> **SUGGESTED ACTIVITY...**
> The effect of UV light on plastics was covered in Chapter 004, page 164. Read this again to refresh your memory.

5   Once all the brackets and fittings have been unclipped, carefully lift out the gutter by twisting the front face of the gutter upwards and out of the brackets.

6   Replace the fittings – ie gutter unions, angles, etc – as necessary, taking care not to alter the fall of the gutter.

7   Measure the distance between the expansion marks of the fittings and cut and deburr the new length of gutter.

8   Install the new gutter by inserting the back edge first and twisting down and away from you.

9   Carefully reclip the gutter into the first fitting and, working towards the second fitting, reclip the gutter into the fascia brackets.

10 When the gutter has been clipped into the last fitting, start testing the gutter with a hosepipe.

11 Check for leaks and clearance of the water from the gutter.

Problems with cast iron gutters

Cast iron gutters present very different problems from PVCu ones. Cast iron gutters, if not regularly painted, will rust from the back edge towards the front, causing weakness of the metal. The rust also attacks the rafter brackets, so they too become weak. When this happens, the weight of the gutter will cause the gutter to drop and become unstable. Occasionally the gutter may even fall without any warning.

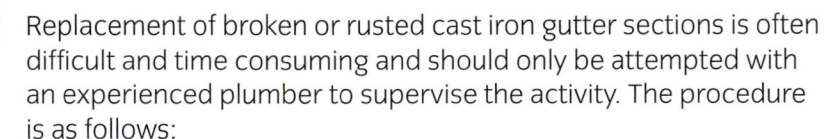

Dropped cast iron gutters

Replacement of broken or rusted cast iron gutter sections is often difficult and time consuming and should only be attempted with an experienced plumber to supervise the activity. The procedure is as follows:

1   Visually inspect the job and assess the risks. A risk assessment should be carried out. Removing sections of cast iron guttering is a two-person job if working from ladders. The guttering is very heavy and this task should not be handled alone.

2   The correct PPE should be worn when attempting this task. Eye protection is essential. It may be beneficial to clean the gutter out beforehand, as this reduces the weight.

3   Carefully cut through the gutter bolts above the nut with a junior hacksaw.

4   Using a nail punch, punch the cut bolts upwards from the cut end. Do not be tempted to punch downwards, as gutter bolts are either large dome-headed or countersunk-style bolts and you risk breaking further lengths of gutter.

5   Once the bolts are removed, carefully break the joints at either end. Be careful, as cast iron gutter often has only one rafter bracket in the centre of the gutter length, and the gutter may suddenly drop.

6 Carefully lift out the gutter by twisting towards you and upwards.

7 With the section of gutter removed, clean the socket and spigot of the gutter either side of the removed length to remove the old jointing material, and paint the inside of the socket and the outside of the spigot using black bitumen paint. The original colour paint of the gutter can be used if the customer requests it.

8 There should be no need to cut the gutter if it is a full length being replaced, as cast iron gutter is supplied in 6ft (imperial) lengths to be compatible with existing systems. Should cutting be required, a hacksaw or angle grinder with an appropriate metal cutting blade can be used. Eye protection is essential.

9 Mark and redrill the bolt hole (if required after cutting).

10 Paint the inside of the socket and the outside of the spigot of the new length with paint and place a 20mm thick bead of soft linseed oil putty in the socket.

11 Place another bead of putty in the existing gutter socket.

12 Carefully lift the new section of gutter to roof height and, ensuring that the spigot meets the socket, lift the new section of gutter into place by inserting the back edge first and twisting down and away from you.

13 Gently press the joints together and insert the gutter bolts at both joints. Retighten the gutter bolts. Do not over-tighten as the gutter may crack.

14 Remove any excess putty from inside and outside of the joints and paint the joints both internally and externally.

15 Once the paint is dry, test the gutter by discharging water from a hosepipe down the guttering, and check for leaks.

Replacing cast iron with PVCu

Replacing cast iron gutters with PVCu is possible with special adapters that convert from cast iron to PVCu (see page 472). When replacing cast iron gutters, do not be tempted to re-use the rafter brackets, as these are not secure enough for PVCu and the new gutter may flap in the wind. Any existing rafter brackets should be removed beforehand and a string line put up between the sockets of the cast iron gutter. This can be done by installing the line between the bolt holes of the existing cast iron to maintain the correct fall. The new fascia brackets can then be installed to the line as previously described.

Cast iron rainwater pipes are easily replaced with PVCu equivalents. The cast iron rainwater pipe should be replaced to the nearest downstream joint or, better still, replace the whole length of cast iron with PVCu pipe.

Blocked gutter

F4 CLEANING AND CLEARING BLOCKAGES

Probably the most common of all maintenance procedures is the cleaning and painting (cast iron only) of gutters. Over a period of time, silt can build up in gutters, especially when the roof tiles are made from concrete – these are rough and tend to be covered with a sand-like coating that washes off during rainfall. Silting can lead to moss growth and eventual blockage, causing gutters to overflow, and this could possibly cause fascia boards and roof joists to rot away and walls to become damp. Cleaning (and painting both inside and out for cast iron gutters) should be carried out during the scheduled preventative maintenance programme on a yearly basis. Where the gutter is found to contain bird droppings, the task should be handled with extreme care, as bird droppings carry disease and should not be inhaled or ingested into the body. A facemask and waterproof gloves should be worn at all times.

CONCLUSION

This chapter has shown the importance of correctly designed and installed gutter and rainwater systems. But this is only half the story. All too often, good, well-installed gutter systems are neglected and left, literally, to the elements. The important points of this comprehensive insight into rainwater management are:

- Think about the design and comply with the Building Regulations Approved Document H3 and recommendations of British Standard BS EN 12056-3.
- Use manufacturers' installation instructions for fall ratios, clipping distances and rainwater pipe positioning.
- Protect the customer's property during installation operations.
- Be aware of health and safety at all times.

009 TEST YOUR KNOWLEDGE

1. What are the four main purposes of guttering systems?

2. What is the correct fall for a guttering system?

3. Name the gutter profiles below:

4. Name four advantages of PVCu as a material for guttering systems.

5. What is the coefficient of linear expansion for PVCu gutters and rain water pipes, and what effect does this have on how rainwater systems are designed and installed?

6. Which type of gutter is manufactured on site?

7. Name the bracket systems below:

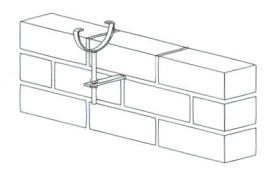

8. What are the purposes of the fittings below?

   a   b

9. Why is the visual inspection one of the most important parts of planned preventative maintenance of guttering systems?

10. Why should a face mask be used when cleaning out gutters?

010 UNDERSTAND AND APPLY DOMESTIC ABOVE GROUND DRAINAGE SYSTEM INSTALLATION AND MAINTENANCE TECHNIQUES

Two hundred years ago, waste water and sewage simply ran down the centre of streets and alleys. These were open sewers, breeding disease that on many occasions caused severe illness and death. Today, the effluent we produce is directed safely away from our homes by a network of pipes known as sanitation systems.

This last chapter investigates domestic sanitation systems. It looks at the many different sanitary appliances available and the systems of above-ground sanitation pipework that they are connected to, which ensure hygienic living conditions in our homes and the surrounding environment.

IN THIS CHAPTER, YOU WILL COVER:

A Legislation and British Standards

B Sanitary appliances
- B1 Materials used for sanitary appliances
- B2 Conventional WCs
- B3 Wash hand basins
- B4 Bidets
- B5 Baths
- B6 Shower trays and cubicles
- B7 Sinks
- B8 Urinals

C Sanitary pipework systems
- C1 Primary ventilated stack system
- C2 Ventilated branch discharge system
- C3 Secondary ventilated stack system
- C4 Stub stack system
- C5 General sanitary pipework requirements

D Traps
- D1 Tubular traps
- D2 Bottle traps
- D3 Anti-vac and resealing traps
- D4 Self-sealing waterless waste valves
- D5 Loss of trap seal

E Installing above-ground drainage systems (AGDS)
- E1 Preparation before installation
- E2 Types of materials
- E3 Waste pipe connections to the soil stack
- E4 Access to pipework
- E5 Soil stack connection to the drain

F Below-ground drainage systems – an overview
- F1 The separate system
- F2 The combined system
- F3 The partially separate system
- F4 Appliance connections to existing below-ground drainage systems

G Bathroom layout specifications

H Sanitary appliance installation
- H1 Preparation
- H2 Assembling the appliances
- H3 The installation process

I Soundness and performance testing of above-ground sanitation systems
- I1 Soundness testing
- I2 Performance testing

J Decommissioning and maintenance
- J1 Simple maintenance tasks
- J2 Dealing with blockages
- J3 Decommissioning sanitary systems

Test your knowledge

A LEGISLATION AND BRITISH STANDARDS

Sanitation systems must comply with Building Regulations Approved Document H1:2002. The general requirements of this document are that a foul water system must:

- convey the flow of foul water to a foul water outfall. This can be a foul or combined foul/rainwater sewer, a cesspool or a septic tank
- minimise the risk of blockage and/or leakage
- prevent foul air from entering the building under working conditions
- be ventilated
- be accessible for clearing blockages
- not increase the vulnerability of the building to flooding
- be large enough to carry the expected flow at any point in the system.

To successfully achieve this, we must consult several documents:

- **BS EN 12056-5:2000: Gravity drainage systems inside buildings – installation and testing, instructions for operation, maintenance and use** – this applies to waste water drainage systems which operate under gravity. It is applicable for drainage systems inside dwellings, commercial, institutional and industrial buildings. Part 5 of the standard gives information that should be followed when installing and maintaining waste water gravity drainage systems as well as the materials that can be used.
- **BS 8000-13:1989: Workmanship on building sites** – the code of practice for above-ground drainage and sanitary appliances. This document gives recommendations on basic workmanship and covers tasks that are carried out in relation to above-ground drainage and sanitary appliance installation.
- **The Water Supply (Water Fittings) Regulations 1999** – These are not strictly relevant with regard to sanitation but sanitary appliances require hot and cold water supplies and, because the appliances are used in connection with foul and waste water, the Water Regulations must be consulted to guard against backflow of contaminated water.

Key British and European standards for the installation of sanitation systems are listed in the table below:

BS EN 12056-2:2000	Gravity drainage systems inside buildings. Sanitary pipework, layout and calculation.
BS EN 12056-5:2000	Gravity drainage systems inside buildings. Installation and testing, instructions for operation, maintenance and use.
BS 8000-13	Workmanship on building sites. Part 13 covers the installation of drainage and sanitation systems.

BS 5627	Plastic connectors for use with horizontal outlet vitreous china WC pans.
BS 6209	Solvent cement for non-pressure pipe systems.
BS EN 1329	Plastic piping systems for soil and waste discharge (low and high temperature) within the building structure – unplasticised polyvinyl chloride (PVCu).
BS EN 1451	Plastic piping systems for soil and waste discharge (low and high temperature) within the building structure – polypropylene (PP) requirements and test methods.
BS EN 1453	Plastic piping systems with structured-wall pipes for soil and waste discharge (low and high temperature) inside buildings – unplasticised polyvinyl chloride (PVCu).
BS EN 1455	Plastic piping systems for soil and waste discharge (low and high temperature) within the building structure – acrylonitrile butadiene styrene (ABS).
BS EN 1519	Plastic piping systems for soil and waste discharge (low and high temperature) within the building structure – polyethylene (PE).
BS EN 1566	Plastic piping systems for soil and waste discharge (low and high temperature) within the building structure – chlorinated polyvinyl chloride (PVCc).
BS EN 12380	Air admittance valves – for use in drainage systems.
BS EN 274	Waste fittings for sanitary appliances.

Before we look at the requirements of these documents, we must first look at the appliances that we install in domestic installations and the materials they are made from. We will also look at urinals which, although they are not strictly domestic sanitary appliances, we may come across during our work as plumbers.

B SANITARY APPLIANCES

There are two purposes of sanitary appliances: to maintain personal hygiene by washing, bathing or showering, and to remove solid and fluid human waste. During this first part of the chapter we will look at the types of sanitary appliances used in dwellings and their working principles.

B1 MATERIALS USED FOR SANITARY APPLIANCES

The materials used in the manufacture of sanitary appliances are listed in the table below. They must be robust, hygienic and easy to clean.

Material	Description	Appliances
Vitreous china	This material is made from white-burning clays and finely grained material mixed with ball clay, a fluxing agent and water into casting clay known as slip. The slip is fired to a high temperature and even in its unglazed state cannot be contaminated by bacteria and remains hygienic in all situations. Glazed vitreous china is stain-proof, burn-proof, rot-proof and non-fading and is resistant to acids and alkalis. Vitreous china is available in many colours and shades.	WC pans and cisterns Wash basins Bidets Urinals
Stainless steel	This is made from 304- or 316-grade stainless steel to European Standard EN 10088-2. It is usually fitted in areas to which the general public has access and is highly resistant to vandalism. All stainless steel sanitaryware conforms to the Department of Health specification.	WCs and cisterns Wash basins Kitchen sinks Urinals
Fireclay	Made from buff-coloured ball clays from Devon and Dorset in the UK, fireclay is very robust to withstand rough treatment but, unlike vitreous china, it is porous. Because of this, it requires firing with a ceramic undercoat to seal the clay before being coated with two coats of white glaze, and then re-fired.	Belfast sinks London sinks Butler's sinks Urinals Heavy-duty WC pans and wash basins for hospitals
High-impact plastic	Usually manufactured by injection moulding techniques.	WC seats WC cisterns Bath panels
Acrylic	Sheet with thicknesses varying between 3mm and 8mm is heated until it becomes soft and pliable and is then placed over an aluminium mould where it is sucked into place. This is known as vacuum forming. Acrylic is warm to the touch and can be moulded into many shapes. It is, however, easily damaged by scratching and abrasive cleaners. Acrylic baths are often strengthened by a baseboard made from chipboard and glass-reinforced polyester (GRP). It is very lightweight and appliances are usually aimed at the domestic market. Acrylic baths require a supporting cradle.	Baths Bath panels Wash basins Shower trays
Enamelled cast iron	Cast iron is extremely robust but is very heavy and very cold to the touch. Because of the nature of cast iron, bath designs tend to be very traditional.	Baths
Porcelain enamelled pressed steel	The steel sheet used in the manufacture of sanitaryware must be of the highest grade low carbon steel. The enamel is sprayed on and then kiln fired. It is rigid but light; very robust, but the enamel is easily damaged.	Baths Wash basins

B2 CONVENTIONAL WCS

The acronym WC stands for 'water closet'. It consists of a WC pan and a flushing cistern. There are three types of WC pans:

- **Wash down type** – This is the most common type of WC fitted in the UK. The pan is cleared by a carefully designed water distribution system, which uses the force of the water flush and the volume of water delivered to the bowl to clear the contents. Wash down-type WC pans are usually around 400mm high, depending on the manufacturer, and have 50mm of water seal in the trap. Smaller versions are available for use in infant schools. The bowl is shaped to provide efficient effluent clearance while maintaining easy cleaning.

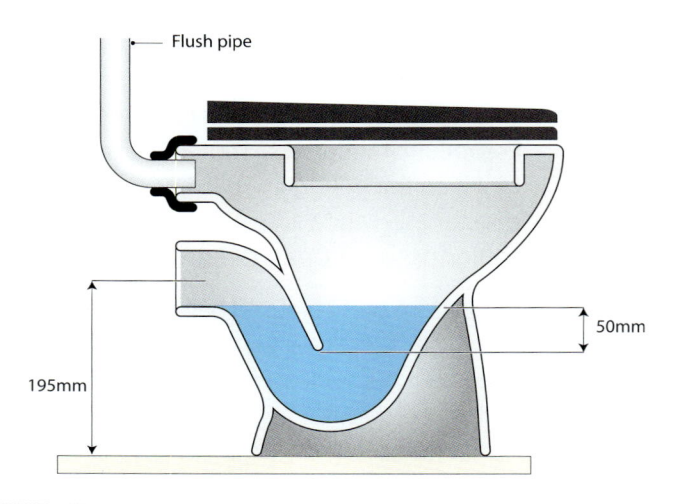

Wash down WC pan

- **Single-trap siphonic type** – There are two types of siphonic pan but both work in a similar manner: the flushing operation creates a vacuum, which contributes to clearing the pan. The first type is the single-trap siphonic WC pan – this has a lower outlet than other pan designs. It is also known as the Malvern-type WC pan. It is usually only installed as a direct replacement, as the design tends to look very dated. It works by restricting the flow of water from the cistern and this allows a build-up of water in the pan, which is then forced through the restricted neck of the trap, creating a vacuum behind it and clearing the pan contents completely.

- **Double-trap siphonic type** – This type is very rarely sold in the UK because the flushing volume of WC cisterns was reduced to 6 litres by the Water Supply (Water Fittings) Regulations 1999. This kind of WC pan is very quiet and extremely efficient at removing the pan contents. Unlike the single-trap siphonic pan, the double-trap siphonic pan has an unrestricted outlet and two water traps. A special pressure-reducing valve called an aspirator (or bomb) is fitted to the bottom of the siphon. When the cistern is flushed, a negative pressure is created in the chamber between the two traps by the aspirator. The aspirator is a venturi device, which sucks out the air from the chamber as the water from the flush passes through it. This causes the contents of the bowl by be sucked through the two traps. The aspirator holds a little water back to refill the second trap after the flush is complete. Double-trap siphonic WCs tend to be longer than wash down types because of the extra water trap.

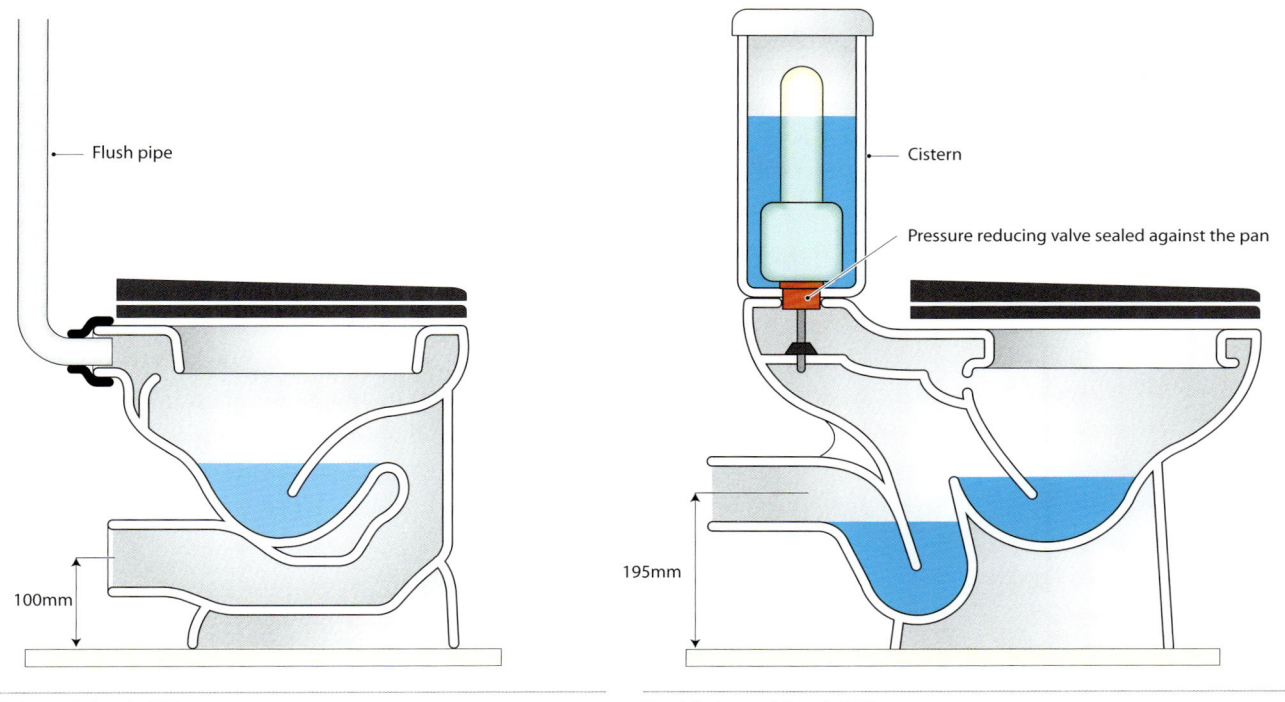

Single-trap siphonic WC pan

Double-trap siphonic WC pan

WC styles

WCs can be manufactured in five main styles:

- **Close-coupled** – The WC pan is designed to have the cistern bolted to the back of the pan to form one unit.

Close-coupled WC suite

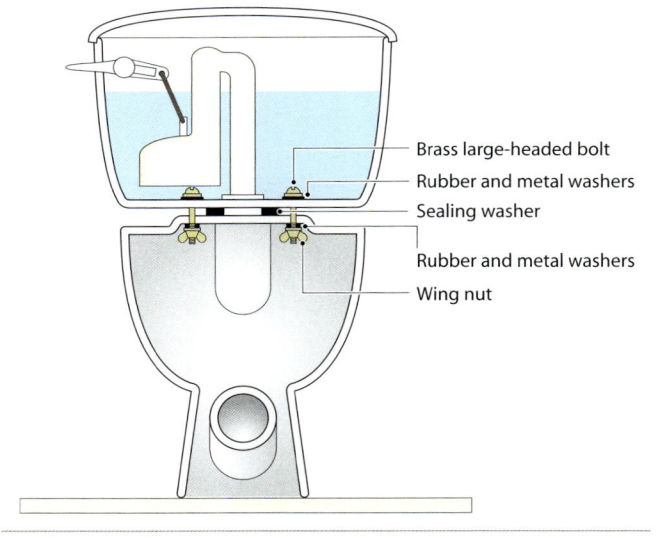

Diagram showing how the cistern is fixed to the WC pan

- **Low level** – The cistern is connected to the WC pan by a short flush pipe to convey the water from the cistern to the WC pan.

- **High level** – Similar to the low level type but the flush pipe is much longer and the cistern is at high level. Usually used when designing period bathroom suites.

- **Back to wall/concealed** – This style is becoming more popular due to the fact that the cistern is concealed in a cabinet or behind a panel. The WC pan sits close to the cabinet or panel.

- **Wall hung** – This type gives the effect of space, as the WC pan is hung on the wall and is completely free of the floor.

In the past, WC pans were manufactured with a variety of P-trap and S-trap configurations (see page 515–6) formed as part of the pan casting, but this proved expensive. As a result, today most WC pans are manufactured with the P-trap configuration, but with the use of an angled WC pan connector they can be made into an S-trap or left or right outlet depending on the installation requirements.

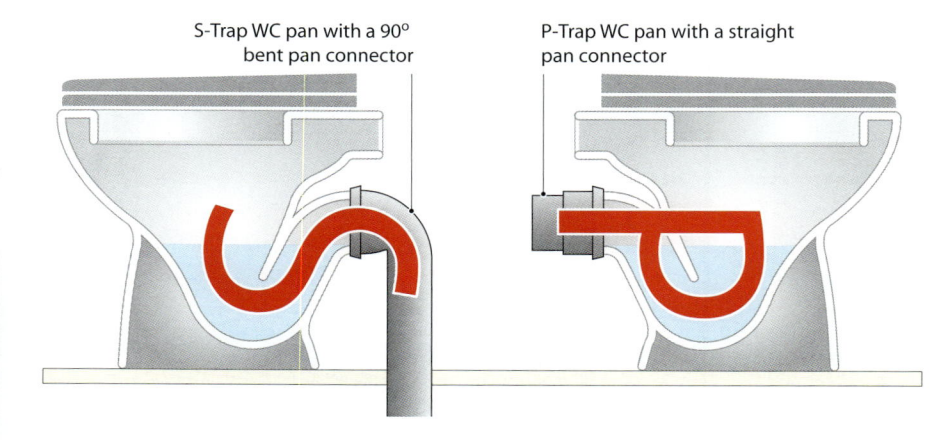

Traditional low-level WC suite

Period Victorian high-level WC suite

Back-to-wall WC suite

Modern wall-mounted WC pan

P-Trap and S-Trap WC suite

The WC cistern

The WC cistern is the method by which the water is discharged into the WC pan. The water can be delivered to the WC pan in several different ways depending on the cistern design:

- **By the use of a siphon** – This is the traditional way to flush a WC cistern. The cistern is flushed using siphonic action (see Chapter 004, page 184). The WC flushing handle is connected to the siphon by a link pin. When the WC cistern handle is depressed, the link pin lifts a plunger in the siphon bell, which has a large thin plastic or rubber diaphragm at the end of it. The diaphragm lifts a column of water up and over the top of the siphon to begin the siphonic action. There are many different styles and sizes of WC siphon available and the correct one must be chosen depending on the cistern size. Some

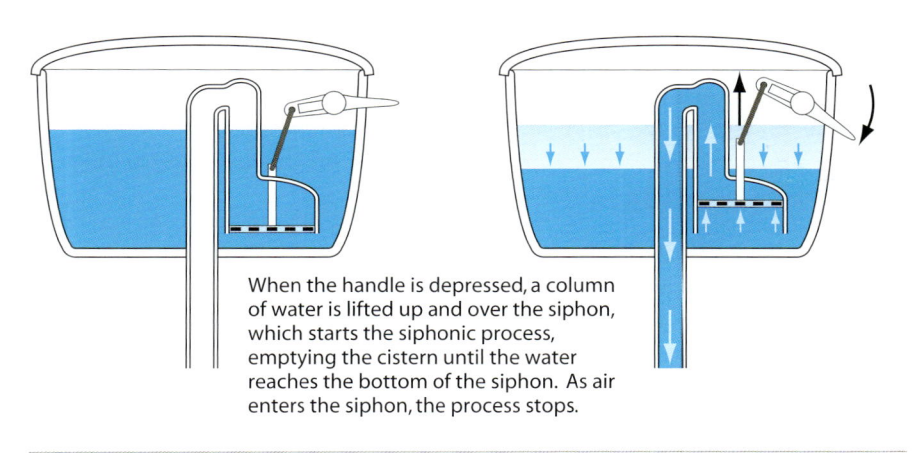

When the handle is depressed, a column of water is lifted up and over the siphon, which starts the siphonic process, emptying the cistern until the water reaches the bottom of the siphon. As air enters the siphon, the process stops.

How a WC siphon works

WC siphon

siphons allow different flushing volumes to be set by adjusting the height at which air is let into the siphon bell to stop the siphonic action.

- **By the use of a dual flush valve** – These can be operated by pressing a button on the top of the WC cistern, or remotely by air blown through a tube when the button is depressed, or by remote infrared sensor linked to a solenoid valve for concealed cisterns. They work by simply opening up a valve when the button is activated; this allows water to flow by gravity to the pan. Siphonic action is not required. Flush valves have a 6-litre and a 4-litre flush action. They also have an integrated overflow that allows water to flow straight to the WC pan should the float-operated valve begin to overflow, so a separate overflow pipe is not required.

- **By the use of a drop valve** – This is also known as a flap valve. It is a very simple valve that allows water to flow by gravity to the pan. In the closed position, it is the weight of the water that makes a watertight seal. When the WC handle is depressed, a link pin simply lifts the valve up. These are not dual flow and will only flush as long as the handle is pressed down. Most flap valves have an integral overflow.

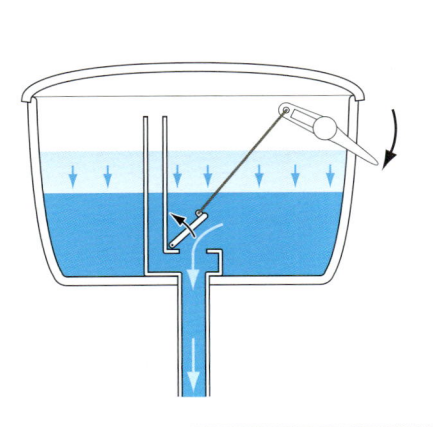

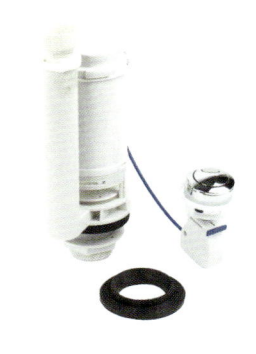

Dual flush valve

KEY POINT

Prior to 1986, the flush volume for a WC cistern was 9 litres. This was lowered in the Model Water By-laws of 1986 to 7.5 litres. Today, the Water Supply (Water Fittings) Regulations 1999 restrict the flushing volumes of new WC cisterns to 6 litres for a long flush and 4 litres for a short flush, but older WC pans will not flush with such a low water volume, so 9- and 7.5-litre cisterns are still available for the replacement market.

When the handle is depressed, the flap lifts allowing water to flow to the pan by gravity

Operation of a drop valve

010 ABOVE GROUND SANITATION SYSTEMS

> **SUGGESTED ACTIVITY...**
> Service valves and float-operated valves were covered in detail in Chapter 006, pages 316–19. Why not check out these chapters and refresh your learning on this important subject?

The water in the cistern is controlled using a float-operated valve conforming to BS 1212 parts 2, 3 and 4. The cistern must also have a service valve fitted as close to the cistern as possible. A separate overflow must be installed with WC cisterns that do not have an integral overflow, and this must discharge safely in a conspicuous position, usually outside the building. Integral overflows discharge directly into the WC pan via the cistern.

WC cisterns can be made from a variety of materials, including vitreous china, plastic and hard rubber, but other materials such as cast iron and lead-lined wood have also been used in the past.

B3 WASH HAND BASINS

There are literally hundreds of different styles of wash hand basin and many of these also come in various sizes and tap arrangements. Corner wash basins are also available. Wash basins should be installed with approximately 800mm from the floor to the front lip of the basin. They can be divided into four basic types:

- **Wall-hung wash basin** – This wash basin is mounted on wall-fixed brackets or bolted directly to the wall. There are several different types of mounting brackets, including towel rail-type or concealed, depending on the wash basin style. The mounting wall must be able to take the weight of the wash basin. If there is any doubt, either a centre leg or a pair of legs should be used.

- **Pedestal wash basin** – There are two different types. A pedestal wash basin is fixed to the wall but relies on the pedestal for its main support. The pedestal is designed to hide the pipework. Semi-pedestal wash basins are becoming increasingly popular. The pedestal does not carry the weight of the basin as it does not reach the floor and it is designed to hide the associated pipework.

Pedestal wash basin

Wall-hung wash basin

Semi-pedestal wash basin

- **Counter-top wash basin** –
  - **Counter-top style** – also known as an inset wash basin. It sits snugly in a worktop surface.
  - **Semi-counter-top style** – also known as a semi-recessed basin. It sits half on and half off a work surface.
  - **Under-counter-top style** – as the name suggests, this type is mounted under a work surface. The work surface is usually marble, agglomerate marble or granite.
- **Vessel wash basin** – This type of basin is designed to be supported by a mounting surface such as a worktop or cabinet.

Wash basins can be made from a variety of materials, including vitreous china, stainless steel and porcelain enamelled pressed steel (see table on page 489).

Counter-top wash basin

Semi-counter-top wash basin

Under-counter top wash basin

Vessel wash basin

Tap hole and waste arrangements for wash basins

There are four main tap hole arrangements for wash basins:

One tap hole basin with monobloc mixer tap 	Specifically designed for use with a monobloc mixer tap.
Two tap hole basin with hot and cold taps 	The traditional tap hole arrangement for use with hot and cold ½-in BSP pillar taps.
Three tap hole basin with remote mixer tap 	This is a little-used tap arrangement where the tap bodies are fitted below the basin with just the wheel heads showing. The spout and the tap bodies are concealed below the wash basin.
No tap hole basin with wall-mounted taps 	Becoming more popular for bespoke bathrooms, these use wall-mounted designer bib taps with concealed pipework.

Wash basins are manufactured with an integral overflow for use with a 1¼-in slotted waste for connection to a 32mm waste trap. There two basic waste types available:

- **Slotted waste, plug and chain** – The old-fashioned method of providing a waste stopper. The slots in the waste are to allow water that has flowed down the integral overflow to find its way safely down to the trap. These are usually made in to the basin with silicone sealant with a plastic poly-washer inserted between the securing nut and the basin. Care should be taken when using gold-plated fittings and silicone sealant as some sealants can discolour the gold plating.

- **Pop-up waste** – These provide a handle, typically designed as part of the tap, which, when pushed down, pops the waste plug up. They tend to have specific sealing washers to seal the waste into the basin. Pop up wastes can also be a spring-loaded type which opens via a spring when the waste plug itself is depressed.

Waste, plug and chain arrangement

Pop-up waste arrangement

B4 BIDETS

Very similar in design to a WC pan, the bidet is often described as a sit-on wash basin. The bidet is a hygienic method of ensuring personal cleansing, especially after using the WC. An important secondary use is that of a footbath. There are two types:

- **The over-rim bidet** – The most common bidet type, installed in the same way as a wash basin. It is available with one or two tap holes, depending on the bidet design, and can be fitted with a variety of taps including monobloc mixers, pillar taps and hand spray-type mixers with a hose connection.

- **The ascending spray bidet** – Very rarely seen in the UK, the ascending spray bidet uses a special tap arrangement to discharge water upwards from inside the bowl of the bidet in a spray similar to a small shower head. Special installation arrangements exist for this appliance because of the risk of contamination of water by backflow through the spray head. It must not be installed on mains pressure systems and the Water Regulations should be consulted for all installations of this type of appliance.

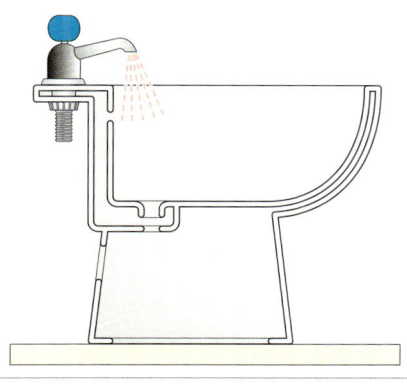

Over-rim bidet

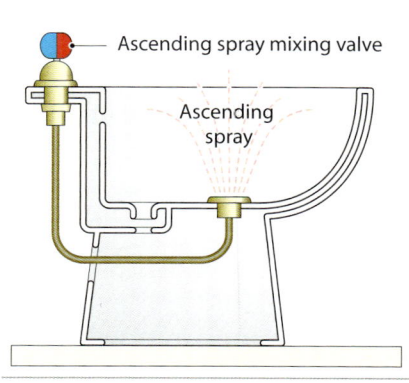

Ascending spray bidet

SUGGESTED ACTIVITY...
The installation of ascending spray bidets was covered in detail in Chapter 007, page 386. Take a look back at that chapter to make sure you are clear on this important topic.

Bidets are usually made from vitreous china. Styles include floor-mounted, back-to-wall and wall-hung types.

Floor-mounted bidet

Back-to-wall bidet

Wall-hung bidet

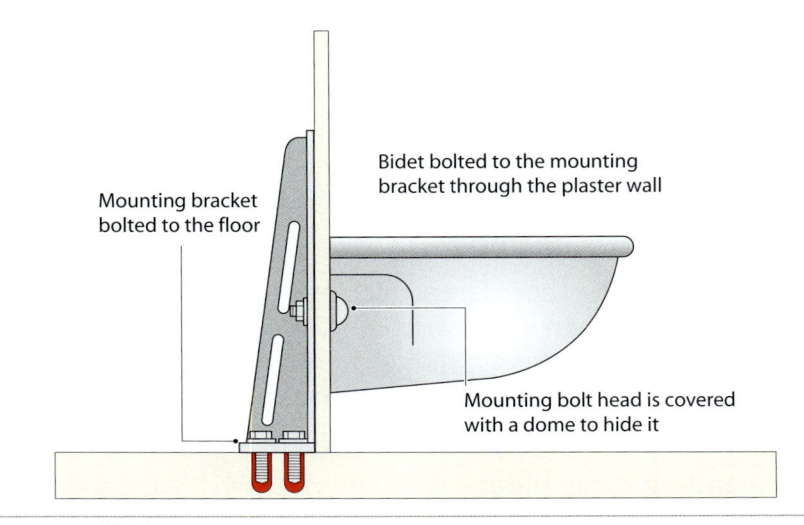

Back-to-wall bidet

B5 BATHS

Baths are manufactured to BS 4305 (EN 198) and can be supplied manufactured from the following materials:

- **Reinforced cast acrylic sheet** – This is the most common material for baths. Some acrylic baths require reinforcement in the form of glass-reinforced polyester, and all types require a steel tubular cradle, a top frame and a base board.

- **Porcelain-enamelled steel** – These tend to be used in commercial situations such as hotels, hostels etc or in housing association and Local Authority housing where durability is important.

In Europe, pressed steel baths are more common than acrylic.

- **Porcelain-enamelled cast iron** – These have a much lower share of the market and tend to be used for the more traditional designs such as roll-top freestanding and rectangular shapes.

Other materials are also available, such as resin-bonded cementatious slurry, known as resinstone, and gel-coated reinforced polyester, but these are generally used in specialist and niche markets.

Each material has its own unique characteristics, which influence the bath design. Baths can be manufactured in a wide variety of styles and designs, including:

- **Standard baths** – These are rectangular-shaped and come in many size and design options. They are usually fitted with a front panel and/or end panels as required.
- **Corner baths** – These fit into the corner of the bathroom. They require a curved bath panel, which is easily cut and trimmed to specific installation requirements.

Standard bath

Corner bath

- **Off-set corner baths** – These are similar to a standard corner bath but they have sides of unequal length. This design utilises the space available while optimising the bathing space. They are available left- or right-handed, depending on the installation requirements.
- **Freestanding baths** – These are designed to stand on their own feet and are usually not fitted against a supporting wall. A range of styles is available, from traditional roll-top and claw-and-ball styles to more contemporary designs.

Off-set corner bath

Freestanding claw-and-ball bath

Double-ended bath   Tapered bath   Shower bath

- **Double-ended baths** – These are usually rectangular in shape but they have two non-tap ends and side-mounted taps. They are designed with two people in mind.

- **Tapered baths** – These are designed where space is at a premium. They are wider at one end and are usually fitted with a shower at the wider end.

- **Shower/baths** – Again, these are usually wider at one end to maximise the space available for showering.

- **Baths for the disabled** – Many baths on the market have been modified for disabled access. Modifications include doors that allow the user to walk in, or chairlifts to lower the user in and out.

- **Spa baths and whirlpools** – These will be covered later in this chapter.

The most common sizes range from 1600mm to 1800mm in length and from 700mm to 800mm in width. The most popular shapes require a front bath panel and, very often, end panels to hide the frame, the cradle and the plumbing.

Tap hole and waste arrangements for baths
Tap holes for baths come supplied in one of three ways:

- **No tap hole** – This type of bath must be drilled so the taps can be installed in the position of the customer's choice, or can be used with wall-mounted taps.

- **Two tap hole** – This is the standard arrangement. The taps may be on either the end of the bath or the side.

- **Three tap hole** – For remote-type taps (one hole for the spout and two for the taps).

Waste connections for baths can be made by:

- **The banjo-type bath waste fitting** – This uses a long, threaded waste fitting with slots on opposite sides near the top. The waste from the overflow comes via a flexible pipe connected by a banjo connection. This is assembled over the waste fitting and is held in place on the underside of the bath by a large 1½-in BSP nut fitted to the bottom of the waste fitting and tightened against the banjo. Silicone sealant should be used at the joint between the bath and the banjo, and the banjo and the nut. This type of bath waste connection is very prone to leakage.

Banjo-type bath waste fitting

Bath waste and overflow fitting

- **Bath waste and overflow kit** – Where there is sufficient space underneath the bath, this is the easiest bath waste connection to fit. It uses a one-piece bath waste connection, which is held in position by a long bolt placed through the centre of the bath waste grille. Both the waste connector and the grille have sealing washers. The bath waste connection incorporates the overflow connection. The bolt pulls the waste connector and the bath waste grille together, and this compresses the washers to make a watertight seal.

- **Bath pop-up waste and overflow fitting** – Pop-up waste systems are becoming increasingly popular. They are fitted in the same way as a bath waste and overflow kit but feature a twist-action overflow (in chrome, gold plate or matching the colour of the bath), which operates a lever to raise or lower the bath waste plug.

- **Combined waste and trap** – This is a fitting that combines the bath waste and overflow with the bath trap.

The bath pop-up waste and overflow fitting

The bath waste and trap combination fitting

Whirlpool and air-spa baths

The whirlpool and air-spa bath is the latest addition to the bath range. The idea of the whirlpool bath to relax the body is not a new one – it was used by the ancient Romans. In the 1950s the idea surfaced again when the Jacuzzi brothers developed the whirlpool bath for domestic use.

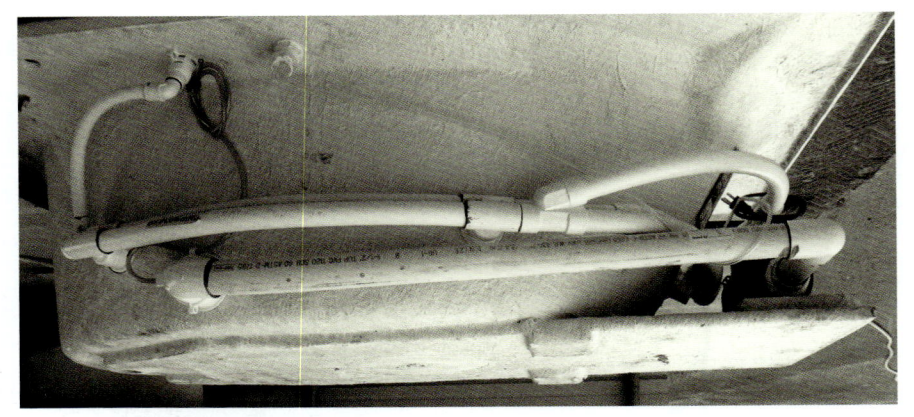

Workings of a whirlpool bath

Today, the whirlpool-type bath can take many forms, such as jetted baths, hydro-pools, hydro-spa and air-spa types, all of which use either the pumping of air and water through nozzles installed into the side or floor of the bath (Jacuzzi type) or the introduction of forced air through smaller nozzles in the bath sides (air-spa type). They can also be retro-fitted to any acrylic or pressed steel bath. The pump is usually situated at one end of the bath.

Most manufacturers use self-draining pipework that incorporates a disinfection system, so that the hidden whirlpool pipework can be cleaned.

All baths of this type incorporate a safety cut-out to stop the pump if anything gets too close to the water suction pipe.

B6   SHOWER TRAYS AND CUBICLES

Shower trays are also known as shower bases. They vary in size from compact square shapes to large rectangular, quadrant and five-sided models. They are made from a variety of materials such as heavy-duty reinforced acrylic sheet, fireclay and resin bonded. The choice of shower tray depends largely on the space and budget available.

Square shower tray

Rectangular shower tray

Five-sided shower tray

Many shower trays have a raised lip which, when placed against the wall, allows tiles to be placed over it to help with the sealing of the tray. Some trays have adjustable feet to assist in levelling the tray. Resin bonded and fireclay trays are bedded on a weak bed of sand and cement.

Waste arrangements for shower trays

The most common waste arrangement for a shower tray is the use of a combined shower waste and trap. Most modern trays are bedded to the floor and, because of the position of the waste on the tray, the trap is, in many cases, inaccessible, making cleaning and clearing of blockages almost impossible. The combined shower waste and trap allows the trap to be cleaned of potential blockages, such as hair, from the top of the waste on the shower tray. The inside of the trap is removable from above.

Combined shower waste and trap

Shower cubicles and enclosures

Shower enclosures are available in three distinct forms:

- **Freestanding shower cubicle** – As the name suggests, a freestanding shower cubicle is one that does not use any of the walls of the building in its construction. However, the cubicle may be fixed to the wall for support.

- **Shower enclosure that uses one or two walls** – This kind of shower enclosure uses either one or two walls to form part of the showering area. This is the most common of all shower enclosures installed.

- **Shower door** – A single shower door is fixed between two opposing walls. This uses three walls of the building to form the enclosure with the shower door being the method of entry and exit.

Freestanding shower cubicle

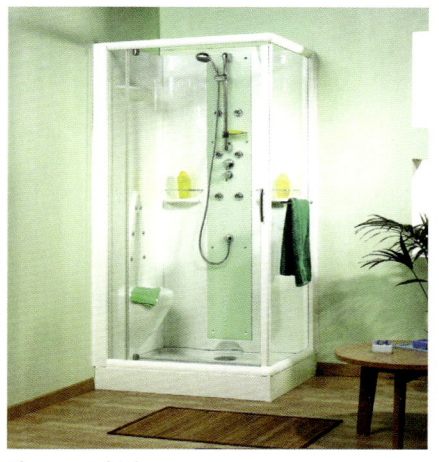

Shower cubicle

Shower door

B7 SINKS

The term sink refers to an appliance fitted in a kitchen or utility room. The ideal sink has to be hardwearing and robust to be able to withstand the abuse it is likely to receive. There are several different types:

- **Kitchen sinks** – These come in a variety of different shapes and sizes. Common arrangements are single bowl and drainer, bowl and a half and double bowl. They are usually set into the work surface and can be made from a variety of materials, such as stainless steel, granite, astro-cast and polycarbonate materials. Vitreous china sinks are also available but these tend to chip easily and will shatter if heavy pans

Styles of kitchen sinks

> **SUGGESTED ACTIVITY...**
> The bonding of metalwork and pipes was covered in detail in Chapter 005, pages 205–6. Take a look at this again to revise this key point.

are dropped into them. It should be noted here that stainless steel sinks will require bonding to the electrical earthing in the property.

- **Cleaners' sinks** – There are three types of cleaners' sinks and all are large, deep, rectangular sinks made of very thick white glazed fireclay. They are usually mounted on cast iron cantilever brackets but modern installations allow them to be fitted into kitchen units.

- **Belfast sinks** – These originate from the early 18th century when they were fitted into the servants' quarters and the butler's area. Today, they are primarily used in utility and cleaners' rooms, although may also be used in period-style kitchens. They are distinguished from all other cleaners' sinks by their integral weir-type overflow. The taps are usually bib types fixed to the wall above the sink.

London sink

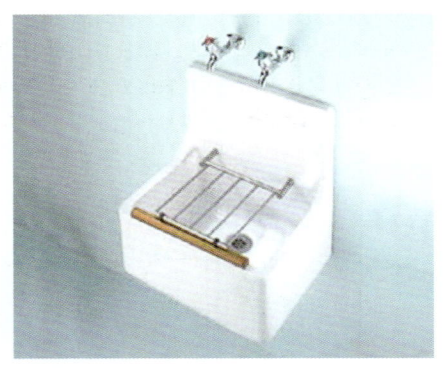

Butler's sink

- **London sinks** – These are visually very similar to Belfast sinks but do not have a weir overflow.

- **Butler's sinks** – The butler's sink is essentially the same as a London sink. There are, however, two main differences: the sink has a high splashback and it also has a bucket grille.

B8   URINALS

Urinals are fitted in non-domestic buildings and come in three different styles:

- **Bowl urinals** – Usually made of vitreous china and stainless steel, these are the most commonly used urinal type and the easiest to install. Dividers may be placed between the urinal bowls to give a little

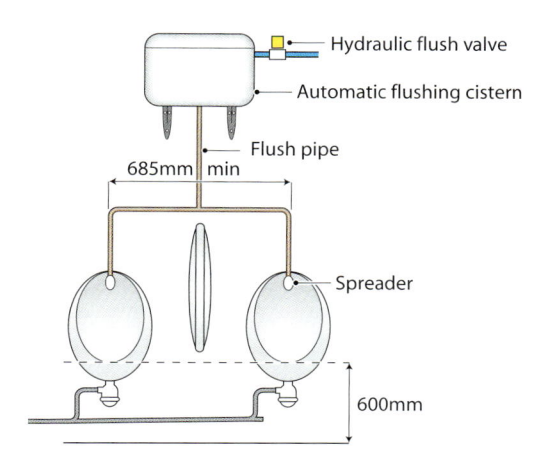

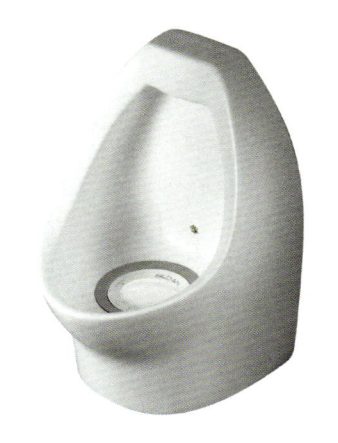

Bowl urinal layout

Bowl urinal

privacy. The bowl should be fixed at around 600mm from the floor to the front lip. This can be reduced for urinals installed in schools.

- **Trough urinals** – These are generally made from stainless steel and installed where the risk of vandalism is high. The trough should be sized according to the number of people that are expected to use it and they are, therefore, available in different lengths. The trough has a waste connection, and the trough floor has a built-in slight fall to allow the urinal to be installed level.

- **Slab urinals** – This type of urinal is manufactured from fireclay and is assembled on site. The channel in the base of the urinal is laid to a slight fall and the waste connection is made directly to the drain via the channel into a trapped gulley.

Stainless steel trough urinal

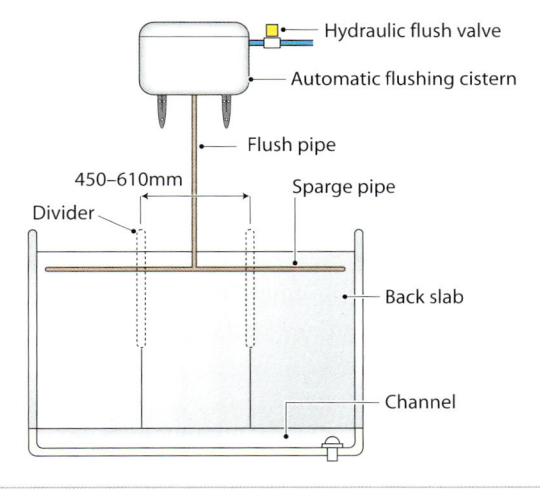

Slab urinal layout

Slab urinal

Flushing the urinal

The Water Regulations state that urinals may be flushed with either:

- a manual or automatically operated cistern

505

- a pressure flushing valve directly connected to a supply or distributing pipe, which is designed to flush the urinal either directly (manually) or automatically, provided that the flushing arrangement incorporates a backflow arrangement or device appropriate to fluid category 5 (see Schedule 2, Paragraph 15).

Clause G25.13 states:

'Where manually or automatically operated pressure flushing valves are used for flushing urinals, the flushing valve should deliver a flush volume not exceeding 1.5 litres per bowl per position each time the device is operated.'

The automatic flushing cistern

As the name suggests, automatic flushing cisterns use an automatic flushing siphon to flush the urinals automatically when the water reaches a predetermined level in the cistern. The Water Regulations stipulate that any auto-flushing cistern must not exceed the following water volumes:

- 10 litres per hour for a single bowl or stall
- 7.5 litres per hour per urinal position for a cistern serving two or more urinal bowls or 700mm of slab.

The maximum flow rate from any automatic flushing cistern must be regulated by the inflow of water from the cold supply. This can be done quite easily by the use of urinal flush control valves such as a hydraulic flush control valve fitted to the incoming water supply. The hydraulic flush control valve allows a certain amount of water through to the cistern when other appliances like taps and WCs are used, rather than have a constant supply of water dripping into the cistern. The sudden reduction in pressure on the mains supply opens the valve to allow a certain amount of water through. The amount of water can be varied depending on the installation requirements and number of urinals. The idea is to prevent the urinals flushing when the building is not being used, thus saving on wasted water.

> **KEY POINT**
>
> When the level of the water reaches the top of the dome, the head of water at point A becomes greater than the pressure at point B. The water pressure in the trap (point C) overcomes the air pressure inside the siphon and this initiates siphonic action, emptying the cistern.

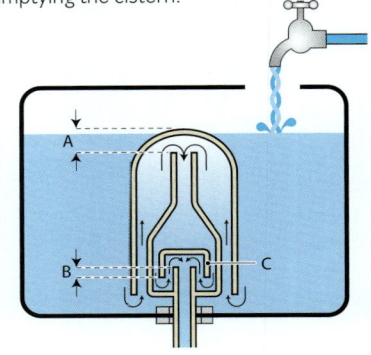

Hydraulic flush control valve

The flushing valve

This is a new method of flushing a urinal that involves the use of a valve, which can either be manual or automatic, that delivers a short 1.5-litre flush to an individual urinal bowl. The water can be supplied either direct from the water main, from a boosted cold water system or at low pressure from a cistern supplied by a distribution pipe.

Manual valves are lever-operated and are located just above the urinal bowl. Automatic valves are activated via an infrared sensor, which senses a person. The sensor must 'see' the person for at least 10 seconds to prevent accidental activation by someone walking by. The sensor activates a solenoid valve and this allows the minimum short flush. Automatic flushing valves require a backflow prevention device to

Manual flushing valve

be included, which prevents backflow of a fluid category 5 contaminated water.

C SANITARY PIPEWORK SYSTEMS

So far, we have looked at the various types of sanitary appliances that we are likely to come across during work, both in domestic and non-domestic situations. Sanitary appliances, however, are only part of the story. Without a system of pipework to take the waste solids and liquids away from the dwelling, sanitary conditions within buildings would not be hygienic and could be potentially damaging to health. In this next part of the chapter, we will look at the various systems of sanitary pipework, often called above ground drainage systems (AGDS), and investigate where these systems should be installed.

All sanitary systems contain two sections. These are:

- **The soil pipe** – Also known as the soil stack, this is the lower, wet part of the system, which takes the effluent away from the building.

- **The vent pipe** – Also known as the vent stack, this is the upper part of the system. It introduces air into the system to help prevent loss of trap seal. Ventilation of a soil and waste system is necessary to prevent water seals in traps being broken due to negative pressure or pressure fluctuations within the system. Broken seals allow foul air and smells to enter the building. The vent pipe is the dry part of the system.

Together the two sections are referred to as the soil and vent pipe.

Before we look at the sanitary systems, we must remember that all sanitary pipework and drainage systems must comply with Approved Document H of the Building Regulations. These requirements will be met if the recommendations of BS EN 12056:2000, which contains recommendations for design, testing, installation and maintenance for all above ground non-pressure pipework systems, are followed.

To comply with Document H, all appliances must be fitted with a water trap seal to prevent foul air from entering the building. Also, the waste pipe diameter and gradient must maintain a water seal in the trap of at least 25mm after the appliance has been used.

C1 PRIMARY VENTILATED STACK SYSTEM

The primary ventilated stack is probably the most common system installed in domestic dwellings. It relies on all the appliances being closely grouped around the stack and therefore does not require an extra ventilating stack, unlike other systems. It is used in situations where the discharge stack is large enough to limit pressure fluctuations without the need for a separate ventilating stack.

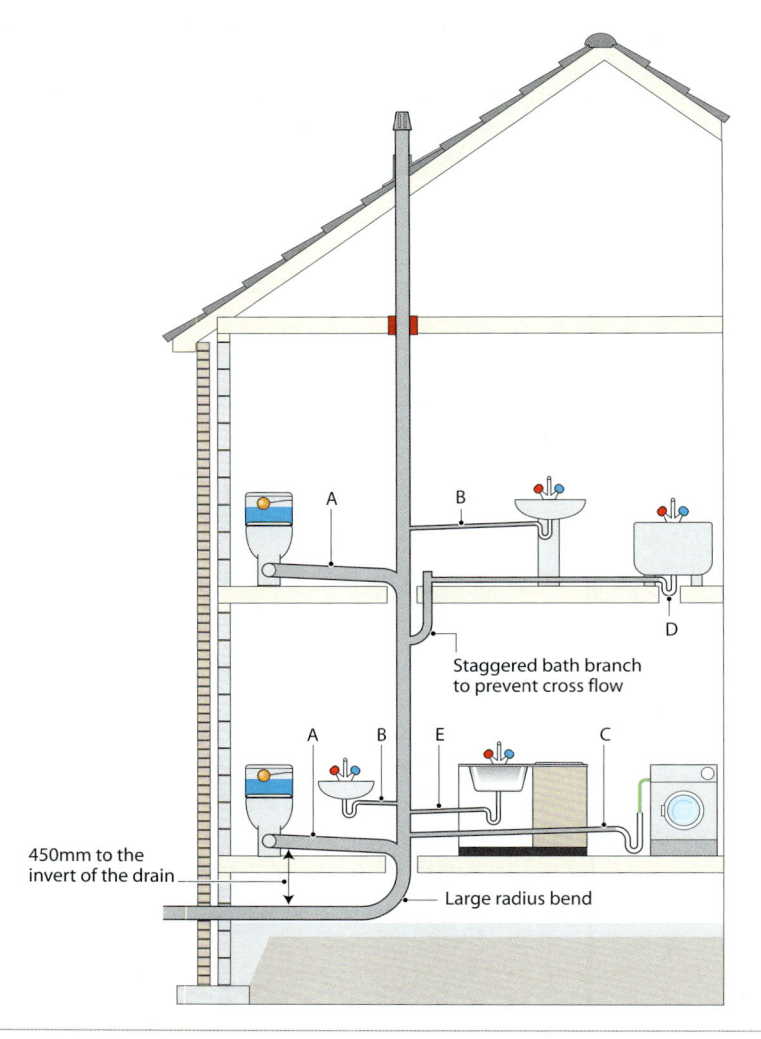

Primary ventilated stack system

Waste pipe sizes and lengths

Waste pipes need to fall away from appliances with enough of a fall for the water to reach what is known as a self-cleansing velocity. The fall is known as the gradient.

The table below shows the size of waste pipe for a given appliance installed on a primary ventilating stack and its maximum length and gradient:

Appliance	Pipe size (mm)	Max. length (m)	Gradient (mm/m)	Trap seal depth (mm)
WC branch (A)	75–100	6	18	50
Wash basin and bidet (B)	32	1.7	18–22 (see graph)	75
Washing machine/dishwasher (C)	40	3	18–90	75
Bath (D)	40	3	18–90	50
Kitchen/utility sink (E)	40	3	18–90	75
Where these lengths are exceeded, the next pipe size up should be used, and 40mm appliances will need to increase to 50mm pipe, the length and gradient of which is listed below.			
Appliances with 50mm waste pipe		4	18–90	75

The rules regarding the gradient for wash basins are slightly different from other appliances. If the maximum length of 1.7m is used, the gradient is 18–22mm/m. For shorter lengths, the gradient can increase and a gradient graph can be used to calculate the gradient needed.

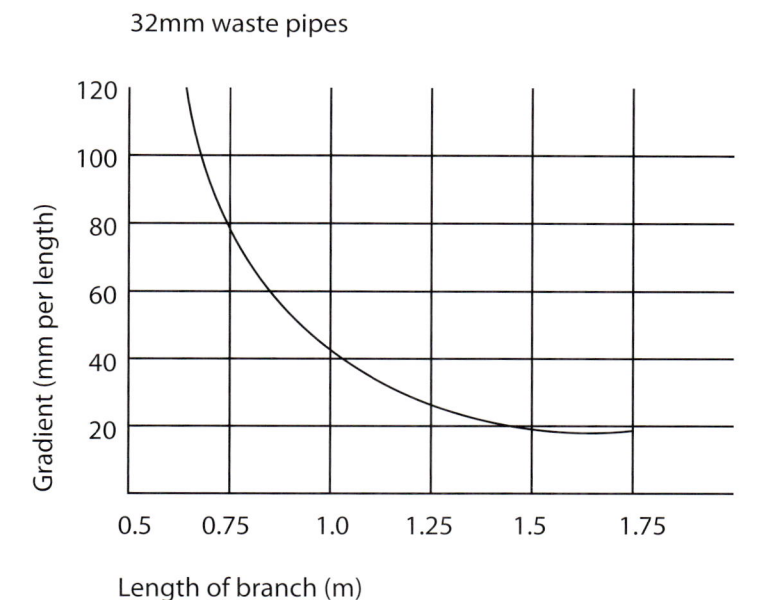

> **SUGGESTED ACTIVITY...**
> Try the gradient curve for yourselves. Determine the gradient of the following waste pipe lengths:
>
> a A waste pipe has a length of 1.5m. What is its gradient?
>
> b A waste pipe has a length of 750mm. What is its gradient?
>
> c A waste pipe has a gradient of 120mm/m. What is its length?
>
> Answers: a. 19mm/m b. 79mm/m c. 0.65m

Gradient graph for wash basin

Reading the graph is a simple task. The vertical line is the length of the waste pipe. The horizontal line is the gradient. So, decide on the length, trace the line up until it meets the curve and follow it across to the left-hand side to read the gradient. For example:

If a 32mm waste pipe is to be installed that is 1m in length, the gradient will be 40mm/m.

Branches at the base of the primary ventilated stack system

For systems up to five storeys high, the distance between the lowest branch connections and the invert of the drain should be at least 750mm. This can be reduced to 450mm for low-rise single dwellings up to two storeys high, such as a house or bungalow. For multi-storey systems, the ground floor appliances should be connected to their own stack or drain but not into the main stack. For buildings that have more than 20 storeys, the ground and the first floors should be connected in this way.

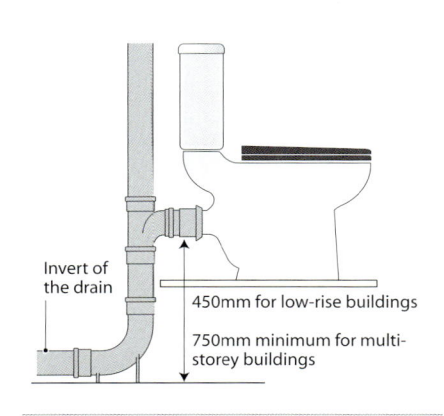

Branch connections at the invert of the drain

Bends and offsets

Bends at the base of discharge stacks should be large-radius with a minimum radius of 200mm. Two 45° bends can be used as an alternative. This is to create a smooth flow of water and solid waste into the drainage system. Tight bends can cause a problem known as compression, where the water hitting the bend forces a shockwave of air upwards, which can blow the water out of waste pipe traps, causing them to lose their seal and let obnoxious smells into the dwelling.

Long radius bend

Double 45° alternative

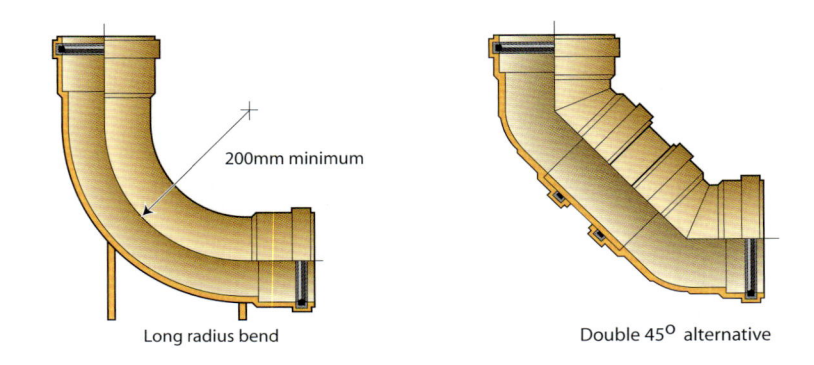

Large-radius bends at the base of the stack

Offsets in the wet part of the stack should be avoided if possible. Where there is no option, large-radius bends should be used with no branch connections within 750mm of the offset. If an offset is to be placed in the wet part of a soil stack in a building of up to five storeys, the stack must be ventilated both above and below the offset.

Branch connections for waste pipes

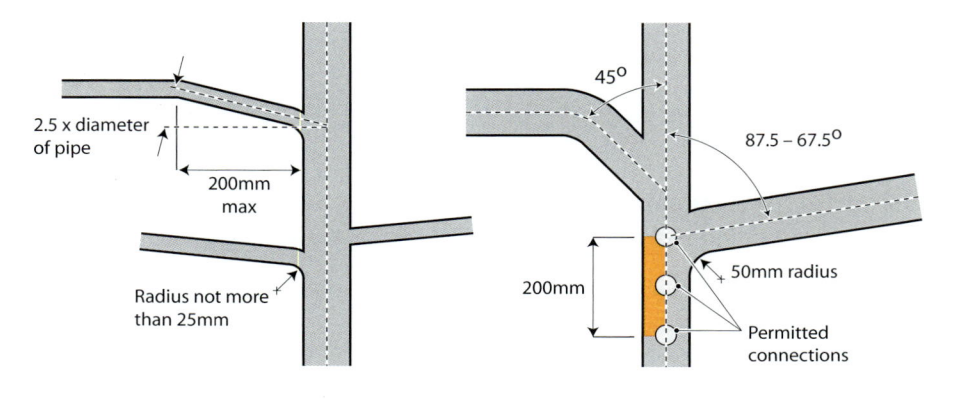

Branch connections

The illustration shows that junctions, including branch pipe connections of less than 75mm, should be made at a 45° angle or with a 25mm bend radius. The prohibited zone shows the distance (opposite the WC connection) in which a branch pipe may not be connected to a distance of 200mm. Branch connection pipes of over 75mm diameter must either connect to the stack at a 45° angle or with a minimum bend radius of 50mm.

Prevention of cross-flow

A branch pipe should not discharge into a stack in such a way that it could cause cross-flow into any other branch pipe, as this could cause loss of trap seal by effluent backflowing up the opposite connection.

The illustration shows the areas of a soil stack where branch connections directly opposite each other are restricted. There are several rules:

(a) Restriction connection area on stack

(b) Examples of permitted connections

(c) Opposing waste pipes

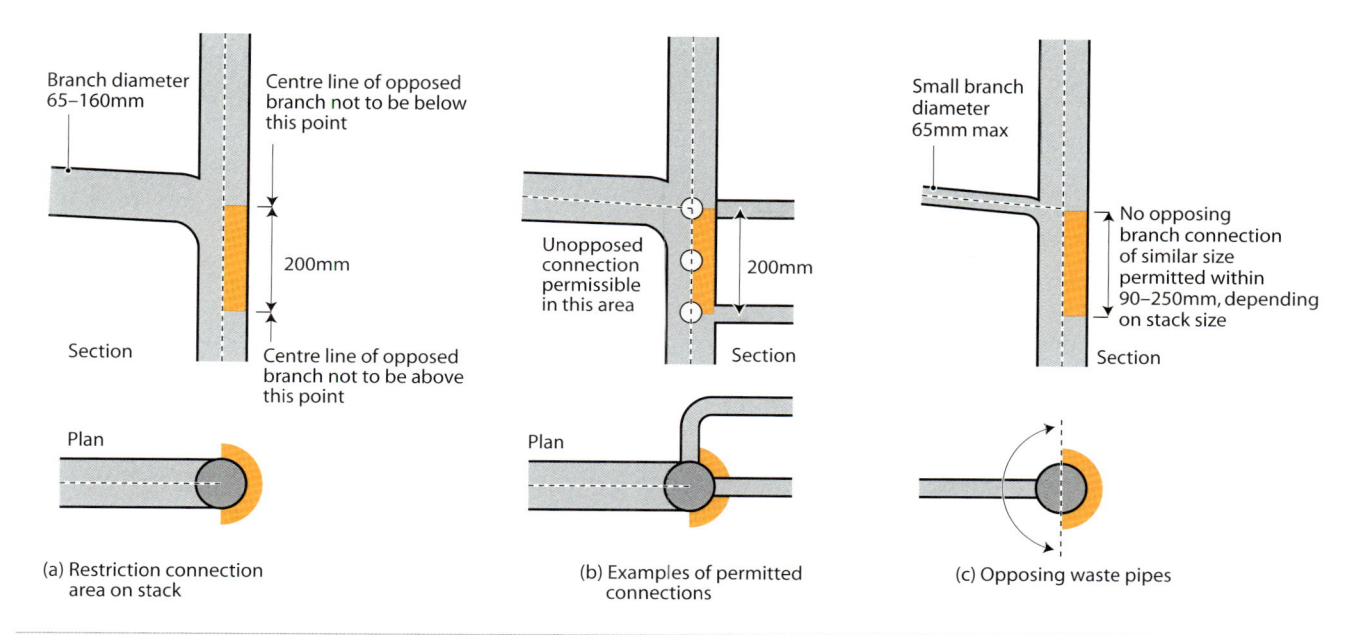

Preventing cross-flow

- Where a branch connection into a stack is between 65mm and 160mm in diameter, such as a WC branch, no other connection may be installed opposite for a distance of 200mm vertically downwards (see diagram (a) above).

- Side connections at 90° to the branch are allowed (see diagram (b) above).

- Where the branches are of similar size (for example, two 40mm connections), the restricted distance will depend on the size of the main stack (see diagram (c) above):

  - on a stack up to 65mm in diameter, no connection is allowed for a distance of 90mm
  - on a stack up to 110mm in diameter, no connection is allowed for a distance of 110mm
  - on a stack up to 160mm in diameter, no connection is allowed for a distance of 250mm.

Where it is not possible to meet the branch requirements of the primary ventilated stack, such as excessive waste pipe lengths, extra ventilation to the system will need to be added to safeguard the trap seal. This can be done in two ways:

- the ventilated discharge branch system, where each waste pipe branch is separately ventilated
- the secondary ventilated stack system, where the waste stack is directly ventilated.

C2 VENTILATED BRANCH DISCHARGE SYSTEM

The ventilated branch discharge system is used on larger systems where there is a risk of trap seal loss because the waste pipe lengths are excessive. Control of the pressure in the waste pipe (the discharge branch) is achieved by ventilating it no further than 750mm from the appliance; this safeguards against trap seal loss by induced or self-siphonage. Alternatively, small air admittance valves may be used at each appliance. These allow air into the system when the appliance is in operation.

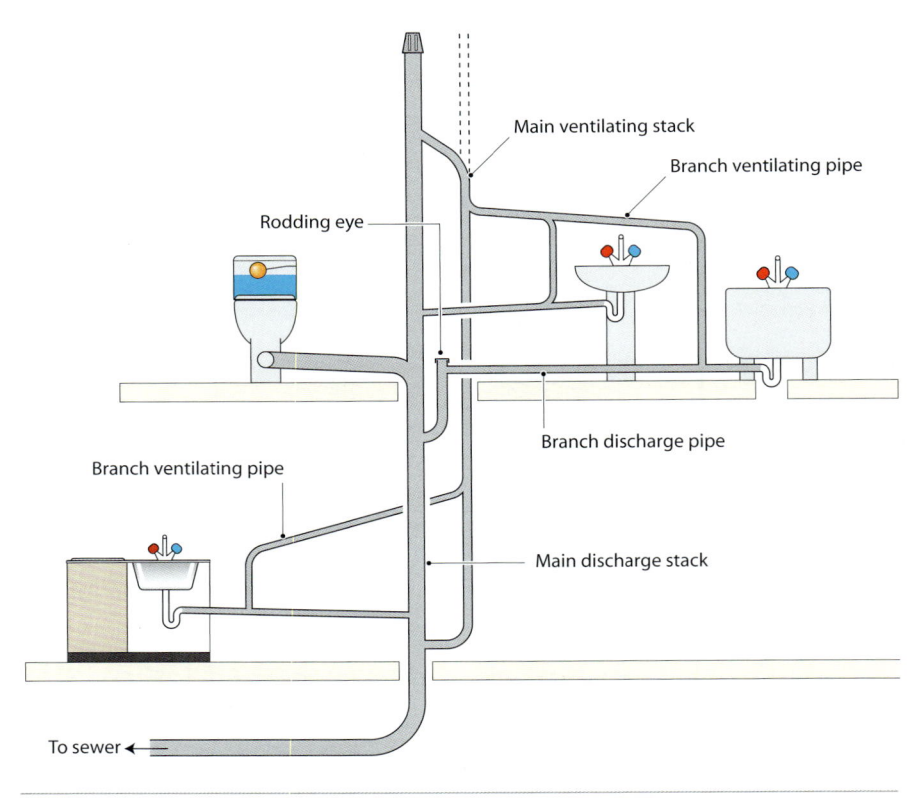

Ventilated branch discharge system

C3 SECONDARY VENTILATED STACK SYSTEM

With this system, only the main discharge stack is ventilated. The secondary ventilated system arrangement safeguards against positive and negative pressure fluctuations.

Where branch ventilating pipes must be installed, the following rules apply:

- Any branch ventilating pipe must be connected to the discharge stack above the spill-over level of highest appliance fitted to the stack. The ventilating pipe must also rise away from the appliance.

- The minimum size of any ventilating pipe to a single appliance is 25mm. However, if is longer than 15m or the ventilating pipe serves more than one appliance, the size must be 32mm.

- The main ventilation stack must be a minimum of 75mm. This also applies to the dry part of the primary ventilating stack.

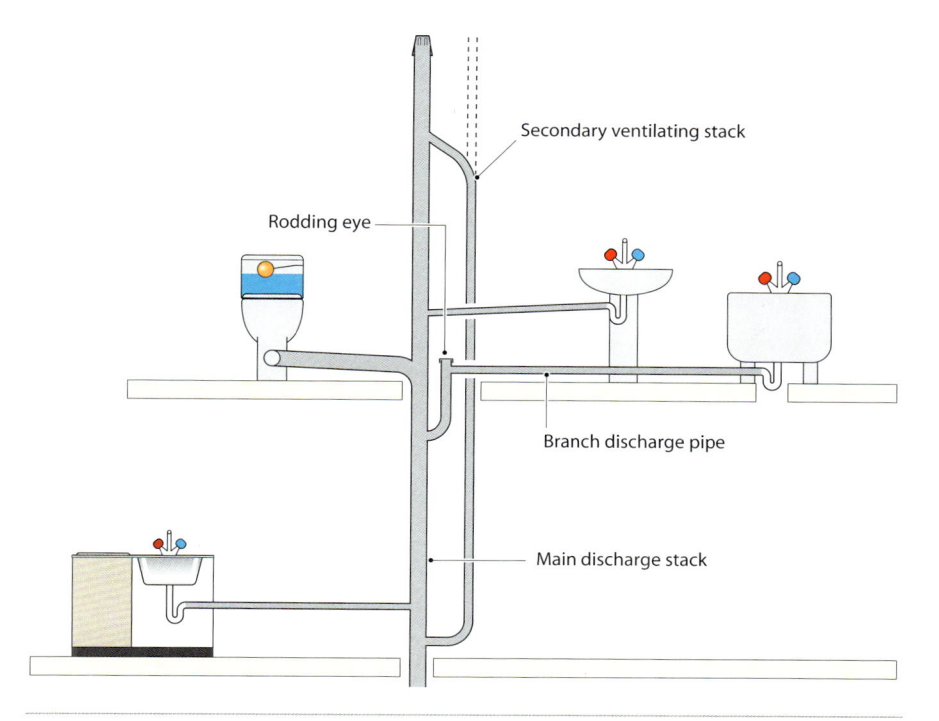

Secondary ventilated stack system

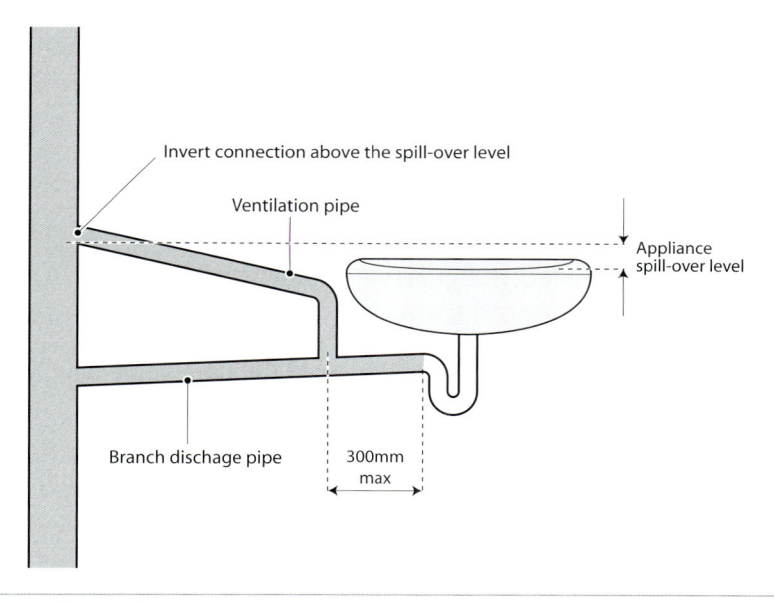

Branch ventilating pipe rules

C4 STUB STACK SYSTEM

When a group of appliances is connected directly to the drain, under certain circumstances, a 110mm stub stack may be used. The illustration overleaf shows a typical ground floor stub stack. Ventilation is required when the connection from the invert of the drain to the highest connection of an appliance to the stack exceeds 2m, or the WC crown connection to the invert of the drain exceeds 1.3m. Ventilation of a stub stack is via an air admittance valve.

An air admittance valve allows air into a stub stack to prevent the loss of trap seals. The subsequent suction action, when an appliance is used, opens the valve. This stabilises the condition of the air in the stack

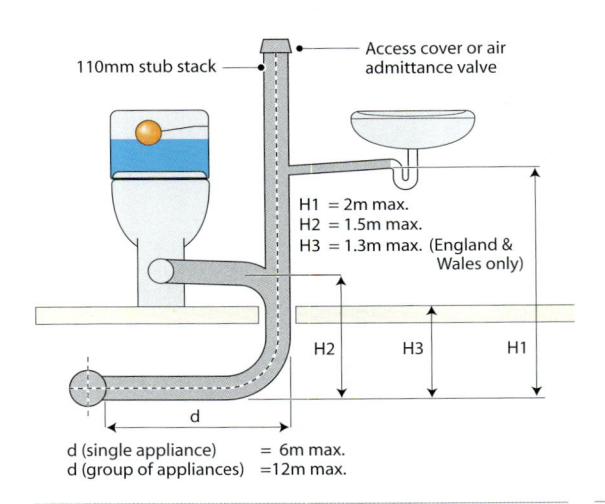

110mm stub stack — Access cover or air admittance valve

H1 = 2m max.
H2 = 1.5m max.
H3 = 1.3m max. (England & Wales only)

d (single appliance) = 6m max.
d (group of appliances) = 12m max.

Stub stack

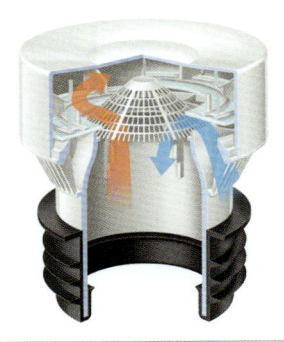

Operation of air admittance valve

because air is sucked into the stack through the valve, which has a secondary effect of not letting smells or foul air out. When the appliance has finished its operation the valve closes, preventing smells escaping into the space where the valve is installed.

Air admittance valves should be fitted in an uninhabited area such as a roof space. This minimises the risk of freezing while still keeping the valves accessible. On no account should they be fitted outside, due to the risk of freezing up in the closed position during cold weather unless the valve is specifically designed and manufactured for outside use. If air admittance valves are installed within a boxing, the boxing must be ventilated. In all cases the valve must be accessible for repair or replacement.

An important point to remember is that air admittance valves are not a substitute for ventilation stacks; any drain where an air admittance valve is fitted will still require conventional venting at some point. This is simply to minimise the effects of back pressure, which could occur if the underground drainage system were to become blocked. The requirements are that one stack in five must be ventilated to the atmosphere using a conventional ventilation stack, and that this should usually be done at the head or start of the drain run. The general rules are:

- Up to four domestic properties of no more than three storeys high can be ventilated using air admittance valves.

- Where an underground drain serves more than four properties fitted with an air admittance valve, the following rules apply:

  - where five to ten buildings exist, additional conventional ventilation stacks must be installed at the head of the drain run

  - where 11 to 20 buildings exist, additional conventional ventilation stacks must be installed at the head of the drain and at the midpoint in the run of the drain

  - all multi-storey domestic properties will require additional conventional ventilation if more than one property is fitted with an air admittance valve and is connected to a common drain, which is not ventilated by a conventional ventilation stack.

C5 GENERAL SANITARY PIPEWORK REQUIREMENTS

As well as the requirements we have already looked at, sanitary pipework systems should follow these general rules:

- Where a ventilation stack is installed within 3m of an opening window, the stack should be installed at least 900mm above the window.

- A cage should be fitted to the top of the vent pipe to prevent birds nesting at the top of the stack. Birds' nests have the effect of blocking off the air supply to the stack, causing waste pipes to lose their trap seal.

- A vent cowl should be fitted in exposed or windy positions. This prevents a condition known as wavering out, whereby the wind blowing across the top of the stack causes the trap water to move from side to side, which could potentially result in trap seal loss by the momentum of the water.

- Access should be provided above the spill-over level of the highest appliance to allow for clearing blockages.

- When installing a soil stack for waste pipes only, the size of the stack must be at least the same size as the largest trap or branch connection to it.

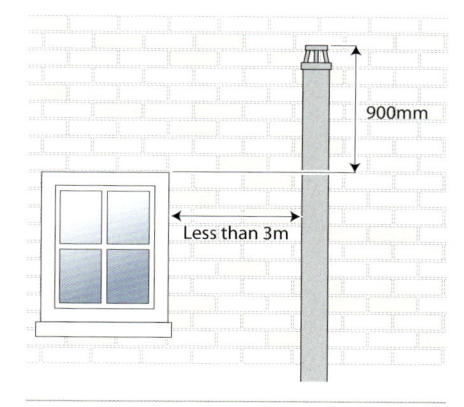

Position of a vent stack next to an opening window

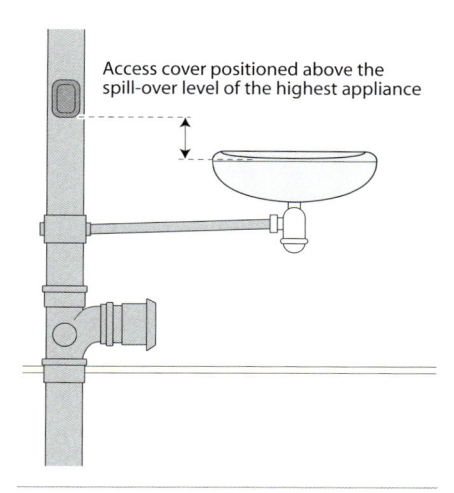

The position of access for clearing blockages

D TRAPS

The purpose of a trap is to stop obnoxious smells from entering the dwelling. There are many different types of traps to suit numerous appliances and applications. Traps are generally manufactured from polypropylene to BS EN 274:2002 for domestic applications but can also be made from copper or brass. Both of these materials can be chrome plated where aesthetics dictate a luxury finish and are usually used with chrome-plated copper waste pipe. Jointing methods include push-fit type joints and compression-type with a rubber compression ring.

Trap depths and sizes

Where a trap diameter is 50mm and above, the trap seal must be 50mm, such as the traps in WC pans. There are two reasons for this and both are reliant on the cohesive quality of water:

- A trap with a diameter of 50mm and over contains more water than, say, a 32mm or 40mm diameter trap. This makes the water much more difficult to move by induced siphonage, wavering out or compression.

- Because of the pipe size, it is unlikely that an appliance will discharge at full bore. If a pipe runs at full bore it will try to pull air along with it. If there is no air to pull, the water in the trap is pulled instead until the trap is empty and the pipe can pull air, thus breaking the siphonic action.

KEY POINT

Water has both cohesive and adhesive qualities and these were explored in Chapter 004, page 167. Take a look back to remind yourself about them.

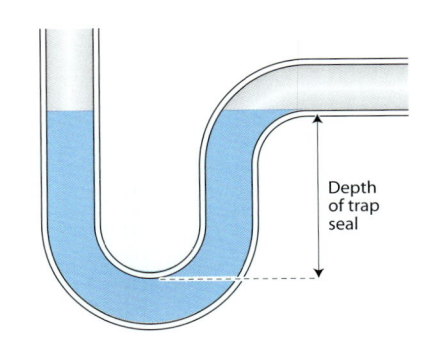

Trap seal depth

Where a waste pipe runs into a hopper head or a gulley, the trap depth can be reduced to 50mm for wash basins, kitchen sinks and electrical appliances such as washing machines and dishwashers, and 38mm for baths and shower trays. The reason for the bath trap difference is that baths and shower trays are large, flat-bottomed appliances, which by their nature discharge water more slowly than a wash basin, for example. The flat bottom of the bath means that the last drops of water run away more slowly than the water from a wash basin, and so the seal is retained.

The trap seal depth required for different appliances is shown in the following table:

Appliance	Waste fitting size	Diameter of trap	Trap seal depth when fitted to a primary ventilated system (BS EN 12056-2)
Wash basin	1¼ in	32mm	75mm
Bidet			
Bath	1½ in	40mm	50mm
Shower			
Bowl urinal		40mm	75mm
Washing machine		40mm	75mm
Dishwasher			
WC pan	N/A	75mm	50mm
		100mm	50mm

D1 TUBULAR TRAPS

Tubular traps can take several different forms:

- **Swivel traps** – These are often used on new work and appliance replacements because they have a union connection in the centre. This allows the trap to swivel through 360°, which facilitates multi-positioning and allows many different pipe connection options. They can be either P-trap, S-trap or running trap types.

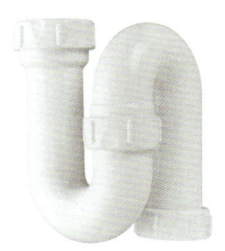

P [left] and S [right] type swivel traps

P-trap and S-trap are so called because of their shapes. A P-trap is used where the waste pipe is installed from the appliance horizontally directly through the wall and into a gulley or stack. The S-trap will allow pipework to be installed vertically downwards from the trap into a waste pipe serving a number of appliances or into an under-floor waste pipe.

- **Running traps** – The idea behind these is that an appliance or group of appliances can be trapped away from the appliances themselves, the trap being installed on the waste pipe run. They are sometimes used where space to install a trap at the appliance is limited.

Running trap

- **In-line traps** – These are specifically designed with wash basins in mind; an in-line trap is basically an S-trap where both the inlet and outlet are in line. They allow wash basin wastes to be totally hidden behind a pedestal, but can be restrictive and tend to block easily.

In-line trap

- **Washing machine traps** – As the name suggests, these are generally used for appliances such as washing machines and dishwashers. Generally of P-trap configuration, they have an extended neck to accommodate a washing machine/dishwasher outlet hose.

Washing machine trap

- **Bath traps** – Two different types are available. One is a swivel type with a 50mm trap seal and the other has a 38mm trap seal. They are specially made to be fitted in the restricted space under a bath or shower tray.

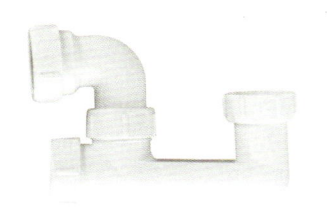

Swivel-type 50mm seal bath trap

38mm seal bath trap

D2 BOTTLE TRAPS

Often used on wash basins because of their neat appearance, bottle traps can be very restrictive to the flow of water. There are certain appliances to which a bottle trap is not suited, such as a kitchen sink or a urinal, as they block so easily. Regular trap cleaning is important to maintain an adequate water flow. There are two main types:

- **Bottle traps** – These are used with wash basins and bidets. Access for cleaning is via the bottom of the trap, which unscrews to facilitate removal of blockages.

Bottle trap

- **Shower traps** – Although not strictly a bottle trap, the operating principle of the shower trap is exactly the same. The main difference is that the trap seal depth is much less than 75mm and access for cleaning is through the grille on the top of the trap rather than underneath.

Shower trap

D3   ANTI-VAC AND RESEALING TRAPS

There is no substitute for a well-planned, well-designed system of sanitary pipework. If the system is designed in accordance with BS EN 12056, problems with trap seal loss should not occur. Anti-vac (anti-vacuum) and resealing traps are not an alternative to a good sanitary pipework system, but they can alleviate the problems that occur with existing systems due to upgrades to appliances and additions to the system. Fitting an anti-vac trap is not recommended on new systems as they have a habit of not holding an air test at the installation and testing stage.

Anti-vac traps

This type of trap uses a small air admittance valve, which is located after the water seal. The valve opens if a drop in pressure occurs when the appliance is used, and this allows air into the system to break any siphonic action that may occur.

Resealing traps

Externally, these are identical to a normal bottle trap. Internally, however, they are quite different. The trap has a bypass within the body of the bottle trap, which allows air to enter the trap via a dip pipe in the event of siphonic action occurring. They are only available as bottle traps, so are not suitable for all installation situations.

D4   SELF-SEALING WATERLESS WASTE VALVES

A self-sealing valve is a waterless valve that uses a thin neoprene rubber membrane to create an airtight seal, stopping foul air from entering the dwelling and maintaining equal pressure within the soil and vent systems.

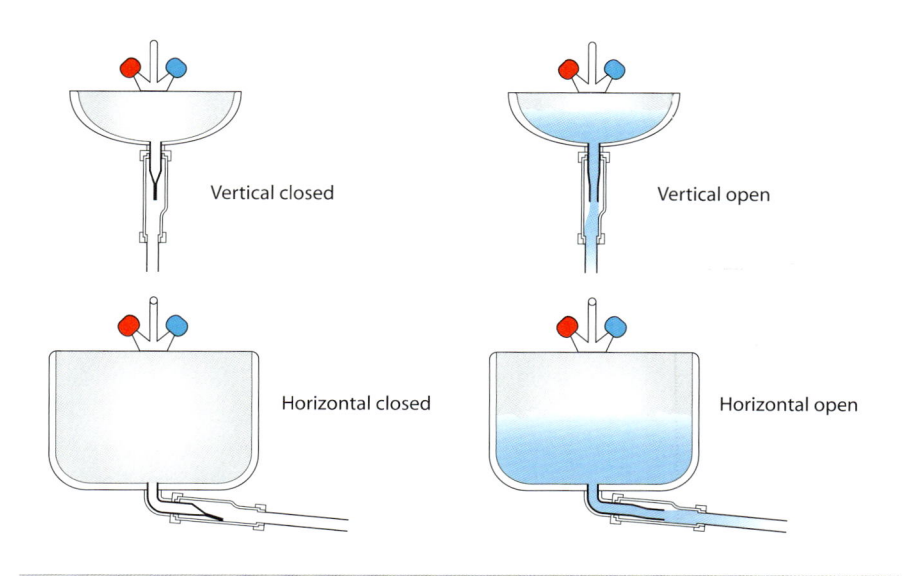

Operation of self-sealing valve

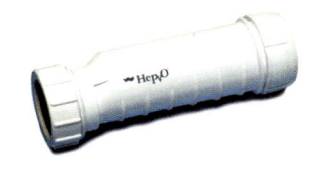

Self-sealing valve

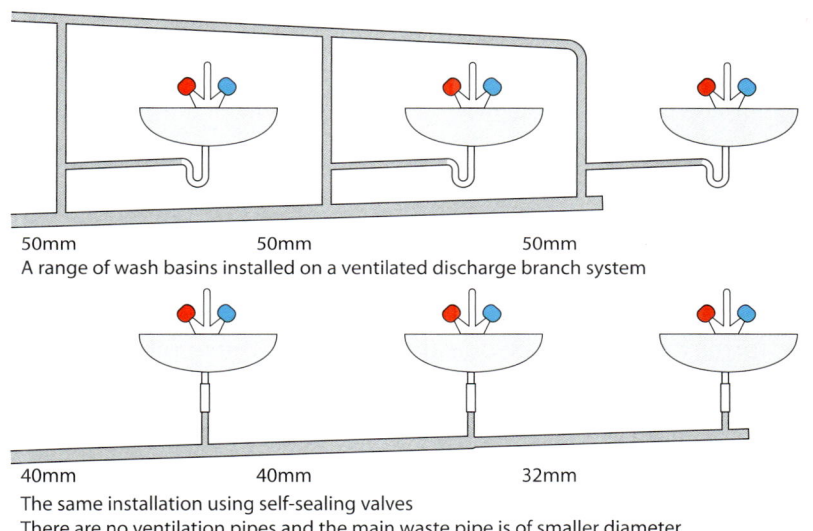

50mm 50mm 50mm
A range of wash basins installed on a ventilated discharge branch system

40mm 40mm 32mm
The same installation using self-sealing valves
There are no ventilation pipes and the main waste pipe is of smaller diameter

Multiple installations of self-sealing valves

The membrane opens under the pressure of water from an appliance, closing again when the water discharge has finished.

It is so effective that it can be used safely on primary ventilated stack systems and ventilated discharge branch systems. A self-sealing valve has certain advantages over a conventional trap:

- The valve removes the problems associated with negative pressure within a system by opening to allow air in, in much the same way as an air admittance valve. This creates a state of equilibrium within the system and means that air admittance valves and extra vent pipes are not required.

- Because there is no water in the valve, the problems of self-siphonage and induced siphonage are eliminated.

- The valve operates silently. This eliminates the noises generally associated with water-filled traps.

- The valve allows a greater number of appliances to be installed on the same discharge system without the risk of compromising system efficiency.
- The valve can withstand back pressures equivalent to ten times greater than those experienced in a typical sanitary pipework system.

D5 LOSS OF TRAP SEAL

Provided we follow the recommendations in BS EN 12056:2002, we should not experience any problems of trap seal loss. Most trap seal problems occur even before water has been let down the trap, simply because they can be attributed to design and installation issues with the sanitary pipework system. When loss of trap seal occurs, obnoxious smells will permeate the dwelling. Most trap seal problems can be traced back to the following faults:

- waste pipes that are too long
- waste pipes that are too small for the appliance
- waste pipes that are laid to an incorrect fall
- incorrect bends at the foot of the soil stack
- too many appliances on the same waste branch
- too many changes of direction.

The next part of the chapter looks at the various ways in which trap seal loss occurs.

Self-siphonage

Self-siphonage occurs when water is discharged from an appliance. The water forms a plug that, as it disappears down the appliance waste, creates a partial vacuum in the waste pipe between the plug of water and the water in the trap. This then pulls the water from the trap.

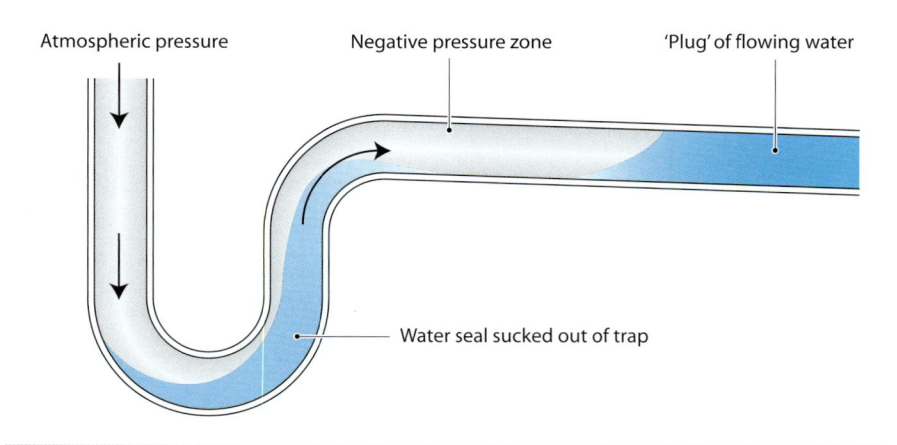

Self-siphonage

In most cases, self-siphonage will not occur if the waste pipe length is kept within the recommended lengths of BS EN 12056. If it does occur, the installation of a vent on the waste pipe branch may be necessary or an anti-siphon trap could be fitted.

Self-siphonage is most common on wash basins due to the rapid evacuation of the water from the bowl and the small size of the waste pipe.

Induced siphonage

Induced siphonage can occur by one appliance causing the loss of trap seal of another appliance connected to the same waste pipe. When water is discharged down an appliance, the water in the previous trap can be drawn out by negative pressure. A similar effect can happen with an appliance further downstream, as the plug of water passes the branch connection.

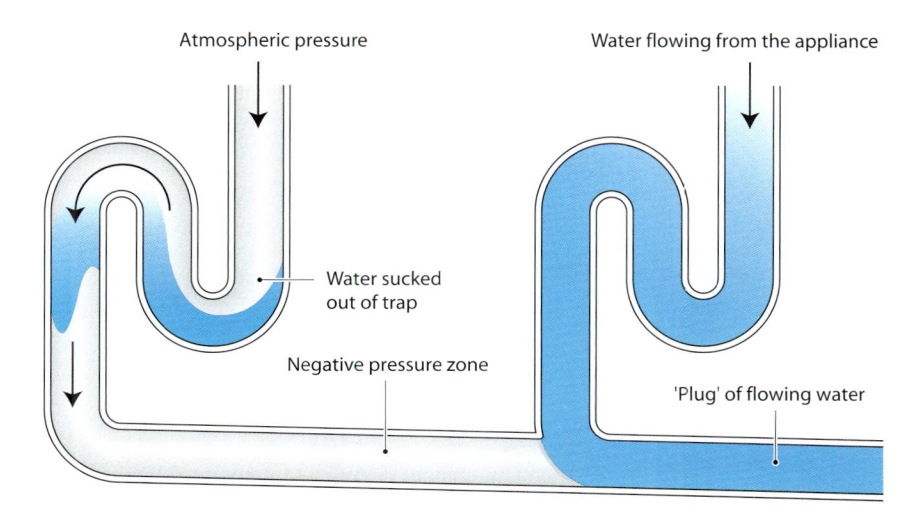

Induced siphonage

Connection of two appliances onto a single internal waste pipe should be avoided. As water is drained from one appliance it can induce a vacuum (through siphonage) that will empty the trap of the connected secondary appliance. This is a regular occurrence where a bath and wash basin are connected to the same waste pipe. For this reason, the pipe size leading to the final branch connection should be increased to 50mm.

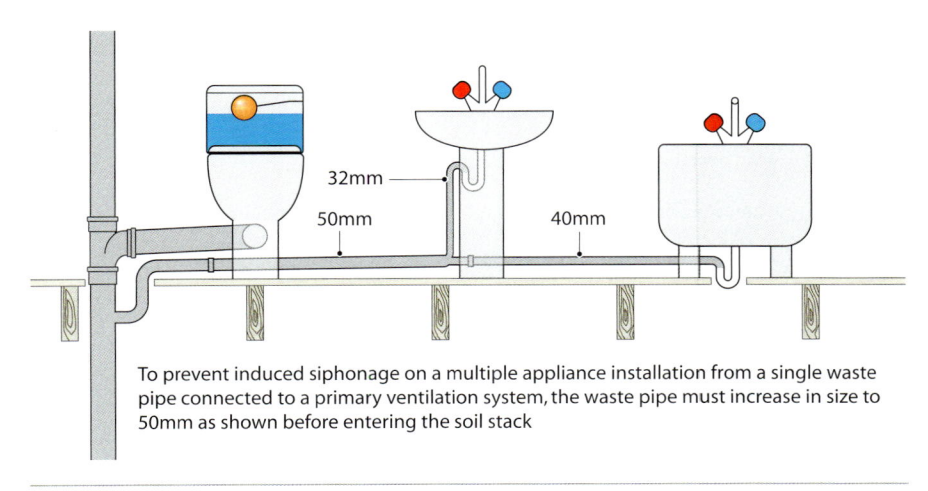

To prevent induced siphonage on a multiple appliance installation from a single waste pipe connected to a primary ventilation system, the waste pipe must increase in size to 50mm as shown before entering the soil stack

Multiple appliance installations

Compression

When water is discharged from a WC at first floor level, it falls rapidly to the base of the stack. If the bend at the base of the stack has a tight radius, the water momentarily stops flowing, causing the water to back up, which creates a back pressure of air. The back pressure travels up the stack and moves through ground-floor waste pipes, eventually blowing the water out of the traps.

The use of large-radius bends or two 45° bends at the base of the stack prevents this from happening by allowing the easy flow of water from the soil stack to the drain, allowing the water to maintain its forward motion and velocity.

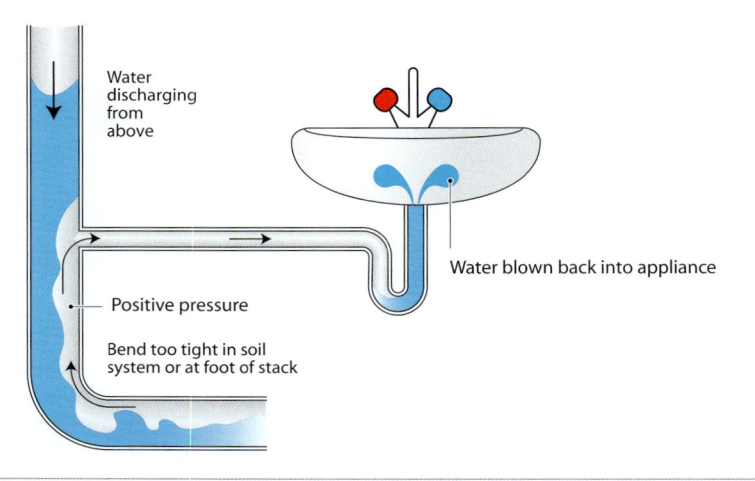

Compression

Wavering out

Wavering out is caused by a natural phenomenon – the wind. In high winds or exposed positions the effect of the wind blowing across the top of the vent pipe will cause the water in the traps of appliances to move with a wavelike motion because of pressure fluctuations. This momentum can often cause water to be pulled over the weir of the trap, resulting in trap seal loss. It can be prevented by fitting a wind cowl to the top of the vent pipe.

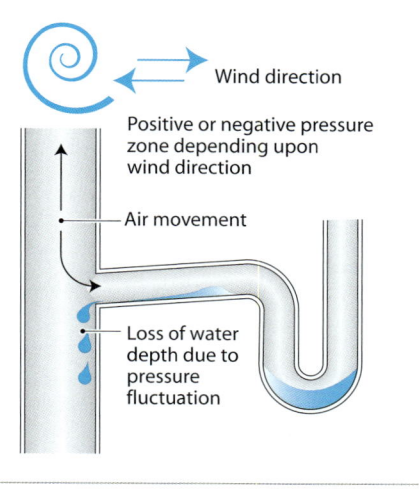

Wavering out

Evaporation

This is a natural form of trap seal loss caused by lack of use of the appliance. Traps, to some extent, rely on the appliance being regularly used to keep them topped up with water. When an appliance is not used, the water in the trap will begin to evaporate until it is all gone. The rate of evaporation can vary but on average it is about 2.5mm of trap seal per week, increasing when the weather is hot and dry.

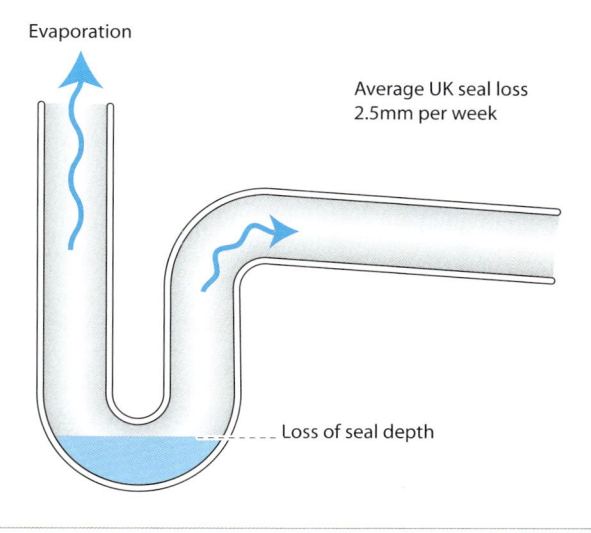

Evaporation loss from trap seal

Capillary action

Capillary action generally only occurs in S-traps when long fibres or long hair gets lodged over the weir of the trap. Capillary action draws the water out of the trap and down the lodged material, and the trap water slowly drips away.

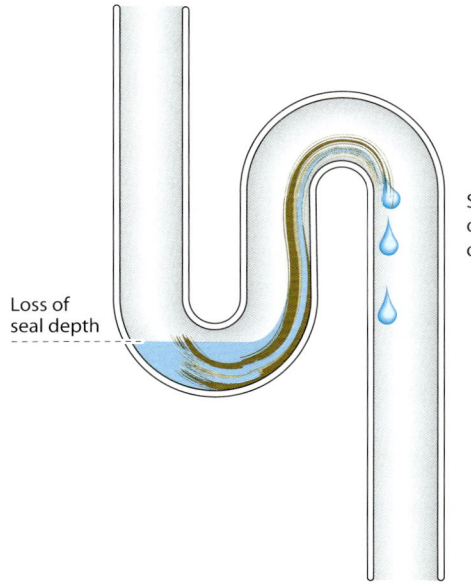

SUGGESTED ACTIVITY...
Remember – capillary action was covered in Chapter 004. To refresh your learning, take another look at pages 166–67.

Capillary action loss from trap seal

Momentum

Loss of trap seal by momentum only occurs when a large amount of water is suddenly discharged down the trap of an appliance. The force of the water will dislodge most of the water in the trap in a similar way to self-siphonage.

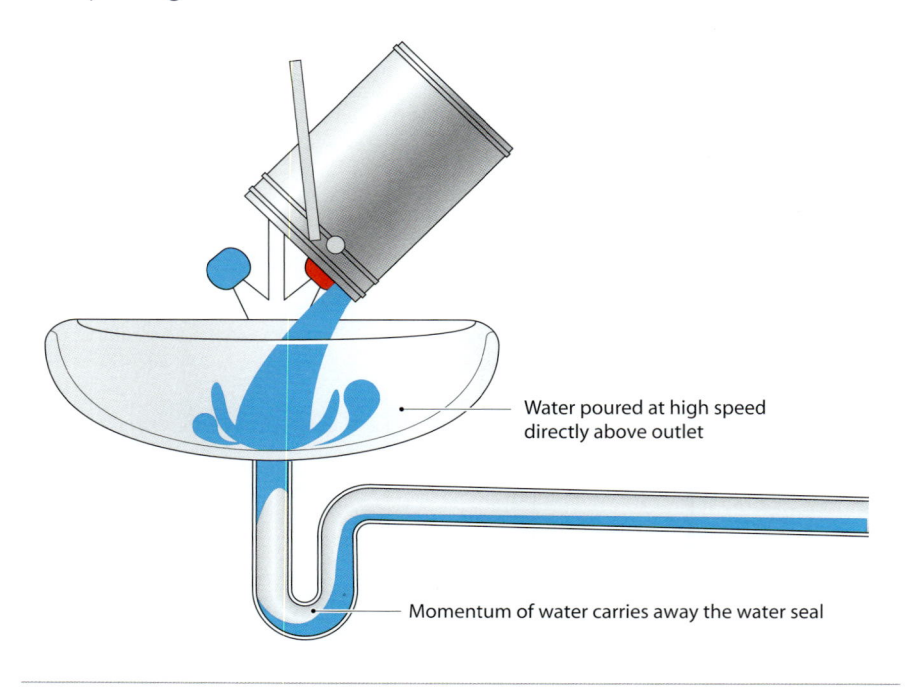

Water poured at high speed directly above outlet

Momentum of water carries away the water seal

Momentum loss from trap seal

Foaming

Foaming is a direct result of too much detergent being used. The resultant excess foam can back up waste pipes and soak away the water in the trap. It can usually be detected by the appearance of foam emerging from traps in appliances.

E INSTALLING ABOVE-GROUND DRAINAGE SYSTEMS (AGDS)

Remember that the installation of AGDS is covered both by the Building Regulations Document H3 and BS EN 12056. These important documents restrict the systems we install and the materials we use to ensure that hygienic conditions are maintained in dwellings and buildings at all times. The way we install them is also an important issue and is subject to a code of practice: BS 8000-13: Workmanship on building sites. Code of Practice for above ground drainage and sanitary appliances: 1989.

E1 PREPARATION BEFORE INSTALLATION

On new-build installations such as multi-dwelling housing developments, the position of the soil and vent pipework will be determined by the drain connection installed to the Architect's drawings. Any preparation work required to allow the installation of the sanitary pipework should be agreed with the relevant trades beforehand. For instance, on some sites, any hole-drilling required in brickwork or timberwork is undertaken by the building or joinery contractors. This must be completed prior to installation to avoid unnecessary and costly delays; it is a good idea to check the preparation work to ensure that pipe and fitting clearances are adequate.

E2   TYPES OF MATERIALS

Generally, the materials used on modern AGDS are made from plastic. The range of plastics used was comprehensively covered in Chapter 005 and is briefly summarised in the table below:

Material	BS number	Characteristics
PVCu	BS EN 1329-1	These three materials can be either solvent welded using solvent cement to BS ISO 6209:2009 for waste pipes from 32mm to 50mm diameter, or push-fit and solvent welded for soil and vent pipes from 82mm to 160mm diameter. Push-fit soil and waste fittings should be to BS EN 1329-1. Pipe is available in lengths of 2.5m, 3m or 4m, either plain-ended or socket- and spigot-ended.
muPVC	BS EN 1566-1	
ABS	BS EN 1455-1	
Polypropylene	BS EN 1451-1	Polypropylene is a push-fit waste system with sizes ranging from 32mm to 50mm diameter. It cannot be solvent welded and is identifiable by a warm but slightly greasy feel to the pipe. It is more flexible than PVCu or ABS and does not break or shatter.

The choice between push-fit and solvent weld waste pipes and fittings is down to personal preference, although on some housing contracts, solvent weld will be specified. Each system has its benefits and drawbacks:

Waste pipe type	Advantages	Disadvantages
Push-fit	Easy to make a watertight joint. Pipe is light and easy to install. Easy to take apart for unblocking and maintenance. Joints allow movement for thermal expansion and multi-positioning. Can be tested immediately after jointing.	Pipe tends to sag if not correctly clipped. Joints can pull apart easily causing unsuspected leaks. Suffers from UV light degradation so may require painting if installed outside.
Solvent weld	The pipe is much more rigid than polypropylene pipe and does not suffer as much from sagging. Neater appearance. Joints will not push apart. Will resist most acids, alkali and chemicals.	Joints are permanent and will not allow for repositioning. Joints must be left for a period of time before testing can begin. Fumes from the solvent cement can be damaging to health. Suffers from UV light degradation so may require painting if installed outside.

> **SUGGESTED ACTIVITY...**
>
> For more information on soil and waste pipes and fittings, revisit Chapter 005 or check out manufacturers' literature. Most manufacturers produce fittings catalogues, and these are available from your local plumbers' merchant or for download as PDF files from the manufacturers' websites. Check out the following:
>
> www.hunterplastics.co.uk
> www.hepworth.wavin.com
> www.osma.co.uk
> www.marley.co.uk
> www.polypipe.com

There is a full range of fittings available for both polypropylene and PVCu and some of these were looked at in Chapter 005. It may be a good idea to keep a fittings catalogue to hand when working on site so that you are aware of the full range of fittings available.

The fixing details for polypropylene and PVCu (ABS and muPVC) were covered previously in Chapter 005, pages 248–49 and 253.

Cast iron

Cast iron was used for many years in both domestic and industrial installations. Now, it is restricted to large installations and public buildings such as hospitals. You still may be required to work on cast iron, especially when refurbishing existing dwellings.

Cast iron has the advantage of being very robust but it is also very heavy and difficult to work with. The jointing system is much easier than it used to be. Today, cast iron is jointed using a special jointing system called Timesaver, which is simply bolted together using special torque wrenches so that the joints are not over-tightened.

E3 WASTE PIPE CONNECTIONS TO THE SOIL STACK

Waste pipe connections to the soil stack can be made in two ways:

- **By the use of a boss pipe** – These can be push-fit or solvent cement types. Each connection for the waste pipe will need to be drilled out using an appropriate-size hole saw beforehand and the correct insert for the waste pipe size used.

- **By the use of a strap boss** – solvent welded onto the soil pipe.

Both of these were featured in Chapter 005, page 247. Care should be taken to ensure that they are installed the right way up because both boss pipes and strap bosses have a slight gradient in the moulding to ensure the correct fall for the waste pipe. The method of installing a strap boss is as follows:

1 Determine where the strap boss is to be installed and mark the centre of the hole.

2 Using the correct size hole saw and a cordless drill, drill the hole for the strap boss, ensuring that the lip on the inside face of the strap boss fits snugly inside the hole. It is important not to have too much play in the hole as this may result in leakage once the solvent cement has set.

3 Clean around the hole and the surface of the strap boss with cleaning fluid.

4 Apply solvent cement to the strap boss first and then around the hole on the soil pipe.

5 Place the strap boss in position, insert the nut and bolt* at the back of the boss and gently tighten.

6 Clean away any excess solvent cement with a clean dry cloth.

Boss pipe adaptor

7 The boss must be left for at least five minutes for the solvent cement to cure enough for testing to be carried out.

   * Not all strap bosses have nuts and bolts to keep them in place. Some just clip together to make a watertight seal. All strap bosses, however, must be solvent welded.

Waste pipes that are to be installed on an internal soil stack can use a waste pipe manifold. This is an adapter that allows multiple waste pipe connections and avoids problems with cross-flow exclusion zones.

Waste pipe manifold

E4   ACCESS TO PIPEWORK

Access to AGDS pipework for cleaning, clearing blockages and maintenance is a requirement of the Building Regulations. There are many ways that we can fulfil this requirement, for example:

- by the use of access plugs inserted into soil pipe junctions and waste pipe tees
- by the use of purpose-designed access covers and fittings.

Access plugs in soil junctions	Access plugs in waste pipe	Purpose-made access fittings

E5   SOIL STACK CONNECTION TO THE DRAIN

The connection to the drain could be in one of several materials depending on the age of the building and its use. Older properties tend to have salt-glazed earthenware drains, and public buildings often use cast iron drains. Connection to these materials is usually by a collar, which is sealed with a sand and cement mortar joint. Modern houses use either PVCu or HepSleve® clay piping. The jointing methods to these materials are shown below:

PVCu	HepSleve®	Salt-glazed earthenware	Cast iron
These two sockets are simple push-fit types. The soil pipe should be chamfered and silicone lubricant applied before inserting into the socket.		These sockets require jointing with a strong 2:1 ratio sand and cement mortar. They should be left for 24 hours before testing is carried out.	

Multi-fit pipe adapters are also available for connecting different pipe materials below ground. The image shows a multi-fit adapter that can be used to connect PVCu to cast iron, earthenware or HepSleve® pipework.

Multi-fit adaptor

The drain connection, as we have already seen, is made to a large-radius bend. On new dwellings the position of the drain connections will be marked on the building plan. On older buildings you will have to use what is already in place, so careful consideration should be made of the method of jointing you are going to use. If the soil stack is external (outside the building), an access pipe can be used as the drain exits the ground. On internal soil stacks, access must be above the spill-over level of the highest appliance.

Because we are connecting to the below-ground drainage system, it is important to have some knowledge of the types of systems that we are likely to come across. The next section of the chapter looks at the various types of below-ground system and their basic layouts.

F BELOW-GROUND DRAINAGE SYSTEMS – AN OVERVIEW

A below-ground drainage system takes the soiled water (also known as black water) and rainwater away from the dwelling and deposits it into a main sewer in the road. From here it will flow to the sewage plant. There are three systems of below-ground drainage:

- the separate system
- the combined system
- the partially separate system (sometimes called the partially combined system).

F1 THE SEPARATE SYSTEM

This is the system that is favoured by the Local Authorities. With this system, foul water and rainwater flow into separate drainage systems. These are then connected to a separate foul water sewer and surface water drain in the road. The foul water from WCs, baths, wash basins and kitchen sinks is conveyed to the sewage treatment plant and the rainwater flows to the nearest water course. The layout of the design is shown below:

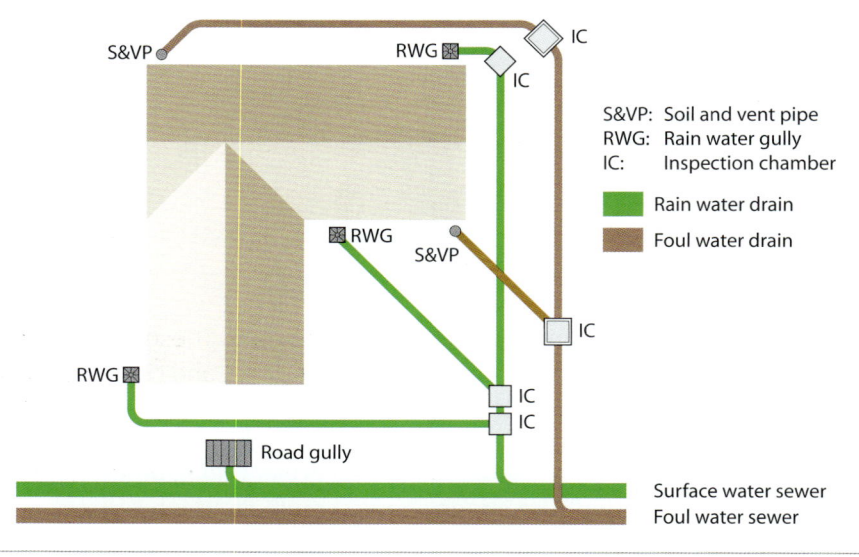

Separate drainage system

Advantages and disadvantages of the combined system

ADVANTAGES	DISADVANTAGES
▪ Because the drains are separate, the sewage plant does not get inundated with water when it rains heavily. ▪ Trapped gullies are not required for the rainwater connections. This helps to identify the drainage system in use.	▪ It is an expensive system to install because two drains are required. ▪ The foul water drain does not get flushed and cleaned out by the rain. ▪ There is a risk of making incorrect connections to the rainwater drain. ▪ The number of inspection chambers required is greater.

F2 THE COMBINED SYSTEM

With this system both foul and rainwater drains discharge into a common sewer. This makes connections to the drains much simpler. It is a simple and economic system to install. However, this system is no longer recognised by the Building Regulations as a viable system on new installations.

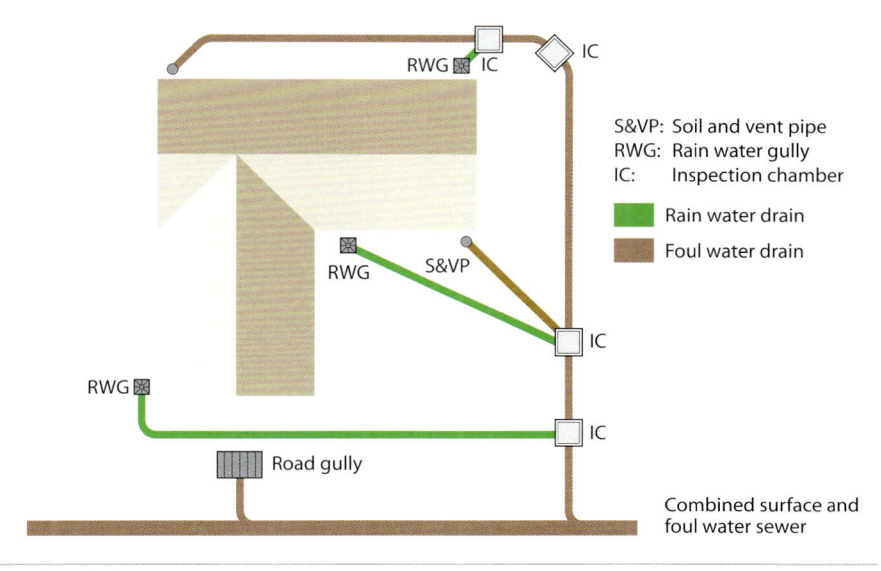

S&VP: Soil and vent pipe
RWG: Rain water gully
IC: Inspection chamber

■ Rain water drain
■ Foul water drain

Combined drainage system

Advantages and disadvantages of the combined system

ADVANTAGES	DISADVANTAGES
▪ Maintenance of the drains is much easier. ▪ It is a cheaper system to install. ▪ It is impossible to connect to the wrong drain. ▪ All drains are flushed out when it rains.	▪ All discharge must pass through the sewage treatment plant, which is expensive and difficult to handle during heavy rainfall.

F3 THE PARTIALLY SEPARATE SYSTEM

This system is a compromise between the separate and the combined systems. Two drainage systems are used: one that carries part of the rainwater discharge from the roof and one that carries foul water and part of the rainwater discharge. Soakaways are also used with this system to collect water from a roof not connected to the surface water drain. It is also known as the partially combined system.

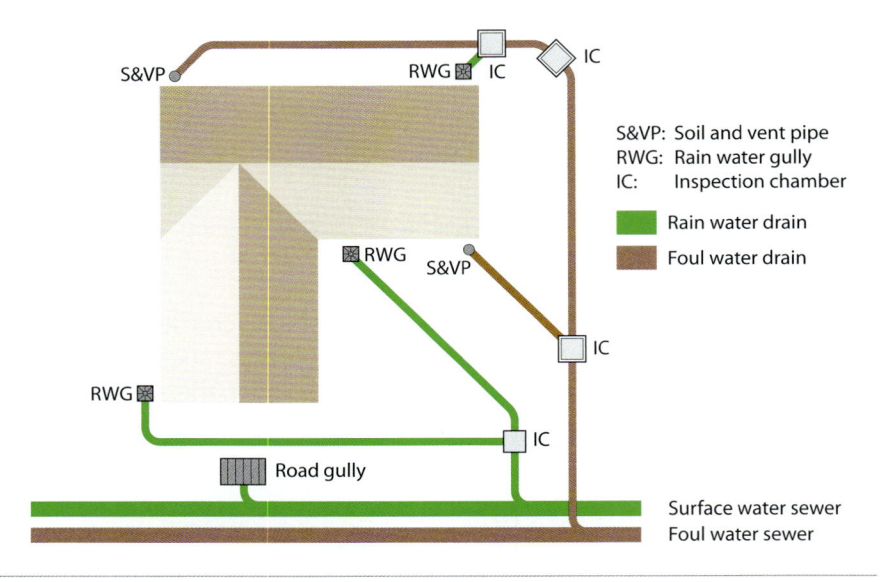

Partially separate drainage system

A soakaway is a pit, usually 1m × 1m × 1m, dug into the ground and filled with gravel where the rainwater pipe discharges into. It allows rainwater to soak naturally away to the water table. Soakaways should be situated at least 5m away from the property.

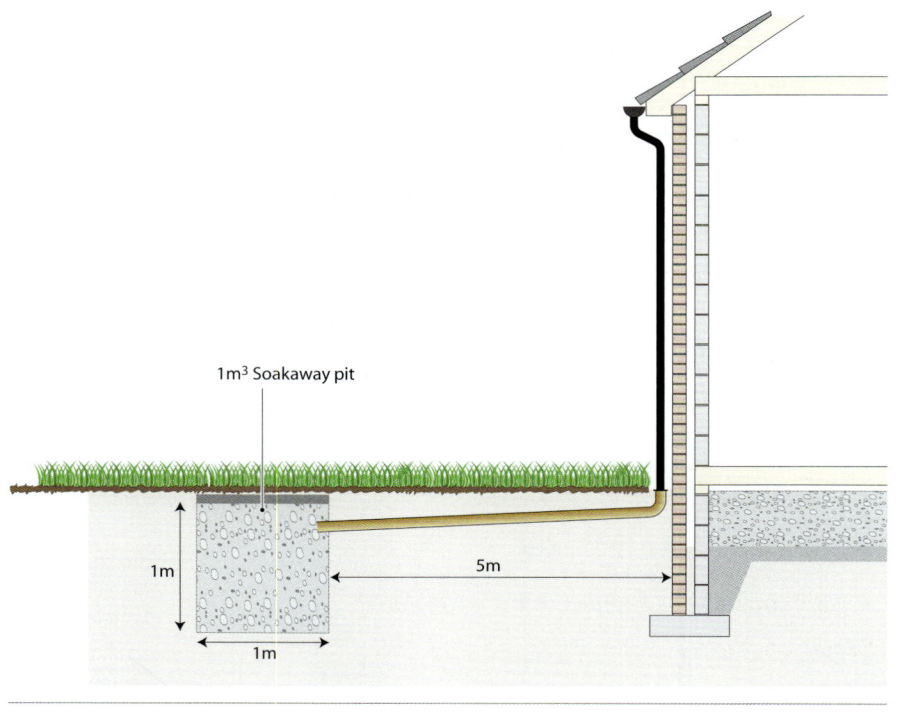

Soakaway

Advantages and disadvantages of the partially separate system

ADVANTAGES	DISADVANTAGES
▪ It can reduce costs by allowing isolated rainwater connections to the foul water drain. ▪ Rodding eyes can be used at strategic points instead of costly inspection chambers.	▪ Care must be taken when installing foul water outlets to ensure the correct system is used.

F4 APPLIANCE CONNECTIONS TO EXISTING BELOW-GROUND DRAINAGE SYSTEMS

The method you use for connecting appliances to below-ground drainage systems will depend on the appliance and the material that the below-ground drainage pipework is made from.

Waste pipes up to 50mm diameter

Appliances such as kitchen sinks, cleaners' sinks and wash basins may discharge directly into the back-inlet gully of a below-ground drainage system. The waste pipe must discharge below the grating but above the water line in the gully. This ensures that an air break is maintained and that no smells can enter the building.

Waste pipes discharging into back inlet gullies

WC connections to ground floor drains

Where a WC is to be connected to a ground floor drain, this can be done simply by the use of a WC pan connector. Pan connectors are available in a number of lengths and outlet sizes to suit 75mm to 110mm drainage systems and to fit both modern and existing WC pans.

WC pan connectors

G BATHROOM LAYOUT SPECIFICATIONS

Sanitary appliances within a dwelling should be installed so that the minimum amount of space is provided for each appliance for both personal use and for an adult to supervise the bathing and washing of children. British Standard BS 6465-2:1996 Code of Practice for space requirements for sanitary appliances recommends the minimum space required by each appliance for adequate usage.

British Standard BS 6465-1:2006+A1:2009 Code of Practice for the design of sanitary facilities informs us that there must be a minimum amount of appliances within a dwelling based on the number of people occupying the property:

Sanitary appliance	Number per dwelling	Notes
WC	One for up to four people Two for five or more people	There should be a wash basin adjacent to every WC in the property.
Wash basin	One	
Bath or shower	One for every four people	
Kitchen sink	One	

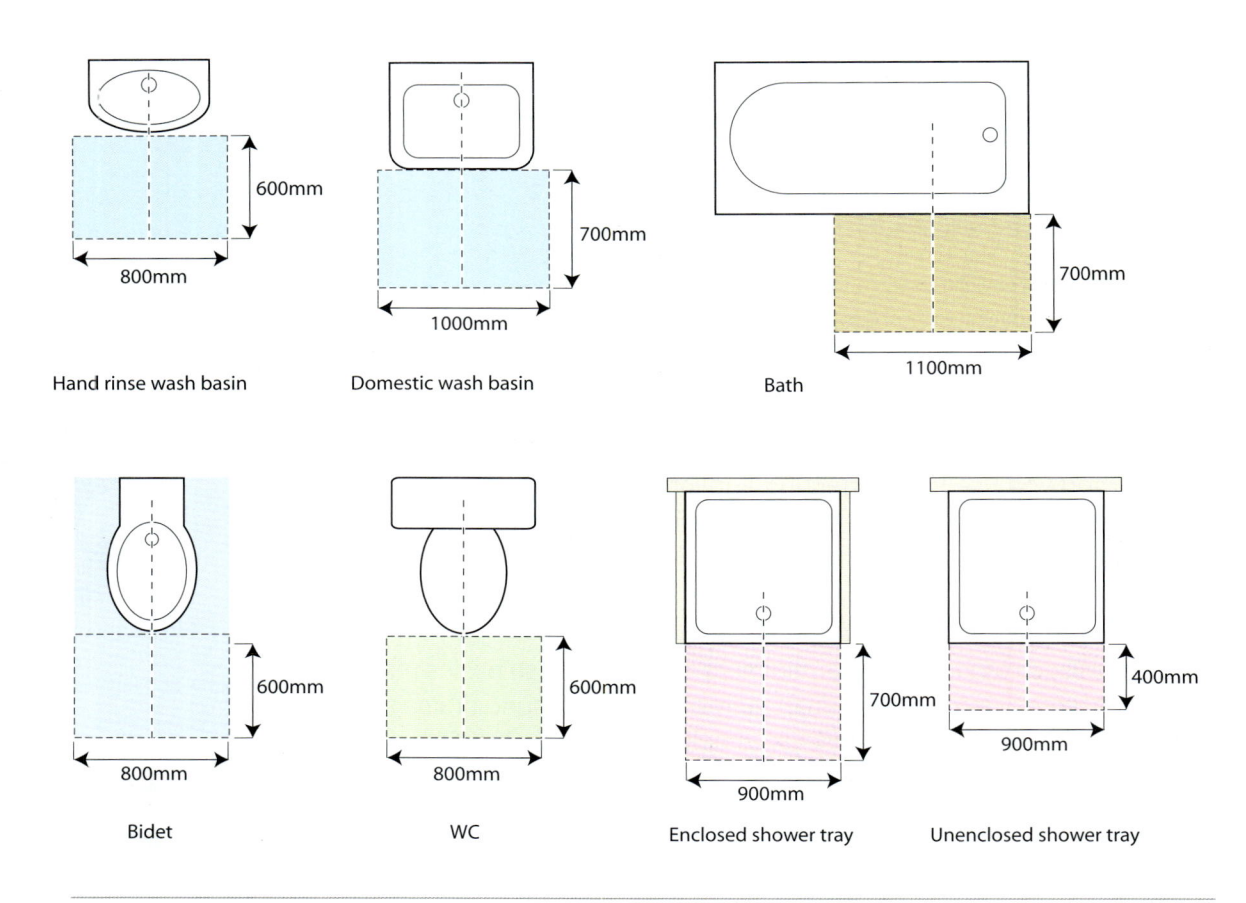

Space provision for sanitary appliances

- **Hand rinse wash basin**: 800mm × 600mm
- **Domestic wash basin**: 1000mm × 700mm
- **Bath**: 1100mm × 700mm
- **Bidet**: 800mm × 600mm
- **WC**: 800mm × 600mm
- **Enclosed shower tray**: 900mm × 700mm
- **Unenclosed shower tray**: 900mm × 400mm

In some cases it is not possible to maintain these distances, especially when the bathroom is small. In these situations, the British Standard allows overlap of the appliance space.

Overlaps also occur in cloakrooms and downstairs WCs.

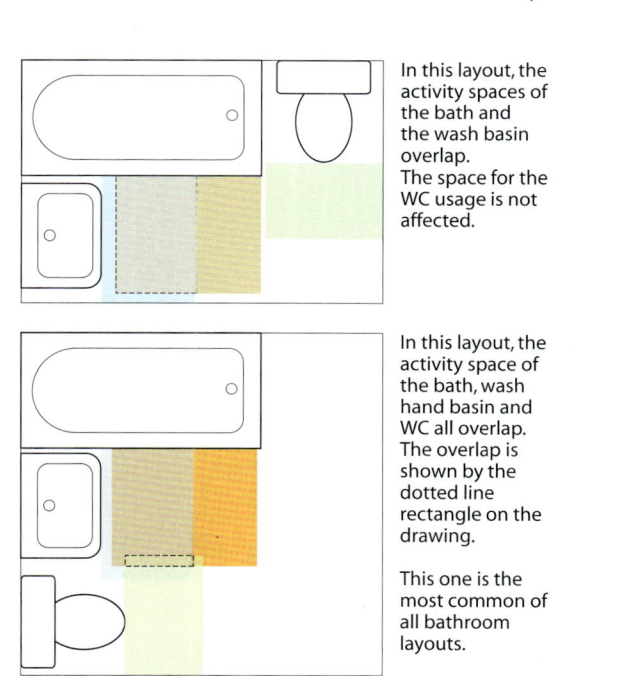

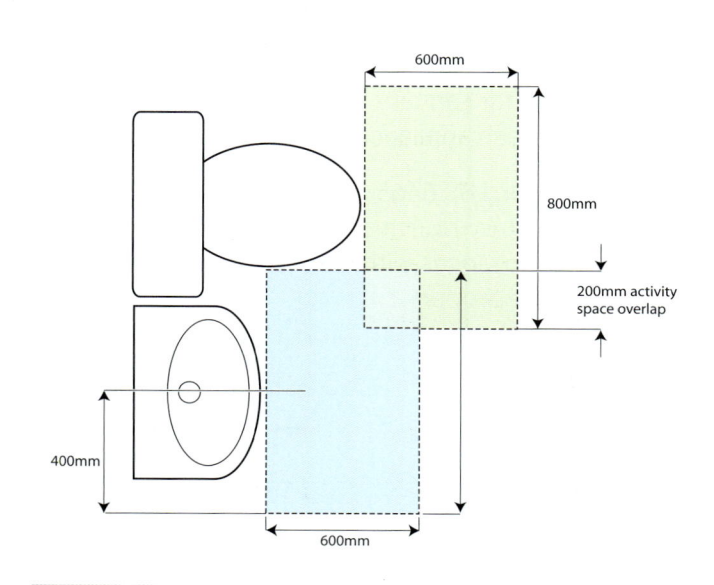

In this layout, the activity spaces of the bath and the wash basin overlap. The space for the WC usage is not affected.

In this layout, the activity space of the bath, wash hand basin and WC all overlap. The overlap is shown by the dotted line rectangle on the drawing.

This one is the most common of all bathroom layouts.

Overlap in a downstairs WC: 600mm × 800mm, 400mm, 600mm, 200mm activity space overlap

Overlap of the appliance space in a bathroom

Overlap in a downstairs WC

H SANITARY APPLIANCE INSTALLATION

Before installing the sanitary appliances, a risk assessment will be required. The nature of bathroom suite installations means that we may come into contact with previously used sanitary equipment, which can be detrimental to health. It is important to use the correct PPE, such as rubber gauntlets and goggles, when removing old sanitary appliances and altering the soil and waste pipework.

H1 PREPARATION

Good preparation for the installation of sanitary appliances is essential as it is probably the most visual of all the installations that we undertake. Customers spend hundreds, sometimes thousands, of pounds replacing their bathroom suites and it is vital to get it right. British Standard BS 8000-13 gives essential guidance on the workmanship that is expected when installing above-ground sanitation systems and sanitary appliances. The preparations made before installing sanitary appliances needs very careful consideration. Good planning includes:

- making sure that the hot and cold pipework has been installed in accordance with drawings
- making sure that any chases and holes necessary have been prepared
- checking that sanitaryware has been delivered on time, is correctly stored and is free from damage. It is a fact that one in four bathroom suites delivered to site are either damaged, incorrect or have parts missing. These delays can be costly in terms of time and repeat customer business.

Remember:

- When materials are ordered, contact the merchant to ensure that the bathroom suite is going to be delivered to the correct address and on the correct day before you start the job.
- Always check the delivery note before signing to ensure that the equipment on the sheet is the same as that being delivered.
- Always handle sanitaryware with care. Most appliances are easily scuffed or damaged.
- When storing materials, ensure that the store is secure and that the materials have been stacked correctly.

Before the job commences:

- Ensure that the work area is completely clear of all debris.
- Ensure that the customers' carpets and furniture are protected.
- Ensure that you have all the manufacturers' instructions to hand. These will need to be left with the customer at the end of the job.
- Consult with the customer to ensure that there are no last-minute changes that may require the intervention of your supervisor.

H2 ASSEMBLING THE APPLIANCES

Assembling sanitary appliances is an activity that should be completed before the appliances are fitted; it is also known as dressing the appliances. Preparation of sanitaryware includes:

- **Installing the taps and wastes to the bath, wash basin and bidet** – Taps are a personal choice and will have been chosen by the customer with a lot of thought. Treat them with care to ensure that they are not damaged during the installation. Taps should always be fitted in accordance with the manufacturer's instructions. The washers provided for sealing the taps to the appliance should always be used and care taken to ensure that they are not over-tightened or you risk cracking the appliance. Never use bare grips to hold the tap, as this will mark the chrome/gold plate. Wastes will either be slotted for appliances with integral overflows, unslotted or a pop-up waste system. Wastes should be made into the appliance with silicone sealant or specific washers if the manufacturer provides them. If silicone sealant is used, try not to use too much as it is difficult to remove from the glaze of the appliance.

Taps being fitted to bath

Waste being fitted to wash basin

- **Assembling the WC cistern** – This means installing the siphon, float-operated valve, overflow (if applicable), flushing handle and close-coupling bracket (if the WC is a close-coupled model).

- **Carefully fixing the bath cradle and feet to the bath** – The bath should be carefully turned upside down on a clean dustsheet for this operation. The bath feet are adjustable to enable the bath to be fitted level and to the correct height. Great care must be taken here as the cradle comes with specific screws for different positions. If the wrong screws are used, you may pierce the bath itself. Always read the manufacturer's instructions beforehand.

Assembling a dual-flush WC cistern

Adjusting the bath feet

H3 THE INSTALLATION PROCESS

The installation process for bathrooms for refurbishments is quite different from new-build installations. We will only deal here with the installation of the three most common appliances:

- bath
- wash basin
- WC suite.

Before work begins, make sure the customer is aware that appliances will be decommissioned and that warning notices have been placed in strategic areas to prevent the accidental turn-on of water supplies and the unintentional use of partially fitted appliances such as WCs.

The protective tape around the sanitary ware should be removed before installation takes place so that the appliances can be checked for damage prior to being fitted.

New-build installations

On new-build installations, the choice of bathroom suite is often not as broad as it is for a private customer, especially on housing contracts where there are only two or three house styles being constructed. The work can become very repetitive, with the same suite types being installed time and again and always in the same positions. The appliance positions are set by the Architect and it is often difficult to deviate from these plans. It is usual for the first fix to have been installed beforehand with hot, cold and waste pipework tails visible. The hot and cold will have been hydraulically pressure tested previously.

Installing the bath

Although there are no set rules for the order in which appliances are installed, it is common practice to install the bath first as this is the largest of the appliances and is much easier to manoeuvre into position in an empty bathroom. The bath should be placed in position and the feet adjusted until it is level on all sides at the correct height to suit the bath panel (if one is being fitted). It may be a good idea to fix timber to the floor where the bath feet will sit as this helps to spread the weight of the bath, the water and the person using it. When you are sure that it is ready to be fixed to the wall, mark the brackets that hold the bath to the wall.

Remove the bath temporarily, and drill the fixing holes. The type of fixings you use will depend on the type of wall it is. For masonry, concrete block and thermalite block walls, wall plugs and brass screws may be used. Plasterboard studded walls will require plasterboard fixings unless wooden noggings have been placed in the wall previously. Fix the bath in its permanent position and, after checking once more to ensure correct level and height, screw the feet to the floor. Make sure that all the feet are screwed down as this is often missed and can cause bath movement later if not done correctly.

Once the bath has been fixed in place, it can be connected to the hot and cold pipework. How this is done will depend on the first fix pipework material. Polybutylene is by far the easiest material to work with but copper adds rigidity to the installation. It is a good idea to install service valves to both taps, as this will facilitate easier maintenance in the

Levelling the bath

future. Make sure that both tap connectors are fully tightened. The waste pipe to the bath can also be installed at this stage while all other appliances are out of position. It is often difficult to work under a bath, especially if the wash basin pedestal or WC pan is in the way. Ensure that the service valves are in the off position prior to commissioning.

When the bath is fixed it is normal practice for the bathroom to be tiled and grouted before any further appliances are installed.

Installing the wash basin

The wash basin often comes next. This can be a tricky installation. The centre line of the basin should be marked lightly in pencil on the tiles. This is usually the centre between the hot and cold pipework. Also, mark the centre on the wash basin itself. This will allow both centre lines to be lined up, ensuring that the basin is in the correct position for the pipework and the drawing specification.

Place the pedestal in position and gently lower the wash basin, complete with the waste trap fitted, onto the pedestal, ensuring that the centre lines match. Do not use any silicone on the pedestal face at this time. It is important first to ensure that the wash basin position is correct, that the basin and pedestal match properly and that the basin is level. Place a spirit level on the top of the wash basin and, once the appliance has been adjusted, mark the fixing holes underneath the basin and mark around the pedestal at floor level. This will ensure that both pedestal and basin go back into the same position once the wall has been drilled.

Carefully drill the tiles and the wall, ensuring that the fixing holes are deep enough to allow the wall plugs to be inserted below the tile surface. Reposition the pedestal in line with the previous floor mark. It is a good idea at this stage to put a thin bead of silicone sealant around the face (lip) of the pedestal where the wash basin sits. This will ensure that both wash basin and pedestal are fixed together once it has cured. Again, carefully reposition the wash basin and screw back to the wall using brass or stainless steel screws. Do not over-tighten the screws, or the fixing holes will break. Once again, check for level and clean any surplus silicone sealant from the pedestal.

The hot and cold pipework is placed behind the pedestal to hide it as much as possible. It is often difficult to install the pipework inside the pedestal itself. Any bends in the pipework need to be as high as possible so that they cannot be seen when a person is standing up. Do not be tempted to solder pipework joints close to the pedestal, or cracking of the pedestal (and wash basin) may occur. It is considered good practice to install service valves on the pipework. Ensure that both the tap connectors (or compression joints if a monobloc mixer tap is used) are fully tightened. The waste pipe can now be finished onto the previously fitted trap. Some pedestals have fixing holes at floor level; if these are present, carefully screw the pedestal to the floor. Ensure that the service valves are in the off position prior to commissioning.

Installing the WC suite

Most WC suites today are close-coupled style. The following procedure is based on this type of installation.

Wash basin fixing holes being marked

When installing the WC suite, the distance between the WC pan outlet and the wall should be measured so that the soil pipe can be trimmed to the correct length. The distance from the wall can also be obtained from the manufacturer's literature. Remember to put the pan connector on the pan outlet first so that an accurate measurement can be taken. Once the soil pipe has been cut to length, insert the pan connector into the soil pipework and carefully place the WC pan into position. Now place the cistern onto the pan and fix it using the nuts, bolts and washers provided. At this stage, it is better to step back from the pan and look to make sure that the pan and cistern sit correctly. Place a spirit level across the back of the cistern to ensure that it is level, and mark the cistern fixing holes with a pencil. If an overflow pipe is required this can also be marked. Remove the cistern and carefully drill the fixing holes, again ensuring that the holes are deep enough for the wall plugs to be pushed below the surface of the tiles. Carefully drill the hole for the overflow (if required).

Refix the cistern to the WC pan, ensuring that the large foam sealing washer that seals the cistern to the pan is in place around the WC siphon tail on the bottom of the cistern. Before screwing the cistern to the wall, it is worth considering putting spacing washers (tap washers will suffice for this) between the cistern and the wall. This helps to prevent the build-up of condensation at the back of the cistern by allowing air movement, and this in turn prevents the build-up of black mould on the tiles where the cistern is fitted. Also, to prevent breaking the cistern, place a tap washer over both brass screws before the cistern is screwed back to the wall. Once this has been done, you can then proceed to screw the WC pan down, again using the correct gauge and length of screw. Brass or stainless steel screws are best used in this situation to prevent the screw from corroding.

The water connection should be installed as neatly as possible as it will be on view all the time. It is a requirement of the Water Supply (Water Fittings) Regulations 1999 that the cold water supply to WC cisterns contains a service valve. Be careful when connecting the tap connector, as it is very easy to strip the thread of the plastic float-operated valve. Do not over-tighten the connector as this may also

Assembling WC pan and cistern

Screwing WC pan to floor

strip the thread. Ensure that the service valve is in the off position prior to commissioning.

Refurbishments of existing bathrooms

This is where you can show your creativity by designing bespoke installations to suit the customer's requirements. The customer may already have an idea of how they want their bathroom to look, so it is important to consider the ideas they may put forward. It may also mean that the original bathroom layout will be altered, with appliances occupying different positions than they did originally. We will presume here that the appliances are returning to their original positions and that the first fix pipework has been completed previously.

Although the method of installing the appliances is identical to the new-build installation, the order in which they are fitted might not be. In this situation, the customer cannot be without a WC, especially if the one you are replacing is the only one in the property. There are two choices:

- Leave the existing WC in place until all other appliances have been fitted – if the appliances are to installed in the same positions as the original bathroom suite, this is probably the better option as the WC will be eventually removed and any damage done to it while installing the bath will not matter. If the new WC was fitted first, damage could be potentially costly in terms of materials and labour charges.

- Replace the WC first – If the bathroom layout is being altered, with the WC occupying a new position, this will obviously be the only option, as the soil stack connection will need to be altered before the bathroom installation can begin. It is pointless installing the original WC on a new soil stack as this wastes precious installation time.

In both cases, the soil pipe to the WC should be blanked off when there is no WC fitted. This will prevent obnoxious smells from entering the work area. This can be done by the temporary use of a drain plug or PVCu cap end, both of which are shown below:

Drain plug	PVCu cap end

SOUNDNESS AND PERFORMANCE TESTING OF ABOVE-GROUND SANITATION SYSTEMS

Once the bathroom suite has been installed, you can think about testing the sanitary pipework. The testing of above-ground sanitation systems is the final part of the installation process. When testing sanitary pipework, there are two elements that we are looking at:

- ensuring that the pipework is sound and does not have any leaks
- ensuring that it performs to the recommendations of BS EN 12056.

I1 SOUNDNESS TESTING

Before testing begins, make a visual check of the system to ensure that it conforms to the British Standards and that you are happy with the clipping distances and that all joints appear to be made correctly. Testing should be completed in accordance with BS 12056-2:2000. If the system is installed in a multi-storey property the system may need to be tested in stages or floor by floor.

1 First, seal the pipework at the top and the bottom either by using drain plugs or drain testing bags. The bottom drain plug can be inserted through the access cover at the base of the stack.

2 Fill all the traps on the system by letting a little down each appliance and let a little water down the WC to cover the bottom plug. This will ensure the plugs' airtightness.

3 On the top plug, a rubber tube is fastened. The tube needs to have a T-piece inserted. On one side of the T is a hand pump and an air inlet valve, on the other side a manometer (a water gauge) is installed. The manometer is measured in millimetres.

4 The hand pump is pumped until a measurement of 38mm is reached and the air inlet valve is turned off. This 38mm is the maximum pressure that should be pumped into the system as the WC only has 50mm of water in the trap and any higher pressure than this will breach the trap. The 38mm test pressure must remain constant for a minimum of three minutes.

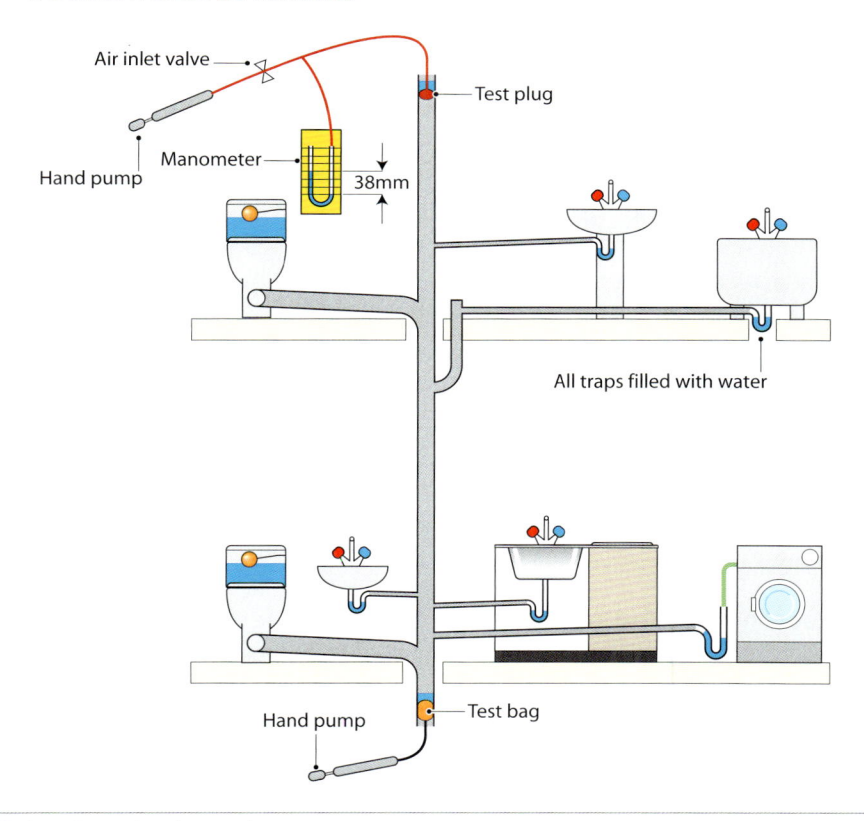

Test equipment and procedure

> **KEY POINT**
>
> The dipstick is a thin piece of wood painted matt black. The dipstick is inserted down the centre of the waste fitting until it reaches the bottom of the trap. When it is withdrawn, the wetness should be measured and the diameter of the trap deducted. What is left is the depth of trap seal.

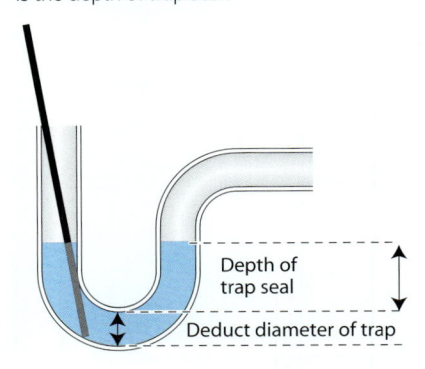

I2 PERFORMANCE TESTING

With the soundness test complete and the test equipment removed, performance testing can begin. This is done to confirm that the system meets the recommendations of the British Standards and the Building Regulations:

1 Fill all of the appliances with water up to their overflow levels and release the water from the appliances simultaneously.

2 At the same time flush the WCs.

3 When all of the appliances have emptied and the WC flushes have finished, the traps of all the appliances can be checked for water seal depth. The trap seal depth after all of the appliances have discharged their water must be at least 25mm. This can be checked with a dipstick.

J DECOMMISSIONING AND MAINTENANCE

Like all of the systems covered in this book, maintenance of AGDS should be carried out periodically to ensure problem-free operation. This is especially important with older systems as some of the materials used in the past, such as cast iron, corrode with time, and others may bring health and safety issues, such as asbestos pipework. On larger systems or housing contracts, periodic maintenance will be carried out to a maintenance schedule, which lists the properties and systems to be checked.

J1 SIMPLE MAINTENANCE TASKS

Simple maintenance tasks that can be performed are:

- **Cleaning out traps** – Traps, especially bath and shower traps, accumulate hair and soap residue that will eventually cause slow discharge of water or even complete blockage. If left they will also begin to smell. Kitchen sink traps collect grease and this can be a constant source of problems, as the grease clings to the waste pipework and makes smooth flow of water less likely in the future. These can be disconnected from the appliance and thoroughly cleaned by hand. Cleaning chemicals may be used but they should be administered with caution as some can burn the skin on contact. Always wear appropriate PPE when using chemicals or cleaning traps such as rubber gloves and eye protection and always read the dosage instructions. Never mix different cleaning chemicals as this could result in dangerous fumes developing and possibly explosive mixtures.

- **Cleaning out the overflows of the appliances** – Belfast sink overflows are notorious for blocking. These can be cleaned with stiff wire and then thoroughly flushed out.

- **Checking access covers** – These should be checked for leakage and tested to ensure that the bolts on the access door are free-moving. A little silicone grease will prevent the bolts from rusting.

Also check the rubber seals to make sure that they are not showing signs of perishing.

- **Checking the pipework** – Pipework is often neglected during periodic maintenance. Always check for signs of leakage and to ensure that the clips are in good order, especially if the soil and waste pipes are external and may be affected by the weather. Direct sunlight is especially damaging to pipework and clips. Damaged or broken clips should be replaced. Also, check the cage on the top of the stack as these often blow off in high winds. These should be replaced as necessary.
- **Checking for signs of overflowing WC cisterns** – Adjust the water levels and check their correct operation.

Make a note of all actions taken on the maintenance report.

J2 DEALING WITH BLOCKAGES

Unblocking drains and soil stack pipework is probably the most unpleasant of all the jobs a plumber undertakes and it can pose a real health risk. In this section we will look at some of the problems we may be asked to deal with and the actions to take. Always wear the correct PPE including rubber gauntlets, eye protection, a face mask, full boiler suit and wellington boots.

Blocked soil pipes

There are a number of reasons why soil pipes and drains block. Often it can be attributed to one of three possible causes:

- **A broken drain** – If this is suspected, there is very little that a plumber can do. The drain will probably need a camera inspection to accurately pinpoint the problem. Broken drains often occur because of ground compression or movement.
- **A tree root growing through the drain** – Again, if this is suspected a camera inspection will be necessary.
- **A physical blockage** – These are usually caused by something being flushed down the toilet and eventually becoming wedged in the drain. These objects may be moved by the use of drain rods, which come with various attachments to deal with a variety of blockage situations.

Drain rods and attachments

Blocked sinks, wash basins and baths

These can often be cleared with a tool known as a force cup (also known as a plunger). The blockage is cleared by filling the appliance with water and pressing down repeatedly on the handle of the force cup. This creates a positive pressure on the downward push and a negative pressure on the upward pull, encouraging movement of water in the waste pipe, which is usually enough to dislodge the blockage. The force of the water when the force cup is removed will move the blockage down the wastepipe, breaking it up as the water flows.

Force cup

Blocked WCs

A blocked WC can often be cleared using a special kind of force cup, designed specifically to unblock WC pans and external gullies, known as a WC plunger.

Blocked waste pipes

These can often be cleared by the use of a hand spinner. An auger at the end of the hand spinner rotates as it enters the waste pipe, breaking up the blockage on contact. Care must be taken if this tool is being used with push-fit waste systems to ensure that the joints are not being forced apart. Thorough testing should be conducted after use to make certain that leaks have not been created.

J3   DECOMMISSIONING SANITARY SYSTEMS

When we remove old sanitary appliances and replace them with new ones, we are decommissioning the above-ground drainage system. In many cases, this will mean the removal of the soil and vent stack and waste pipes too. These are procedures that need careful consideration. The last part of this chapter assesses the most effective way of decommissioning an existing system of above-ground drainage.

The old sanitary appliances

Removal of the old appliances should be carried out with care. You must use appropriate PPE as you will be handling sanitaryware that has been used for personal ablutions and will most definitely be carrying disease. A risk assessment must be carried out, as some of the appliances, such as cast iron baths, will be heavy and may need the assistance of a second person, especially if the bath is to be carried out of the property in one piece. Ensure that the area outside the property is clear and free from obstacles.

Try not to damage vitreous china sanitaryware such as WCs and wash basins when you are removing them. The vitreous china is extremely sharp when broken. Always wear rubber gauntlets and eye protection during these operations.

Cast iron baths are often broken into four pieces before being carried outside. This is quite a dangerous task, as the enamel on the bath is glass and will fly in all directions when hit. It is best to use a club/lump hammer for this. Start at the waste hole as this is the weakest and thinnest point and work down the spine of the bath, then work across the bath. Eventually, the bath will break into four almost identically sized pieces, which are relatively easy to carry. You must be aware that this task is extremely noisy and can be alarming to the customer, so reassure them beforehand of your intentions. The required PPE is eye protection, gloves and ear defenders. It may be that the bath has more value in one piece, as there is a market for second-hand cast iron baths. In this case, help will be needed to manoeuvre the bath safely outside.

When all of the appliances have been removed from the property, they should be stripped of any scrap metal as this is recyclable.

The old sanitary pipework

Old sanitary pipework can be made from a variety of materials, including cast iron, lead and asbestos. Each of these materials has its own health and safety issues, which must be observed.

Cast iron

This is a heavy material. Invariably you will be working at height when removing this kind of pipework, so precautions must be taken to block off the pathway around where you are working and post signs warning of the danger. The most common types of fixings for cast iron were nails and small wooden or plastic pipe spacers (known as bobbins) fixed through lugs on the cast iron pipe sockets (known as ears). It is usual practice to break the ears off the pipe to free it from the wall. Care should be taken, as these can fly off when being broken. Breaking the pipe in sections, working from the top, is the best way of removing it but you must ensure that pieces of broken pipe do not enter the drain and block it. The correct PPE should be worn during this process, including hard hat, goggles, gauntlets and eye protection. A risk assessment should also be carried out. Cast iron has scrap value and should be recycled.

Asbestos

If you suspect that the soil and vent pipes are made from asbestos, you must seek advice from your supervisor. On no account must you break the pipe, or you risk releasing potentially dangerous fibres into the atmosphere. If the material is asbestos, it must be removed by a specialist asbestos removal company by law (see Chapter 001, page 29).

Lead

Traditionally, lead pipe was used for WC branches and waste pipes. Occasionally soil and vent pipes made from lead may also be found but this is extremely rare. As with all lead, pipes should be handled with great care. Lead can sometimes corrode, leaving a fine white powder residue known as lead oxide. This material is extremely dangerous as it offers the quickest way of being ingested into the body by breathing in the powder. Always wear the correct PPE when handling lead, such as barrier cream on the hands, or wear gloves, a face mask and goggles. Lead is also a heavy material; take care when lifting it. A risk assessment must be carried out. Like all metals, lead has a scrap value and should be recycled.

General points about decommissioning

When working at high level, place barriers and warning notices around where you are working. If the system is being decommissioned for a short period, ensure that warning notices are placed at the appliances to prevent accidental usage while you are working on the system. Inform the customer of the length of time you expect the system to be out of action, and always wear an appropriate level of PPE and conduct a risk assessment.

CONCLUSION

We have seen as we have worked through this chapter just how important above-ground sanitation systems are with regards to both personal and environmental hygiene. Correctly installed and functioning sanitary appliances and pipework protects us from the diseases that were rife in the UK only 200 years ago and still continue to cause severe illness in other parts of the world to this day.

Properly installed sanitary appliances and pipework are a visual reminder of how well we can portray our plumbing skills, while providing a necessary, hygienic environment for ourselves, the customer and the environment at large.

010 TEST YOUR KNOWLEDGE

1. Which Building Regulation Document must sanitation systems comply with?

2. There are seven general requirements of above-ground drainage systems (AGDS). What are they?

3. There are two British Standards that must be consulted when installing sanitation systems. What are they?

4. Name the three types of WC pan.

5. Identify the following WC pan:

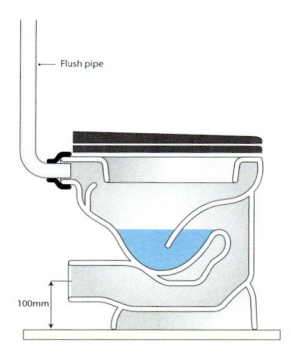

6. Name three types of wash basin.

7. What three materials are baths commonly manufactured from?

8. What is the most common type of bidet used in the UK?

9. There are two ways a urinal can be flushed. What are they?

10. Name the system shown below:

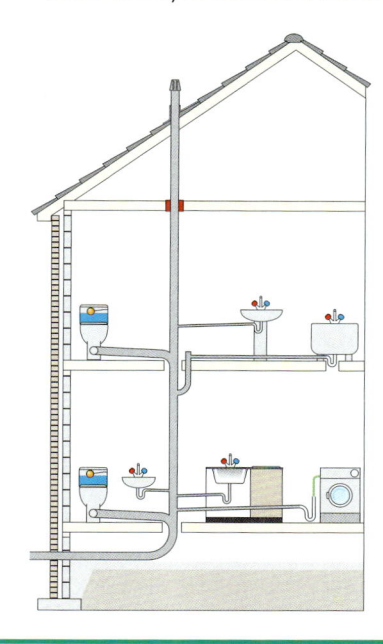

11. Where would a ventilated discharge branch system be used?

12. Where would you use a bottle trap?

13. Which kind of trap seal loss can be caused by water from one appliance displacing the trap of another appliance downstream?

14. What are the two basic methods of jointing PVCu soil and waste pipes?

15. What is the test pressure for above-ground drainage systems?

16. How is trap seal loss measured?

17. What precautions must we take when decommissioning sanitary pipework installed at high level?

18. List three operations that can be carried out during periodic routine maintenance of above-ground drainage systems.

TEST YOUR KNOWLEDGE ANSWERS

001 Understand and carry out safe working practices in building services engineering

1. The Health & Safety At Work Act 1974.

2. It is the duty of every employer, so far as is reasonably practicable, to ensure the health, safety and welfare at work of their employees.

3.
   - Improve the planning and management of projects from the very start.
   - Identify hazards early on, so they can be eliminated or reduced at the design planning stage and the remaining risks can be properly managed.
   - Target effort where it can do the most good in terms of health and safety, and discourage unnecessary red tape.

4. The employer.

5. The Control of Substances Hazardous to Health.

6. The Reporting of Injuries, Diseases and Dangerous Occurrences Regulations. These regulations cover the legal duty to report certain work-related accidents, diseases and dangerous occurrences.

7. Any TWO from the following:
   - The Water Supply (Water Fittings) Regulations
   - The Gas Safety (Installation and Use) Regulations
   - The 17th Edition IEE Regulations

8. The Health and Safety Executive (HSE).

9. Yellow triangle — Danger; Blue circle — Mandatory.

   Red square — Fire; Green square — Safe condition.

10. Absorption, ingestion and inhalation.

11. The three main types of asbestos.

12.
    - Minor cuts.
    - Minor burns.
    - Objects in the eyes.
    - Exposure to fumes.

13. Cardiopulmonary resuscitation.

14.

25V	Violet
50V	White
110V	Yellow
230V	Blue
400V	Red

15. It should have a combined inspection and testing every three months, as well as a formal visual inspection on a monthly basis. This also applies to any IT, movable, portable and hand-held electrical equipment used on site.

16. To provide a continuous earth for the pipework, to prevent an electric shock in the event of any electrical fault.

17. **Propane** — Signal red
    Acetylene — Maroon
    Oxygen — Black

18. True. This is important point when dealing with LPG as it means that it can collect in places below ground such as cellars and confined spaces where it becomes an unseen potential explosion hazard.

19

Class A	SOLIDS such as paper, wood, plastic etc
Class B	FLAMMABLE LIQUIDS such as paraffin, petrol, oil etc
Class C	FLAMMABLE GASES such as propane, butane, methane etc
Class D	METALS such as aluminium, magnesium, titanium etc
Class E	Fires involving ELECTRICAL APPARATUS
Class F	Cooking OIL and FAT etc

20 75°

21 1.2m

002 Understand how to communicate with others within building services engineering

1 The Architect.

2 The client. The client is the reason the building is being constructed and so indirectly employs everyone on and off site. The client can be a person, company or Local Authority.

3 The Quantity Surveyor.

4 The Clerk of Works.

5
- The Civil Engineer
- The Structural Engineer
- The Building Services Engineer.

6 The Heating and Ventilation Engineer.

7 The Building Control Officer. He or she is responsible for ensuring regulations on public health, safety, energy conservation and disabled access are met. They work to the Building Regulations.

8 The Equality Act 2010. This came into force from October 2010 and provides a modern, single legal framework to tackle disadvantage and discrimination.

9 The Freedom of Information Act 2000.

10 Timesheet. These are completed by employees and detail hours worked and work done for wages and invoicing purposes.

11 Letter. A letter on headed paper, with the company name or logo, is formal recognition that a notice, request or instruction from the company representative is being given.

12 A responsible person.

13 Diagrams and visual aids.

14 Your immediate supervisor.

15
- Meetings
- Negotiation/mediation sessions
- Other dispute resolving methods.

003 Understand how to apply environmental protection methods within building services engineering

1 The Climate Change Act 2008.

2 Document L in Wales and England, Document J in Scotland and Document F in Northern Ireland.

3 Domestic Building Services Compliance Guide 2010

4 Seasonal Efficiency of Domestic Boilers United Kingdom (SEDBUK). This is a banded scale from A to G, which gives energy efficiency ratings, with band A being the most efficient. It is likely to be replaced in the near future with minimum percentage efficiency figures.

5 Electricity. Wind power is classed as a zero carbon, renewable energy source.

6 Any TWO from the following:
- Solar thermal
- Biomass
- Hydrogen fuel cells
- Heat pumps

- CHP
- CCHP

7 Kerosene. This is the fuel oil used in the majority of domestic boilers.

8 Natural gas. The largest component of this is methane.

9. Photovoltaics.

10 The Energy Saving Trust or the Carbon Trust. The Energy Saving Trust is considered the leading UK organisation in helping to reduce carbon emissions.

11 An Energy Performance Certificate (EPC) gives information on the current energy efficiency and advice on reducing energy use and carbon dioxide emissions for a building. Although they are required for the majority of buildings, only public buildings are required to have them on display.

12 Soldering fittings, because this requires the use of a blowtorch and soldering equipment, contributing to CO_2 emissions.

13 An illegal activity.

14 Asbestos.

15 Recyclable material.

16 G (regulation 17.K).

17 125 litres.

18 Rainwater harvesting.

19 Grey water recycling.

20 About 30%.

004 Understand how to apply scientific principles within mechanical engineering services

1

Measure of	Base SI unit	Symbol
Length	metres	m
Mass	kilograms	Kg
Time	second	s
Electric Current	ampere	A
Thermodynamic temperature	kelvin	K

2 Any FOUR from the following:
- copper
- aluminium
- lead
- zinc
- tin
- iron

3 A mixture of two or more metals.

4 It is 0.6–0.7 times lighter than air (which is used as a standard substance to compare the relative density of gases).

5 Thermoplastics.

6 The tensile strength of a material is a measure of how well or badly it reacts to being pulled or stretched until it breaks.

7 Temporary hard water and permanently hard water. Temporary hard water precipitates calcium carbonate when boiled, which can cause scaling in pipes.

8 The formula for this is:

L × Δt × SHC of water

140 × 76 × 4.187 = 44,549.68kJ

9 The volume of gas is halved.

10 When the water changes its physical state from boiling water to steam.

11 Conduction, convection and radiation.

12 The pressure exerted by water is due to its mass.

13 Newton (equal to kgm/s^2).

14
- Changes of direction of pipe.
- Size of pipe.
- Pressure of water.
- Length of pipe.
- Frictional resistance of the internal bore of the pipe.
- Constrictions such as valves and taps.

15 …an equal but opposite reaction.

16 A moment of a force is the measure of the turning effect (or torque) produced by a force acting on a body. An example of this in plumbing is a spanner tightening a nut.

17 A circuit where the electrons always flow from the negative (–) pole towards the positive (+) pole.

18 Resistivity is the ability of a material to resist the flow of electricity where the resistance to electricity generates usable heat. It is the opposite to conductivity.

19 30 seconds.

20 A system where all metal fixtures in a domestic property, such as hot and cold water pipes, central heating pipes and gas pipes, radiators, stainless steel sinks, steel and cast iron baths and steel basins are bonded to earth and all have an equal potential.

005 Understand and carry out site preparation and pipework fabrication techniques for domestic plumbing and heating systems

1
a Scissor bender
b Adjustable spanner
c Plastic pipe cutters
d A circular saw
e A core drill
f A hacksaw

2 Always check the tool, the cord and the plug for any signs of wear or damage. Always check to make sure that the tool is the correct voltage for the power supply.

3 R220, R250 and R290. R250, also know as half-hard tempered, is the most widely used grade for plumbing and heating applications.

4 End feed – requires solder to be fed into the end of the fitting.

Integral solder ring – has a band of lead-free solder inside the fitting, so additional solder is not needed.

5 Fluxes are used to clean oxides from the surface of the copper and to help with the flow of solder into the fitting.

6 Type A fittings are non-manipulative – neither the tube nor the fitting need working to make the joint.

Type B fittings are manipulative – these require that the end of the tube is worked, or flared, to make a successful joint.

7 Low carbon steel.

8 A standard thread cut onto the ends of pipes and blackheart malleable, male fittings to ensure a watertight, gas tight or steam tight joint. The thread tapers towards the end of the tube, ensuring that the tube tightens the further it is screwed into the fitting.

9 PVCu and ABS.

10 Acrylonitrile butadiene styrene (ABS). In contrast, unplasticised polyvinyl chloride (PVCu) has a good resistance to UV light, but suffers from photodegradation.

11 120mm

12 Walk around the house with the customer. Point out any existing damage to furniture, fixtures, carpets and wall coverings. This will prevent any misunderstandings regarding damage and marks already in place.

13 25mm

14 Start   350mm
    Finish   1.25m

15 This is a collar that is placed around a pipe passing between floors, and it expands in the presence of heat to stop the spread of fire.

16 The benchmarking certificate and the Building Regulations Compliance certificate.

17 Where a system or appliance is permanently disconnected and/or removed.

18 Where systems and equipment are serviced and checked at regular intervals to ensure optimum performance.

19 —(M)—

20 ▷|◁

006 Understand and apply domestic cold water system installation and maintenance techniques

1 The sun warms the Earth causing water on its surface to evaporate. The evaporated water vapour rises with the air and is carried by the prevailing winds. As the water vapour passes over land, some of it condenses to form clouds and as more water vapour is attracted, the cloud becomes saturated to the point where it can no longer hold the moisture and the vapour is released in the form of rain, sleet, snow or hail, which falls back to the Earth. Here, some of the water runs into rivers, streams, lakes and the oceans allowing the process to begin again.

2 Any THREE from the following:
   - deep wells
   - shallow wells
   - upland surface water
   - springs
   - rivers
   - canals
   - aquifers
   - artesian wells and springs
   - boreholes

3 Water that is fit for human consumption.

4 Chlorine.

5 A grid system of pipework for supplying wholesome water. They are installed and maintained by the water undertaker.

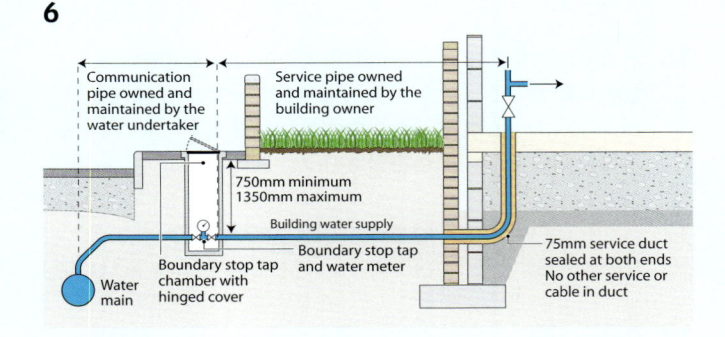

6

7 A drain-off valve.

8 A system where all cold water outlets are supplied direct from the mains cold water supply.

9 An indirect system of cold water supply.

10 A cold water distribution pipe (22mm minimum) is fitted to the cistern in the roof space and distributes cold water from the cistern to the wash basin, WC and bath.

11 Direct from the mains supply.

12 230 litres.

13 It acts as a warning that the float-operated valve has developed a problem and the cistern is in danger of overflowing.

14 The flowing of water in the wrong direction due to loss of system pressure.

15 A verifiable single check valve.

16 A double check valve.

17 A part 2 is made from brass and a part 3 is made from plastic.

18 In a quarter-turn tap.

19 Draw off a small supply of drinking water for the period of isolation.

20 Any TWO from the following:
- noise
- corrosion
- air locks
- leakage

007 Understand and apply domestic hot water system installation and maintenance techniques

1
- The size of the property and the distance from the outlets.
- The number of occupants and the amount of hot water required.
- The number of hot water outlets.
- The type(s) of fuel to be used.
- Installation and maintenance costs.
- Running costs and fuel efficiency.
- The pressure and flow rate of the incoming mains supply.

2 Centralised systems are those where the source of hot water is sited centrally in the property for distribution to all of the hot water outlets.

3
- Indirect system with a double-feed indirect cylinder.
- Indirect system with a single-feed self venting cylinder.
- Direct system.

4 a A maximum normal operating temperature of 60°C.

b
- To avoid problems with scaling.
- To prevent the growth of legionella bacteria.

5 One-pipe circulation, often called parasitic circulation.

6 A coil (or, in some older systems, a smaller cylinder called an annular).

7 Air entrapment.

8 To prevent dead legs.

9 A pump made from bronze.

10 To prevent galvanic corrosion.

11 A Indirect system with a double-feed indirect cylinder.

B Indirect system with a single-feed self-venting cylinder.

C Direct system.

12 Single-point system.

13 A heater that serves more than one point or draw-off.

14 The thermostatic mixing valve is positioned on the hot water outlet and connects the hot water to a branch from the incoming cold supply. Its purpose is to blend the hot and cold supplies to a specific temperature so that the hot water supply is not too hot.

15 To prevent backflow.

16 1 metre.

17 Twin impeller inlet shower boosting pump.

A single impeller outlet pump.

18 They must have equal flow rate and pressure. This is important because it ensures the correct mixing of both hot and cold water, thus preventing scalding.

19 The customer.

008 Understand and apply domestic central heating system installation and maintenance techniques

1. The simultaneous heating of all spaces in a dwelling to maintain specified temperatures based on calculated heat losses. It is also referred to as full central heating.

2. The one pipe semi-gravity system.

3. The C-Plan has independent thermostatic control over both hot water and heating temperatures. The two-pipe semi-gravity system does not.

4. Fully pumped systems with mid-position valve (commonly referred to by the Honeywell trade name Y-Plan).

    Fully pumped systems with 2 x two-port valves (commonly referred to by the Honeywell trade name S-Plan).

5. Fully pumped system with mid position valve (commonly referred to by the Honeywell trade name Y-Plan).

6. The expansion vessel.

7. Via a temporary filling loop.

8. A manifold is a multi-connection fitting that is fitted to the central heating flow and return pipes of a microbore system of central heating.

9. It is the means by which water enters the system for filling and top-up and it allows space for the system water to expand to when it is heated.

10. The use of an air separator helps in the positioning of the feed and vent by ensuring that the neutral point is built into the system.

11. Any FOUR from the following:
    - panel radiators
    - column radiators
    - low-surface-temperature radiators
    - fan convectors
    - tubular towel warmers
    - tubular towel warmers with integral panel radiator
    - skirting heating
    - underfloor heating

12. In order left to right:
    - single panel radiator
    - single panel radiator with convector
    - double panel radiator with single convector
    - double panel radiator with double convector.

13. The automatic bypass valve controls the flow of water across the flow and return circuit of fully pumped heating systems by opening automatically as other paths for the water close.

14. The AB port. The other two ports are as follows: the A port connects to the heating circuit, and the B port to the hot water circuit.

15. Automatic air valves are fitted where air is expected to collect in the system, usually at high points.

16. It should be placed close to the vulnerable parts of the system especially if it is fitted in unheated garages and roof spaces. The frost thermostat should be used in conjunction with a pipe thermostat.

17. The boiler interlock is not a single control device but the interconnection of all of the controls on the system, such as room thermostats, cylinder thermostats and motorised valves. The idea behind the boiler interlock is to prevent the boiler firing up when it is not required.

18. This will ensure that no debris enters the pump.

19. 1.5 times normal working pressure.

20. Black oxide sludge is made up of minute particles of steel that have been 'robbed' by galvanic corrosion. It causes problems in central heating systems including:
    - black water appearing at the air release valve when the system is bled
    - coating of circulators
    - blocked pipework, preventing heat from

- reaching radiators
- cold spots in radiators when sludge collects at the bottom
- formation of hydrogen gas in a radiator
- noise from blocked heat exchangers.

009 Understand and apply domestic rainwater system installation and maintenance techniques

1.
   - To protect the buildings foundations.
   - To reduce ground erosion.
   - To prevent water penetration and damp in the building structure.
   - To provide a means for collecting rainwater for later use, ie rainwater harvesting.

2. 1:600

3. Answers from left to right:

   Half round

   Ogee

   Square section

4. Any FOUR from the following:
   - It is easy to install.
   - It is lightweight and easy to handle.
   - Minimal maintenance is required.
   - It requires no painting.
   - It is economical.
   - It is corrosion free.
   - It has a smooth internal bore.
   - It has a life expectancy of 50 years.

5. The coefficient of linear expansion for PVCu is 0.06mm/m/°C. This means that for every metre of gutter, PVCu expands by 0.06mm for every degree rise in temperature. To counteract this expansion, fittings are manufactured and should be installed with a 10mm expansion gap.

6. Extruded aluminium.

7. Top: Top-fitted rafter bracket

   Bottom: Drive-in rise and fall bracket

8. a   To convert a half-round profile gutter to ogee profile gutter.

   b   To allow a rainwater pipe to be connected to a drain.

9. Visual inspections help in establishing the overall condition of the gutter and rainwater pipe installation, joints and fittings and in pinpointing specific problems.

10. In case the gutter contains bird droppings. These should be handled with extreme care as they carry disease and should not be ingested into the body by breathing in.

010 Understand and apply domestic above ground drainage installation and maintenance techniques

1. Building Regulations Approved Document H1: 2002.

2. AGDS must:
   - convey the flow of foul water to a foul water outfall. This can be a foul or combined foul / rainwater sewer, a cesspool or septic tank
   - minimise the risk of blockage and/or leakage
   - prevent foul air from entering the building under working conditions
   - be ventilated
   - be accessible for clearing blockages
   - not increase the vulnerability of the building to flooding
   - be large enough to carry the expected flow at any point in the system.

3. BS EN 12056-5:2000: Gravity drainage systems inside buildings. Installation and testing, instructions for operation, maintenance and use.

   BS 8000 Part 13:1989: Workmanship on building sites.

4. Wash down, single trap siphonic and double trap siphonic.

5. Single trap siphonic.

6 Any TWO from the following:
- Wall-hung wash basins
- Pedestal wash basins
- Semi-pedestal wash basins
- Counter-top style
- Under-counter-top style

7
- Reinforced cast acrylic sheet
- Porcelain-enamelled steel
- Porcelain-enamelled cast iron

8 Over-rim bidet.

9 a Manual or automatically operated cistern

b Pressure flushing valve directly connected to a supply or distributing pipe

10 Primary ventilated stack system.

11 The ventilated branch discharge system is used on larger systems where there is a risk of trap seal loss because the waste pipe lengths are excessive.

12 On a wash basin or bidet.

13 Induced siphonage.

14 Push-fit and solvent weld.

15 38mm for 3 minutes with no pressure loss.

16 By using a dipstick.

17 When working at high level, place barriers and warning notices around where you are working.

18 Any THREE from the following:
- Cleaning out traps.
- Cleaning out the overflows of the appliances.
- Checking access covers.
- Checking the pipework.
- Checking for signs of overflowing WC cisterns.

GLOSSARY

17th edition IEE Regulations (BS 7671)
The National standard to which all wiring, industrial or domestic, should conform.

Abutment
The junction between a pitched roof and a vertical wall.

Acceleration
A measure of the rate at which a body increases its velocity.

Acceleration due to gravity
The rate of change of velocity of an object due to the gravitational pull of the earth.

Acetylene (C_2H_2)
A flammable gas used in conjunction with oxygen for welding.

Acrylonitrile butadiene styrene (ABS)
A type of thermoplastic used for waste pipes, soil pipes, underground drainage, gutters and rainwater pipes. Can be solvent welded.

Acts of Parliament
These create a new laws or change an existing one.

Adhesion
The way that water tends to stick to whatever it comes into contact with.

Air admittance valve
Allows air into a stub stack to prevent the loss of trap seals.

Air changes
The amount of air movement within the building.

Air gap
A physical unrestricted open space between the wholesome water and the possible contamination.

Air separator
A fitting designed to correctly position the feed and vent pipes on a central heating system to ensure that the neutral point is automatically built in the system.

Air temperature
The temperature of the air within a building.

Air velocity
The speed at which air travels through a building.

Alloy
A mixture of two or more metals.

Alternating current (AC)
An electrical current that reverses its direction of travel constantly and uniformly throughout the circuit.

Ampere
The unit of electrical current.

Annealing
A process that involves heating the copper to a cherry-red colour and then quenching it in water. This softens the copper tube so that the copper can be worked without fracturing, rippling or deforming.

Anodic corrosion protection
A form of corrosion protection that uses a sacrificial anode to distract the corrosion away from vulnerable parts of the system.

Anodising
Coating one metal with another by electrolysis to form a protective barrier from corrosion.

Anti-gravity valves
A valve used in older central heating systems to stop unwanted gravity hot water circulation. Often called a dumb ball valve.

Approved Codes of Practice (ACoP)
Documents giving practical guidance on complying with the Regulations.

Aquifers
Water-bearing rocks below the Earth's surface.

Architect
The designer of a building or structure.

Artesian wells and springs
Water that rises from underground water-bearing rock layers under its own pressure.

Asbestos
A naturally occurring fibrous material that has been a popular building material since the 1950s, now known to cause serious and fatal illness.

Atmospheric pressure
The amount of force or pressure exerted by the atmosphere on the earth and the objects located on it.

Atom
A fundamental piece of matter made up of three kinds of particles called subatomic particles: protons, neutrons and electrons

Automatic bypass valve
A spring-loaded valve used on fully pumped heating systems; it is designed to automatically open when other paths for water flow begin to close.

Automatic urinal flushing cistern
Used to flush urinals.

Back boiler
A boiler made from a non-ferrous metal that is situated behind a real fire. Used with a direct cylinder.

Back siphonage
A vacuum that can suck water backwards causing contamination of the water supply.

Backflow
The flowing of water in the wrong direction due to loss of system pressure.

Backflow prevention device
A mechanical device, usually a fitting, designed to prevent contamination of water through backflow or back siphonage.

Banjo-type bath waste fitting
A type of waste fitting fitted to a bath that connects an overflow to the waste trap.

Barbed shanked nail
A nail with grooves cut into the shank. This makes the nail difficult to pull out once it has been driven into the wood.

Batch feed boiler
A solid fuel boiler where the fuel is fed by hand.

Bill of Quantities (BOQ)
A document used in tendering in the construction industry in which materials, parts and labour (and their costs) are itemised. It also (ideally) details the terms and conditions of the construction or repair contract and itemises all work to enable a contractor to price the work for which he or she is bidding.

Biomass
Any plant or animal matter used directly as a fuel or that has been converted into other fuel types before combustion.

Black water
Water and effluent from WCs and kitchen sinks that can only be treated by a water undertaker at a sewage works.

Blackheart fittings
A type of fitting for low carbon steel pipe with a tapered female thread.

Boiler cycling
This happens when a heating system has reached temperature, and the boiler shuts down. A few minutes later the boiler will fire up again to top the temperature up as the system loses heat, and after a few seconds shuts down again. This constant firing up and shutting down as the system water cools slightly wastes a lot of fuel energy.

Boreholes
Man-made wells that are drilled directly to a below-ground water source.

Boyle's law
A gas law that states that the volume of a sample of gas at a given temperature varies inversely with the applied pressure.

Branch ventilating pipe
Used on the ventilated branch discharge system of sanitary pipework to ventilate excessively long waste pipe runs.

BS 1566-1:2002
The British Standard for copper indirect hot water storage cylinders.

BS 6700:2006+A1:2009
The main British standard for the installation of hot and cold water installations in dwellings.

BS 8000-13:1989
The Code of Practice for the workmanship on site relating to the installation of sanitation systems.

BS EN 12056-3:2000
The British and European Standard for the installation of rainwater and guttering systems.

BS EN 12056-5:2000
The British and European Standard for the installation of sanitary pipework.

BS EN 12588:2006
The British and European Standard for rolled (milled) sheet lead.

BSP or BSPT
Stands for British Standard pipes and British Standard pipe threads, and relates to the type of thread we use on screwed low carbon steel pipes and fittings. Although the pipe is measured in mm, it is universally referred to in imperial measurements, eg ½-inch BSPT (½-inch British Standard pipe thread).

Building Control Officer
Responsible for ensuring that regulations on public health, safety, energy conservation and disabled access are met.

Building Regulations Approved Document F: Ventilation
Document dealing with indoor air quality to ensure buildings are properly ventilated.

Building Regulations Approved Document H3
The main document concerning the installation of rainwater discharge systems.

Building Regulations Approved Document L: Conservation of fuel and power: 2010 (Part J in Scotland and Part F in Northern Ireland)
Document controlling the insulation values of building elements, the heating efficiency of boilers, the insulation and controls for heating appliances and systems together with hot water storage, lighting efficiency and air permeability of the structure.

Building Services Engineer
Designer of the internal services within the building such as heating and ventilation, hot and cold water supplies, air conditioning and drainage. Many Building Services Engineers are members of the Chartered Institution of Building Services Engineers (www.cibse.org).

Calorific value
The amount of energy released when a known volume of gas, oil or coal is completely combusted under specified conditions. Solid and liquid fuels are measured in megajoules per kilogram (MJ/kg) and gases are measured in megajoules per cubic metre (MJ/m^3).

Capillary attraction
The process where water (or any fluid) can be drawn upwards through small gaps against the action of gravity.

Capillary fitting
A fitting for copper tubes that uses the principle of capillary attraction to draw solder into the joint when heated.

Carbon Trust
An independent, non-profit organisation set up by the UK Government with support from businesses to encourage and promote the development of low carbon technologies.

Carburising flame
A sooty flame containing too much acetylene.

Celsius (°C)
A common unit of temperature that has a zero point (0°C), which corresponds to the temperature at which water will freeze.

Central Heating System Specifications (CHeSS) 2008 CE51
Produced by the Building Research Energy Conservation Support Unit (BRECSU) to create a set of common standards for energy efficiency which

domestic heating installers and manufacturers should work towards.

Centralised hot water systems
Those systems where the source of hot water is sited centrally in the property for distribution to all of the hot water outlets.

Centre to centre
Measuring from the centre line of one pipe to the centre line of another so that all the tube centres are uniform. This ensures that the pipework will look perfectly parallel because all of the tubes will be at equal distance from each other.

Ceramic discs
Two thin close-fitting, slotted ceramic plates that control the flow of water from a tap.

Chamfer
To take off a sharp edge at an angle. If we chamfer a pipe end, we are taking the sharp, square edge off the pipe.

Charles's law
A gas law discovered by Jacques Charles which states that the volume of a quantity of gas, held at constant pressure, varies directly with the Kelvin temperature.

Chlorine
A chemical added to water for sterilisation purposes.

Cistern
A vessel for storing cold water that is only subjected to atmospheric pressure.

Civil Engineer
Designer of roads into and out of the building along with any bridges, tunnels etc that may be required.

Clerk of Works (CoW)
The Architect's representative on site. He or she ensures that the building is constructed in accordance with the drawings while maintaining quality at all times.

Climate Change Act 2008
Sets a target for the UK to reduce carbon emissions to 80 per cent below 1990 levels by 2050.

Coal
A heavy hydrocarbon that releases high content of sulphur dioxide and carbon dioxide when burnt.

Cohesion
The way in which the water molecules 'stick' together to form a mass rather than staying as individuals.

Coke
Produced by heating coal in an oven which reduces both sulphur and carbon dioxide content. Known as a smokeless fuel.

Combination boiler
A boiler that supplies both instantaneous hot water and central heating from the same appliance.

Combined cooling, heat and power (CCHP)
Uses the excess heat from electricity generation to achieve additional building heating or cooling.

Combined heat and power (CHP)
A plant where electricity is generated and the excess heat generated is used for heating.

Combined storage and feed cistern
Stores water for the domestic hot water system and the indirect system of cold water to the appliances, wash hand basin, bath, WC, washing machine etc.

Combined system
A system of below-ground drainage where both rainwater and foul water discharge into the same drain.

Combustion
A chemical reaction in which a substance (the fuel) reacts violently with oxygen to produce heat and light.

Commissioning
The process of bringing a system or appliance into full working operation through a system of checks to ensure correct operation to the design specification.

Communication pipe
A pipe connecting the water main to the customer's external stop valve. Owned by the water undertaker.

Compression
Back pressure of air created by water discharging down a soil pipe travelling up the stack blowing the water out of the traps.

Compression fitting
A mechanical fitting that requires tightening with a spanner to make a watertight joint.

Compressive strength
The maximum stress a material can sustain when being crushed.

Condensation
A process where steam turns to water.

Condensing boiler
A boiler that extracts all usable heat from the combustion process, cooling the flue gases to the dew point. The collected water is then evacuated from the boiler via a condensate pipe.

Conduction
Heat travelling through a substance with the heat being transferred from one molecule to another.

Conductivity
The property that enables a metal to carry heat (thermal conductivity) or electricity (electrical conductivity).

Construction (Design and Management) Regulations 2007
The principal piece of Health and Safety legislation specifically written for the construction industry.

Control of Asbestos Regulations 2006
Legally enforceable document prohibiting the importing, supplying and use of all forms of asbestos.

Control of Lead at Work Regulations 2002
Legally enforceable document that applies to all work which exposes any person to lead in any form whereby the lead may be ingested, inhaled or absorbed into the body.

Control of Substances Hazardous to Health (COSHH) Regulations 2002
Legally enforceable document intended to protect people from illness caused by exposure to hazardous substances.

Convection
Heat transfer through the movement of a fluid substance, which can be water or air.

Corrosion
Any process involving the deterioration or degradation of metal components.

Coulomb
Unit of electrical charge.

Creep
A term that is used to describe the effects of thermal movement whereby the lead fails to return to its original position after expansion has taken place.

Cross-connection
When one fluid category connects with another, for example, within a mixer tap.

Data Protection Act 2018
Gives people the right to know what information is held about them.

Delivery note
A document that lists the type and amount of materials that are delivered to site.

Deposition
The process where steam passes directly to ice.

Dew point
The temperature at which the moisture within a gas is released to form water droplets. When a gas reaches its dew point, the temperature has been cooled to the point where the gas can no longer hold the water and it is released in the form of water droplets.

Dezincification
A form of selective corrosion (often referred to as de-alloying) that happens when zinc is leached out of brass.

Direct current (DC)
An electrical current where the polarity or direction of the electron flow never reverses.

Direct hot water storage cylinder
A hot water storage vessel that does not contain a heat exchanger.

Direct system of cold water
A cold water system where all cold water outlets are connected to the main cold water supply.

Disability Discrimination Act 1995
Applies to companies who employ over 20 people. They are required to accommodate the needs of the disabled.

District heating
A system for distributing heat generated in a centralised location for residential and commercial heating requirements.

Domestic Building Services Compliance Guide 2010
Lays down rules for minimum boiler energy efficiency requirements. Often abbreviated to DBSC Guide.

Double-feed indirect hot water storage cylinder
A hot water storage vessel that contains a heat exchanger in the form of a coil or an annular.

Ductility
A mechanical property that describes by how much solid materials can be pulled, pushed, stretched and deformed without breaking.

Dynamic pressure
The pressure of water while it is in motion.

Economy 7 electricity
A UK tariff that provides for seven hours of cheaper-rate electricity, usually between 1 am and 8 am in the summer and 12 am and 7 am in the winter (although times may vary between regions and suppliers).

Effort arm
In mechanics, the arm where the force is applied.

Electrolyte
A fluid that allows the passage of electrical current, such as water. The more impurities (such as salts and minerals) there are in the fluid, the more effective it is as an electrolyte.

Elevation
A drawing showing one side of a building.

End feed fitting
A capillary fitting for copper tubes that requires solder to be fed into it during the soldering process.

Energy Performance of Buildings (Certificates and Inspections) (England and Wales) Regulations 2007
States the requirements for clients and landlords to produce energy performance certificates when buildings are constructed, rented out or sold.

Energy Saving Trust (EST)
An independent non-profit organisation set up after the 1992 Rio 'Earth Summit' that attempts to reduce energy use in the UK.

Equality Act 2010
Implemented by the Equality and Human Rights Commission (EHRC) to provide a single legal framework with clear, streamlined law that will be more effective at tackling disadvantage and discrimination.

Equipotential bonding
A system where all metal fixtures in a domestic property such as hot and cold water pipes, central heating pipes and gas pipes, radiators, stainless steel sinks, steel and cast iron baths and steel basins are connected together through earth bonding so that they are at the same potential voltage everywhere.

Erosion corrosion
Corrosion that occurs in tubes and fittings because of the fast flowing effects of fluids and gases.

Estimate
A costing for a piece of work that is not a fixed price but can go up or down if the estimate was not accurate or the work was completed ahead of schedule.

Expansion vessel
A vessel divided by a membrane with air one side and water the other that allows the expansion of water to take place safely.

Fan assisted boiler
A boiler that uses a fan to evacuate the products of combustion.

Fascia bracket
A clip for securing a gutter to a fascia board.

Feed and expansion cistern
Used to feed a vented central heating system and also allows expansion of water into the cistern when the system is hot.

Feed cistern
Only holds the water required to supply the hot water storage vessel.

Ferrous metal
A metal that contains iron and is susceptible to corrosion through rusting.

Filling loop
A method of filling sealed central heating systems direct from the water main.

Fireclay
A malleable clay used for heavy-duty sanitary appliances.

Flame arrester
A device fitted to lead welding equipment to prevent a dangerous situation known as flame blowback.

Flange
A projecting flat rim or collar, which is designed to strengthen or attach to another object. Flanges can also be found on large industrial pipe installations.

Flashings
A term given to a small weathering, usually at an abutment.

Flow rate
The amount of fluid or gas that flows through a pipe or tube over a given time.

Fluid category
A method of water classification from 1 to 5 according to its potential level of contamination, with 5 being the most dangerous.

Flushing valve
A method of flushing a urinal and WCs fitted in industrial premises using water direct from the mains supply without the need for a cistern.

Flux
A paste used to clean oxides from the surface of the copper and to help with the flow of solder into the fitting.

Footing a ladder
Standing with one foot on the bottom rung, the other firmly on the ground.

Force
The influence on an object which, acting alone, will cause the motion of the object to change. It is measured in newtons (kgm/s^2).

Forced draught
Any flue that uses a fan to help evacuate the products of combustion.

Fossil fuels
Formed by anaerobic decomposition of buried dead carbon-based plants, these fuels are known as hydrocarbons and release a high carbon dioxide content when burnt.

Freedom of Information Act 2000
Gives people the right to ask any public body for all the information they have on any subject.

Fully pumped heating systems
A heating system where both hot water circulation and central heating are pumped by a central heating circulator.

Galvanic corrosion
Corrosion that occurs when two dissimilar metals are in contact with each other in the presence of an electrolyte, usually water.

Gantt chart
Otherwise known as a programme of work, it is used on site to illustrate dates and lengths of time to complete particular jobs. It includes start and finish dates, labour and materials required and overall progress.

Gas Safety (Installation and Use) Regulations 1998
These cover the safe installation, maintenance and use of gas and gas appliances in private dwellings and business premises, aimed at preventing carbon monoxide (CO) poisoning, fires and explosions.

Gradient curve
A method of determining the fall of a 32mm waste pipe.

Gravity feed boilers
A solid fuel boiler where the fuel is automatically fed to the fire bed via gravity.

Grey water recycling
A method of collecting water used for bathing from baths, showers and wash basins and using it for other purposes such as WC flushing.

Guardrails
Erected to stop a person falling from a scaffold.

Gutter profile
The shape of a gutter when viewed from the side.

Hardness
The property of a material that enables it to resist bending, scratching, abrasion or cutting.

Hazard
Anything that may cause harm, such as chemicals, electricity, gas, or working from ladders.

Hazardous waste
Waste that is harmful to human health, or to the environment, either immediately or over an extended period of time.

Health and Safety at Work Act 1974
The principal piece of legislation covering occupational health and safety in the UK.

Health and safety file
A document held by the client by which health and safety information is recorded and kept for future use.

Health and Safety Inspectors
Persons employed by either the Health and Safety Executive or the Local Authority to enforce health and safety legislation

Heat exchanger
A device or vessel that allows heat to be transferred from one water system to another without the two water systems coming into contact with each other. The transfer of heat takes place via conduction.

Heat pumps
An electrical device with reversible heating and cooling capability. It extracts heat from one medium at a low temperature (the source of heat) and transfers it to another at a high temperature (called the heat sink), cooling the first and warming the second.

Hertz (Hz)
The SI unit of frequency, measuring the number of cycles per second in alternating current.

Home Energy Conservation Act (HECA) 1995
Places obligations on Local Authorities to draw up plans to increase domestic energy efficiency in their area by 30% over 10–15 years.

Hopper head
A large bucket type fitting for collecting rainwater from two or more rainwater pipes.

Humidity
The amount of moisture in the air.

Hydroelectric power
Electricity generated by turbines driven by the gravity movement of large amounts of water.

Ice
Water in its solid state when subjected to temperatures below its freezing point.

Immersion heater
A hot water heater that uses an electrical heating element to heat the water. Controlled by a thermostat.

Independent boiler
A freestanding boiler, usually solid fuel.

Independent scaffold
A scaffold which does not require the building to support it because it has two rows of vertical standards.

Indirect system of cold water supply
A cold water system where only the kitchen sink is connected to the mains cold water supply. All other cold water outlets are fed from a protected cistern.

Induced siphonage
An appliance causing the loss of trap seal of another appliance connected to the same waste pipe.

Instantaneous hot water systems
A system of hot water supply that heats cold water directly from the cold water main via a heat exchanger. There is no storage capacity.

Integral solder ring fitting
A capillary fitting for copper tubes with a ring of lead-free solder in the joint.

Job specification
A description of the installation that is being quoted for, complete with the types of materials and appliances that the installation must contain.

Joule
Unit of heat. 4.186 joules of heat energy (equals one calorie) is required to raise the temperature of 1g of water from 0°C to 1°C.

Jumper plate
A circular plate that holds a tap washer in place. It can be fixed or loose depending on the type of tap in which it is fitted.

Kelvin (K)
A unit of temperature where the lowest point, 0 Kelvin, corresponds to the point at which all molecular motion would stop. 0 Kelvin is −273° Celsius or absolute zero.

Kerosene fuel oil (grade C2 28-second viscosity oil to BS 2869)
A medium hydrocarbon liquid fuel. It is a residual by-product of crude oil, produced during petroleum refining. It has a high carbon content and is clear or very pale yellow in colour.

Ladders
Used to gain access to scaffolds or light work at high levels. There are three main classes: class 1, 2 and 3.

Latent heat
A change of state as a result of temperature rise.

Lead welding
A type of fusion welding to join two sheets of lead.

Legionella bacteria (legionella pneumophila)
Bacteria which breed in stagnant water. They can give rise to a lung infection called legionnaire's disease, which is a type of pneumonia.

Legislation
A law or group of laws that have come into force. Health and safety legislation for the plumbing industry includes the Health & Safety at Work Act and the Electricity at Work Regulations.

Level
When pipework is perfectly horizontal.

Lever
A rigid object that can be used with a pivot point or fulcrum to multiply the

mechanical force that can be applied to another, heavier object.

Liquid petroleum gas (LPG)
The generic name for the family of carbon based flammable gases that are found in coal and oil deposits deep below the surface of the earth. They include propane, butane, methane and ethane.

Local Authority
Ensures that all works carried out conform to the requirements of the relevant planning and building legislation.

Localised hot water systems
Systems of hot water supply that are installed at the place where they are needed.

Locking out
A process by which a thermostat protects the boiler from overheating by shutting it down when a temperature of around 85°C is reached. High limit thermostats are manually resettable by pushing a small button on the boiler itself.

Low-pressure, open-vented central heating systems
A central heating system that is fed via a feed an expansion cistern and contains an open vent pipe.

Low surface temperature radiator (LST)
A radiator designed to give full heat output whilst being cooler to the touch.

Low water content boiler
A boiler that contains only a small amount of water for quick water heating.

Lubricant
A substance, often a liquid or a grease, introduced between two moving surfaces to reduce the friction.

Malleability
The property of a material, usually a metal, to be deformed by compressive strength without fracturing.

Manifold
A manifold, in systems for moving fluids or gases, is a junction of pipes or channels, typically bringing one into many or many into one.

MCB (miniature circuit breaker)
A type of fast-reacting, resettable fuse.

Microbore system
A central heating system using very small pipework, usually 8mm and 10mm, to feed the heat emitters.

Molecule
The smallest particle of a specific element or compound that retains the chemical properties of that element or compound.

Momentum
Trap seal loss caused by a large amount of water is suddenly discharged down the trap of an appliance.

Multi-point hot water heater
A water heater that serves more than one hot water outlet.

Natural gas
A light hydrocarbon fuel found naturally wherever oil or coal has formed. Predominantly contains five gases – methane, ethane, butane, propane and nitrogen.

Neutral water
Water that is neither hard nor soft that has a pH value of 7.

Newton
A unit of measurement of force. (kgm/s^2)

Nogging
A term often used on site to describe a piece of wood that supports or braces timber joists or timber-studded walls. They are particularly common in timber floors as a way of keeping the joists rigid and at specific centres, but they can also be used as supports for appliances such as wash hand basins and radiators that are being fixed to plasterboard.

Non-ferrous metal
Metals that do not contain iron.

Non-rising spindle
Mainly found in taps, a non-rising spindle is connected to a hexagonal barrel holding the washer. It does not rise when the tap is opened.

Ohm
The unit of electrical resistance.

One-pipe central heating system
A simple ring circuit of pipework to and from the boiler and as such, there are no separate flow and return pipes.

Open flue
A flue that is open to the room where the appliance is fitted and relies on heat from the combustion process to create an updraught to evacuate the products of combustion. Often called natural draught.

Open-vented central heating systems
Systems fed from an F&E cistern in the roof space that contains a vent pipe, which is open to the atmosphere.

Open-vented direct hot water storage systems
A hot water storage system containing a direct cylinder.

Open-vented hot water systems
Systems fed from a cistern in the roof space that contains a vent pipe which is open to the atmosphere.

Open-vented indirect hot water storage systems
A hot water storage system containing an indirect type cylinder.

Outriggers
Tubes or special units that connect to the bottom of tower scaffolds at the corners, giving a greater overall base measurement and, therefore, an increase in height.

Overflow pipe
A method of warning of float-operated valve malfunction.

Overheads
On a building site, costs that include those of the site office and site/administration staff salaries.

Oxygen (O$_2$)
A very powerful oxidising agent used in gas form with acetylene when welding.

Parallel threads
A screw thread of uniform diameter used on fittings such as sockets.

Partially separate system
A system of below-ground drainage where the foul water and some of the rainwater discharges into the foul water drain and all other rainwater discharges in a rainwater drain or soakaway.

Peat
A poor quality fossil fuel that has a high carbon content but much less than coal with large amounts of ash produced during combustion.

Permanently hard water
Water that contains magnesium and calcium chlorides and sulphates in the solution. Cannot be softened by boiling. Alkaline, with a pH value above 7.

Permit to Work
A document that gives authorisation for named persons to carry out specific work within a nominated time frame.

Personal protective equipment (PPE)
A garment or piece of equipment worn by a person designed to create a barrier against workplace hazards.

Photovoltaic
A method of generating electricity from the power of the sun. Also known as solar arrays.

Pitting corrosion
The localised corrosion of a metal surface and is confined to a point or small area that takes the form of cavities and pits.

Planned preventative maintenance
Planned maintenance, usually to a schedule, so that systems and equipment can be serviced and checked at regular intervals to ensure optimum performance.

Planning Officer
Responsible for processing planning applications, listed building consent applications and conservation area consent applications.

Plumb
When pipework is perfectly vertical.

Polybutylene
A type of thermoplastic used to manufacture pipes for cold water, hot water and central heating systems.

Polyethylene
A type of thermoplastic used to manufacture mains cold water pipes.

Polypropylene
A type of thermoplastic used to manufacture cold water cisterns, WC siphons and push-fit waste and overflow pipe.

Polyurethane foam
A sprayed form of insulation applied to hot water storage cylinders.

Portable appliance testing (PAT)
A method of testing portable electrical appliances and tools to ensure that they are safe to use.

Potable
Water that is fit to drink. It is pronounced 'poe-table'.

Power shower
A cistern-fed shower mixing valve that uses a boosting pump to increase flow rate and pressure.

Press-fit fittings
Fitting for copper tubes that require a special electrical press tool, which crimps the fitting onto the tube to make a secure joint.

Pressure
Defined as force per unit area. It is measured in pascals (newtons per square metre – N/m^2).

Pressure jet burner
An oil burner found on oil burning central heating boilers that atomises the fuel prior to combustion.

Pressure relief valve
A safety valve that safeguards against over-pressurisation by allowing excess water pressure to safely discharge to drain.

Primary open safety vent
A pipe on a central heating system that is open to the atmosphere to provide a safety outlet should the system overheat.

Primary ventilated stack
A system of sanitary pipework that relies on all the appliances being closely grouped around the stack and therefore does not need an extra ventilating stack.

Private water supply
Drinking water source which is not provided by a licensed water undertaker.

Propane (C$_3$H$_8$)
A flammable gas that is heavier than air. One of the five principal gases in natural gas.

Proprietary trench support
A specially designed support to prevent trench collapse.

Pure metal
Derived directly from the ore and containing very little in the way of impurities.

Push-fit fittings
Simple push-on fittings for copper tubes or polybutylene pipe.

Putlog scaffold
A scaffold which is not self-supporting and has only one row of vertical standards.

Quantity Surveyor
A financial consultant or accountant who advises as to how the building can be constructed within the client's budget.

Quotations
A fixed price for a job, which cannot vary.

Race Relations Act 1976
An Act of Parliament that makes discrimination on grounds of race illegal.

Radiation
Heat transfer as thermal radiation from infrared light, visible or not, which transfers heat from one body to another without heating the space in between travelling in straight lines.

Rafter bracket
A bracket fixed to the roof members of a dwelling for securing a gutter when no fascia board is available.

Rainwater cycle
A natural process where water is continually exchanged between the atmosphere, surface water, ground water, soil water and plants. The scientific name is the hydrological cycle.

Rainwater harvesting
A method of collecting rainwater and using it for other purposes such as WC flushing.

Residual current device (RCD)
A fast-reacting type of fuse that detects fluctuations in current flow.

Refrigerant
Fluorinated chemicals which are used in both liquid and gas states to create both heating and cooling effects.

Regulations
Rules, procedures and administrative codes set by authorities or governmental agencies to achieve an objective. They are legally enforceable and must be followed to avoid prosecution.

Relative density
The ratio of the density of a substance to the density of a standard substance under specific conditions.

Reporting of Injuries, Diseases and Dangerous Occurrences Regulations 1995 (RIDDOR)
Places a legal duty on your employer, the self-employed and people in control of work premises to report some work-related accidents, diseases and dangerous occurrences.

Resistance arm
In mechanics, the arm where the load is concentrated.

Reverse osmosis
A water filtration process whereby a membrane filters unwanted chemicals, particles and contaminants out of the water.

Reversed central heating return system
A central heating system where the return travels away from the boiler in the same direction as the flow before looping around to be connected to the return at the boiler.

Rising spindle
Mainly found in taps, a rising spindle is connected to the washer and jumper plate. It rises as the tap is opened.

Risk
The chance, no matter how high or low, that somebody could be harmed by a hazard, together with an indication of how serious the harm could be.

Risk assessment
A detailed examination of any factor that could cause injury.

Room-sealed appliance
An appliance where the combustion process and flue gas evacuation is sealed from the space where the boiler is fitted. These can be natural draught or forced draught.

S-Plan central heating system
A fully pumped heating system that uses a two two-port zone valves.

Saddle
The top piece of an abutment flashing.

Sand cast lead sheet
Lead sheet produced by traditional casting on a bed of sand.

Sealed, pressurised central heating systems
A central heating system fed direct from the cold water main and incorporating an expansion vessel.

Secondary circulation
A method of hot water circulation to prevent dead legs of cold water in hot water systems.

SEDBUK
The Seasonal Efficiency of Domestic Boilers in the United Kingdom. A list of boiler efficiency ratings.

Self-siphonage
Water from a sanitary appliance usually a wash basin discharging a plug of water, which creates a partial vacuum in the waste pipe between the plug of water and the water in the trap. This then pulls the water from the trap.

Semi-gravity heating system
A system of central heating where the hot water circulation is via gravity and the heating is pumped.

Sensible heat
A temperature rise without a change of state.

Separate system
A system of below-ground drainage where rainwater and foul water discharge into separate drainage system.

Service pipe
A pipe that connects the external stop valve to the dwelling.

Sex Discrimination Act 1975
An Act of Parliament that protects employees against discrimination on the grounds of gender.

Shear strength
The stress state caused by opposing forces acting along parallel lines of action through the material. The action of ripping or tearing.

Sick building syndrome (SBS)
A combination of ailments associated with an individual's place of work or residence.

Single-feed self-venting indirect hot water storage cylinder
A hot water storage vessel that contains a heat exchanger that uses air entrapment to separate the primary water from the secondary water.

Single point hot water heater
A hot water heater that serves only one outlet. Also known as a point-of-use water heater.

Siphonic WC pan
A WC pan that uses a vacuum to clear the contents of the pan.

Soakaway drain
A specifically designed and located pit, sited away from the dwelling, which allows the water to soak away naturally to the water table.

Soaker
A small piece of code 3 lead used as part of an abutment weathering on a plain tiled or slated roof.

Soft water
Water with a high content of carbon dioxide (CO_2). Acidic, with a pH value below 7.

Soil stack
The lower, wet part of a sanitary pipework system, which takes the effluent away from the building.

Solar collector
Used with solar hot water heating, the solar collector collects the sun's warmth and transfers it, through a heat exchanger, to the hot water storage vessel.

Solar thermal
Technology that utilises the heat from the sun to generate domestic hot water supply.

Solenoid valve
A solenoid valve operates with the aid of an electromagnet. When electricity is supplied to the electromagnet of the valve, the valve becomes magnetised and snaps open, allowing water to flow. Once the electricity has been switched off, the valve is no longer magnetised and a spring snaps the valve shut.

Specific heat capacity
The amount of heat required to change a unit mass of that substance by one degree in temperature. Measured in kJ/kg/°C.

Spigot
Another name for the plain end of a pipe. If the fitting we buy has a plain pipe end, we call this a spigot end.

Steam
Water that has undergone a change of state in the presence of heat.

Storage cistern
Designed to hold a supply of cold water to feed appliances fitted to the system.

Structural Engineer
Calculates the loads (wind, rain, the weight of the structure itself) and the effects of the loads on the structure.

Stub stack system
A system of sanitary pipework where an air admittance valve replaces the vent pipe.

Sublimation
A process in which ice passes directly to steam.

Surface water drain
Used to collect rainwater and discharge it away from a dwelling direct to a water course, river or stream.

Surveyor
The person responsible for ensuring that the Building Regulations are followed in the planning and construction phases of a new building and extensions and conversions to existing properties.

System boiler
A central heating boiler that contains an expansion vessel and pressure relief valve in a single unit.

Tapered threads
A standard thread cut onto the ends of pipes and blackheart malleable, male fittings to ensure a watertight, gas tight or steam tight joint. The tube tightens the further it is screwed into the fitting.

Temper
The temper of a metal refers to how hard or soft it is.

Temporary continuity bonding
Provides a continuous earth to prevent an electric shock in the event of any electrical fault while removing or replacing metal pipework.

Temporary hard water
Water that contains minerals such as calcium carbonate (limestone). Can be softened by boiling. Alkaline, with a pH value above 7

Tensile strength
A measure of how well or badly a material reacts to being pulled or stretched until it breaks.

Thermal envelope
The part of a building that is enclosed within walls, floor and roof, and that is thermally insulated in accordance with the requirements of the Building Regulations.

Thermistor
A resistor that varies with temperature.

Thermo-mechanical cylinder control valves
A non-electrical method of controlling secondary hot water temperature. Works in a similar way to a thermostatic radiator valve.

Thermocouple
A connection between two different metals that produces an electrical voltage when subjected to heat.

Thermometer
A device for measuring temperature.

Thermoplastic
A type of plastic made from polymer resins that becomes liquid-form when heated and hard when cooled.

Thermosetting plastic
Rigid plastics, resistant to higher temperatures than thermoplastics.

Toe boards
A board placed around a platform or

a sloping roof to prevent personnel or materials from falling.

Torque
The property of force that is exhibited when an object rotates around its axis.

Tower scaffold
A small, temporary structure for holding workers and materials during the construction or repair of a building. They can be static or mobile.

Trade foreman
The leader of the tradesmen on site. For instance, a plumbing foreman is the plumber who is running the plumbing installation on site. The plumbing supervisor would have many sites to visit, and each one would have a plumbing foreman.

Two-pipe central heating system
A system having two pipes, a flow and a return, which are connected to the boiler.

Underfloor heating
A method of using concealed underfloor pipework to warm a dwelling.

Unplasticised polyvinyl chloride (PVCu)
A type of thermoplastic used for waste pipes, soil pipes, underground drainage, gutters and rainwater pipes. Can be solvent welded.

Unvented hot water storage systems
Systems fed directly from the cold water main that are not open to the atmosphere and contain an expansion vessel or expansion bubble.

Vaporising burner
An oil burner found on some oil fired appliances that warms the fuel to vaporise it prior to combustion.

Velocity
The measurement of the rate at which an object changes its position.

Vent stack
The upper part of a sanitary pipework system that introduces air into the system to help prevent loss of trap seal.

Ventilated discharge branch system
A sanitary pipework system used on larger installations where there is a risk of trap seal loss because the waste pipe lengths are excessive.

Venturi boost mixing valves
A shower valve using principle of a venturi tube for mixing hot and cold water to produce a showering temperature.

Venturi tube
A pipe that is suddenly reduced in size creating a reduction in pressure but an increase in velocity, in accordance with Bernoulli's principle.

Vitreous china
Clay material with an enamelled surface used to manufacture bathroom appliances.

Volt
The unit of electrical potential.

Waste carrier's licence
A licence required by the Local Authority for anyone transporting waste materials.

Waste management duty of care code of practice
Legislation that aims to ensure that producers of waste take responsibility for making sure that their waste is managed without harm to human health or to the environment.

Water
A compound constructed from two hydrogen atoms and one oxygen atom. The most abundant compound on earth. It can be fresh water or saline (salt) water.

Water course
A river, stream or other flowing natural water source.

Water Supply (Water Fittings) Regulations 1999
These relate to the supply of safe, clean, wholesome drinking water to properties and dwellings, specifically targeting the prevention of contamination, waste, undue consumption, misuse and erroneous metering.

Water undertaker
A water company in the UK. A supplier of treated, wholesome water.

Watt
SI unit for power. It is equivalent to one joule per second (1 J/s), or in electrical units, one volt ampere (1 V·A).

Wavering out
Trap seal loss caused by wind blowing across the top of a vent stack.

WC cistern
Used to flush a WC.

Whiteheart fittings
A type of fitting for low carbon steel pipe with a parallel female thread.

Wind turbine
A method of generating electricity from a turbine connected to a large propeller driven by the wind.

Working drawings
All plans, elevations and details needed by the contractor and trades to complete the building.

Y-Plan central heating system
A fully pumped heating system that uses one three-port mid-position valve.

Zero carbon fuel
A fuel where the net carbon dioxide emissions from all the fuel used is zero.

INDEX

A

ABS (acrylonitrile butadiene styrene) 157, 245–246
acceleration 180
access, restricting 19
accidents
    prevention 16–19
    reporting 20
    see also first aid
acetylene 54–55
acid rain 115
action and reaction 192
air change rates 393, 395
airlocks
    hot water systems 388
    low-pressure systems 342–343
amperage 195–197
anodising 164
antifreeze 168
Architect 83
asbestos
    disposal 29, 140–141
    hazards 27
    legislation 13
    removal 29, 543
    types of 27
    working with 26–29
atmospheric pressure 184

B

back boilers 419–420
    see also solid fuel appliances
back siphonage 308–310, **308**, 385
backflow 308–314, 385–386, **386**
bathroom layout specifications 531–532
    see also sanitary appliances; sanitation systems
baths 498–502, 534–536, 541
batteries, disposal of 143
benchmarking 271
bending pipework
    calculations for 266–267
    copper tube 223–230, **224–229**
    low carbon steel pipe 240–241, **240**, **241**
    pipe bending tools 215
    polybutylene pipe 252
Bernoulli Effect 185, 323, 371
bidets 497–498, 534
biomass boilers 116, 119–120, **120**
black oxide 457
black water 287
blowtorches 60

boilers
    combination (non-condensing) 395, 424
    condensing 396, 414, **414**, 422, 424–425, **425**
    condensing combination boilers 372–374, **373**, 414–415, **415**
    efficiency of 109
    gas 421–425, **423**
    oil-fired 109, 112, 426–428, **427**
    see also central heating systems;
boreholes 286–287
Boyle's Law 171–172
brass fittings
    corrosion 343
    dezincification 161
British Standards 92
    BS EN 12056 487–488
    BS EN 12828 393
    BS EN 12831 393, 395
    BS EN 14336 393
    BS 6700 293
    BS 8000 487–488
Building Contractor 85
building regulations
    conservation of fuel and power 107–110
    ventilation 107
Building Regulations Compliance certificate 271
Building Services Engineer 84
butane 114, 169

C

calorific value 114–115
cancellation rights, of customer 95
capillary attraction 166–167
capillary fittings 230–232, **231**
carbon dioxide 9, 108, 111, 126, 170
carbon footprint 108
cast iron pipework, removal of 542
Celsius 173
central heating systems
    air separator **408**
    balancing 449
    combination boiler (non-condensing) 395, 424
    commissioning 449
    condensing boilers 396, 414, **414**, 422, 424–425, **425**
    condensing combination boilers 414–415, **415**

    corrosion protection 450–452
    decommissioning 450
    designing 444–445
    draining down 455
    electrical central heating controls 440–446
    fault finding 456–457
    feed and expansion cistern 408–409
    filling 448–449
    flue systems 428–430, **429**, **430**
    fully pumped systems 395, 403–409, **403**, **405**
    gas boilers 421–425, **423**
    heat emitters 430–437
    heat exchanger 395, 422–423
    installation 446–449
    maintenance 452–456
    microbore system 394, 416–417, **416**
    oil-fired boilers 426–428, **427**
    open-vented low-pressure systems 395–409
    pipework 394, 447
    power flushing 455
    pump replacement 452–453
    pumped central heating only systems 396–397, **397**
    reversed return system 417–418, **418**
    sealed (pressurised) heating systems 409–415, **414**, **415**
    semi-gravity systems 397–402, **398**, **399**, **400**, **401**
    solid fuel appliances 401–402, **401**, 419–421, **420**, **421**
    temperature controls 437–440, 443–444
    testing 448
    thermal comfort 393–394
    see also radiators
centre of gravity 191
ceramic disc tap, replacing 340
Charles's Law 171
chasing 264
circuits see electrical circuits
cisterns
    cold water storage 300–306, 348–349
    commissioning 332–333
    corrosion of 343
    feed and expansion 408–409, **409**
    inlet requirements 302
    installing 301–303
    interconnecting 304–305, **304**, **305**

stagnation 303
Civil Engineer 83
Clerk of Works 84
clients, duties of 6, 82
Climate Change Act 2008 105
clipping
   copper tube 238–239
   low carbon steel pipe 244
   plastic pipes 248, 249, 253
coal 111
coke 112
cold water systems 293–314
   backflow/back siphonage 308–314, **308**
   commissioning 331–333
   decommissioning 341
   direct system 295–297, **295**, **296**
   domestic supply 295–307
   existing installations 333–334
   fault-finding 342
   frost protection 306–307
   indirect system 297–299, **298**
   installing 327–331
   insulating 328, 331
   isolation valves 314–317, **315**, **317**
   leakage 343–344
   maintenance 334–341
   planning 328–330
   stop valve 294, 314–315, **315**
   testing 331
   see also cisterns
combustion 60–61
commissioning 270–271
   central heating systems 449
   cisterns 332–333
   cold water systems 331–333
communication 97–102
   verbal 98
   written 95–96, 98
company policies 96–97
compression fittings 232–233, **232**, **233**, 243–244, 249–250
compressive strength 158
condensing combination boilers
   backflow protection 314
   central heating 414–415, **415**
   hot water systems 372–374, **373**
conductivity 159–160, 176–177, 195, 199
confined spaces, working in 77–78
conflicts, in the workplace 100–101
Construction (Design and Management) Regulations 2007 5–7
construction site
   documentation 92–95
   electricity 46–47
   inspections 88–90
   management 81–90
   on-site trades 87–88
   safety 16–22
   security 19

contract of employment 96
contractors, duties of 7
Control of Substances Hazardous to Health (COSHH) Regulations 2002 8–10
convection 177
copper
   properties of 154, 155, 222
   thermal conductivity 176
   copper tube 327
   bending 223–230, **224–229**
   clipping 238–239
   corrosion 343
   insulating 307
   jointing 230–237
   red band 333
   uses of 223
corrosion 156, 160, 343
   air infiltration 450–451
   in cisterns 343
   electrolytic 451–452
   in hot water storage cylinders 367
   of metals 161–163
   of plastics 163–164
   preventing 164
CPR (cardiopulmonary resuscitation) 43–44
   see also first aid
cuPVC 245, 334
customer care 259, 268
   handover information 96, 129, 271–272
   policy 94
   protecting property 268–269, 478–479

D

decommissioning 272
   central heating systems 450
   cold water systems 341
   hot water systems 389
   sanitation systems 542–543
Designers, duties of 6
Diaphragm-type FOV, repairing 341
domestic properties
   electrical hazards 48–49
   electricity supply 52
   plumbing installation 257–269
   protection of 268–269, 478–479
   see also customer care
ductility 158
dust 9
dynamic pressure 184

E

ear defenders 33
earthing 205
efficiency ratings 109

electric shock 47–48, 205
   first aid for 42–43
electrical
   circuits 201–205
   conductivity 160, 199
   current 197–199, 201–205
   inspections 90
   installations 48–49, 52
electrical appliances, energy ratings 127–128
electrical equipment
   disposal of 142–143
   safety of 10–11
electricity
   hazards 47–49
   isolation procedures 52–53
   measuring 195–197
   non-renewable energy sources 115
   principles of 193–195
   renewable energy sources 117
   safety 45–53
electricity supply
   construction sites 46–47
   domestic properties 52
Electricity at Work Regulations 1989 10–11, 45–46
emergencies
   fire evacuation 45
   raising the alarm 44–45
   see also first aid
employer responsibilities
   first aid 37–38
   health and safety 4–5, 7, 20–21
employment legislation 90–91
energy
   conservation 105–110, 125–126
   efficiency 128–130
   sources of 111–117
energy performance, of buildings 105–106
Energy Performance Certificates (EPCs) 126–127
environmental policy 95
equilibrium 192–193
equipment
   hand tools 209–214, 217–219
   pipe cutting/bending 214–215
   pipe threading 217
   plumbing-specific 214–218
   power tools 219–222
   soldering 216
equipotential bonding 53, 205–206
erosion corrosion 162
estimates 95
excavations *see* trenches
eye protection 31

F

fan convectors 435

fault-finding
    central heating systems 456–457
    hot water systems 387–388
    cold water systems 342
filtration 284–285
fire
    classification 61
    extinguishers 61–62
    safety 60–62
fire stopping 269
fireclay 157
first aid 36–45
    CPR (cardiopulmonary resuscitation) 43–44
    injuries 38–44
    legislation 37
    recovery position **44**
fittings
    copper tube 230–237
    low carbon steel pipe 242–244
    plastics 246–253, **247**, **248**
    proprietary 237
    recognition **236–237**
fixings 254–257
flashback 56
float-operated valves 317–319
floorboards, lifting 260
flow rate 180, 185–196
flue systems 428–430, **429**, **430**
flushing procedure, cold water system 331–333
flushing systems 490–493, **493**, 505–507
fluxes 25, 232
force 181, 190–192
fossil fuels 111
frost protection, cold water systems 306–307
fuel
    fossil 111
    gases 54–60
    high carbon 111–115
    low carbon 116
    zero carbon 117
fumes, exposure to 40
fusion welded fittings 250

G

galvanic corrosion 162, 367
galvanisation 164
gas
    boilers 109, 421–425, **423**
    bottled 54–60
    composition of 113
    equipment 55–60
    legislation 14
    natural 113, 169
    see also gases; liquid petroleum gas (LPG)

Gas Safe Register 10, 14
Gas Safety (Installation and Use) Regulations 1998 268
gases
    air 169
    carbon dioxide 170
    and corrosion 160–161
    liquid petroleum gas (LPG) 54–55, 59–60, 113–114, 169
    measuring 170–172
    pressure 170
    steam 169
gloves 32–33
glycol 168
gravity
    and acceleration 180
    centre of 191
    and force 181
gravity circulation 177
green technology
    biomass solid fuel 116, 119–120, **120**
    combined heat and power 121–123
    heat pumps 116, 120–121, **121**
    solar photovoltaic panels 117, 124, **125**
    solar thermal systems 116, 118–119, **118**
    wind turbines 117, 123–124, **124**
greenhouse gases 105, 111, 115
grey water 287
    recycling 149–150, **149**
guttering systems 461–484
    cast iron systems 470–471, 482–483
    extruded aluminium systems 472–473
    fittings **469**
    installation 473–477, **474–476**
    leaking 480–481
    maintenance 479–484
    profiles 468, **468**, 472–473
    PVCu systems 467–470, 481–482
    rainfall intensity 462–463
    roof area 463–465
    running outlet position 465–466, **465**

H

hand tools 209–214, 217–219
handover to customer 96, 271–272
hardness see Mohs hardness scale
hazardous substances
    categories of 22, 309
    disposal of 140–143
    legislation 8–10
hazards, on site 16–18
health and safety
    inspections 15–16, 89

legislation 3–15
of the public 19–20
Health and Safety Executive (HSE) 6, 15–16, 20
Health & Safety at Work Act 1974 3–5
hearing protection 33
heat
    exchangers 352, 356, 368
    pumps 116, 120–121, **121**
    transfer 172, 175–178
heating see central heating systems
height, working at see working at height
high carbon fuels 111–115
high density polyethylene (HDPE) 334
hot water systems
    centralised instantaneous systems 368–374
    centralised open-vented systems 348–368, 379
    combination open-vented cylinders 358–360, **359**
    combined storage unit (CPSU) system 370–371, **370**
    condensing combination boilers 372–374, **373**
    decommissioning 389
    direct systems 349–352, **351**
    fault finding 387–388
    gas-fired water heater 363–365, **364**
    heat exchangers 352, 356, 368
    immersion heaters 350–351, 361–363, **362**, 379, 387
    indirect systems 352–358, **353**, **355**, **358**
    instantaneous gas-fired multi-point heaters 371–372, **371**
    instantaneous point-of-use heaters 374–375
    localised systems 374–378
    maintenance 389
    pilot light 371
    pipework 367–368, 374, 378–379, 380–381
    secondary circulation 366–367, **366**
    storage cylinders 365–367
    storage point-of-use heaters 376–378, **376**, **377**
    temperature control 379–380
    testing and commissioning 387
    thermal stores 368–369, **369**
    thermostatic mixing valves 379–380
    see also showers
humidity 393
hydroelectric power 117
hydrogen fuel cells 116

I

immersion heaters 350–351, 361–363, **362**, 379, 387
improvement notice 16
injuries 34
   first aid 38–44
   reporting 10
in-situ installations 268
inspections
   construction site 88–90
   electrical 48–49, 90
   health and safety 89
   water 89–90
insulation 160, 195
   cold water pipes 328, 331
   hot water pipes 380
   pipework 306–307
isolation valves 314–317, **315**, **317**

J

jointing
   copper tube 230–237
   low carbon steel pipe 242–244
   plastics 245–253, **245**, **246**
joists, notching and drilling 262–264, 328, **329**

K

Kelvin temperature 171, 173
kerosene 112–113, 168
kettling 457
kick-space heaters 435
Kyoto Protocol 105

L

ladders
   lifting and carrying 67–68
   raising **66**, 66–67
   safety 69–70
   securing **67**
   types of 63–65
lead 155
   disposal of 142
   legislation 13
   poisoning 24
   thermal conductivity 176
   welding 59
   working with 23–25
lead pipework 333
   removal 543
leakage, cold water systems 343–344
legislation
   building services 14
   employment 90–91
   health and safety 3–15
   sanitation systems 487–488
   waste management 131–132
levers 187
lifting *see* manual handling
limescale 283, 325
limits to personal authority 96–97
liquid petroleum gas (LPG) 54–55, 59–60, 113–114, 169
liquids
   capillary attraction 166–167
   properties of 164–168
Local Authority 84–85
low carbon fuels 116
low carbon steel pipe 239–244
low-pressure systems, airlocks 342–343
LPG *see* liquid petroleum gas
lubricants 168

M

machines, and mechanical advantage 186–190
maintenance 272–273
   central heating systems 452–456
   cold water systems 334–341
   hot water systems 389
   sanitation systems 540–541
making good 273–274
malleability 159
management, of construction projects 81–90
manual handing 34–36
   one-person lift **35**
   two-person lift **36**
Manual Handling Operations Regulations 2005 12
manufacturers' guidance 92
materials
   delivery of 258–259
   properties of 154–160
matter, states of 174
mechanical advantage 187–190
mechanical lifting gear 36
metals 155–156
   ferrous 156
   non-ferrous 156
   oxidisation 160
   recycling 137–138
method statement 18–19
Mohs hardness scale 159
monobloc mixer tap 312, **313**
muPVC 244–246

N

natural gas 169
Newton's third law of motion 192
nitrogen 9
noise
   central heating systems 446–447
   cold water systems 342
   hot water systems 388
   pipes expanding 328
   protection from 33
   water hammer 319
non-rising spindle tap, re-washering and reseating **338**–**339**
notifiable projects 82

O

offset bends **225**, **241**
   see also bending pipework
Ohm's Law 195, 200–201
oil-fired boilers 109, 112, 426–428, **427**
oxidative degradation 163–164
oxidisation, of metals 160
oxyacetylene equipment 55–59, **56**
oxygen cylinders 54–55

P

passover bends **226**
peat 112
perimeter fencing 19
permits to work 19
personal
   authority 96–97
   conduct 17, 96, 259
   hygiene 13
Personal Protective Equipment at Work Regulations 1992 8
pH value, of water 282
photo degradation 163–164
pipework
   bending tools 215
   central heating systems 394, 447
   cutting tools 214
   hot water systems 367–368, 374, 378–379, 380–381
   prefabrication 265–267, **266**
   sanitation systems 507–515, **508**, **512**, **513**, **514**, 525–528, 543
   sleeving 268
   surface-mounted 265
   temporary continuity bonding 53
   threading equipment 217
   see also bending pipework; clipping
pitting corrosion 162–163
plasterboard fixings 256
plastics
   clipping pipes 248, 249, 253
   degradation of 163–164
   jointing and fittings 244–253, **247**, **248**
   recycling 138–139
   thermoplastics 156–157
   thermosetting 157

INDEX

plumbing appliances, drawing symbols **274**
   see also sanitary appliances
plumbing systems
   designing 257–258
   installation (domestic) 257–269
   preparation 259–264
   see also central heating systems; cold water systems; hot water systems
policy documents 94–95
polybutylene 157, 250–253, 334, 328
polyethylene 156, 176, 249–250
polypropylene 154, 157, 248–249
Portable Appliance Testing 49–51
Portsmouth-type FOV tap, repairing 340–341
power, measuring 197
power tools 219–222
   battery-powered cordless tools 51–52
   Portable Appliance Testing 49–51
   safety 49–52
PPE see protective equipment
press-fit fittings 235, **235**
pressure 181–184
   of gases 170
   of water 182–183
pressure-jet burners 426–427, **427**
prohibition notice 16
Project Manager 84
propane 54, 114, 169
proprietary fittings 237
protective equipment 8, 29–33
public, health and safety of 19–20
pulleys 188–190
push-fit
   fittings 233–234, **234**, 250
   joints 246, 252–253
PVCu 154, 156, 244
   fittings 246–247
   jointing 245–246, **245**, **246**

Q

qualifications, and authority 96–97
Quantity Surveyor 83
quotations 95

R

radiation 177–178
radiators 430–434
   column 434
   connections 431
   dressing 432
   faults 456–457
   hanging 257, 432–433
   low surface temperature 434
   panel 430–432
   positioning 432–433, 447
   removing 269
   replacement 453–454
   valves 437–440
radon gas 9
rainfall intensity 462–463
rainwater
   cycle 279–280, 282
   harvesting **148**, 148–149, 287
   recycling 135–140
   grey water **149**, 149–150
   metals 137–138
   plastics 138–139
   water 287
reduced voltage system 46–47
refrigerants 167–168
   disposal of 141–142
refrigeration cycle **167**
relative density 154
renewable energy 117, 355
Reporting of Injuries, Diseases and Dangerous Occurrences Regulations (RIDDOR) 1995 10
resistance 195–196, 199, 201–205
respirators 31–32
rising spindle tap, re-washering and reseating **336–337**
risk assessment 4–5, 17–18
rusting 160–161, 164
   see also corrosion

S

sacrificial anodes 164, 367
safety
   helmets 30
   signs 20–22
   see also health and safety
Safety Signs and Signals Regulations 1996 12–13
sanitary appliances
   bathroom layout specifications 531–532
   baths 498–502, 534–536, 541
   bidets 497–498, 534
   installation 533–538
   layout specifications 531–532
   materials 488–489
   removal 542
   shower trays and cubicles 502–503
   sinks 503–504, 541
   urinals 504–507, **505**
   wash hand basins 494–497, 534, 536, 541
   WCs 490–494, 536–538, 541–542
sanitation systems
   below-ground drainage 528–531
   blockages 541–542
   decommissioning 542–543
   flushing systems 490–493, **493**, 505–507
   installation 524–528
   legislation 487–488
   maintenance 540–541
   pipework 507–515, **508**, **512**, **513**, **514**, 525–528, 543
   soil stack connection to drain 527–528
   testing 538–540
   traps 515–524, **519**
   waste pipe connection to soil stack 526–527
scaffolds 11–12, 70–75
scaling 325
SEDBUK ratings 109
sedimentation 284
shear strength 158
showers
   booster pumps 382–383, **383**
   cistern-fed 382–384, **382**, **383**
   electric 375
   mains hot and cold fed 384, **384**
   mixing valves 312–313, **313**, 322–325, **323**
   trays and cubicles 502–503
   unbalanced supply pressures 385, **385**
SI system of measurement 153, 172–173, 178, 180, 195
sick building syndrome 107
signs see safety signs
sinks 503–504, 541
siphonic action 184
skirting heating 436
sleeving pipework 268, 329–330
socket tools 216
solar photovoltaic power 117, 124, **125**
solar thermal technology 116, 118–119, **118**, 177–178
soldering
   equipment 216
   gases 54
   safety 268–269
solid fuel appliances 401–402, **401**, 419–421, **420**, **421**
solvents 26
specific heat capacity 172
spring bends 227–230, **228**, **229**
stainless steel tube 333–334
steam 169
sterilisation 285
stop valve 294, 314–315, **315**
Structural Engineer 84
surface-mounted pipework 265
Surveyor 83
sustainable homes 106

T

taps 320–322, **321**
   ceramic disc tap 322
   replacing 340
   non-rising spindle 321
   re-washering and reseating **338–339**
   rising spindle 320–321, **321**
   re-washering and reseating **336–337**
temperature, measuring 173–174
temporary continuity bonding 53
tensile strength 158
terminal fittings *see* taps
testing 270
   cold water systems 331
   Portable Appliance Testing 49–51
   pre-testing checks 269
   sanitation systems 538–540
thermal
   comfort 393–394
   conductivity 159, 176
   degradation 163
thermometers 173–174
thermostatic radiator valves 437
thread direction, fuel gases 54
threading tools 217
tidal electricity generation 117
time sheets 93
tools *see* equipment
torque 188, 190–191
towel warmers 436
trades, on-site 87–88
trenches, working in 75–77

U

ultraviolet degradation 164
unconsciousness, first aid 41–42
underfloor heating 437
units of measurement 153, 172–173, 178, 180, 195
urinals 504–507, **505**

V

valves
   drawing symbols **275**
   float-operated 317–319
   isolation 314–317, **315**, **317**
   stop 294, 314–315, **315**
vaporising burners 427–428, **428**
velocity 180
vitreous china 157
voltage 195–196, 201–205
   power tools 219
   reduced voltage system 46–47

W

wash hand basins 494–497, 534, 536, 541
waste
   disposal 130–135, 140–143
   hazardous materials 140–143
   legislation 131–132
   pipes 246–249
   recycling 135–140
water
   boiling point 165
   boreholes 286–287
   conditioners 325
   contamination 144–145
   distribution of 288–292
   efficiency calculations 145–147
   filters 326–327
   fluid categories 309
   grey water recycling 149–150, **149**
   hardness 282–283
   inspections 89–90
   legislation 14
   meters 292
   pH value 166, 282
   pressure 182–183
   properties of 165–166
   rainwater cycle 279–280, 282
   recycling 287
   reducing consumption 148
   softeners 283, 326
   sources of 280–282
   treatment 283–285, 325–327
   wastage 145
   see also cold water systems; grey water; rainwater harvesting
water main, connection to 289–292
water saving devices 137, 148
Water Supply (Water Fittings) Regulations 1999 292–293, 297, 309, 329, 487
WCs 490–494, 536–538, 541–542
welding equipment 55–59
wetting 25
wheel and axle, principle of 188
wind turbines 117, 123–124, **124**
Work at Height Regulations 2005 11–12
work wear 30
working at height 11–12, 63–75, 478
working drawings 93, **274**, **275**

Z

zero carbon fuels 117

NOTES

NOTES

NOTES